Natural populations of plants show intricate patterns of variation. European botanists of the eighteenth and nineteenth centuries used this variation to classify different 'kinds' into a hierarchy of family, genus and species. Although useful, these classifications were based on a belief in the fixity of species and the static patterns of variation. Darwin's theory of evolution changed this view; populations and species varied in time and space, and were part of a continuing process of evolution. The development of molecular techniques is transforming our understanding of microevolution and the evolutionary history of the flowering plants. This new edition reviews recent progress in its historical context, showing how hypotheses and models developed in the past have been critically tested. The authors consider the remarkable insights that molecular biology has given us into the processes of evolution in populations of cultivated, wild and weedy species, the threats of extinction faced by many endangered species and the wider evolutionary history of the flowering plants as revealed by cladistic methods.

PLANT VARIATION AND EVOLUTION

PLANT VARIATION AND EVOLUTION

D. BRIGGS
Department of Plant Sciences, University of Cambridge

S. M. WALTERS
Former Director of the University Botanic Garden, Cambridge

3rd EDITION

PUBLISHED BY THE PRESS SYNDICATE OF THE UNIVERSITY OF CAMBRIDGE
The Pitt Building, Trumpington Street, Cambridge, United Kingdom

CAMBRIDGE UNIVERSITY PRESS
The Edinburgh Building, Cambridge CB2 2RU, UK http://www.cup.cam.ac.uk
40 West 20th Street, New York, NY 10011–4211, USA http://www.cup.org
10 Stamford Road, Oakleigh, Melbourne 3166, Australia
Ruiz de Alarcón 13, 28014 Madrid, Spain

First published by Weidenfeld and Nicolson 1969
Second edition published by Cambridge University Press 1984
Reprinted 1986 (with corrections), 1988, 1990
Third edition 1997
Reprinted 2000

Printed in the United Kingdom at the University Press, Cambridge

Typeset in Times 10/13 pt [VN]

A catalogue record for this book is available from the British Library

Library of Congress Cataloguing in Publication data

Briggs, D. (David), 1936–
Plant variation and evolution / D. Briggs & S. M. Walters. – 3rd ed.
p. cm.
Includes bibliographical references and index.
ISBN 0 521 45295 3 (hb). – ISBN 0 521 45918 4 (pbk.)
1. Botany – Variation. 2. Plants – Evolution. I. Walters, S. M.
(Stuart Max) II. Title.
QK983.B73 1997
581.3′8–dc21 96-39293 CIP

ISBN 0 521 45295 3 hardback
ISBN 0 521 45918 4 paperback

'The standing objection to botany has always been, that it is a pursuit that amuses the fancy and exercises the memory, without improving the mind, or advancing any real knowledge: and, where the science is carried no farther than a mere systematic classification, the charge is but too true. But the botanist that is desirous of wiping off this aspersion should be by no means content with a list of names; he should study plants philosophically, should investigate the laws of vegetation, should examine the powers and virtues of efficacious herbs, should promote their cultivation; and graft the gardener, the planter, and the husbandman, on the phytologist. Not that system is by any means to be thrown aside; without system the field of Nature would be a pathless wilderness; but system should be subservient to, not the main object of, pursuit.'

GILBERT WHITE

from *The natural history of Selborne* 1789, letter written on 2 June 1778

'... there is no better method for scientists of one period to bring to light their own unconscious, or at least undiscussed, presuppositions (which may insidiously undermine all their work) than to study their own subject in a different period. And ... when the writings of an earlier author have apparently been taken as the basis of subsequent work, constant scrutiny is necessary to prevent his presuppositions becoming fossilized, so to speak, in the subject and producing unnoticed inconsistencies when modifications have been made as a result of subsequent work.'

A. J. CAIN, 1958

To Daphne and Lorna

Contents

Preface to the Third Edition		*page* xv
Acknowledgements		xix
Note on names of plants		xxi
1	**Looking at variation**	1
	'Kinds', species and natural classification	1
	Individual variation	5
	The nature of species	5
2	**From Ray to Darwin**	7
	Ray and the definition of species	7
	The Chain of Being	9
	Linnaeus	10
	Buffon and Lamarck	17
	Darwin	20
	Tests of specific difference	29
3	**Early work on biometry**	33
	Commonest occurring variation in an array	35
	Estimates of dispersion of the data	36
	Histograms, frequency diagrams and the normal distribution curve	38
	Other types of distribution	41
	Comparison of different arrays of data	42
	Complex distributions	43
	Local races	46
	Correlated variation	48
	Problems of biometry	50
4	**Early work on the basis of individual variation**	52
	Phenotype and genotype	53
	Transplant experiments	55

	The work of Mendel	58
	Pangenesis	64
	Mendelian ratios in plants	65
	Mendelism and continuous variation	66
	Physical basis of Mendelian inheritance	73
5	**Post-Darwinian ideas about evolution**	80
	Experimental investigation of evolution	80
	The mutation theory of evolution	84
	Neo-Darwinism	85
6	**Modern views on the basis of variation**	88
	Molecular basis of heredity	88
	Mutation	94
	Cytological differences	98
	Non-Mendelian inheritance	102
	Modern techniques used in studying genetic variation	103
	Electrophoretic studies of enzymes	104
	Analysis of DNA	107
	Use of DNA in studies of variation	111
	Patterns of variation	114
	Phenotypic variation	114
	Developmental variation	116
	Phenotypic plasticity	120
7	**Breeding systems**	124
	Outbreeding	124
	Late-acting self-incompatibility systems	133
	Self-fertilisation	133
	Apomixis	134
	Consequences of different reproductive modes	143
	Advantage and disadvantages of different breeding systems	145
	Breeding systems in wild populations	148
	Environmental control of facultative apomixis	157
	The use of molecular markers in the study of the reproductive behaviour of apomictic plants	158
	Evolution of breeding systems	160
	Concluding remarks	165
8	**Infraspecific variation and the ecotype concept**	167
	Turesson's pioneer studies and other experiments	167
	Experiments by American botanists	174
	The widespread occurrence of ecotypes	183

Clines 183
Factors influencing the variation pattern 186
The refining of genecological experiments 190
Sampling populations 190
Cultivation experiments 196
The designed experiment 201
The interpretation of experiments 206
9 **Recent advances in genecology** 208
Variation in populations 209
Plant populations 211
Gene flow 212
Gene flow: early ideas 212
Gene flow: agricultural experiments 213
Gene flow: insights from the movement of pollen 215
Gene flow: studies of seed dispersal 217
Gene flow: studies using molecular tools 218
'Neighbourhoods' in wild populations 221
Effects of chance 223
Founder effects in introduced species 225
Selection in populations 226
Fitness 228
Studies of single factors 228
Studies of several interacting factors: *Lotus* and *Trifolium* 229
Reciprocal transplant experiments 237
Experimental evidence for disruptive selection 239
Co-selection in swards 244
The speed of microevolutionary change: agricultural experiments 244
Rapid change in polluted sites 246
Microevolution in arable areas 250
Adaptive and non-adaptive characters 252
Patterns of variation in response to seasonal or irregular extreme habitat factors 255
Concluding remarks 257
10 **Species and speciation** 259
The species concept 259
Other species definitions 260
Gradual speciation 263
Abrupt speciation 264

11 Gradual speciation and hybridisation 270
Evidence for gradual speciation 271
Crossing experiments with species of *Layia* 272
The interpretation of crossing experiments 273
Studies of *Layia* using molecular methods 274
Uncertainties about the concept of gradual speciation 276
Natural hybridisation 277
The consequences of hybridisation: some theoretical considerations 284
Results of disruptive selection in polymorphic populations 287
Introgression and other patterns of hybridisation 290
Genetic investigations of hybridisation 295
Chemical studies of hybridisation 297
Critical tests of the hypothesis of introgression 299
Recent studies of introgression using molecular tools 302
Introgression in Louisiana Irises 304
Concluding remarks 307
12 Abrupt speciation 309
How common is polyploidy? 309
Experimental studies of polyploids 312
Early cytogenetic studies 312
Resynthesis of wild polyploids 313
Auto- and allopolyploidy 314
Genome analysis 318
Genome analysis: uncertainties about ancestry 322
Genetic control of chromosome pairing: the implications for genome analysis 325
Studies of karyotypes 327
Chemical studies 330
Autopolyploids: reassessment of their evolutionary potential 331
Polytopic multiple origin of polyploids 332
The origin of new polyploids: the role of somatic events and unreduced gametes 338
The persistence of polyploids 340
Gene flow between diploids and polyploids 344
Polyploids: their potential for evolutionary change 346
Distribution of polyploids 348
How important is polyhaploidy? 351
The delimitation of taxa within polyploid groups 351

Abrupt speciation 352
Changes in chromosome number 353
Chromosome repatterning 357
Speciation following hybridisation 358
Minority disadvantage 359
13 The species concept 361
The biological species concept 362
The views of botanical taxonomists 364
14 Evolution: some general considerations 367
The fossil record 368
Diversification of the angiosperms 370
Microevolution and macroevolution 372
The devising of phylogenetic trees 379
The use of computers in taxonomy 382
The influence of numerical taxonomy 384
Cladistics 385
A critique of cladistic approaches 389
Transgenic plants 396
15 Conservation: confronting the extinction of species 399
What are the threats to biodiversity? 400
What classes of evidence are available for assessing claims concerning threats of extinction? 400
The threats induced by changes in land use 402
Threats to native biota from introduced plants and animals 402
The effects of pollution 404
How many species are there in the world? 406
How many species are threatened with extinction? 408
Processes involved in the extinction of species 411
Demographic stochasticity 412
Effects of fragmentation 414
Genetics of small populations 414
Minimum viable populations 416
What priorities should be set in attempting to reverse the decline of endangered species? 419
Ex situ conservation 421
The role of protected areas in countering the threat of extinction 423
Managing resources to prevent extinction of species 426
Restoration ecology 427

Manipulating and creating populations of endangered species in an attempt to prevent extinction 428
Arguments for conservation 432
Glossary 434
References 438
Index 499

Preface to the Third Edition

When it was first proposed to establish laboratories at Cambridge, Todhunter, the mathematician, objected that it was unnecessary for students to see experiments performed, since the results could be vouched for by their teachers, all of them of the highest character, and many of them clergymen of the Church of England. (*Bertrand Russell*, 1931)

Although experimental science is firmly established, introductory books in biology are often influenced by the 'Todhunter' attitude – the teacher has all the answers. In many text books a judicious mixture of concepts, mathematical ideas and selected results of laboratory and field experiments is combined in an elaborate pastiche to provide a more or less complete edifice. Perhaps one or two areas of uncertainty may be indicated, but the general impression is of a house well built but awaiting the placing of the last few roof-tiles. Conversations with research biologists, however, quickly reveal a different picture. Almost nothing is settled: current views represent a provisional framework, and even some parts of the subject long held to be clarified are suddenly overturned by new discoveries.

Teaching experience reinforces our view that students of science should be shown the way in which, slowly and painstakingly, our present partial pictures have been built up, how and to what extent they are testable by experiment and observation, and in what way they remain vague or defective. A healthy scepticism in the face of the complexity of organic evolution is the best guarantee of real progress in understanding its patterns and processes.

The reception given to the two editions of our book has encouraged us to prepare this new, largely rewritten, edition. The general shape and intention of the book remains the same: it aims to provide an authoritative, introductory university text, while at the same time satisfying a general reader with a real interest in the subject.

Our aim is to show how one branch of biology – namely the study of variation and evolution in plants – has developed over the last 300 years. The development has been increasingly scientific, leading to a realisation of the crucial importance of hypothesis and experiment. Throughout the book, in which we are uncommitted and even sceptical about neat explanations and simple formulations, we have tried to provide a critical but concise account of current excitements and advances in the subject – especially those arising from the increasing use of molecular tools – while, at the same time, paying attention to the difficulties and uncertainties. Moreover, we have tried to engender a critical attitude in the mind of the reader.

In this account we have intentionally introduced and shown the connections between many complex subjects, and have therefore provided references to important research papers and books in order that the student may build on our framework. Our survey concentrates on four main themes:

1. We have examined the logical and historical framework of early observation and experiment, which in our experience is almost wholly neglected in university courses.
2. While readily admitting the importance of laboratory-based observations and experiments, we believe that biologists who wish to understand evolution must study in detail the variation in plants out of doors, whether in wild, cultivated or weedy habitats. Accordingly, our book reveals how molecular studies, building on the work of the past, are providing new insights into patterns and processes in plant populations.
3. Taxonomists were the first biologists to study plant variation. An important theme of this book, therefore, is the impact evolutionary investigations have had on the theory and practice of taxonomy, especially studies of phylogenetic relationships, which use cladistic methods to examine morphological and molecular data sets.
4. Our subject has a peculiar attraction in that it is developing rapidly and at many levels. However, the knowledge of wild populations is so incomplete that a keen student may make a very real contribution. We believe that the biologist interested in field observations has a very great advantage over other scientists, to whom the possibility of significant discovery is greatly restricted, or even denied, by the nature of their material.

Finally, we stress the relevance of our subject to the twenty-first century. As we shall see, our insights into microevolution have come through the study of both cultivated and wild plants. There should be no schism between the pure and applied aspects of biology. Also, it is encouraging that many professional scientists are investigating the living plant in its outdoor environment, for we believe that increased food production, enlightened nature conservation and a proper sustainable use of biological resources can only come if people appreciate and understand the patterns of plant and animal diversity all around us. We hope this book, in its new edition, will continue to make a small but significant contribution to filling what we have found in our experience to be a very real gap in the available biological literature.

D. Briggs
S. M. Walters

The second edition of our book, published in 1984, has sold steadily, apparently filling a need felt by students both at University and sixth-form level. This is very gratifying to the authors.

During the twelve years since we wrote the second edition, I have myself been in retirement and to a large extent detached from the demands – and satisfactions! – of undergraduate teaching. It is therefore appropriate that I express my admiration of my co-author's work, since it was early apparent to both of us that the brunt of the necessary revision and up-dating would fall on him, and that my role would be largely supportive and subsidiary.

In spite of considerable revision, which has included some rearrangement, we believe that we have retained the general flavour of this book, a flavour which has not radically altered since the first edition in 1969. We remain convinced that 'students of science should be shown the way in which ... our present partial pictures have been built up ...' and we have accordingly retained, especially in the first few chapters, the historical approach that many reviewers particularly commended.

S. M. Walters

December 1996

Acknowledgements

In the preparation of the first edition of this book we received assistance from many people. We record our thanks to Alex Berrie, Arthur Chater, Charles Elliott, Martin Lewis (photographs), Don MacColl, Hugh McAllister, David Ockendon, and Robin Sibson (photographs). Diagrams were drawn by K. G. Farrell, J. Messenger and Design Practitioners Ltd.

For help with the second edition we are grateful to John Akeroyd, May Block, Christopher Chalk (photographs), David Coombe, Joachim Kadereit, Joseph Pollard, Duncan Porter, Peter Sell, Lorna Walters (preparation of the index), Suzanne Warwick, Harold Whitehouse, and David Valentine.

In preparing the third edition we acknowledge helpful discussions with Jayne Armstrong, Elizabeth Arnold, John Barrett, Tony Bradshaw, Arthur Cain, Quentin Cronk, Harriet Gillett, Mark Gurney, John Harper, Peter Jack, Joachim Kadereit, Vince Lea, Elin Lemche, Roselyne Lumaret, Pierre Morisset, Chris Preston, Donald Pigott, Ada Radford, Andrew Theaker, John Thompson, Suzanne Warwick, John West, Alison Smith, Kerry Walter and Peter Yeo. In thanking them we should make it clear that, although we have not in every case acted on their advice, we have always valued it.

We thank Daphne Briggs, Alastair Briggs, Jonathan Briggs, Nicholas Oates and Gina Murrell for help in preparing and checking the manuscript, and Stephen Drury-Morris for taking the photographs.

Some of the research for the third edition was carried out when David Briggs was a visitor to Agriculture Canada, Ottawa. I record my special thanks to Suzanne Warwick and Chris Reynolds for their generous hospitality in making this sabbatical leave so memorable and stimulating. Also, I would like to thank my parents, my wife Daphne and my family for their encouragement, understanding and support.

We have consistently appreciated the support given by the Publisher, and would wish to thank particularly Maria Murphy and Karen Binks.

Finally, we wish to acknowledge our appreciation of the support of successive Professors of Botany in Cambridge. To Sir Harry Godwin in particular we owe the original stimulus to write a book on this subject and from his successors Percy Brian, Richard West and Tom ap Rees we have received sympathetic encouragement throughout.

Note on names of plants

Scientific names are generally in accordance with Stace: *New Flora of the British Isles* (1991) for British plants, and Tutin *et al.*: *Flora Europaea* (1964–1993) for European plants not in the British flora. In the cases not covered by either work we have used the name we believe to be correct. Where an author has used a non-current name, this is noted in brackets.

While some botanists continue to use long-standing names for important families, others use names derived from the name of the type genus (e.g. Compositae/Asteraceae, Cruciferae/Brassicaceae, etc.). Thus, the reader is warned that different legitimate names are in current use in the taxonomic literature. We give alternative names in the text.

1
Looking at variation

The endless variety of organisms, in their beauty, complexity and diversity, gives to the biological sciences a fascination which is unrivalled by the physical world. Some recognition of these different 'kinds' of organisms is a feature of all primitive societies, for the very good reason that man had to know, and to distinguish, the edible from the poisonous plant, or the harmless from the dangerous animal. We now distinguish more than a quarter of a million species of plants and many more than a million species of animals. What do we mean by 'species'? This is one of the questions we intend to explore in the chapters that follow.

'Kinds', species and natural classification

Suppose we assemble and examine an array of living organisms. We find breaks in the pattern of variation and can recognise a number of different 'kinds'. It is not, however, a question of simple discontinuity in particular features of the form of the organism. We can appreciate this if we think of a concrete example. In an Oak–Ash wood, the trees with simple, lobed leaves are Oak (*Quercus robur*) and others with pinnately compound leaves are Ash (*Fraxinus excelsior*). Looking more closely at these two 'kinds' of trees, we can find differences in bark, twig, flower or fruit – indeed in the characters of any part of the tree. It is clearly not only discontinuity which characterises the gross variation of organisms, but also the fact that the discontinuously varying characters are highly significantly correlated. Thus we see in the case of Oak that lobed leaves, alternately arranged on the twig, go together with an acorn fruit, whilst pinnate leaves are correlated with opposite arrangement and a winged 'key' fruit in Ash. Of course, this simple distinction may not hold in every case. It is possible, for example, to see in Botanic Gardens a peculiar Ash tree with simple leaves

(*F. excelsior* forma *diversifolia*) and we know it is an Ash because it conforms in every other character. Nevertheless, the correlations of characters are strong enough to make broad agreement on the delimitation of 'kinds' a reasonably satisfactory aim for plant (or animal) taxonomy. Moreover, if we look at the 'folk taxonomies' developed independently by primitive peoples, we find a fair measure of agreement between the 'kinds' recognised, and the genera and species scientifically named and classified by botanists.

There is a further dimension to the variation pattern, however. 'Kinds' of organisms can be arranged in a hierarchy, each higher group containing one or more members of a lower group. A study of primitive biological classifications reveals that this hierarchical classification is also a widespread feature of languages in general, and the particular hierarchy of genus and species which modern biology uses is really only a special case of a general linguistic phenomenon. The inescapable conclusion from such comparative studies is that hierarchical taxonomic classifications arise in human societies wherever those societies develop, and that the detail of the treatment which we find in folk taxonomies reflects the importance of the organisms concerned in the life of the particular tribe or culture. This view of taxonomy as a product of man's need to understand, describe and use the plants and animals around him can in fact be applied to the Latin biological classification which we use at the present day.

Continuing our example, we find that the common European Ash tree, *F. excelsior*, is now grouped with some 65 other North American, East Asiatic and European species in the genus *Fraxinus* (Mabberley, 1987). This genus is in the Oleaceae, a family of 24 genera with 400 species, amongst which is also Olive (*Olea europaea*). Likewise, *Q. robur* belongs to a genus of some 600 species in the family Fagaceae. Other notable plants in the Fagaceae include Beech (*Fagus*), Southern Beech (*Nothofagus*) and Chestnut (*Castanea*).

Although we may talk of associating species into genera and genera into families, this is not what happened in the early days of biological classification. It is indeed arguable that the ordinary man's idea of a 'kind' of plant or animal corresponds in the history of classification more closely to a modern genus than it does to a species. The reason for this is that the classical and medieval ideas of 'kinds' of plants were available in the eighteenth century to Linnaeus, who stabilised the scientific names in the 'binomial' form in which we still use them; and so the Linnaean genera (*Quercus* = Oak, *Fraxinus* = Ash, etc.) indicate the level of recognition of 'kinds' in the botany of medieval Europe. This is beautifully illustrated by the Carrot family (Umbelliferae/Apiaceae), many of which were familiar plants in classical times in Europe, mainly because they were cultivated for food or flavouring (for example, *Daucus* = Carrot, *Pastinaca* = Parsnip) or because they were poisonous (for example, *Conium* = Hemlock). All these familiar European plants were accurately described and given what were later to become their generic names, long before Linnaeus. Figure 1.1 shows a page of illustrations of the fruits of Umbelliferae from the earliest monograph of a family of plants, published by Morison in Oxford in 1672. With closer examination, and particularly as the exploration of the plants of the world proceeded, Linnaeus and his successors then distinguished other 'kinds' of Oak, Ash, Carrot, etc., retaining the name of the genus for all the species so distinguished (Walters, 1961, 1962, 1986*b*).

One other important element in our understanding of the history of biological classification must now be introduced: the idea of a 'natural' classification. Looking back as we can now do, equipped with our modern evolutionary picture of the diversity of living organisms, we are apt to feel that a 'natural' classification must be one which accurately reflects the evolutionary history of the plants or animals concerned. As we shall see in later chapters, this was a view held by Charles Darwin and indeed others before him, although not widely expressed until after Darwin's ideas were generally accepted. Yet it is abundantly clear in the history of botanical and zoological thought that so-called 'natural' classifications were used and discussed long before evolution was an accepted picture. What did a 'natural' classification mean to John Ray in the seventeenth century or to Linnaeus in the eighteenth? This is a complex subject, but two things can be said. First, Ray, Linnaeus and all earlier biologists believed that, in describing and naming different 'kinds' of organisms, they were discovering a divine pattern in the created world, a pattern in which the different 'kinds' (modern genera or species) showed patterns of 'affinities' or relationships with each other. This was the 'natural order' awaiting man's discovery and it had been fixed by God, who had separately created all the different 'kinds' of plants and animals. Secondly, the idea of 'natural' classification should be logically contrasted with 'artificial' classification; in the former, the sum total of 'affinity' or resemblance is taken into account in classifying groups together, whereas in 'artificial' classifications a single character (or a small set of characters) defines the group. Thus, as we shall see, Linnaeus developed a so-called 'Sexual System' to classify his genera of plants into what we would now call 'families', but in spite of the initial success of this system, which was frankly artificial, it was superseded by the 'Natural System' of families which we use today.

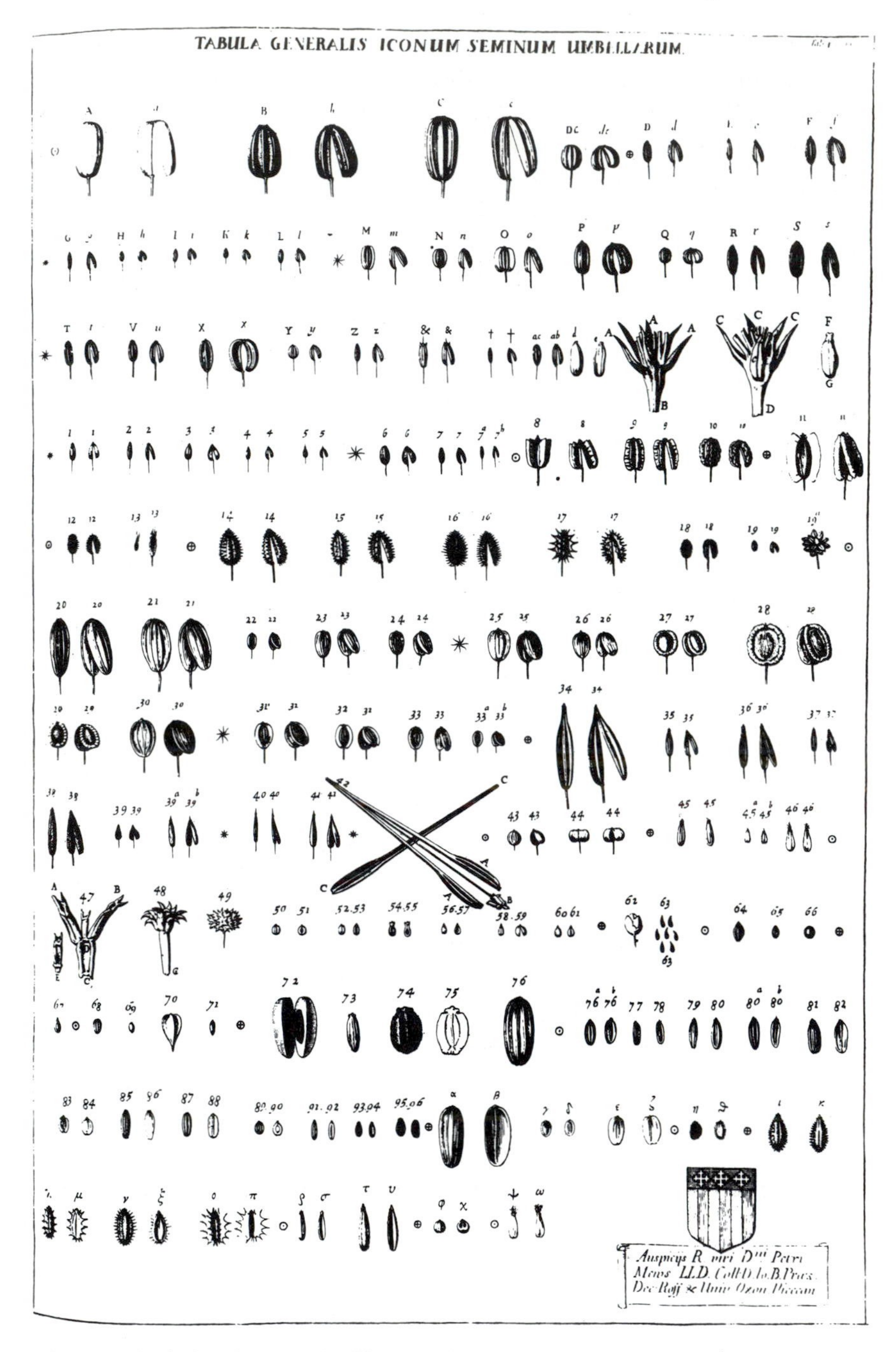

Fig. 1.1. The fruits of the Umbelliferae (Apiaceae), from Morison's monograph of the family published in 1672.

Individual variation

To introduce the main themes of this book let us first look at individual variation. If we examine carefully a group of plants of any one species it is soon clear that not all the individuals are alike. For instance, in a seedling Ash the seedling leaves are simple, quite unlike the pinnate leaves of mature individuals. Here we have to consider developmental variation. Another source of variation between individuals can be plausibly attributed to factors of the external environment, such as coppicing by man or grazing by animals. Further, we may note that adjacent specimens of ordinary Ash and 'Weeping' Ash remain distinct in cultivation. Thus we can distinguish three main types of differences between individuals, which we might call 'developmental', 'environmentally induced' and 'intrinsic'. For many purposes we may very usefully distinguish, in a study of variation, a component, fixed and heritable, which is the intrinsic genetic character of the 'kind' (ordinary growth habit versus 'weeping' habit in the case of Ash), from a component which is environmentally induced, non-heritable and imposed, as it were, from outside. In addition to these we have the phenomenon of developmental variation, by which the adult differs, often strikingly, from the immature individual.

In many species of plants, including the Oak and Ash, individuals at all stages from the seedling to the mature plant may be studied. However, in other species, for example the grasses in a lawn, the delimitation of the individual presents problems. Many species spread freely, in some cases exclusively, by vegetative means, the parts of the plant being, at first, connected, but ultimately becoming separated to produce clonally derived individuals. In such cases the simple concept of the individual is inapplicable. Looked at from this viewpoint, there is more in common between the higher animals (including man) and the Ash, than the Ash trees and certain grass species in the turf. We are, perhaps inclined to think that man and Ash are 'normal', for in these species individuals are born, mature and die, while the clonally reproducing plants once established would appear to be 'odd' in being potentially 'immortal'. We ought to see this for what it is – an anthropocentric distortion. In later chapters we shall examine recent work on the nature of individual variation, including some of the properties of clones.

The nature of species

The main problems to be examined in this book concern the nature of species. Anyone familiar with the vegetation of an area has to face a number of questions which have puzzled biologists increasingly in the last 100 years. For instance, why are certain species clearly distinct, while in other cases we find a galaxy of closely similar species, often difficult to distinguish from each other? Is there, in fact, any objective way of delimiting species? Why is it that hybridisation occurs in certain groups of plants and not in others? Further, how can one account for the different degrees of intraspecific variation found in species?

In the early nineteenth century an examination of these questions, as we shall see in Chapter 2, produced a static picture of variation. Since 1859, however, with the publication of *On the origin of species*, all such studies have been made in the light of Darwin's profound generalisation of evolution by natural selection. Even though this theory has not always been accepted, it has had a tremendous impact on all fields of biology. Nowadays, the fact of evolution is taken for granted, in part because of the wealth of evidence assembled by Darwin and other scientists. There is often at the same time an uncritical acceptance of the theory – a tendency to say 'it must be true, for it is in all the books'. Implicit in Darwin's ideas is the assumption that evolution is still taking place. Thus in this book we shall not only look at the problems of species and patterns of variation but also the evidence for evolution, particularly experimental evidence for evolution on a small scale, often called 'microevolution'.

In discussing variation and microevolution it is essential to realise that the basic raw material for our studies exists in every country of the world. Even though we use mainly European and North American examples, because in these regions variation has been most carefully examined, similar examples can be found in countries where the flora is comparatively unknown. There is a further point of importance. It is not only 'natural', unspoiled vegetation which we can usefully study; equally illuminating results may be obtained from the study of communities radically altered by man, and in fact some of the most important insights into microevolution have come from studies of introduced plants, agricultural crops, weeds, and the vegetation of areas subject to pollution. Such studies, as we shall see, have revealed how certain species have adapted successfully in agricultural and urban industrial landscapes. Botanists have also been concerned with extinctions, for it is now believed that 10% or more of the world's flora is endangered by man's activities. Guided by our knowledge of pattern and process of microevolutionary change, we will consider the steps that are being taken to prevent this catastrophic loss of biodiversity.

2

From Ray to Darwin

In 1660, Robert Sharrock, Fellow of New College, Oxford, wrote a book entitled *History of the propagation and improvement of vegetables by the concurrence of art and nature.* He was concerned in its early pages to debate a live issue of the day (Bateson, 1913), namely:

> It is indeed growen to be a great question, whether the transmutation of a species be possible either in the vegetable, Animal or Minerall Kingdome. For the possibility of it in the vegetable; I have heard Mr Bobart and his Son often report it, and proffer to make oath that the Crocus and Gladiolus, as likewise the Leucoium, and Hyacinths by a long standing without replanting have in his garden changed from one kind to the other.

The Bobarts were both professional botanists. Sharrock investigated their claim, and found '...diverse bulbs growing as it were on the same stoole, close together, but no bulb half of the one kind, and the other half of the other'. In this age we find it hard to understand a belief in the possibility of transformation of Crocus into Gladiolus. Our reason for disbelief is partly concerned with the nature of evidence; we are not satisfied with the test for the alleged transmutation and would not have been content merely to examine the crowded underground parts. Another reason, however, relates to current ideas of the nature of species. We have a different notion of species from that of the seventeenth century.

Ray and the definition of species

It was the English naturalist John Ray (1628–1705) who was probably the first man to seek a scientific definition of species (Raven, 1950). In his definition is an implied rejection of the sort of transmutation of species claimed by the Bobarts of Oxford, although in other passages in Ray's work he does not wholly dismiss the possibility of transmutation. For instance, he cites as reliable the case of Cauliflower seed supplied by a London dealer, which on germination produced Cabbage. Richard Baal, who sold the seed, was tried for fraud and ordered by the court at Westminster to refund the purchase money and pay compensation (De Beer, 1964).

Ray's views on species were published in 1686 in *Historia plantarum.* He wrote (trans. Silk in Beddall, 1957):

> In order that an inventory of plants may be begun and a classification of them correctly established, we must try to discover criteria of some sort for distinguishing what are called 'species'. After a long and considerable investigation, no surer criterion for determining species has occurred to me than distinguishing features that perpetuate themselves in propagation from seed.

He is concerned to define species as groups of plants which breed true within their limits of variation. This definition of species, based as it is partly upon details of the breeding of the plant, was a great advance upon older ideas, which relied entirely upon consideration of the external form.

Ray was also very interested in intraspecific variation. In his letters to various friends (collected by Lankester, 1848), he noted several striking variants of common plants discovered on his journeys around Britain. For example, at Malham in Yorkshire he noticed white-flowered as well as the normal blue-flowered Jacob's Ladder (*Polemonium caeruleum*), and from other localities he reported white-flowered Foxglove (*Digitalis purpurea*), double-flowered specimens of Water Avens (*Geum rivale*) and white-flowered Red-Rattle (*Pedicularis palustris*). Ray also made observations on a prostrate variant of Bloody Cranesbill (*Geranium sanguineum* var. *prostratum*). He wrote to a friend: 'Thousands hereof I found in the Isle [Walney] and have sent roots to Edinburgh, York, London, Oxford, where they keep their distinction'. This report on the constancy of this distinct variant of *G. sanguineum* in cultivation is of particular interest, and is referred to again in Chapter 4.

We may learn more of Ray's ideas on the nature of species and intraspecific variation by examining a discourse given to the Royal Society on 17 December 1674 (Gunther, 1928). In this, he expresses his concern that great care should be taken in deciding what constitutes a species and what variation is insufficient for specific distinction. He shows, for instance, that within a species there might occur individuals different from the normal in one or more of the following characters: height, scent, flower colour, multiplicity of leaves, variegation, doubleness of flower, etc. Plants differing by such 'accidents', as Ray calls them, should not be given specific status. He records the origin of one notable variant in his own garden: 'I found in my own garden, in yellow-flowered Moth-Mullein (*Verbascum*), the seed whereof sowing itself, gave me some plants with a white flower'. Concerning other variants Ray suggests that they are caused by growing plants under unnatural conditions, for example a rich or a poor soil, extreme heat and so on.

He concludes his analysis of specific differences and the problem of intraspecific variation as follows:

> By this way of sowing [rich soil, etc.] may new varieties of flowers and fruits be still produced in infinitum, which affords me another argument to prove them not specifically distinct; the number of species being in nature certain and determinate, as is generally acknowledged by philosophers, and might be proved also by divine authority, God having finished his work of creation, that is, consummated the number of species, in six days.

Ray's views on the origin of specific and intraspecific variation are here laid bare. Given sufficient regard for the variation patterns of a particular group of plants, a botanist should be able to avoid elevating 'accidental' variants to the level of the species. Species themselves were, for Ray, all created at the same time and all therefore of the same age. That new species can come into existence, Ray denies, as this is inconsistent with the account of the Creation given in Genesis. This idea is again expressed in a passage written towards the end of his life: 'Plants which differ as species preserve their species for all time, the members of each species having all descended from seed of the same original plant' (Stearn, 1957).

Ray, an ordained minister himself, firmly upholds the doctrine of special creation. This view was almost universally accepted in the seventeenth century, Protestants being particularly influenced by the works of Milton. Indeed, a fundamentalist approach to the Biblical account of the Creation was characteristic of most biologists until the middle of the last century.

The Chain of Being

A very powerful idea underlay the attitudes of philosophers and theologians throughout Classical and Medieval Europe when they attempted to understand the world. This world-view was Platonic in origin, and consisted of two parts which, at the risk of oversimplification, could be presented as an all-powerful but ultimately unknowable God, and a fixed order of lower creatures, including man whose position on the scale or 'Chain of Being' was 'lower than the angels' but higher than the 'brute beasts'. Lower organisms including plants were at the base of the chain. The relationship between God and Nature was one of the great philosophical questions which underlay the rise of natural science in the sixteenth and seventeenth centuries. The acceptance by great pioneer biologists such as Ray of the natural world as rational and available to observation and experiment marked the beginning of modern science, and the '*Scala Naturae*' or 'Chain of Being' was, in time, transformed from the rigid timeless Plan to a blue-print of biological evolution. There is one outstanding contribution to this subject to which interested readers must be referred: Lovejoy (1936, repr. 1966); further valuable comment can be found in the definitive biography of Ray by Raven (1942, 1950, 1986) and, in a wider context, in Morton (1981, especially Chapter 6).

Linnaeus

In our examination of historical aspects of the subject, we must next study Linnaeus (Carl von Linné, 1707–78), the great Swedish systematist, who made extremely important contributions (Goerke, 1989). He too, in *Critica botanica* (1737), championed the idea of the fixity of species:

> All species reckon the origin of their stock in the first instance from the veritable hand of the Almighty Creator: for the Author of Nature, when He created species, imposed on his Creations an eternal law of reproduction and multiplication within the limits of their proper kinds. He did indeed in many instances allow them the power of sporting in their outward appearance, but never that of passing from one species to another. Hence today there are two kinds of difference between plants: one a true difference, the diversity produced by the all-wise hand of the Almighty, but the other, variation in the outside shell, the work of Nature in a sportive mood. Let a garden be sown with a thousand different seeds, let to these be given the incessant care of the Gardener in producing abnormal forms, and in a few years it will contain six thousand varieties, which the common herd of Botanists calls species. And so I distinguish the species of the Almighty Creator which are true from the abnormal varieties of the Gardener: the former I reckon of the highest importance because of their Author, the latter I reject because of their authors. The former persist and have persisted from the beginning of the world, the latter, being monstrosities, can boast of but a brief life.
> *(trans. Hort, 1938)*

The approaches of both Ray and Linnaeus were typological; they upheld the Greek philosophical view that beneath natural intraspecific variation there existed a fixed, unchangeable type of each species. It was the job of botanists to see these 'elemental species'; 'natural variation' was in a sense an illusion.

We see also in the passage quoted above that Linnaeus had a very similar attitude to intraspecific variation to that of Ray. Stearn (1957), in an interesting analysis of the origin of Linnaeus' views, draws attention to his love for gardening and his experience as personal physician and superintendent of gardens to George Clifford, a banker and director of the East India Company. For some years Linnaeus, working on his great illustrated book on the plants in Clifford's gardens – the *Hortus cliffortianus* – lived at Hartekamp, near Haarlem, in the centre of the Dutch bulb-growing area. Here thousands of varieties of Tulips and Hyacinths were grown. Linnaeus wrote the *Critica botanica* during this period, and no doubt his personal observations at the time prompted the following outburst: 'Such monstrosities, variegated, multiplied, double, proliferous, gigantic, wax fat and charm the eye of the beholder with protean variety so long as gardeners perform daily sacrifice to their idol: if they are neglected these elusive ghosts glide away and are gone'.

Other observations of Linnaeus in the *Critica botanica* show his familiarity with variation in wild plants and his experimental approach to problems. For instance, he studied flower colour, noting that purple flowers tend to fade after a few days, turning to a bluish colour; but '... sprinkle these fading flowers with any acid, and you will recover the pristine red hue'. Concerning aquatic plants he notes: 'Many plants which are purely aquatic put forth under water only multifid leaves with capillary segments, but above the surface of the water later produce broad and relatively entire leaves. Further, if these are planted carefully in a shady garden, they lose almost all these capillary leaves, and are furnished only with the upper ones, which are more entire.' As an example, Linnaeus gives *Ranunculus aquaticus folio rotundo et capillaceo*, the aquatic species of *Ranunculus* to which we refer later.

Linnaeus was particularly interested in cultivation and its effect upon plants:

> *Martagon sylvaticum* is hairy all over, but loses its hairiness under cultivation.
>
> Hence plants kept a long while in dry positions become narrow-leaved, as *Sphondylium, Persicaria*... Hence broad-leaved plants, when grown for a long while in spongy, fertile, rich soil have been known to produce curly leaves, and have been distinguished as varieties, ... the following have been distinguished as '*crispum*': *Lactuca, Sphondylium, Matricaria*, etc.

The early botanical work of Linnaeus is extremely important in the history of ideas about species and variation. He championed firmly the reality, constancy and sharp delimitation of species. He was also concerned to refute the Ancient Greek idea of transmutation of species, which was still widely believed in his day. In *Critica botanica* he wrote:

No sensible person nowadays believes in the opinion of the Ancients, who were convinced that plants 'degenerate' in barren soil, for instance, that in barren soil Wheat is transformed into Barley, Barley into Oats, etc. He who considers the marvellous structure of plants, who has seen flowers and fruits produced with such skill and in such diversity, and who has given more credence to experiments of his own, verified by his own eyes, than to credulous authority, will think otherwise.

Linnaeus is immortalised for botanists by his great work *Species plantarum* (1753), in which are described in a concise and methodical fashion all the approximately 5900 species of plants then known to man. In classifying these species, Linnaeus grouped species into genera and genera into classes on the basis of the number and arrangement of their stamens. These groups were then subdivided into orders based on the number of pistils. This classification, which Linnaeus acknowledged as artificial, was to be a preliminary to a more natural classification – of which he only produced a fragment – based on overall resemblance. Linnaeus' classification of the plant kingdom did not yield a 'Chain of Being' but a hierarchical classification which he likened to the pattern of countries on a map (Jonsell, 1978; Bowler, 1989*b*).

Linnaeus' concept of the species seems to have been subject to change as his experience grew. In early works, and most explicitly in the theoretical *Philosophia botanica* (1751), he stresses the clear distinction between *species*, which were constituted as such by the Creator from the beginning, and mere *varieties*, which may be induced by changed environmental conditions, or raised by the art of gardeners. Nevertheless, not infrequently in *Species plantarum* there are comments which show that Linnaeus did not always find specific distinctions clear: for example, under *Rosa indica* we find that 'the species of *Rosa* are with difficulty to be distinguished, with even greater difficulty to be defined; nature seems to me to have blended several or by way of sport to have formed several from one' (Stearn, 1957). It is even true that Linnaeus speculates, in a few cases, on the possible evolutionary derivation of one species from another in the pages of *Species plantarum*. Thus, under *Beta vulgaris* we find (Fig. 2.1), after a list of seven agricultural crop varieties, the fascinating statement: 'Probably born of *B. maritima* in a foreign country'. *B. maritima*, the Wild Beet, is given separate treatment as a distinct species! This and several other cases are interestingly discussed by Greene (1909), who points out that there is good evidence to support the view that the dogmatic 'special creation' statements of *Philosophia botanica*

BETA.

maritima. 1. BETA caulibus decumbentibus.
Beta caulibus decumbentibus, foliis triangularibus petiolatis. *Mill. dict.*
Beta ſylveſtris maritima. *Baub. pin.* 118. *Raj. angl.* 4. *p.* 127.
Habitat in Angliæ, Belgii *littoribus maris.*

vulgaris. 2. BETA caule erecto.
Beta. *Hort. cliff.* 83. *Hort. upſ.* 56. *Mat. med.* 113. *Roy. lugdb.* 220.
rubra. α. Beta rubra vulgaris. *Baub. pin.* 118.
β. Beta rubra major. *Baub. pin.* 118.
γ. Beta rubra, radice rapæ. *Baub. pin.* 118.
δ. Beta lutea major. *Baub. pin.* 118.
ε. Beta pallide virens major. *Baub. pin.* 118.
Cicla. ζ. Beta alba vel palleſcens, quæ Cicla officinarum. *Baub. pin.* 118.
η. Beta communis viridis. *Baub. pin.* 118.
Habitat - - - - -, ♂, forte a maritima, in exoticis, prognata.

Fig. 2.1. Treatment of *Beta* in Linnaeus' *Species plantarum*, 2nd edn, 1763. (Now called *Beta vulgaris* with subspp. *maritima*, *cicla* and *vulgaris*.)

and similar writings of Linnaeus did not, even in his earlier days, represent Linnaeus' real views, but were diplomatic writings to satisfy the 'orthodox ecclesiastics who, in his day, ruled the destinies of all seats of learning in Sweden'.

If he was orthodox on these matters in the main works, which established his academic and scientific reputations, Linnaeus allowed himself much more freedom in several of the 186 'dissertations' which his research students, following the medieval rules of disputation, had to 'defend' in Latin. It is clear from these writings that Linnaeus came to believe less rigidly in the fixity of species. For instance, in 1742 a student brought to him, from near Uppsala, an unusual specimen of Toadflax (*Linaria vulgaris*). The flower was not of the usual structure but had five uniform petals and five spurs. Experiments showed that the plant bred true and Linnaeus called it *Peloria* (Fig. 2.2). After close study Linnaeus decided that *Peloria* was a new species which had arisen from *L. vulgaris* (Linnaeus, 1744). He also considered that certain other species might have arisen as a result of hybridisation (Linnaeus (1749–90; Erikkson, 1983). In

(*a*)

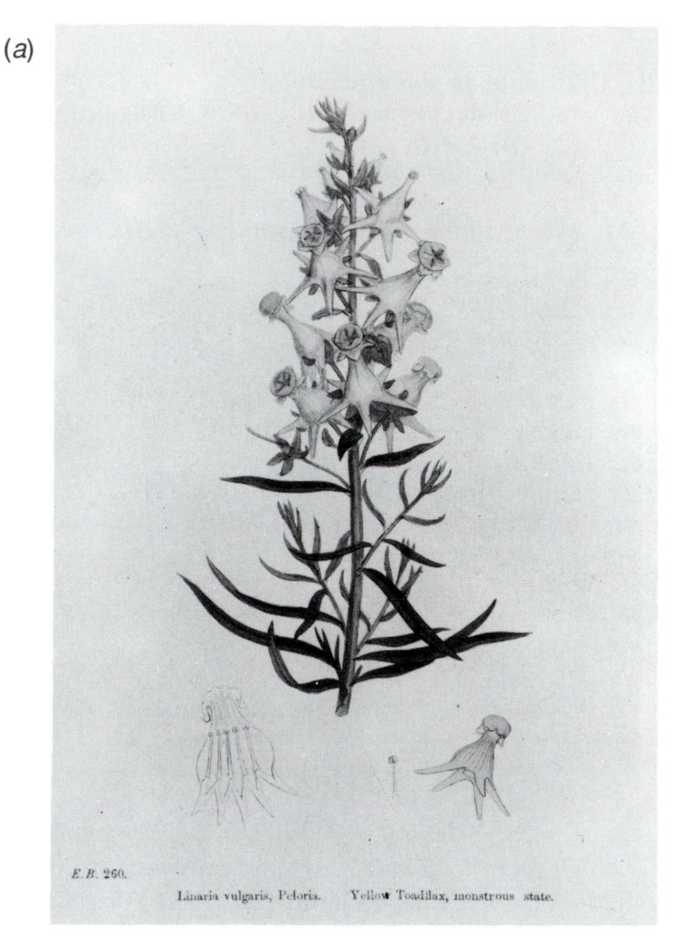

Fig. 2.2. *Linaria vulgaris* (*a*) and its *Peloria* variant (*b*). (Illustrations by Sowerby in Boswell Syme, 1866.)

Plantae hybridae (1751) records are given of 100 plants which might be regarded as hybrids. In *Somnus plantarum* (1755) we read:

The flowers of some species are impregnated by the farina [pollen] of different genera, and species, inasmuch that hybridous or mongrel plants are frequently produced, which if not admitted as new species, are at least permanent varieties.

Later, in the summer of 1757, Linnaeus made what might be considered to be the first scientifically produced interspecific hybrid, between the Goatsbeards, *Tragopogon pratensis* (yellow flowers) and *T. porrifolius*

(*b*)

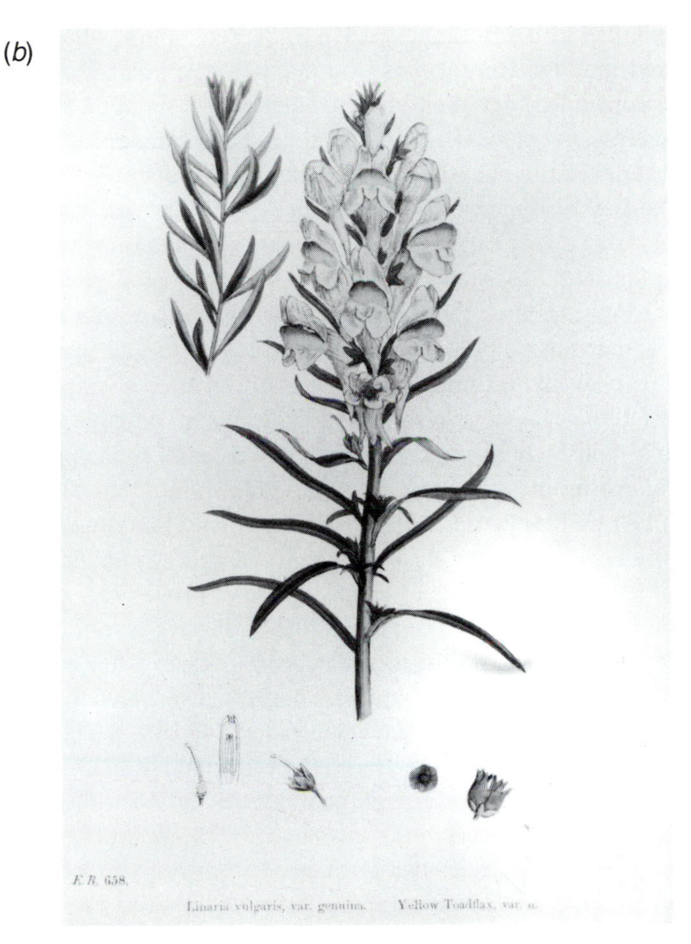

(violet flowers). Ownbey (1950), who studied *Tragopogon* in America, gives the following details of Linnaeus' experiment. After rubbing the pollen from the flower-heads of *T. pratensis* early in the morning, Linnaeus sprinkled the stigmas with pollen of *T. porrifolius* at about 8 a.m. The flower-heads were marked, the seed eventually harvested and subsequently planted. The first generation hybrid plants flowered in 1759, producing purple flowers yellow at the base. Seed of the cross, together with an account of the experiment and its bearings upon the problems of the sexuality of plants, formed the basis for a contribution to a competition arranged by the Imperial Academy of Sciences at St Petersburg. Linnaeus was awarded the prize in September 1760. It is of great historical interest that the seed sent by

Linnaeus was planted in the Botanic Garden in St Petersburg, where the progeny flowered in 1761. Here it was examined by the great hybridist Kölreuter, who concluded that 'the hybrid Goatsbeard ... is not a hybrid plant in the real sense, but at most only a half hybrid, *and indeed in different degrees*'. It is also interesting that the second generation progeny produced by the inter-crossings of Linnaeus' hybrid plants clearly showed segregation of different types, a very early record of genetic segregation which we discuss in Chapter 4.

We see how Linnaeus came to believe that, as in the case of *Peloria*, certain species had arisen from others in the course of time, and that new species could arise by hybridisation. There is, however, contemporary evidence against Linnaeus' views (Glass, 1959). Adanson, an eighteenth-century French botanist whose originality has only recently been appreciated, tested *Peloria* more fully than Linnaeus. He found that *Peloria* specimens supplied by Linnaeus to the Paris Jardin des Plantes were not stable, producing flowering stems with both 'peloric' and normal flowers. Germination of seed of these plants often gave normal progeny as well as 'peloric'. Adanson concluded that the plant was a monstrosity, not a new species. He came to similar conclusions in two other cases, after experiments with an entire-leaved Strawberry (*Fragaria*) discovered by the horticulturalists Duchesnes and son at Versailles in 1766, and the famous laciniate plant of *Mercurialis annua* discovered by Marchant in 1715. There was also evidence against the origin of new species by hybridisation. Kölreuter made a large number of crosses in Tobacco (*Nicotiana*) and other genera. True-breeding new species were not produced by hybridisation; indeed the hybrids were often almost completely sterile, and even when they were fertile there was great variation in the progeny.

Returning to the writings of Linnaeus, we find that in later life he also gave further thought to the origins of the patterns of variation in plant groups. He speculated on the Creation as follows (*Fundamenta fructificationis*, 1762, trans., quoted from Ramsbottom, 1938):

> We imagine that the Creator at the actual time of creation made only one single species for each natural order of plants, this species being different in habit and fructification from all the rest. That he made these mutually fertile, whence out of their progeny, fructification having been somewhat changed, Genera of natural classes have arisen as many in number as the different parents, and since this is not carried further, we regard this also as having been done by His Omnipotent hand directly in the beginning; thus all Genera were primeval and constituted a single Species. That as many Genera having arisen as there were individuals in the beginning, these plants in course of time became fertilised by others of different sort

and thus arose Species until so many were produced as now exist ... these Species were sometimes fertilised out of congeners, that is other Species of the same Genus, whence have arisen Varieties (Fig. 2.3).

Linnaeus ascribes here almost an evolutionary origin to present-day species, genera having been formed at the Creation, species-formation being a more recent process (Erikkson, 1983; Linroth, 1983). This most important change in Linnaeus' views relates to his hybridisation studies. He appears to have been convinced in later life that species can arise by hybridisation, and moved away from the idea of a fixed number of species all created at the same moment in time. Linnaeus' early views on the fixity of species received wide circulation in Europe in his main works, *Critica botanica*, *Systema naturae* and *Species plantarum*, while his more mature views, presented in the dissertations, did not have such a wide readership. So it is not surprising that even today he is often credited with rigid views on the question.

Buffon and Lamarck

In the mid-eighteenth century zoologists, too, were considering special creation. Linnaeus' contemporary, the French zoologist Buffon (1707–88), had also started his career with orthodox beliefs: 'We see him, the Creator, dictating his simple but beautiful laws and impressing upon each species its immutable characters'. Later, in 1761, however, he speculated on the mutability of species: 'How many species, being perfected or degenerated by the great changes in land and sea, ... by the prolonged influences of climate, contrary or favourable, *are no longer what they formerly were*?' (Osborn, 1894).

The speculative ideas of Buffon and others remained untested by experiment; the majority of botanists and zoologists, engaged as they were in the late eighteenth century on the naming and classification of the world's flora and fauna, believed in the fixity of species. This belief was indeed so firmly held by naturalists that Cuvier (1769–1832), who had studied many fossil animals, accounted for extinct species by postulating a series of great natural catastrophies, which wiped out certain intermediate species. Cuvier believed that there had been only one Creation and that after each disaster the earth was repopulated by the offspring of the survivors. The last catastrophe was the Great Flood recorded in Genesis.

The doctrine of fixity of species was not without its critics in the nineteenth century. Lamarck (1744–1829), in his *Philosophie zoologique*

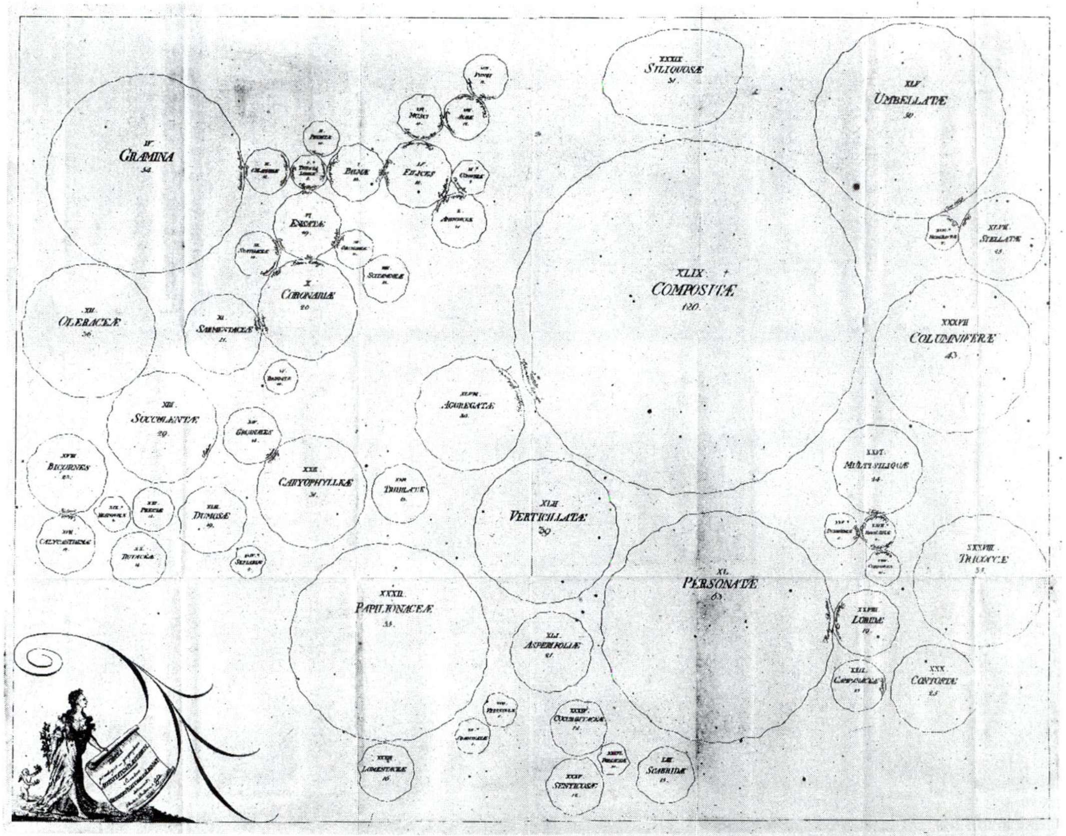

Fig. 2.3. Representation of Linnaeus' idea on the relationships between plant families, by his pupil Giseke in 1792. The Roman numerals given for each 'natural order' – i.e. family – are as in Linnaeus' *Genera plantarum* (6th edn, 1764), whilst Arabic numerals give the number of genera. The size of each circle is roughly proportional to the size of the family expressed as number of genera. At the edge of certain circles appear the names of genera which bridge the gap between groups. Such groups were intended to be a well-hidden secret, hence the more or less illegible names. Jonsell (1978) sees this as yet another example of Linnaeus' view that new forms could arise by hybridisation.

(1809), attacked the belief that all species were of the same age, created at the beginning of time in a special act of Creation. He believed, much as Ray and Linnaeus did, that species could be changed by growth in different environments, but he also believed that modifications in plant structure brought about by environmental change were inherited (Elliot, 1914):

> In plants, ... great changes of environment ... lead to great differences in the development of their parts ... and these acquired modifications are preserved by reproduction among the individuals in question, and finally give rise to a race quite distinct from that in which the individuals have been continuously in an environment favourable to their development... Suppose, for instance, that a seed of one of the meadow grasses... is transported to an elevated place on a dry, barren and stony plot much exposed to the winds, and is there left to germinate; if the plant can live in such a place, it will always be badly nourished, and if the individuals reproduced from it continue to exist in this bad environment, there will result a race fundamentally different from that which lives in the meadows and from which it originated.

Thus Lamarck believed that a normally tall plant, dwarfed by growth at high altitude, would produce dwarf offspring. His belief in such an inheritance of acquired characters, which is closely paralleled in the writings of Erasmus Darwin (1731–1802), formed the basis of his evolutionary speculation – one species evolved into another as hereditary changes arose in a plant under the impact of environmental variation. Lamarck, who suffered ill-health at the end of his life and was totally blind for the last 10 years, did not make any experimental investigations in search of evidence for his hypothesis (Jordanova, 1984). He did, however, cite a number of possible cases of apparent change of species brought about by environmental agency. For example:

> So long as *Ranunculus aquatilis* is submerged in the water, all its leaves are finely divided into minute segments; but when the stem of this plant reaches the surface of the water, the leaves which develop in the air are large, round and simply lobed. If several feet of the same plant succeed in growing in a soil that is merely damp without any immersion, their stems are then short, and none of their leaves are broken up into minute divisions, so that we get *Ranunculus hederaceus*, which botanists regard as a separate species.

In this interesting quotation we see that Lamarck puts quite a different interpretation upon variation exhibited by aquatic *Ranunculus* species, from that of Linnaeus, who considered such changes in leaf characters part of intraspecific variation. We consider modern interpretations of this variation in Chapter 6.

Darwin

The views of Lamarck, at least in their simple form, are now rejected by most biologists. Our ideas on evolution are based on the work of Charles Darwin (1809–82). In an attempt to understand observations made on the voyage of the *Beagle*, Darwin began a series of note books on transmutation, the first dated July 1837 (De Beer, 1960–61), and he also wrote a sketch of his views in 1842 and a longer 'essay' in 1844 (Francis Darwin, 1909*a*, *b*). Recently a new edition of all these note books has been published (Barrett *et al.*, 1988) and the first complete edition of Darwin's correspondence is being produced (Burkhardt & Smith, 1985–).

Many historians have investigated the development of Darwin's ideas, as revealed by his writings and annotations in his books (Smith, 1960; Schweber, 1977; Kohn, 1985; Bowlby, 1990; Desmond & Moore, 1991; Browne, 1995), research which suggests the crucial role of such writers as Lyell, Comte, Adam Smith, Quetelet and Malthus. Darwin delayed publication of his work, possibly because he wished to collect further information relevant to a theory of evolution (De Beer, 1963). It has also been suggested that he was concerned about the likely social consequences of his theory (Desmond & Moore, 1991). In 1844, *Vestiges of the natural history of Creation*, a famous book advocating an evolutionary interpretation of nature, was published anonymously. It received strong condemnation in reviews and this may have contributed to the delay in the publication of Darwin's ideas (Schweber, 1977). In 1858, Darwin received an essay from the naturalist Wallace (1823–1913), which set out a hypothesis almost identical to his own. It has been suggested that Darwin plagiarised Wallace's ideas (Brackman, 1980; Brooks, 1983), but this has been firmly rebutted by Kohn (1981). The whole question is discussed at length in the Introduction to Volume 7 of *The correspondence of Charles Darwin* (Burkhardt & Smith, 1991). Bowler (1989*b*) may be consulted for a discussion on whether Wallace has been fairly treated as co-discoverer of natural selection or whether, as many authors have hinted, 'history has been less than just to him'.

Darwin's friends helped to resolve the delicate question of priority (Burkhardt & Smith, 1991) and, at a meeting of the Linnean Society on 1 July 1858, at which Wallace's paper was read, Darwin's views were represented by unpublished extracts from his papers and a letter he wrote to Professor Asa Gray in 1857. The contributions of Darwin and of Wallace, which were prefaced by a letter from Lyell and Hooker explaining the historical background, have been reprinted in Jameson, 1977.

The main strands of the hypothesis of Darwin and Wallace may be summarised as follows:

1. Plants and animals vary. Darwin recognised two sorts of intraspecific variation: discontinuous variants (sports, monstrosities, jumps, saltations) and continuous variations (small, slight or individual differences, deviations or modifications). In his letter to Gray, Darwin wrote '*Natura non facit saltum*', making clear his view that it was individual differences and not saltations which were important in evolution.
2. Because of the fecundity of organisms there would be a geometrical increase in numbers unless checked. Such natural checks occur. Darwin and Wallace both acknowledged a debt to Malthus in their understanding of natural checks to population increase (Habakkuk, 1960; Pantin, 1960).
3. As a consequence of these checks, only those individuals survive which have an inherent advantage over others in the population.
4. These better-fitted organisms, surviving this 'natural selection', pass on their 'advantage' to a proportion of their offspring.
5. Selection continues over thousands of generations and in a rapidly changing environment new variants take the place of the original organisms.

The principal ideas of Darwin's hypothesis are set out in the following quotations from the extract read at the Linnean Society meeting:

De Candolle, in an eloquent passage, has declared that all nature is at war, one organism with another, or with external nature... It is the doctrine of Malthus applied in most cases with tenfold force... Reflect on the enormous multiplying power *inherent and annually in action* in all animals; reflect on the countless seeds scattered by a hundred ingenious contrivances, year after year, over the whole face of the land; and yet we have every reason to suppose that the average percentage of each of the inhabitants of a country usually remains constant. Finally, let it be borne in mind that this average number of individuals (the external conditions remaining the same) in each country is kept up by recurrent struggles against other species or against external nature (as on the borders of the Arctic regions, where the cold checks life), and that ordinarily each individual of every species holds its place, either by its own struggle and capacity of acquiring nourishment in some period of its life, from the egg upwards; or by the struggle of its parents ... with other individuals of the *same* or *different* species. But let the external conditions of a country alter... Now, can it be doubted, from the struggle each individual has to obtain subsistence, that any minute variation in structure, habits or instincts, adapting that individual better to the new conditions, would tell upon its vigour and health? In the struggle it would have a better *chance* of surviving; and those of its offspring which inherited the variation, be it ever so slight, would also have a better *chance*. Yearly more are bred than can survive; the smallest grain in the balance, in the long run, must tell on which death shall fall, and which shall survive. Let this work of selection ... go on for a thousand generations, who will pretend to affirm that it would produce no effect...

Darwin then goes on to give an example:

If the number of individuals of a species with plumed seeds could be increased by greater powers of dissemination within its own area (that is, if the check to increase fell chiefly on the seeds), those seeds which were provided with ever so little more down, would in the long run be most disseminated; hence a greater number of seeds thus formed would germinate, and would tend to produce plants inheriting the slightly better-adapted down.
(Darwin & Wallace, 1859)

After the meeting of the Linnean Society, Darwin spent the next few months writing the text which was eventually published in 1859 under the title *On the origin of species by means of natural selection.* Darwin saw the *Origin* as an introduction to a series of works. As Vorzimmer (1972) has shown, the book, which provided insights into so many facets of biology, provoked very considerable controversy and, reworking the material to accommodate the views of critics and the development of his own ideas, Darwin produced six editions in all. Sometimes substantial changes were made, as can be seen in the *variorum* text of the *Origin* produced by Peckham (1959).

We rightly give credit to Darwin for establishing, against considerable opposition, a plausible mechanism in natural selection to explain organic evolution. As is usually the case in the history of ideas, however, a careful reading of the literature reveals a number of statements of 'selectionist' ideas long before Darwin and Wallace. Indeed Zirkle (1941), in a remarkably interesting and little-quoted paper, provides abundant evidence of such ideas, tracing them back to the writings of Empedocles (495–435 BC) and Lucretius (99–55 BC). In the sixth edition of the *Origin* Darwin himself provides a short historical review in which he makes it clear that the idea of natural selection had occurred to others and indeed had been published, for example by Dr W. C. Wells (1818) in 'An account of a white female, part of whose skin resembles that of a Negro' and by Patrick Matthew (1831) in his book *On naval timber and arboriculture*.

Many of the difficulties raised by contemporary critics, almost all of which are of interest to modern biologists, are very involved. Mivart (1871) produced a fascinating book which reviews all the contemporary difficulties in accepting Darwin's views, and Vorzimmer (1972) may be consulted for full details of many problems; here is a brief treatment of some of the most important.

The role of saltations in evolution

As Darwin pointed out, discontinuous variants – sports, monstrosities, etc. – are often sterile and therefore of little or no consequence in evolution. However, various biologists including Harvey (who studied a mutant *Begonia*, 1860), Huxley (in his review of the first edition of the *Origin*) and Asa Gray all suggested that saltations may be important in evolution.

> When you suppose one species to pass, by insensible degrees into another, so many facts of variation support your view that it does not seem very improbable; but where a generic limit has to be passed, bearing in mind how *persistent* generic differences are, I think we require a *saltus* (it may be a small one) or a real break in the chain, namely, a sudden divarication.
> *(Unpublished letter from Harvey to Darwin, quoted by Vorzimmer, 1972)*

The mechanisms of heredity

In many ways this was the most important problem raised by Darwin's critics. For evolution to take place there must be selection of favoured varieties, such variants on crossing leaving better-adapted offspring. Darwin was unable to understand the mechanism of heredity, believing in a type of blending inheritance (see Chapter 4). Fleeming Jenkin, Professor of Engineering at Edinburgh University, writing anonymously in *The North British Review*, June 1867, showed a very serious weakness in Darwin's argument. He pointed out that if a rare variant favoured by natural selection appeared in a population, it would cross with the more abundant less-favoured plants in the population. Its hereditary advantage would then be lost in blending inheritance. How did favoured genetic variants ever become abundant? Darwin was never able satisfactorily to answer this criticism. Apparently Jenkin was not alone in appreciating this particular difficulty, for Vorzimmer (1972) argues that before Jenkin's review several biologists, including Darwin himself, were aware of the implication of blending inheritance.

The effect of chance

Jenkin also drew attention to another important problem – the effect of chance on the survival of favoured variants. He discusses his ideas in relation to hares:

> ... let us here consider whether a few hares in a century saving themselves by this process [burrowing] could, in some indefinite time, make a burrowing species of hare. It is very difficult to see how this can be accomplished, even when the sport is very eminently favourable indeed; and still more difficult when the advantage gained is very slight, as must generally be the case. The advantage, whatever it may be, is utterly out-balanced by numerical inferiority. A million creatures are born; ten thousand survive to produce offspring. One of the million has twice as good a chance as any other of surviving; but the chances are fifty to one against the gifted individuals being one of the hundred survivors. No doubt, the chances are twice as great against any one other individual, but this does not prevent their being enormously in favour of *some* average individual. However slight the advantage may be, if it is shared by half the individuals produced, it will probably be present in at least fifty-one of the survivors, and in a larger proportion of their offspring; but the chances are against the preservation of any one 'sport' in a numerous tribe.

The limits of variation

While not the first to raise this problem, Jenkin, in the following quotations, points to what many biologists saw as an important difficulty in Darwin's theory:

> If we could admit the principle of a gradual accumulation of improvements, natural selection would gradually improve the breed of everything, making the hare of the present generation run faster, hear better, digest better, than his ancestors; ... Opinions may differ as to the evidence of this gradual perfectibility of all things, but it is beside the question to argue this point, as the origin of species requires not the gradual improvement of animals retaining the same habits and structure, but such modification of those habits and structure as will actually lead to the appearance of new organs. We freely admit, that if an accumulation of slight improvements be possible, natural selection might improve hares as hares, and weasels as weasels, that is to say, it might produce animals having every useful faculty and every useful organ of their ancestors developed to a higher degree; more than this, it may obliterate some once useful organs when circumstances have so changed that they are no longer useful, for since that organ will weigh for nothing in the struggle of life, the average animal must be calculated as though it did not exist.
>
> We will even go further: if, owing to a change of circumstances some organ becomes pre-eminently useful, natural selection will undoubtedly produce a gradual improvement in that organ, precisely as man's selection can improve a special organ... Thus, it must apparently be conceded that natural selection is a true cause

or agency whereby in some cases variations of special organs may be perpetuated and accumulated, but the importance of this admission is much limited by a consideration of the cases to which it applies:... Such a process of improvement as is described could certainly never give organs of sight, smell, or hearing to organisms which had never possessed them. It could not add a few legs to a hare, or produce a new organ, or even cultivate any rudimentary organ which was not immediately useful to an enormous majority of hares... Admitting, therefore, that natural selection may improve organs already useful to great numbers of a species, does not imply an admission that it can create or develop new organs, and so originate species.

The origin of complex organs and structures

Many biologists, assuming a useless incipient stage in the development of complex organs (for example, the eye) could not see how such structures could have evolved as a consequence of natural selection either of small individual differences or of saltations. Similar difficulties are encountered, for example, in the evolution of complex floral structures. For instance, the extraordinary adaptations shown by many Orchids, in which cross-pollination is brought about by particular insect visitors to the flower, were used by Darwin as evidence for the variety and complexity of adaptation to be explained by natural selection (Darwin, 1862, 1877*b*). Writers such as Mivart (1871), however, who were opposed to any completely selectionist interpretation of evolution, were inclined to turn Darwin's 'evidence' against him. Darwin had mentioned the remarkable Orchid, *Coryanthes speciosa*, in which the labellum, modified to a bucket with a lip, is filled with water secreted by special glands. Pollination involves the visiting bees in an involuntary bath from which they can only rescue themselves by crawling through the narrow passage at the lip and thus effect pollination. As Mivart observes (p. 62): 'Mr Darwin gives a series of the most wonderful and minute contrivances ... structures so wonderful that nothing could well be more so, except the attribution of their origin to minute, fortuitous and indefinite variations'. More than a century of intervening research has done little to remove this particular difficulty; the evolution of highly complex structures still presents a real problem of interpretation. Yet, Mivart's offering – the role of a Creator God, manifest as innate forces, in meticulous detailed design – is quite unacceptable to modern science.

What has been called 'the creative power of selection' remains a difficult concept, providing material for arguments essentially similar to those which Darwin faced but often, of course, in the more recent literature, based on more sophisticated examples of claimed 'adaptations'. Williams (1966) provides a particularly interesting and well-argued treatment of these themes.

The role of isolation

Whilst it may be argued from Darwin's writings (Mayr, 1963) that he saw isolation as providing a 'favourable circumstance' for speciation, it was the German naturalist Wagner (1868), pointing to the probable loss of favourable variants in 'blending' inheritance, who argued that spatial isolation was a necessary condition of speciation. This important difference of opinion is discussed in some detail later.

The age of the earth

The modern biologist may find it difficult to realise the extent to which controversies about the age of the earth and its rocks preceded and eventually made possible the Darwinian revolution. The study of fossils, or 'figured stones' as they were first called, undoubtedly produced questioning about the age of the earth as early as the middle of the sixteenth century, when, for example, Conrad Gesner wrote an illustrated work on fossils. As the botanist H. Hamshaw Thomas wrote in an excellent short paper on 'The rise of geology...' published in 1947: 'fossils provided the first challenge to the orthodox cosmogony of the day'. By the end of the eighteenth century, when Hutton published his *Theory of the earth*, geology was recognised as a separate and quite well-based science, and the Genesis account of the origin of the earth was already under great strain. After overcoming naïve Church opposition, which offered a date of 4004 BC for the Biblical Creation, Darwin's ideas were, however, faced with much more sophisticated criticism from the eminent physicist, Lord Kelvin, who estimated the probable age of the earth from calculations of the average rate of heat loss based upon measurable temperature increases down bore-holes and mines. In his famous paper (Kelvin, 1871) 'On geological time', read at a meeting of the Geological Society of Glasgow in 1868, Kelvin calculated that the consolidation of the crust of the earth took place at a maximum of 400 million years ago, and expressed the view that this would have allowed insufficient time for the slow evolutionary processes postulated by Darwin (Burchfield, 1975). Full details of this interesting controversy are given by Mivart (1871). We now know that Kelvin's estimate was far too small; using evidence from the rate of decay of radioactive minerals, and in other ways, modern estimates of the age of the earth are of the order of about 4.6 billion years (Futuyma, 1975).

The nature of specific difference

Darwin's view of species, based as it was on a thorough study of living organisms as well as on the pertinent literature, is very different indeed from that of Ray and Linnaeus.

In Chapter 2 of the sixth edition of the *Origin* we read:

Hence, in determining whether a form should be ranked as a species or a variety, the opinion of naturalists having sound judgement and wide experience seems the only guide to follow. We must, however, in many cases, decide by a majority of naturalists, for few well-marked and well-known varieties can be named which have not been ranked as species by at least some competent judges.

He was impressed in his study of the variability of plants and animals by how difficult it was, in many groups, to delimit species, and gives many examples. In polymorphic groups he notes: 'With respect to many of these forms, hardly two naturalists agree whether to rank them as species or as varieties. We may instance *Rubus*, *Rosa* and *Hieracium* amongst plants'. He also considered the opinion of such great taxonomists as de Candolle, who, completing his monograph of the Oaks of the world, wrote:

They are mistaken, who repeat that the greater part of our species are clearly limited, and that the doubtful species are in a feeble minority. This seemed to be true, so long as a genus was imperfectly known, and its species were founded upon a few specimens, that is to say, were provisional. Just as we come to know them better, intermediate forms flow in, and doubts as to specific limits augment.

Darwin concludes:

Certainly no clear line of demarcation has as yet been drawn between species and sub-species – that is, the forms which in the opinion of some naturalists come very near to, but do not quite arrive at, the rank of species: or, again, between sub-species and well-marked varieties, or between lesser varieties and individual differences. These differences blend into each other by an insensible series...

In Chapter 9, on hybridism, he notes, after examining the extensive writings of Gärtner, Kölreuter, etc., that even though exceptions are known: 'First crosses between forms, sufficiently distinct to be ranked as species, and their hybrids, are very generally, but not universally, sterile'. In the section on intraspecific crosses he writes: 'It may be urged, as an overwhelming argument, that there must be some essential distinction between species and varieties, inasmuch as the latter, however much they may differ from each other in external appearance, cross with perfect facility, and yield perfectly fertile offspring'. Notwithstanding these views on the crossing of different groups, one is impressed on reading the *Origin* by the absence of any definition of species incorporating both morphological and crossing information. Why Darwin provides no definition of species is very clear from the discussion of species in the concluding chapter:

When the views advanced by me in this volume, and by Mr Wallace, or when analogous views on the origin of species are generally admitted, we can dimly foresee that there will be a considerable revolution in natural history. Systematists will be able to pursue their labours as at present; but they will not be incessantly haunted by the shadowy doubt whether this or that form be a true species. This, I feel sure and I speak after experience, will be no slight relief. The endless disputes whether or not some fifty species of British brambles are good species will cease. Systematists will have only to decide (not that this will be easy) whether any form be sufficiently constant and distinct from other forms, to be capable of definition; and if definable, whether the differences be sufficiently important to deserve a specific name. This latter point will become a far more essential consideration than it is at present; for differences, however slight, between any two forms, if not blended by intermediate gradations, are looked at by most naturalists as sufficient to raise both forms to the rank of species. Hereafter we shall be compelled to acknowledge that the only distinction between species and well-marked varieties is, that the latter are known, or believed, to be connected at the present day by intermediate gradations whereas species were formerly thus connected. Hence, without rejecting the consideration of the present existence of intermediate gradations between any two forms, we shall be led to weigh more carefully and to value higher the actual amount of difference between them. It is quite possible that forms now generally acknowledged to be merely varieties may hereafter be thought worthy of specific names; and in this case scientific and common language will come into accordance. In short, we shall have to treat species in the same manner as those naturalists treat genera, who admit that genera are merely artificial combinations made for convenience. This may not be a cheering prospect; but we shall at least be freed from the vain search for the undiscovered and undiscoverable essence of the term species.

From this passage it is abundantly clear that Darwin considered the taxonomist's task in recognising and naming species as a severely practical one, to be decided on criteria of degree of discontinuity and clarity of descriptive diagnosis. Undoubtedly he was influenced by his close study of domesticated plants and animals, and his views were not very acceptable to most practising taxonomists working with wild plants and animals, who continued to be impressed by the high proportion of apparently clear-cut entities to be described and named in the natural world. To them, species were 'real' in a way that genera were not.

Classification

Darwin's ideas about evolution provoked him to a new and radical reappraisal of classification and the results of this are set out in Chapter 14 of the *Origin*. First, Darwin notes that 'organic beings, like all other objects, can be classed in many ways, either artificially by single characters, or more naturally by a number of characters'. Clearly, specimens in a diverse collection may be grouped on flower colour, habit or any other single characters. As we have seen Linnaeus' 'Sexual System' was clearly artificial. While his system was highly successful in practical terms, Linnaeus himself recognised that his work would be superseded by natural classifications, in which plants were classified on the basis of many characters. In the period between the *Species plantarum* and the *Origin* much progress was made in constructing 'Natural Systems', major contributions being made by Bernard Adanson and Antoine Laurent de Jussieu, Robert Brown, de Candolle, and others (Davis & Heywood, 1963; Cronquist, 1988).

As a prelude to stating his own views, Darwin notes that: 'Some authors look at it [the Natural System] merely as a scheme for arranging together those living objects which are most alike, and for separating those which are most unlike... But many naturalists think that something more is meant by the Natural System; they believe that it reveals the plan of the Creator'. Darwin then presents his own ideas at length, of which the following is a brief quotation:

> the Natural System is founded on descent with modification; ... community of descent is the hidden bond which naturalists have been unconsciously seeking, and not some unknown plan of creation, or the enunciation of general propositions, and the mere putting together and separating objects more or less alike... Thus, the natural system is genealogical in its arrangement, like a pedigree: but the amount of modification which the different groups have undergone has to be expressed by ranking them under different so-called genera, sub-families, sections, orders and classes.

The implications of these ideas about branching phylogenetic trees will be discussed in later chapters.

Tests of specific difference

We have examined so far in this chapter a number of ideas about what constitutes a species. Linnaeus stressed the morphological difference between species whilst Darwin, considering both external morphology and the results of hybridisation experiments, found the species difficult to define.

Many other botanists were interested in the species problem in the mid-nineteenth century and tests of specific rank were devised.

At first, it seemed that hybridisation experiments might provide an objective guide as to whether a plant was a species or a variety. A number of scientists supported this view, in particular Professor Godron of the University of Nancy. In 1863, he published the following opinion (*fide* Roberts, 1929). If two given plants could be crossed without difficulty giving fertile offspring, they were to be called varieties of one species. If, on the other hand, two plants crossed with difficulty, if sterility barriers existed between different plants, then such plants were to be considered different species. Further, crossing between plants of different genera was impossible. The categories of variety, species and genus were therefore to be determined by crossing experiments. Godron's rigid ideas, which were based upon his own work as much as on the extensive publications of earlier hybridists, contrast sharply with the cautious views of Darwin. Of other botanists interested in the problems of defining species experimentally one must mention Professor von Nägeli of Vienna, with whom Mendel fruitlessly corresponded (see Chapter 4). He published a massive review of hybridisation in 1865, noting in particular the difficulties in the sort of ideas published by Godron.

A second test of species is associated with the name of Alexis Jordan of Lyons in France. He considered that cultivation experiments, with progeny testing, provided an objective means of distinguishing species. In 1864, he published a great many of his results. He is perhaps best remembered for his work on *Erophila verna*, in which he described 53 'elementary species', each retaining its distinctive characters in cultivation and coming true from seed. An even greater number could easily have been described, for he indicates that he had more than 200 distinct lines of *E. verna* in cultivation (Fig. 2.4). His experiments were not confined to this taxon, and it is of considerable interest to note the large number of 'elementary species' he described in several common genera, as shown in Table 2.1.

Jordan was followed by others in his practice of describing 'elementary species' within Linnaean species; for instance, Wittrock in Sweden working with *Viola tricolor*, and later De Vries (1905) experimenting with *Oenothera* species. The practice was condemned by many botanists as it led to an inordinate number of new plant names.

The idea of using only a single line of experimental evidence as a test of specific rank did not meet with universal approval. For example, Hoffmann, Professor of Botany in the University of Giessen, carried out a large number of experiments with many taxa. He observed plants closely in the

Table 2.1. *Numbers of 'elementary species' in various taxa published by Jordan in 1864*

Arabis	23	*Iberis*	23
Biscutella	21	*Ranunculus*	25
Erophila	53	*Thalictrum*	47
Erysimum	26	*Thlaspi*	21

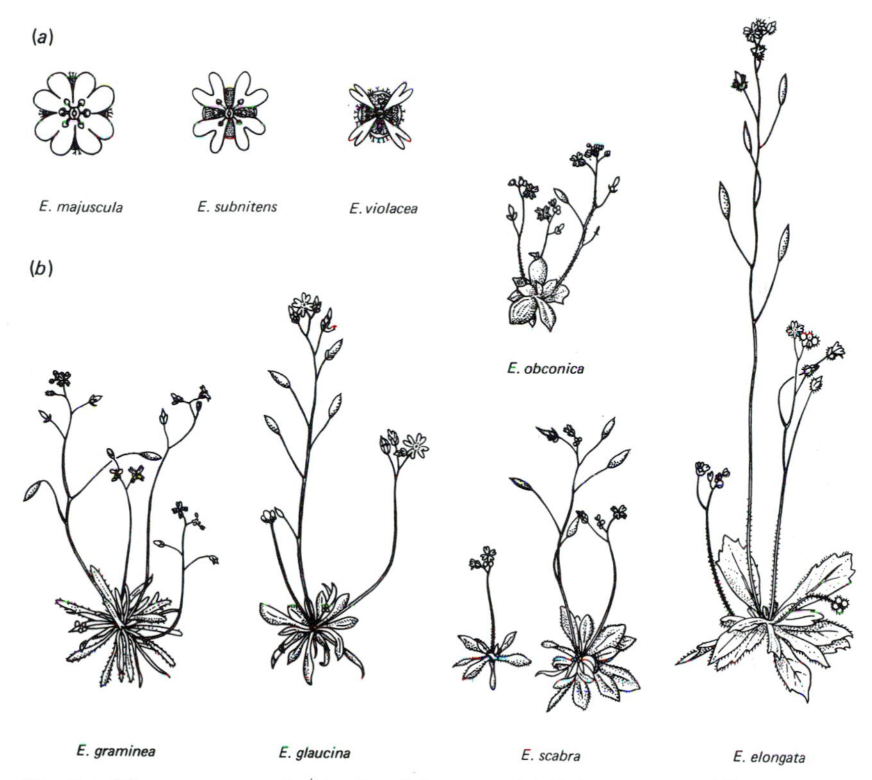

Fig. 2.4. 'Elementary species' in *Erophila verna*. (*a*) Enlargements of flowers showing petal variation (× 1.6). (*b*) Habit variation (× 0.75). (From Rosen, 1889.)

wild, and also carried out cultivation and crossing experiments. In a review of his researches (1881), he considered the many different lines of evidence which could be used in judging specific rank. Not only did he study the performance of plants in cultivation and the results of crossing plants, but he also took into account geographical distribution and the extent of hybridisation *in the wild*.

This historical review of species and their variation has brought us almost to the end of the nineteenth century, and it is in this period that there emerged two new aspects of the study of variation. First, the statistical examination of biological variation: some of the results of this work are the subject of the next chapter. Secondly, following the epoch-making rediscovery in 1900 of Mendel's work on heredity, the science of genetics made its appearance.

3

Early work on biometry

In the second half of the nineteenth century, as Darwinism was making its impact upon biology, an interesting new approach to biological variation, especially intraspecific variation, was being examined. Instead of trying to describe variation patterns in words, the investigators, examining large samples of organisms, collected numerical data and subjected them to statistical analysis. In the account which follows we discuss selected themes. For more detail on the history of the subject see Pearson & Kendall (1970), Kendall & Plackett (1977) and Porter (1986).

The first worker to study natural variation statistically, a science which became known as biometry, was probably the Belgian Quetelet (1796–1874). He wrote a famous series of letters on the subject to his pupil, the Grand Duke of Saxe-Coburg and Gotha. Later in the century, Darwin's cousin, Francis Galton (1822–1911), made notable contributions to the statistical investigation of variation and inheritance. Like Quetelet he was particularly fascinated by human variation, especially social questions.

Darwin acknowledged that in human societies the effect of natural selection, in weeding out the unfit, is greatly reduced. Galton and others became very concerned about the implications of this view for the health of society (Bowler, 1989*b*; Keynes, 1993). In his book *Hereditary genius*, published in 1869, Galton concluded from a survey of prominent families that the intellectual ability of fathers was inherited by their sons. He also noted that parents of lower mental ability often produced the most children. These observations suggested to him that prominent families should be encouraged to produce many offspring, while the parents of lower ability should be prevented from passing on their infirmities. Many reacted against Galton's views, for example claiming that the children of prominent families may owe some of their success to a better upbringing. Cowan (1972) is of the opinion that Galton was so convinced in his eugenic ideas that his biometrical and statistical techniques were developed in the hope of persuading the scientific community of the importance of eugenics. Likewise, both Pearson (Mackenzie, 1981) and Fisher (Bennett, 1983; Norton, 1983) were actively concerned with eugenics, and this may have provided strong motivation for their biometrical and statistical investigations.

It is clear from these comments why much of the early work on biometry was concerned with measuring groups of criminals, students, etc. It is not the intention in this book, which deals with plant variation and evolution, to explore the details of this work. However, we should acknowledge the major concern behind the measurement of variation, and note that the very interesting work carried out on plants was something of a sideline to the early interest in eugenics.

In presenting the botanical findings, it is important to examine first the general characteristics of this new approach. Instead of contenting themselves with the study of a few herbarium specimens or cultivated plants, the early biometricians took large samples, often using living material of common species. These samples were then carefully scrutinised and measured, as we read in Davenport (1904): 'Having settled upon the general conditions, of race, sex, locality, age, which the individuals to be measured must fulfil, take the individuals methodically at random and without possible selection of individuals on the basis of the magnitude of the character to be measured'. Finally, having collected the samples and obtained the numerical data, the worker performed the analysis, observing Quetelet's precept that statistics must be collected without any preconceived ideas and without neglecting any numbers (Quetelet, 1846).

These early studies of biological material established the important point that there are two main kinds of intraspecific variation. First, much of the variation is discontinuous. If, for instance, one is examining the number of chambers in a capsule, the number of seeds in a fruit, the number of leaves on a plant – in fact any variation in the number of parts – then the numbers found must be integers. One never discovers 14.5 undamaged peas in a pod. Often, in considering variation in the number of parts – so-called meristic variation – a more or less complete series of members is found. For instance, Pearson (1900) noticed, in a cornfield in the Chiltern Hills, England, that Poppies (*Papaver rhoeas*) had different numbers of stigmatic bands on the capsule (Fig. 3.1). He collected a very large sample of 2268 capsules: the frequency of different numbers of stigmatic bands is given in Table 3.1. In other instances of discontinuous variation, however, only two or a few strikingly different variants are found. For instance, the Opium Poppy, *Papaver somniferum*, may or may not have a dark spot at the base of the

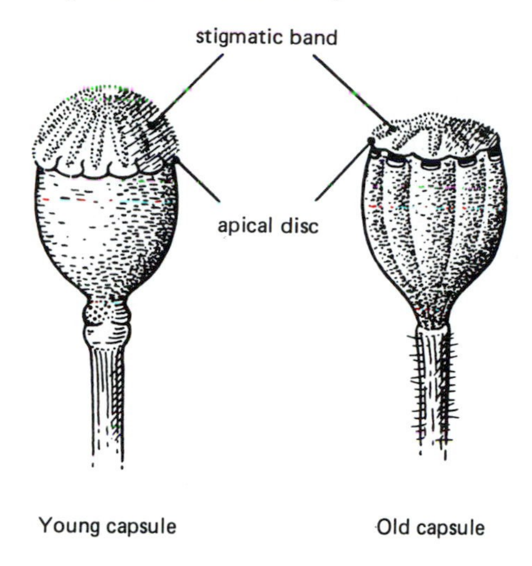

Fig. 3.1. Young and old capsules of the Poppy, *Papaver rhoeas* (× 0.5).

petal; Groundsel (*Senecio vulgaris*) has either radiate or non-radiate flowers; and Foxglove (*Digitalis purpurea*) may have white or red flowers, and hairy or glabrous stems. On the other hand, a second type of variation – continuous variation – is also common in plants. In considering variation in such characters as height, weight, leaf length and root spread, any value is possible within a given range. There are no breaks in the variation for particular characters.

Strikingly discontinuous variation patterns, as in white or red-purple flower colour in *D. purpurea*, presented little difficulty in examination or classification. The analysis of arrays of data, however, whether of discontinuous or continuous variates, posed somewhat more complex problems. With an array of data, how was it possible to show numerically where the bulk of the variation lay; how could a numerical estimate or spread of the data within the sample be obtained and, further, how could the variability of two samples be compared? Using Pearson's data for variation in stigmatic band number in *Papaver rhoeas* (Table 3.1), we may now briefly examine some of the statistics employed by the early biometricians.

Commonest occurring variation in an array

Sometimes a knowledge of the mode, or most frequent class, and the median, or middle value of an array, is a useful indication of where the bulk of the variation lies in a sample. These are, however, less useful than the

Table 3.1. *Calculation of mean, variance, standard deviation and coefficient of variation in number of stigmatic bands in capsules of the Poppy,* Papaver rhoeas *(Data from Pearson, 1900)*

Number of bands x	Frequency, f	fx	Difference from mean $x - \bar{x}$	Square of difference $(x - \bar{x})^2$	$f \times$ square of difference $f(x - \bar{x})^2$
5	1	5	− 4.8	23.04	23.04
6	12	72	− 3.8	14.44	173.28
7	91	637	− 2.8	7.84	713.44
8	295	2360	− 1.8	3.24	955.80
9	550	4950	− 0.8	0.64	352.00
10	619	6190	+ 0.2	0.04	24.76
11	418	4598	+ 1.2	1.44	601.92
12	195	2340	+ 2.2	4.84	943.80
13	54	702	+ 3.2	10.24	552.96
14	25	350	+ 4.2	17.64	441.00
15	5	75	+ 5.2	27.04	135.20
16	3	48	+ 6.2	38.44	115.32
	2268	22 327			5032.52

$$\text{Mean} = \frac{22\,327}{2268} = 9.8$$

$$s^2 = \frac{\sum f(x - \bar{x})^2}{n - 1} = \frac{\sum d^2}{n - 1} = \frac{5032.52}{2267} = 2.2 \quad s = 1.49$$

$$\text{Coefficient of variation} = \frac{s}{\bar{x}} \times 100 = \frac{1.49}{9.8} \times 100 = 15.2\%$$

arithmetic mean, $\bar{x}$. This is calculated quite simply by summing (Σ) the observed values (x) and dividing by the number of observations, n.

$$\text{Mean} = \bar{x} = \frac{\sum x}{n} \qquad (1)$$

Estimates of dispersion of the data

Values of the mean give no indication of the variation within a sample. Identical means may be obtained if the data are all clustered very closely to the mean or if the data are markedly above and below the mean (Fig. 3.2). It is clearly very important to have an estimate of the degree of dispersion of the data within a particular sample.

There are several possible ways of examining dispersion. Early biomet-

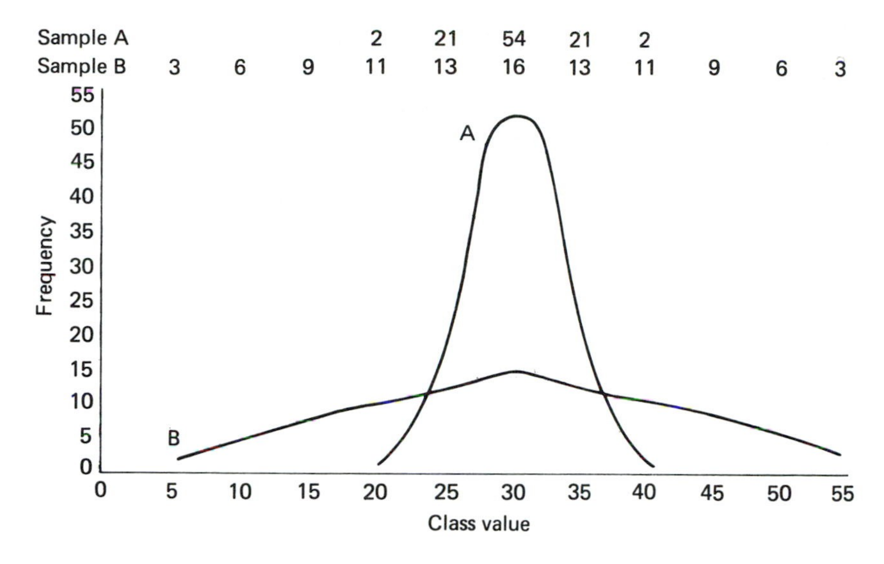

Fig. 3.2. Hypothetical frequency distribution for two population samples with the same mean. Sample B is much more variable than sample A. (From Srb & Owen, 1958.)

ricians often noted the extreme values of the array, or alternatively they calculated how much each value of the array differed from the mean, and after summing the differences, calculated an *average deviation from the mean*. They also used a statistic, now seldom if ever calculated, called the *probable error*. (Details of this calculation may be found in statistics books.) In more recent times, however, dispersion has been estimated by calculating the variance, s^2, and *standard deviation*, s.

The variance, s^2, is calculated by summing the squares of the deviations of all the observations from their mean (d^2) and dividing by $n-1$.

$$s^2 = \frac{\sum d^2}{n-1} \tag{2}$$

Except where samples are small, d^2 is more readily calculated by employing the equations:

$$\sum d^2 = \sum(x - \bar{x})^2 \tag{3}$$

$$\sum(x - \bar{x})^2 = \sum x^2 - \frac{\left(\sum x\right)^2}{n} \tag{4}$$

In calculating the variance, s^2, it is important to note (and statistics books should be consulted for justification) that the divisor is $n - 1$. The standard deviation, s, is found by obtaining the square root of the right-hand side of Equation (2). The calculation of variance and standard deviation for Pearson's Poppy data is given in Table 3.1.

The variance is a valuable statistic, giving a measure of the dispersion of the data about the mean. It is used a good deal in more complex statistics, where different populations are being compared. The standard deviation too is a useful measure of dispersion, especially as the 'spread' of the data is here expressed in the same units as the mean. (The probable error – the statistic estimating dispersion which was often calculated by early biometricians – is 0.6745 times the standard deviation.) Now that the variance and standard deviation values have been calculated, how are they to be interpreted? Before we examine this point, let us look at early work on the visual representation of arrays of data.

Histograms, frequency diagrams and the normal distribution curve

Most people find it easier to comprehend the significance of data expressed visually rather than numerically. The variation in *Papaver rhoeas* may be expressed as: $\bar{x} = 9.8$, $s^2 = 2.22$, $s = 1.49$, or it may be represented in the form of a diagram. Histograms and plotted curves were frequently employed in early biometrical studies. The distribution of the values for stigmatic band number in *P. rhoeas* has been plotted as a histogram in Fig. 3.3; the distribution is roughly bell-shaped, being almost symmetrical about the mean value. Small irregularities in the distribution are the result of small sample size; a closer fit to a bell-shaped curve would result from an even larger set of data for stigmatic band number. The results for *P. rhoeas* are an example of a very common frequency distribution in biological material – the 'normal' or Gaussian distribution, the latter after Gauss (1777–1855), one of the investigators of this type of distribution.

In the last decades of the nineteenth century, approximately normally distributed variation was demonstrated in a great range of biological materials. Davenport (1904), in the second edition of his book *Statistical methods with special reference to biological variation*, first published in 1889, provides a very valuable survey of early biometrical results, giving references and details of scores of botanical and zoological examples. Figure 3.4 shows some typical cases.

As approximately normal distributions are frequently encountered in biological material, it is important to look at some of their properties. First, Fig. 3.5 shows that in a normal curve the median, mode and mean of the

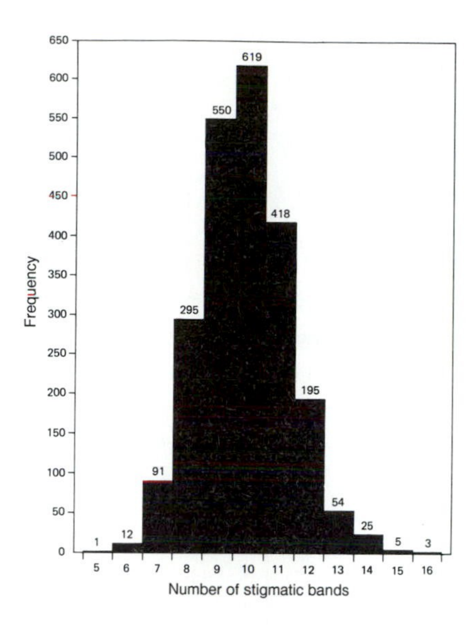

Fig. 3.3. Histogram of Pearson's data (Table 3.1) for the variation in the number of stigmatic bands in a sample of capsules of *Papaver rhoeas*. Such histograms were often used in early biometrical studies. Campbell (1967), discussing the use of histograms, suggests that they should be used only in cases of continuous variation. For examples of meristic or other discontinuous variation, frequency diagrams are preferable; in these, the frequency of each class is indicated by a vertical line on the graph.

array fall at the same point. Secondly, and of great importance, is the relation of standard deviation to the curve. We have outlined above how to calculate variance and standard deviation. Now, how precisely does knowledge of the standard deviation help us to understand the dispersion of the data within the sample? Examining Fig. 3.5 we see that about two-thirds (68.26% to be exact) of the total variation under a normal curve falls within the range 'mean $\pm$ one standard deviation'. Twice the standard deviation on each side of the mean excludes about 5% of the variation, 2.5% in each tail of the normal curve. For different sets of data which are normally distributed, different values of the standard deviation will be found. Thus, if we have a large amount of variability in a sample, a wide curve corresponding to the large standard deviation will be obtained. Whatever the width of the curve, however, the 'mean $\pm$ one standard deviation' always contains 68.26% of the variation. We can see now how the standard deviation is so useful in indicating dispersion. One last point remains to be considered. How is the appropriate normal curve to be fitted to a histogram? It cannot, of course, be drawn 'by eye'. Recourse to a

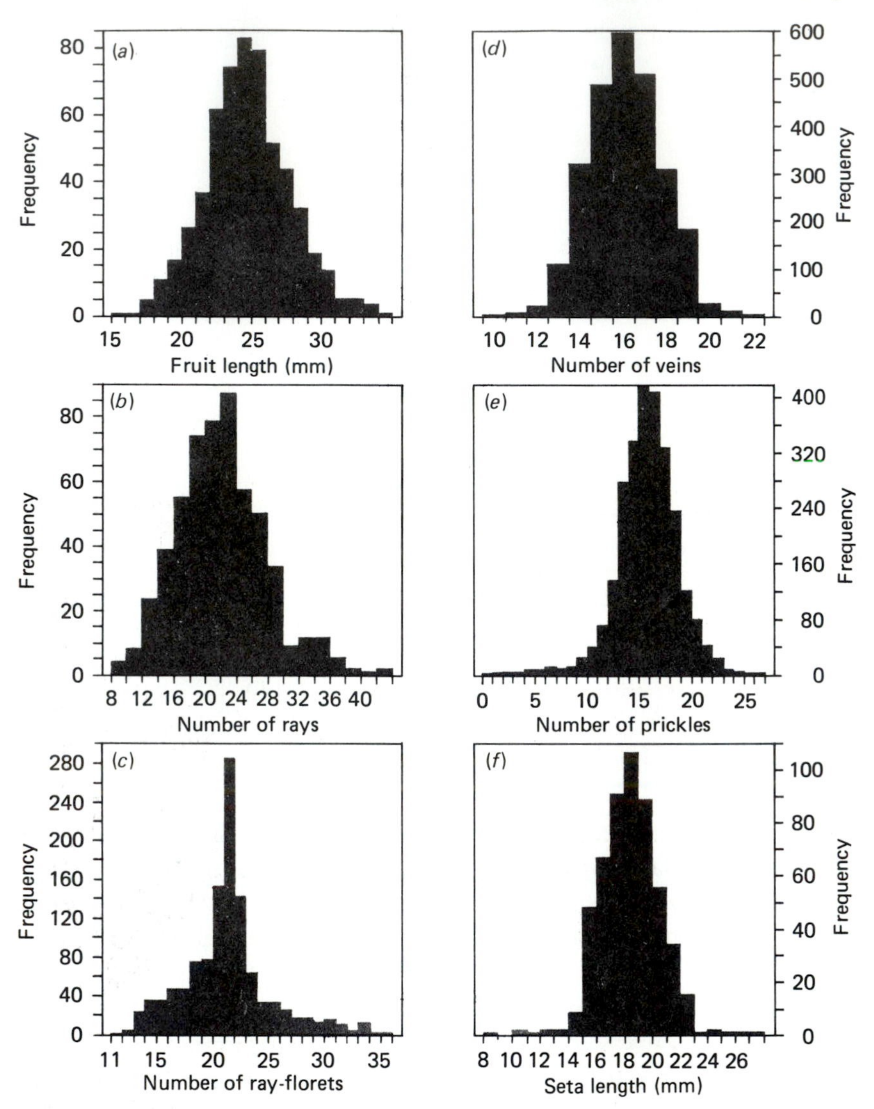

Fig. 3.4. Early botanical results showing approximately normal distributions. (*a*) Fruit length in the Evening Primrose, *Oenothera glazioviana* (*O. lamarckiana*), for 568 plants collected in October 1893 (De Vries, 1894). (*b*) Number of primary umbel rays in *Anethum graveolens* for 552 plants collected in July 1893 (De Vries, 1894). (*c*) Number of ray-florets in Ox-Eye Daisy, *Leucanthemum vulgare* (*Chrysanthemum leucanthemum*), collected from 1133 heads in Keswick, England, July 1895 (Pearson & Yule, 1902). (*d*) Number of main-branch veins in leaves of Beech (*Fagus sylvatica*) from 2600 leaves (Pearson, 1900). (*e*) Number of prickles on 2600 leaves of Holly (*Ilex aquifolium*) from trees in Somerset, England (Pearson *et al.*, 1901). (*f*) Seta length in the moss, *Bryum cirrhatum*, for 522 plants (Amann, 1896).

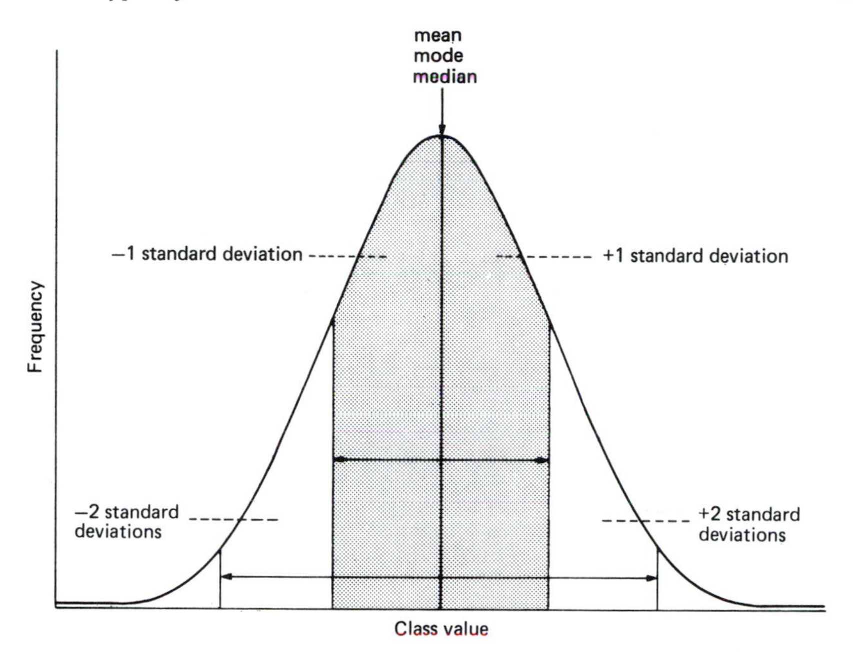

Fig. 3.5. A normal distribution, showing proportions of the distribution that are included between ± 1 standard deviation and ± 2 standard deviations with reference to the mean. (From Srb & Owen, 1958.)

statistics book will indicate full details, and it is sufficient for our purposes to note that it involves the substitution of values of the mean and standard deviation into the equation for the normal curve.

Other types of distribution

Not all the biometrical studies of plant materials gave normally distributed variation, however. De Vries (1894) was one of the first to point to deviant distributions, calling attention to what he called 'Half-Galton' curves. Table 3.2 gives sets of data for the number of compartments in the fruit of Sycamore (*Acer pseudoplatanus*), and petal number in Marsh Marigold (*Caltha palustris*) and Silverweed (*Potentilla anserina*), which in each case approximates to half a normal curve. In *Caltha* and *Acer*, the 'right-hand half' of the curve is represented, whilst in *Potentilla* only the 'left-hand half' is found.

Other researches of this period, especially those dealing with numbers of plant parts, revealed further asymmetrical and deviant frequency distribu-

Table 3.2. *Half-Galton curves (De Vries, 1894)*

Caltha palustris	Petal number	5	6	7	8
	Frequency	300	87	25	4
Acer psudoplatanus	Number of fruit compartments	2	3	4	
	Frequency	50	17	3	
Potentilla anserina	Petal number	3	4	5	
	Frequency	6	537	1819	

Table 3.3. Ranunculus repens *(Data of Pledge, 1898, in Vernon, 1903)*

	3	4	5	6	7	8	9	10	11	12	13
Sepal frequency	1	20	959	18	2						
Petal frequency		8	706	145	72	38	15	7	7	1	1

tions. For instance, examining the figures of Pledge (1898) in Table 3.3 for petal frequency in Creeping Buttercup (*Ranunculus repens*), we see that the frequency distribution when plotted would have a long tail to the right: such a curve is described as 'positively skewed'. (A curve with a long tail to the left is said to be 'negatively skewed'.) The data for sepal numbers, collected in the same study, also depart from a normal distribution, in this case by being too tightly bunched together. Such a distribution is said to be leptokurtic (high-peaked). Sometimes flat-topped curves (platykurtic) have been discovered (Bulmer, 1967; Rayner, 1969; David, 1971).

Comparison of different arrays of data

By visual inspection, it is often possible to see that a group of plants is more variable in, say, height than in flower size, and the problem of investigating this biometrically particularly fascinated Pearson. In the late 1890s, he first devised a statistic known as the *coefficient of variation*. Easy to calculate, it is merely the ratio of the standard deviation to the mean. In order to have a scale of reasonable-sized numbers, the resulting coefficient is usually expressed as a percentage:

$$C \text{ (coefficient of variation)} = \frac{s}{\bar{x}} \times 100\%$$

An important property of the coefficient is that, as it is calculated as a *ratio*,

Table 3.4. *Variation in human height and weight; means* = $\bar{x}$, *coefficients of variation* = C *(Data of Pearson and others, in Davenport, 1904)*

		n	$\bar{x}$	*C*
Height				
English upper middle class	male	683	69.215 in	3.66
English criminals		3000	166.46 cm	3.88
US recruits		25 878	170.94 cm	3.84
Cambridge University	male	1000	68.863 in	3.66
students	female	160	63.883 in	3.70
English newborn babies	male	1000	20.503 in	6.50
	female	1000	20.124 in	5.85
Weight				
Cambridge University	male	1000	152.783 lb	10.83
students	female	160	125.605 lb	11.17
English newborn babies	male	1000	7.301 lb	15.66
	female	1000	7.073 lb	14.23

direct comparison of different coefficients is possible. This even applies when the original figures were calculated in different units, as in metres, inches, grams, etc. Table 3.4 shows some data for human height and weight. Taking the figures for height first, a comparison of means is impossible in certain cases, as some of the measurements are in inches and others in centimetres. Direct comparison of coefficients of variation is, however, possible, and we can see that English criminals and US recruits show a similar degree of variation in height. There are small differences in height between male and female students, and between males and females at birth. Considering the information for variation in weight, again we have differences between male and female at birth, and at college. Finally, as high values of the coefficient of variation indicate greater variation for a particular character, we can see that there is a much greater variation in weight than in height in the samples examined.

Coefficients of variation continue to be very useful in the study of variation. Figure 3.6 shows the coefficients calculated by Gregor (1938) for different parts of Sea Plantain (*Plantago maritima*). These data illustrate convincingly a fact known before Linnaeus, namely that floral parts are generally less variable than vegetative parts.

Complex distributions

Other biometrical studies in the 1890s revealed more complex frequency distributions. Some of the results of Professor Ludwig (1895) of Greiz in

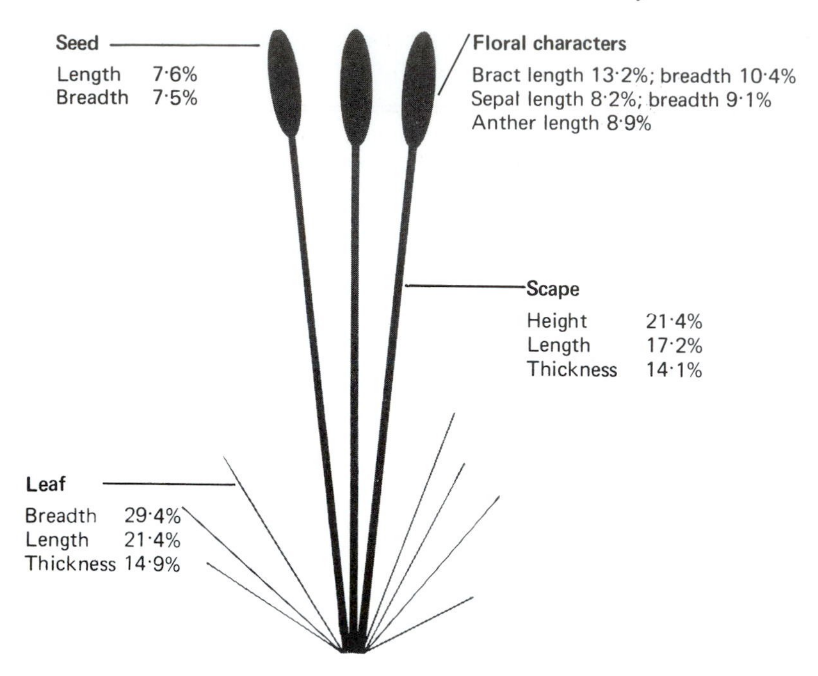

Fig. 3.6. Coefficients of variation in *Plantago maritima*: typical mean values from a number of populations. (Data from Gregor, 1938.)

Germany may be used as an illustration. He counted the numbers of ray-florets in 16 800 heads of the Ox-Eye Daisy, *Leucanthemum vulgare* (*Chrysanthemum leucanthemum*), collected from Greiz, Plauen, Altenberg and Leipzig between the years 1890 and 1895. The frequency distribution he obtained was not of the 'normal' type but had several peaks (Table 3.5). He obtained similar multimodal distributions for many species: for example, in number of ray-florets in Daisy (*Bellis perennis*), number of disc-florets in Yarrow (*Achillea millefolium*) and number of flowers in the umbel of Cowslip (*Primula veris*).

In collecting his *Leucanthemum* (*Chrysanthemum*) data, Ludwig records some interesting differences in ray-floret numbers in different localities. Mountain plants showed 'peaks' at 8 and 13, while lowland plants on the other hand had a 'peak' at 21 ray-florets. In fertile soil, Ludwig often found a strong 'peak' at 34. He considered that these variations between plants from different areas were the result of nutritional factors. What interested Ludwig most of all about his data was the presence (in the results) of clear peaks at 8, 13, 21 and 34 ray-florets. These numbers, he pointed out, belong to the famous Fibonacci sequence of numbers discovered by

Table 3.5. *Variation in the number of ray-florets in* Leucanthemum vulgare (Chrysanthemum leucanthemum)

Number	Ludwig (1895)[a]	5 July 1901 Tower (1902)[b]	30 July 1901 Tower (1902)[b]	Total Tower (1902)[b]
7	2			
8	9			
9	13			
10	36			
11	65			
12	148		1	1
13	427		8	8
14	383		3	3
15	455		6	6
16	479	1	8	9
17	525	0	9	9
18	625	0	8	8
19	856	2	12	14
20	1568	8	19	27
21	3650	17	26	43
22	1790	23	11	34
23	1147	22	10	32
24	812	21	10	31
25	602	22	8	30
26	614	19	5	24
27	375	16	4	20
28	377	14	6	20
29	294	12	4	16
30	196	10	2	12
31	183	16	4	20
32	187	18	2	20
33	307	29	1	30
34	346	20	1	21
35	186	6	0	6
36	64	6	0	6
37	28	0	0	0
38	16	0	0	0
39	16	2	0	2
40	14			
41	0			
42	3			
43	2			
Total	16 800	284	168	452

[a]Ludwig (1895): Plants from Greiz, Plauen, Altenberg and Leipzig, 1890–5
[b]Tower (1902): Plants from Yellow Springs, Ohio, USA (two collections and total from same locality)

Leonardo Fibonacci of Pisa in the twelfth century. The sequence runs 0, 1, 1, 2, 3, 5, 8, 13, 21, 34, 55, 89, 144..., each term being the sum of the two terms which precede it. It represents a set of whole numbers which satisfy almost exactly an exponential growth curve. Not all Ludwig's results gave such clear peaks at the Fibonacci numbers, and he was hard-pressed to explain peaks at 11 and 29, which he discovered in certain plants. Nevertheless, he believed that the Fibonacci sequence of numbers was important in understanding complex patterns of variation.

Certain other biometricians, notably Weldon (1902*b*), were sceptical about Ludwig's claim for the Fibonacci sequences. They pointed to the fact that plants from different areas had been amalgamated in collecting the data. Plants from a single locality often gave a different picture of the variation; for instance, counts for *L. vulgare* ray-florets from Keswick, England, in 1895 gave an approximately normal distribution (Fig. 3.4). Weldon also pointed out that sampling at different times could have an important influence upon the results. This is well illustrated in the results of Tower (1902). He collected, at the beginning and end of July 1901, two sets of *L. vulgare* plants from a locality at Yellow Springs, Ohio, USA. His results show clearly that early flowers have more ray-florets than those produced later in the season. Tower went on to show that it was not a question of different plants in flower at the beginning and end of July; marked plants continued to flower throughout the summer, producing flowers with different numbers of parts at different times in the season. Different peaks for ray-floret numbers are found in early July (22, 33) compared with those found later in the month (13, 21). It is interesting to note that it is only in the amalgamated data that these peaks are found, and that the highest peak is not at the Fibonacci number of 34, as found by Ludwig, but at 33. Such results as these cast some doubt upon the importance of the Fibonacci numbers, indicating that the location of peaks in a complex distribution, far from conforming to a mathematical sequence, was greatly influenced by the method of collecting the data. The precise results obtained would depend upon whether all the plants were collected at the same stage of maturity, a point particularly difficult to ascertain if data for plants from widely different localities and ecological conditions were amalgamated.

Local races

Close study of local variation in the species occupied the attention of many early biometricians. For instance, Ludwig (1901) made a special analysis of variation in Lesser Celandine (*Ranunculus ficaria*). He showed that plants

Table 3.6. *Mean number of stamens and carpels in* Ranunculus ficaria *(Ludwig, 1901)*

	Mean number		Standard deviation
Gais	Stamens	23.8250	2.8872
(80 plants)	Carpels	18.1125	4.2885
Trogen	Stamens	20.3682	3.8234
(385 plants)	Carpels	13.2635	3.0606

Table 3.7. *MacLeod's data for seasonal variation in floral parts in a population of* Ranunculus ficaria *(Lee, 1902)*

	Mean number		Standard deviation
Early flowers	Stamens	26.7313	3.7609
(268 plants)	Carpels	17.4478	3.8942
Late flowers	Stamens	17.8633	3.2984
(373 plants)	Carpels	12.1475	3.3878

from different localities had different numbers of carpels and stamens. Details of the two most dissimilar populations, from Gais and Trogen, are given in Table 3.6. Clearly Gais has plants with more carpels and stamens than Trogen.

Ludwig called these local populations, characterised by different mean numbers of floral parts, '*petites espèces*' or 'local races'. Until this time the term 'local race' had been used rather loosely for plants from particular areas used for biometrical study or experiments, but Ludwig sought to demonstrate the reality of 'local races', using biometrical evidence. In his view, these races could be distinguished on the basis of the mean number of floral organs, amalgamation of data for a number of races giving a multimodal distribution curve.

Ludwig's views were again challenged by British and American biometricians, particularly by Lee (1902) who, using the data of MacLeod on *R. ficaria*, pointed to the great seasonal variation in floral parts (Table 3.7). Her criticism of Ludwig's 'local races' is particularly telling as the variation in early and late flowers from a single locality covers almost the entire range between the Gais and Trogen plants.

This criticism of Ludwig's results did not clinch the issue, however, as there was earlier work by Burkill (1895) on two dissimilar *R. ficaria* populations in

Table 3.8. *Variation in* Ranunculus ficaria *(Burkill, 1895)*

	Date of collection	Number of flowers	Mean number of stamens	Mean number of carpels
Cambridge	3 March	32	22.87	13.41
(under trees)	16 April	75	19.49	11.95
Cayton Bay				
(open field,	31 March	100	38.24	32.32
top of cliffs)	4 May	43	30.67	25.72

which large differences in mean numbers of floral parts were maintained (although not completely) on later sampling on the same site (Table 3.8).

The reality of 'local races' was an important issue in the early volumes of the journal *Biometrika*, which was launched in 1901. In an editorial (**1**: 304–6, 1902) it was contended that the polymorphism found in most results was spurious. It was difficult to defend the notion that each peak of a complex distribution represented a 'local race', especially as peaks often disappeared as sample size was increased. Another important point concerned sampling techniques. It was stressed that random sampling was essential, a point perhaps neglected by early workers. Further, the problem of what constitutes a locality was raised and the validity of putting together data for samples taken from different areas was questioned. Finally, the editorial stressed the difficulties of seasonal variation and environmental effects, and concluded that a species is not broken up into 'local races'.

Returning once again to variation in *R. ficaria*, we find the same conclusion is reached by Pearson *et al.* (1903) in a paper in *Biometrika*, which draws together published records, together with new results of variation in floral parts in different areas of Europe. The tables of data are too large for inclusion here, but the following conclusions were drawn from the extensive statistics. 'Local races' could not be distinguished by the number of floral parts, and the influence of the environment and seasonal variation would seem to be sufficient to mask any difference due to 'local races'. The problem of how to eliminate seasonal and environmental variables from experimental studies was not seriously investigated until later, as we shall see in Chapter 4.

Correlated variation

Many early biometricians examined closely a further aspect of variation, namely the simultaneous variation in pairs of characters. For instance,

Pearson was interested in the relation of measurements of different parts of the human body. Suppose we consider body height and its relation to forearm length. It may be that there is some relation between the two variables or they may be independent. Three different situations are possible:

1. The taller the person, the longer the forearm.
2. The taller the person, the shorter the forearm.
3. A tall person is as likely to have a long or a short forearm as is a short person.

The first situation is one of positive correlation, the second of negative correlation, whilst if the last were discovered we should conclude that there was no correlation between the traits.

In investigating correlation, a statistic called the correlation coefficient (r) is often calculated. It is not necessary for our purposes to give the formulae and details of calculation, which may be found in any statistics book. What is important is the way in which r values indicate correlation or lack of it. $r = +1$ indicates complete positive correlation; $r = -1$ signifies complete negative correlation. If $r = 0$, then correlation is absent.

In biological material, perfect correlation – either positive or negative – is very rare; the various degrees of positive and negative correlation which are often found are indicated by figures which lie between $r = +1$ and $r = -1$.

Examining the relation between stature and forearm length, Pearson demonstrated positive correlation: in one case $r = +0.37$. A number of botanical situations were also studied at this time. Among the problems investigated was the correlation in the size of leaves in the same rosette in *Bellis perennis* (Verschaffelt, 1899), correlation between pairs of measurements of leaves and fruits of various species (Harshberger, 1901) and correlation between various parts in the Desmid, *Syndesmon thalictroides* (Kellerman, 1901). The sort of figures obtained for correlation in the floral parts of plants may be illustrated with data, summarised from various authors, on *Ranunculus ficaria* (Table 3.9). Clearly there is a stronger correlation between numbers of stamens and carpels than between other organs.

Correlation coefficients – and a further method of studying the association of pairs of measurements known as regression analysis – were used, particularly by Galton and Pearson, for studying heredity. It is a matter of common experience that tall fathers tend to have tall sons and that short fathers usually have short sons. The association is by no means complete,

Table 3.9. *Correlation coefficients in* Ranunculus ficaria *(Davenport, 1904)*

Numbers of	Values of r
Sepals to petals	+ 0.34 to − 0.18
Sepals to stamens	+ 0.06 to + 0.02
Sepals to carpels	+ 0.25 to + 0.03
Petals to stamens	+ 0.38 to + 0.22
Petals to carpels	+ 0.35 to + 0.19
Stamens to carpels	+ 0.75 to + 0.43

however. Galton examined the situation biometrically, analysing data from a large number of human families (Galton, 1889): 'Mr Francis Galton offers 500L in prizes to those British Subjects resident in the United Kingdom who shall furnish him, before May 15 1884, with the best Extracts from their own Family Records'. Galton sifted through particulars of 205 couples of parents with their 930 adult children of both sexes. He examined his data carefully, looking for association between the characteristics of parent and offspring. In many cases r values proved to be positive – as high as $r = +0.5$ for height of parents and offspring. We shall examine Galton's interpretation of these results in Chapter 4.

Problems of biometry

In this short survey of early biometrical work a number of problems remain to be examined. In our opening remarks we indicated that there are two main types of variation found on sampling. Arrays of data may be obtained showing either discontinuous or continuous variation. Also there may be found markedly discontinuous patterns of variation, with two or more very distinct non-overlapping categories. The reality of these distinct groups is important, as they figure widely in genetic work. As we shall show in Chapter 4, Mendel's work on genetics, published in 1866 and rediscovered in 1900, involved crossing Garden Peas (*Pisum sativum*) with different coloured cotyledons (green or yellow), or plants of different height (tall or dwarf). Early geneticists crossed glabrous and hairy plants of *Biscutella laevigata* (Saunders, 1897), and *Silene* spp., especially *S. dioica* and *S. latifolia* (*S. alba*) (De Vries, 1897; Bateson & Saunders, 1902). Among the biometricians it was Weldon (1902*a*), an opponent of Mendelism, who pointed out a certain ambiguity in defining discontinuities. For instance, he showed that if a large range of cultivated Pea stocks was examined it was found that there was a continuous range of cotyledon colour from green to yellow. It was impossible to sort into green and yellow categories. Similarly, he also showed that there was an enormous range of hairiness in *Silene* species and that it was very hard to accept a classification into glabrous or hairy variants. The important point to bear in mind, however, is the scale of the operation; it may be that general discontinuities do not occur, but marked discontinuities in limited collections and in the progeny from carefully controlled crosses certainly exist. When we read of Mendel crossing tall and dwarf Peas or yellow and green Peas, it is as well to remember that he deliberately chose stocks with markedly contrasting characters and that, even though there would have been variation in, say, height in his tall and dwarf stocks – perhaps normally distributed variation – there was no overlap in the distribution curves of tall and short plants.

A further problem raised by early biometrical work is that of the significance of differences between sets of numerical data. For instance, the coefficient of variation for weight in Cambridge University students (Table 3.4) shows that females ($C = 11.17\%$) show greater weight variation than males ($C = 10.83\%$). The difference in values is, however, quite small. Now, is this result due to differences in sample size? There were few female students in Cambridge in the 1890s and there was difficulty in getting even 160 measurements. Or is the variation due to chance? Would further samples taken in different years give the same basic pattern of greater weight variation in female students?

This type of problem is widespread in biometry. Is there any statistically significant difference in the frequency distributions of two sets of data? Do the peaks in a multimodal distribution reveal a true polymorphism or is it the result of sample size or chance? Questions of this type are now tackled by applying statistical significance tests. In Chapter 8 we shall go further into these problems; it is sufficient at this point to note that most of these tests came into being because biometricians wrestled with the problems of interpreting data from biological material.

Finally, we must return to another issue: the vexed question of the underlying basis of variation, which fascinated and puzzled early workers. What part of the variation was due to environmental variation and what part was genetic? In the next chapter we examine this problem.

4

Early work on the basis of individual variation

In the last chapter we saw how the early biometricians found great difficulty in analysing some of their data because they were unable to decide which part of the variation had a genetic basis and which part was environmentally induced. For animal studies it was Galton (1876) who appreciated the unique value of twins in investigations of the relative roles of nature and nurture in the development of the individual. To study genetic and environmental effects in plants, specimens selected for comparison may be cultivated under a standard set of environmental conditions. Experiments, both historical and recent, have been performed on the assumption that residual differences between plants of the same species, collected from the same or different habitats and grown under such standard conditions, might be considered to have a genetic basis. What follows is a brief survey of early studies. In Chapter 8 we will discuss in some detail the design and interpretation of garden experiments.

It is very interesting to see how cultivation techniques have developed as methods of analysing variation in plants. Experimental cultivation of plants undoubtedly arose as an adjunct to gardening and horticulture, and in Chapter 2 we saw how Ray, collecting the striking prostrate variant of *Geranium sanguineum* from Walney Island, demonstrated its constancy by cultivating plants in different gardens. The most valuable of these experimental tests were undoubtedly those of a comparative nature. For instance, Mendel cultivated two variants of *Ranunculus ficaria*, which he called *Ficaria calthaefolia* and *F. ranunculoides*, and reported to Dr von Niessl that each remained distinct (Bateson, 1909).

In a paper of quite remarkable scope, Langlet (1971) has reviewed the extent to which foresters in the eighteenth and nineteenth centuries were using experimental cultivation to study adaptive variation in some of the widespread forest trees of Europe. He cites, for example, the neglected (and largely unpublished) work of Duhamel du Monceau, Inspector-General of the French Navy, who, around the time that Linnaeus published his *Species plantarum* (1753), brought together an impressive collection of samples of Scots Pine (*Pinus sylvestris*) from Russia, the Baltic countries, Scotland and Central Europe, and established the first experimental provenance tests for any wild plant. This early development of what we could now call 'genecology' is understandable because of the economic and military importance of the timber supply, but the neglect by most modern writers of the further expansion of such studies in the nineteenth century is less easy to explain and probably, as Langlet suggests, is in part due to the fact that much of this forestry research was published in German. Darwin himself, of course, was greatly interested in the variation of cultivated plants; but forestry differed from agriculture and horticulture, as Langlet shrewdly observes, because its source material was almost entirely the wild species not already subject to artificial selection.

These examples show the importance of simple cultivation of carefully examined material, comparing performance in the wild with that in culture, and comparing also the behaviour of samples of the same or closely related species in the same garden. The method of comparative cultivation, whether seeds or plants are collected from the wild, permits us to investigate the basis of variation patterns. It is easy with hindsight to get a false impression of the ideas of the past and here is a case in point. Even though ideas about the balance between genetic and environmental variation are implicit in some of the writings of the nineteenth century and even discernible in the work of Linnaeus, an explicit statement came only with the research of the Danish botanist Johannsen carried out during 1900–7 and published in 1909. He worked with dwarf beans of the species *Phaseolus vulgaris*, which is naturally self-fertilising.

Phenotype and genotype

Johannsen obtained commercial seeds of the variety 'Princess' and grew 19 of them, each from a different source, in an experimental garden. The progeny from each of these beans had a different mean seed weight, and Johannsen inferred that these differences were genetic. From each of these 19 original beans, he established a separate line by self-fertilisation, growing up to six generations of daughter beans. For each line he raised a sub-line by selecting heavy seeds at each generation and a separate sub-line in which light seeds were selected. Very great care was taken to label the plants, and in each generation the mean seed weight for a line was calculated separately

Table 4.1. *Two pure lines of* Phaseolus vulgaris *(Johannsen, 1909)*

Mean weight (grams) of selected small seeds	Mean weight of progeny		Mean weight (grams) of selected large seeds	Year	Mean weight (grams) of selected small seeds	Mean weight of progeny		Mean weight (grams) of selected large seeds
30			40		60			70
	36	35		1902		63	65	
25			42		55			80
	40	41		1903		75	71	
31			43		50			87
	31	33		1904		55	57	
27			39		43			73
	38	39		1905		64	64	
30			46		46			84
	38	40		1906		74	73	
24			47		56			81
	37	37		1907		69	68	
	Pure line 'A'					Pure line 'B'		

for progeny from heavy and light mother beans. Table 4.1 gives the results for two lines. Johannsen found that for a particular line in any one year the mean seed weight for progenies from light and heavy beans did not differ significantly. From each of the 19 original beans a pure line was established, selection having no effect upon mean seed weight. The implication of these results may be more readily understood later, when it will be shown that habitual self-fertilisation leads to genetic invariability. Thus, genetically identical plants were produced from the progeny of a single bean. Even though the pure lines from the 19 beans were each genetically uniform, Johannsen found great differences in individual bean weights, approximately normally distributed, giving slightly different mean values for a line in different years. He attributed these differences to the effects of the environment.

These experiments led him to define clearly the distinction between genetic and environmental effects upon an organism. Of first importance were the hereditary properties of an individual – the *genotype* – which were largely fixed at fertilisation. The appearance, or *phenotype*, of particular individuals of the same genotype might, however, be different because of environmental factors, e.g. two seeds may have the same genotype but very different weights because of the position in which they developed in the pod. Even though Johannsen's results were obtained for a habitually self-fertilising species, there is no reason to doubt that the concept of genotype and phenotype is of general validity.

Transplant experiments

Besides the rather simple cultivation experiments we have examined so far, nineteenth-century botanists also investigated, through transplant and transfer experiments, the degree of adaptation that a plant showed when placed in a habitat different from that in which it was collected in the wild. Not only were they interested in what we now call changes in phenotype of a plant but also in the persistence of any changes which occurred during the experiment.

As part of a general study of adaptation Bonnier studied many European plants. His experimental technique is of special interest as he used cloned material. Experimental plants were allowed to grow to a convenient size. They were then divided into pieces, and these pieces or 'ramets' were transplanted into experimental beds at different altitudes in the Alps, the Pyrenees and in Paris. His alpine sites were not gardens – ramets were planted into natural vegetation, protected sometimes by fencing. No fertiliser was added and no watering of the plants took place. In the first reports of his experiments (begun in 1882) he showed how 'alpine' ramets grew into very dwarf compact plants with very vivid flowers, in comparison with 'lowland' ramets (Fig. 4.1). In the 1890s, in a series of largely neglected papers, he published a great deal about the physiological and anatomical adaptation of these plants.

In 1920, Bonnier presented a summary of his researches and claimed that in the course of his experiments certain lowland species became modified to such a degree that they were transformed into related alpine and subalpine species or subspecies. This claim, which, Bonnier notes, supports the ideas of Lamarck, is of very great interest and if true would have a profound effect

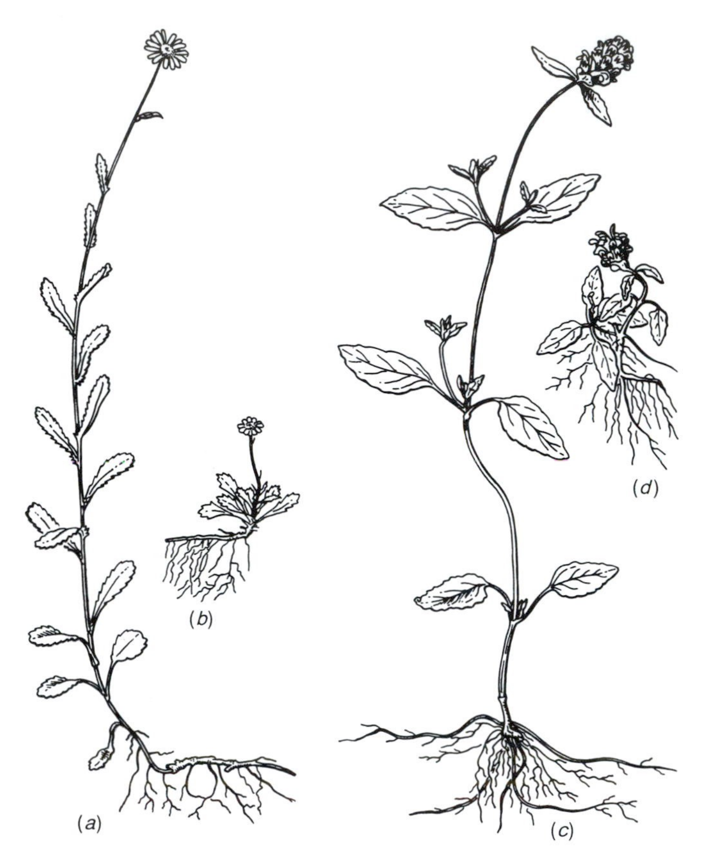

Fig. 4.1. Two examples of Bonnier's transplant experiments, showing the dwarfing effect of cultivation of ramets of the same clone at high altitudes. (*a*) & (*b*) Lowland and mountain *Leucanthemum vulgare* (*Chrysanthemum leucanthemum*). (*c*) & (*d*) Lowland and mountain *Prunella vulgaris*. (From Bonnier, 1895.)

upon the interpretation of natural variation patterns. It is worthy of note that Bonnier did not publish his conclusions in his *earlier* papers. Writing in 1890, he does not mention any transmutation of *Lotus corniculatus* into *L. alpinus*, although he had grown plants for some years both in the Pyrenees and in the Alps. He merely reported the dwarfing of the alpine clones in comparison with lowland ones.

Bonnier's claims were supported by the researches of Clements working in Colorado and California. He made a large series of clone-transplant experiments. In these experiments, too, it was asserted that lowland species had been transformed into alpine ones by growth at high altitude. *Epilobium angustifolium* was considered to have been changed into *E.*

latifolium, and Clements claimed that the grasses *Phleum alpinum* and *P. pratense* could be reciprocally converted (Clausen, Keck & Hiesey, 1940).

Before examining the alleged transformations we should note that a number of Central European botanists, notably Nägeli and Kerner, had been carrying out similar experiments and had come to different conclusions. Nägeli was one of the first to study alpine populations in experimental gardens. He brought a wide range of alpine plants into cultivation at the Botanic Garden in Munich and many changed their appearance greatly. This was particularly true of species of the genus *Hieracium*. Small alpine plants grown at Munich on rich soil became very large, much-branched plants. Nägeli was most interested to discover, however, that the acquired characters disappeared when plants were transplanted to gravelly soil within the garden and the specimens again assumed the appearance of alpine plants.

Kerner, Professor of Botany in Vienna, carried out many transplant and reciprocal sowing experiments using an alpine garden at Blaser at 2195 metres in the Tyrol, and the Botanic Gardens at Vienna and Innsbruck. He discovered that, for many species, if seeds were grown in two contrasting environments, dwarf plants with more vivid flowers were produced in alpine conditions. He noted the parallel case of more vivid colours in snails and spiders transferred to alpine conditions from the lowlands. Writing of his experiments in his famous book *The natural history of plants, their forms, growth, reproduction and distribution* (1895), Kerner noted:

> *in no instance was any permanent or hereditary modification in form or colour observed*... They [the modifications] were also manifested by the descendants of these plants *but only as long as they grew in the same place as their parents.* As soon as the seeds formed in the Alpine region were again sown in the beds of the Innsbruck or Vienna Botanic Gardens the plants raised from them immediately resumed the form and colour usual to that position. [author's italics].

Kerner, therefore, came to very different conclusions from Bonnier and Clements as to the nature of the changes which had taken place in the material planted at high altitude. Since Kerner's experiments, thousands of experimental plantings have been carried out, deservedly the most famous being those of Clausen, Keck & Hiesey (1940) in California. No evidence of transformation of the kind claimed by Bonnier and Clements has been discovered. The most reasonable explanation for their anomalous results is that their experimental areas became invaded by the related alpine species which were growing naturally at these high altitudes.

From these observations it can be seen that experimental cultivation can be of very great value in investigating variation in plants. Simple cultivation tests, in which a range of material is grown under standard conditions, in conjunction with crossing experiments, may reveal genetic differences between the stocks under investigation. Transfer and transplant experiments, properly carried out with special care in labelling and organisation, will give information upon the plasticity and adaptation of different plants. Especially useful are clone-transplants, as the performance of material of a single individual is investigated in different environments. In this respect, Bonnier's experiments were to be preferred to those of Kerner who often used seeds. Seeds, except in special circumstances (see Chapter 7), will be genetically heterogeneous and raise difficulties in interpretation not present in the clone-transplant method.

The work of Mendel

Let us now suppose that cultivation and transplant experiments in a particular instance have established a *prima facie* case of genetic difference between two plants. What is the nature of this difference? Our present knowledge of heredity stems from the various experiments of Mendel which he carried out over many years. Mendel, an Augustinian monk of the monastery of St Thomas at Brünn (now Brno in the Czech Republic), reported his work on crossing Peas in two papers to the Natural History Society of Brünn on 8 February and 8 March 1865, and the proceedings of these meetings were subsequently published in the *Transactions* of the Society in 1866.

Even though Mendel may be credited with the discoveries leading to the establishment of genetics, in many elementary textbooks the accounts of his work lack historical perspective. There is a wealth of pre-Mendelian experiments in hybridisation (Roberts, 1929), although it is true that early workers often had different objectives from those of Mendel. Kölreuter and Linnaeus investigated the phenomenon of sex in plants. Others, such as Laxton and the de Vilmorins, tried to improve varieties of plants of horticultural and agricultural importance. Another group of hybridists, as we saw in Chapter 2, were trying to find criteria for the experimental definition of species, using the data from experimental and natural hybridisation. Darwin was extremely interested in all aspects of hybridisation, and published a book on the effects of self- and cross-fertilisation in plants.

Many findings of Mendel were in fact anticipated by earlier hybridists, although they were not connected into a coherent theory (see Zirkle, 1966, for examples). Kölreuter, for example, discovered that *Nicotiana paniculata* × *N. rustica* and the reciprocal cross gave identical hybrids. He also had crosses which showed dominance: *Dianthus chinensis* (normal flowers) × *D. hortensis* (double flowers) resulted in dominance of double flowers. The phenomenon of segregation was also known long before Mendel's day.

Turning now to discuss the main points of Mendel's contribution, we might ask what are the ways in which his approach to the problem of heredity differed from those of his predecessors? The contemporary preoccupation with species led to many interspecific crossing experiments. For the purpose of elucidating the mechanisms of heredity, species-crosses are not very helpful because species differ in innumerable characters and in a number of generations a bewildering array of hybrid variants may appear. Before Mendel, hybridists did not in general concern themselves with the numbers of progeny of different sorts and sometimes they did not even keep separate the progeny from different plants or different generations. In Mendel's paper we see that he is aware of previous work in the field – the experiments of Kölreuter, Gärtner, Herbert, Lecoq, Wichura and others – and the defects of past experiments (trans. Bateson, 1909):

> Those who survey the work done in this department will arrive at the conviction that among all the numerous experiments made, not one has been carried out to such an extent and in such a way as to make it possible to determine the number of different forms under which the offspring of hybrids appear, or to arrange these forms with certainty according to their separate generations, or definitely to ascertain their statistical relations.

In selecting Peas for his work, Mendel knew that they are usually self-fertilising and that different cultivated varieties differ from each other in a number of respects. First, he tested a selection of stocks (34 in all) and, in a two-year trial, found a number to be true-breeding. This is one of the most important facets of Mendel's work. Then, after carefully removing the unopened stamens of selected flowers, he crossed Pea plants that differed in a pair of contrasting characters. Using the useful terms devised by Bateson & Saunders (1902) for the first and second filial generations (F_1, F_2), we may take as an example the cross between unpigmented plants (white seeds, white flowers, stem in axils of leaves green) and pigmented plants (grey or brownish seeds – with or without violet spotting, flowers with violet standards and purple wings, stem in axils of leaves red). Here Mendel discovered that the first generation of hybrids, the F_1, were all pigmented plants: 'pigmented' Mendel spoke of as 'dominant', the character 'unpigmented' he termed 'recessive'. He obtained the same result when pigmented plants were the seed or pollen parent – in other words reciprocal crosses gave the same results. Following natural self-fertilisation, in the next, F_2, generation he discovered that the recessive character (unpigmented) reappeared along with pigmented plants in a numerical ratio of 3 pigmented: 1 unpigmented (Fig. 4.2). In the next, F_3, generation, Mendel discovered that unpigmented plants bred true, whereas only one-third of the pigmented plants did so. On selfing, the other two-thirds of pigmented plants gave pigmented to unpigmented plants in a 3:1 ratio. The 3:1 ratio in the F_2 was in reality a ratio of 1 true-breeding pigmented: 2 non-true-breeding pigmented: 1 unpigmented.

Mendel obtained essentially similar results in crossing other Peas differing in single characters:

Character	*Dominant*	*Recessive*
Stature	Tall	Dwarf
Seed shape	Round	Wrinkled
Cotyledon colour	Yellow	Green
Pod shape	Inflated	Constricted
Unripe pod colour	Green	Yellow
Flower position	Axillary	Terminal

Particulate inheritance

To explain his results Mendel postulated the existence of 'factors' as he called them. Using current nomenclature, the dominant character, in our example 'pigmented', may be denoted by a factor C and the recessive 'unpigmented' by c. True-breeding parental stocks, CC and cc, produced C and c gametes, respectively, which at fertilisation gave an F_1 of constitution Cc. These F_1 plants, in appearance pigmented, produced in equal numbers two sorts of gametes, C and c, which (mating events being at random) gave three kinds of plants in the F_2 generation in the proportion $1CC$: $2Cc$: $1cc$ – a ratio of 3 pigmented: 1 unpigmented. Mendel realised that, owing to the operation of chance, an exact 3:1 ratio would not be achieved in practice. His results came close to expectation: in our example his F_2 consisted of 705 pigmented: 224 unpigmented plants, giving a ratio of 3.15:1.

Mendel's hypothesis of factors which can coexist in an F_1 and which segregate intact at gamete formation was subject to a further test, that of back-crossing the F_1 (Cc) to the recessive parent (cc). As expected, the progeny were in the ratio 1 pigmented: 1 unpigmented. These confirmatory results of Mendel vindicated his explanation of the earlier crosses.

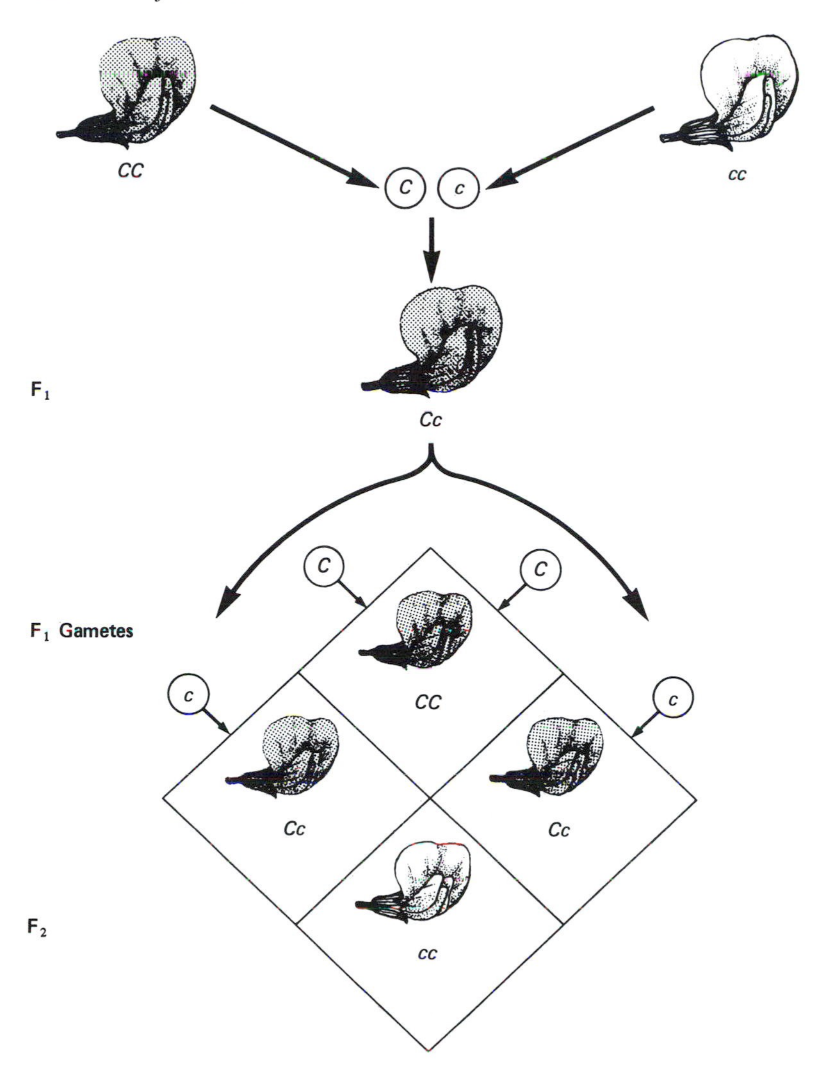

Fig. 4.2. One of Mendel's single-factor crosses, in which 'pigmented' (allele *C*) is dominant to 'unpigmented' (allele *c*). Note the 3:1 ratio of dominant to recessive phenotype in the F_2 progeny.

Mendel's two-factor crosses

We must now examine what happened when Mendel made 'two-factor' crosses. One of his experiments, incorporating his theory of determinants, may be represented by the following outline (Fig. 4.3). Mendel confirmed the genetic constitution of each category of plants by examining the

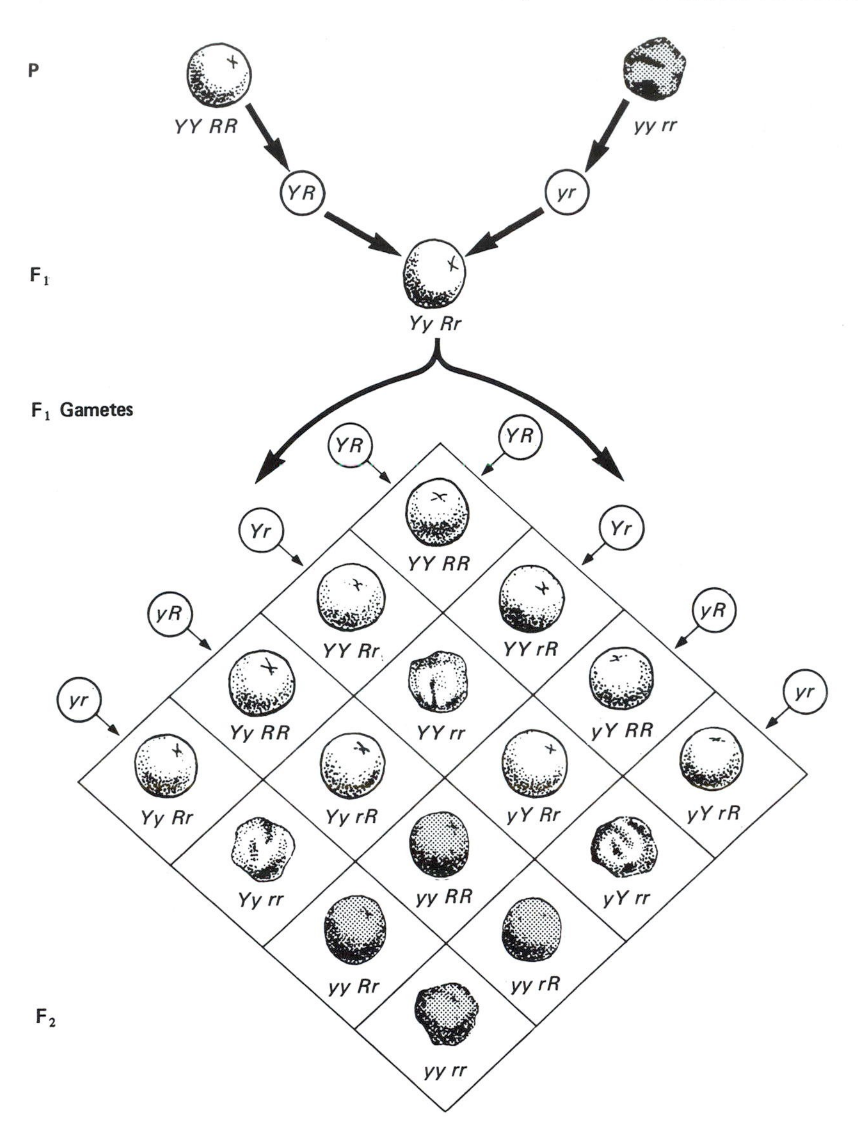

Fig. 4.3. A Mendelian two-factor cross: pure breeding yellow, round Peas (*YYRR*) × green, wrinkled (*yyrr*). The appearance of the F_2 progeny was:

	Yellow round	Yellow wrinkled	Green round	Green wrinkled
Theoretical phenotype	9	3	3	1
Mendel's result	315	101	108	32
Experimental ratio	9.8	3.2	3.4	1

progeny of selfed F_2 individuals. The important principle he discovered in these experiments was that the F_1, besides producing *YR* and *yr* gametes as did its parents, also produced gametes *Yr* and *yR* in numbers equal to those of the parental type. This equality of numbers of the four types of gametes established the *independent assortment* of the pairs of factors. Verification of independent assortment was made by Mendel when he crossed the F_1 (*YyRr*) with the double recessive parent (*yyrr*). As he predicted from his earlier results, four classes of offspring were produced in a 1:1:1:1 ratio.

Further work, in three-factor crosses in Peas, and crosses in French Beans, is reported in Mendel's paper.

A series of letters from Mendel to Nägeli have survived (trans. Stern & Sherwood, 1966) which reveal many facets of Mendel's work. First, Nägeli did not appreciate the significance of Mendel's findings and he questioned some of the results, for instance whether unpigmented plants produced in the F_2 were true-breeding. He thought that the progeny of hybrids must be variable and that in the case of unpigmented F_2 Peas, repeated selfing would eventually lead to segregation. Secondly, it was fortunate that Mendel's experiments on Peas were more or less complete before 1864, for Mendel writes that in the following summers infestations of Pea Beetle (*Bruchus pisi*) made cultivation of Peas difficult and finally impossible in Brno. The letters also reveal that Mendel carried out hybridisation experiments on more than 20 other genera of plants, e.g. *Aquilegia*, *Cirsium*, *Mirabilis*, *Verbascum*, *Viola* and *Zea* (see Iltis, 1932).

Historians of the early history of genetics have critically examined what might be called 'the traditional account' of Mendel's work using many primary sources of information, including Mendel's published papers, his letters to Nägeli and annotations in the books in his library, etc. (Orel & Matalovä, 1983; Orel, 1984; Bowler, 1989*a*). No potential source has been neglected. There is even a published illustration of Mendel's table-cloth with its embroidered monogram (Anon., 1965).

First, scholars have critically examined the aim of Mendel's work. It has been argued that he was not only concerned with studying heredity but also whether hybrids were constant or variable, in the context of a major concern of his day, namely, the role of hybrids in the origin of new species (Olby, 1979). Callender (1988) has proposed that he was an opponent of the theory of descent with modification. In traditional accounts of Mendel's work, Nägeli is 'blamed' for encouraging Mendel to study hybrids in the genus *Hieracium* (Mendel, 1869; trans. Stern & Sherwood, 1966). It is now known that reproduction in this genus is aberrant (we shall see why in Chapter 7). However, Mendel had begun his studies of *Hieracium* before he started his correspondence with Nägeli (Iltis, 1932). If Mendel's primary purpose was to study different sorts of hybrids in relation to the development of new species, then it is quite understandable that he should study not only the segregating hybrids of Pea but also the 'constant' hybrids of the genus *Hieracium*. Thus, it has been suggested that Mendel's experiments on *Hieracium* were not undertaken in the expectation that they would yield the same results as those on Peas (Callender, 1988).

It is important to realise, therefore, that Mendel did not derive a generalised scheme of heredity for all organisms from his experimental results on Peas. Indeed, it was only in the twentieth century that geneticists stated Mendel's findings as 'laws' applicable in plants and animals, including man. He thought of his work as demonstrating the method by which the laws of heredity could be worked out. This point is clearly stated towards the end of his paper on Peas when Mendel wrote: 'It must be the object of further experiments to ascertain whether the law of development discovered for *Pisum* applies also to the hybrids of other plants'.

Secondly, historians have discussed at length whether Mendel's notion of paired factors or elements was conceptually equivalent to the paired alleles of classical genetics (Brannigan, 1979; Olby, 1979).

Thirdly, there is another point of great interest, analysed in detail by Fisher (1936) and Edwards (1986). Mendel's data, taken as a whole, fit expected ratios far too well and consistently do not deviate as much as would be expected by the operation of the laws of probability. Fisher argues cogently that Mendel probably knew what his results would be before he started his experiments and that in reality his experiments were a confirmation or demonstration of a theory he had already formulated. Doubtful individuals were classified to fit expectation, perhaps by an assistant, or aberrant families may have been excluded from the final results (Sturtevant, 1965). The excessive goodness of fit of Mendel's results does not seem to be in dispute, but the conclusion that deliberate falsification was involved has not been accepted by Wright (1966).

Pangenesis

Perhaps we should now compare the ideas current at the end of the nineteenth century with those of Mendel which superseded them. Darwin, in his astonishingly productive later years, gave a great deal of thought to the problems of heredity and, in 1868, in *The variation of plants and animals under domestication*, he put forward his theory of 'pangenesis'. This theory, in many ways derived from Hippocratean ideas about the direct inheritance of characters, suggested that cells of plants and animals threw off minute granules or atoms (Darwin called them gemmules), which circulated freely within the organism. It was these gemmules which were transmitted from parent to offspring. Blending of gemmules occurred in the progeny. The phenomenon of 'segregation' of a recessive plant in an F_2 or subsequent generation, Darwin could account for only by suggesting that sometimes the gemmules were transmitted in a dormant state.

The theory of pangenesis, with its notion that gemmules came to the reproductive cells from all parts of the body, provided a mechanism for the inheritance of acquired characters, an idea favoured by Darwin. This view was challenged by Weismann (1883), who, in the words of Whitehouse (1959), disputed the idea that 'something from the substance of each organ was thought to be conveyed to the reproductive elements'. Weismann pictured just the converse situation: 'that the potentially immortal reproductive lineage carried in some mysterious way an exceedingly complex inheritance, which in each generation gave rise to mortal somatic offshoots – the individuals'. According to Weismann, 'the stream [of life] flowed direct from the reproductive cells of one generation to those of the next, and the individuals themselves (apart from their sexual elements) represented side-streams from which there was no return to the main stream'.

It is not necessary to go farther into Darwin's ideas, as they received no support from experiments. Galton, searching for evidence of gemmules, intertransfused blood of different-coloured rabbits and studied the colour of their offspring. There was no evidence that the presence of 'foreign' blood in a female rabbit made any difference to the colour of her progeny (Darwin, 1871; Galton, 1871). For a detailed account of this fascinating episode in the history of genetics, see Pearson's *The life, letters and labours of Francis Galton* (1924).

Galton himself had many ideas about heredity. Those he developed most forcibly were based upon a belief in blending inheritance. He did not carry out any breeding experiments as did Mendel but, as we saw in Chapter 3, he analysed records of human families and developed the 'law of ancestral heredity'. This 'law' was a statistical statement of general patterns in samples, rather than a genetic analysis.

Mendelian ratios in plants

Limitations of space prevent us from presenting a full account of the awakening of interest in Mendel's paper at the end of the nineteenth century. There had only been a few references to Mendel's work before 1900

(Olby & Gautry, 1968), but this is perhaps to be expected as the paper was published in an obscure journal. Then, tradition has it that De Vries, Correns and Von Tschermak, who had all been conducting breeding experiments, independently rediscovered Mendel's laws. Mendel's paper anticipated their findings, providing a confirmation and an interpretation of their own results. However, the notion of independent rediscovery has recently been dismissed as a 'myth' (see Corcos & Monaghan, 1990, for a full review of the evidence).

When Mendel's results became available in 1900, it was soon realised that his hypothesis of segregating factors, or 'genes' as they came to be called, could explain the results obtained for many plants and animals. Table 4.2, compiled mostly from Bateson (1909), gives a representative list of plants in which Mendelian inheritance was discovered. The characters involved range from those of the general growth habit, to details of the leaf, flower, fruit and seed. It is very interesting to see that variants known for many years were investigated. For instance, white flower colour in a variety of *Polemonium caeruleum* (described by John Ray in the seventeenth century) was shown to be recessive to the normal blue colour, in experiments of De Vries. Not only did morphological characters show Mendelian inheritance, but so did physiological traits. An example is disease resistance in Wheat (*Triticum*) infected with the fungus *Puccinia glumarum*, where susceptibility was shown by Biffen to be dominant.

Cases of independent segregation in two-factor crosses were also discovered. For example, in crossing a white-flowered, 'three-leaved' *Trifolium pratense* with a red-flowered, 'five-leaved' variant, De Vries (1905) obtained an approximate fit to an expected 9:3:3:1 ratio, the characters 'red-flowered' and 'five-leaved' being dominant.

Gradually Mendelian explanations for many single discontinuous variation patterns were accepted by most botanists and a number of useful terms were introduced. The alternative factors *A* and *a*, as, for example, tall and dwarf in peas, were spoken of as *alleles* (allelomorphs) of a gene by Bateson & Saunders (1902), who also introduced the term *heterozygous* (*Aa*) to describe a zygote or individual with two unlike alleles and *homozygous* (*AA*, *aa*) for one with two alike.

Mendelism and continuous variation

Notwithstanding the success of Mendelian explanations of familiar patterns of variation, universal acceptance did not follow. The biometricians, led by Pearson, remained loyal to the 'law of ancestral inheritance' of Galton, which we have shown is based upon blending inheritance.

Among the criticisms of Mendelism, one of great weight was that in certain crossing experiments no clear-cut segregation occurred in the F_2 generation. As an example, East's (1913) data for corolla length in F_1 and F_2 hybrids of *Nicotiana forgetiana* (female) × *N. alata* var. *grandiflora* (male) are given in Table 4.3. Here, a short-flowered plant was crossed with a long-flowered plant; the F_1 was of intermediate corolla length and the F_2, showing wider variation, did not segregate with Mendelian ratios. Is such a situation an example of blending inheritance? Pearson and his school of biometricians considered blending inheritance to be the general rule, Mendelian inheritance only applying in special circumstances. In the early years of the twentieth century, the problem of explaining continuous variation patterns was very urgent. An initial difficulty in understanding continuous variation was in estimating the environmental and genetic components of the variation pattern. This difficulty was largely removed by the work of Johannsen, to which we have already referred.

Yule (1902) was probably one of the first to suggest that many genes were involved in continuous variation. To show what he had in mind, we may take as an example human height, which follows a typical normal distribution and, even though nutritional factors are highly important in determining the height of a person, the fact that Pearson and Galton showed a positive correlation (r about 0.5) between the height of parent and offspring provided a *prima facie* case of genetic control of height. Yule considered that a number of genes might be involved in determining continuous variation patterns, and in this case different genes might determine leg length, trunk length, neck length, etc. In order to make this hypothesis credible, it was necessary to demonstrate that the genetics of a single character could be controlled by at least two genes.

Such a situation was discovered in 1909 by Nilsson-Ehle, who studied hybrids between Wheats with brown and white chaff (Fig. 4.4). In the F_1 of the cross, brown chaff was dominant. Inter-crossing of the F_1 gave an F_2 generation, not in the expected 3:1 ratio of brown: white, but in the ratio of 15 brown:1 white. This result was confirmed in a second experimental cross. Nilsson-Ehle considered that in this case two different genes were involved in chaff colour and that the 15:1 ratio was in reality a modified 9:3:3:1 ratio. The presence of a single dominant in an individual was sufficient to give brown chaff; only one-sixteenth of the progeny (of genotype *aabb*) had white chaff. Here is a clear case of two genes affecting the same character.

These experiments of Nilsson-Ehle, which were paralleled by the independent work of East, provide the necessary basis for an understanding of

Table 4.2. *Plants in which Mendelian inheritance was demonstrated before 1909; in some cases dominance is incomplete (Data mostly from Bateson, 1909; R.E.C. = Royal Society Evolution Committee Reports)*

	Gene dominant	Recessive	Material and author
Growth habit	Tall	Dwarf	Pea (*Pisum*): Mendel; von Tschermak. Sweet Pea (*Lathyrus*): R.E.C. Runner and French Bean (*Phaseolus*): von Tschermak.
	Branched	Unbranched	Sunflower (*Helianthus*): Shull. Cotton (*Gossypium*): Balls.
	Straggling	Bushy	Sweet Pea (*Lathyrus*): R.E.C.
	Biennial	Annual	Henbane (*Hyoscyamus*): Correns.
Leaves	Much serrated	Little serrated	Nettle (*Urtica*): Correns.
	'Palm'	'Fern'	Primula (*Primula sinensis*): Gregory.
	Normal	Laciniate	Greater Celandine (*Chelidonium majus*): De Vries.
	Yellow sap	White sap	Mullein (*Verbascum blattaria*): Shull.
	Rough	Smooth	Wheat (*Triticum*): Biffen.
Stems	Hairy	Glabrous	Campion (*Silene*): De Vries; R.E.C.
Flowers	Beardless	Bearded	Wheat (*Triticum*): Spillman; von Tschermak.
	Long pollen with 3 pores	Round pollen with 2 pores	Sweet Pea (*Lathyrus*): R.E.C.
	Normal pollen	Sterile	Sweet Pea (*Lathyrus*): R.E.C.
	Yellow	Brown	*Coreopsis tinctoria*: De Vries.

Table 4.2. (*cont.*)

	Gene dominant	Recessive	Material and author
	Purple	White	Thorn-apple (*Datura*): De Vries.
	Purple disc	Yellow disc	Sunflower (*Helianthus*): Shull
	Black palea	Straw palea	Barley (*Hordeum*): von Tschermak; Biffen.
	Purple spot	No spot	Opium Poppy (*Papaver somniferum*): De Vries.
	Red chaff	White chaff	Wheat (*Triticum*)
	Red flower	White flower	Clover (*Trifolium pratense*): De Vries (1905)
	Blue flower	White flower	Jacob's Ladder (*Polemonium caeruleum*): De Vries (1905).
	Blue-purple flower	White flower	Sea Aster (*Aster tripolium*): De Vries (1905).
Fruits	Prickly	Smooth	Buttercup (*Ranunculus arvensis*): R.E.C.
	Blunt pods	Pointed pods	Pea (*Pisum*): von Tschermak: R.E.C.
	Two-celled	Many-celled	Tomato (*Lycopersicum esculentum*): Price and Drinkard.
	Dark	Light	Deadly Nightshade (*Atropa belladonna*): De Vries; Saunders.
Seeds	'Long staple'	'Short staple'	Cotton (*Gossypium*): Balls.
	Round	Wrinkled	Pea (*Pisum*): Mendel; Correns; von Tschermak; Lock; Hurst; R.E.C.
	Starchy	Sugary	Maize (*Zea*): De Vries; Lock; Correns.
	Yellow endosperm	White endosperm	Maize (*Zea*).

Table 4.3. *Frequency distribution of corolla length in the cross* Nicotiana forgetiana × N. alata *var.* grandiflora (*Data of East*, 1913)

	Length of corolla (mm)														
	20	25	30	35	40	45	50	55	60	65	70	75	80	85	90
N. forgetiana	9	133	28	—	—	—	—	—	—	—	—	—	—	—	—
N. alata var. *grandiflora*	—	—	—	—	—	—	—	—	—	1	19	50	56	32	9
F_1	—	—	—	3	30	58	20	—	—	—	—	—	—	—	—
F_2	—	5	27	79	136	125	132	102	105	64	30	15	6	2	—

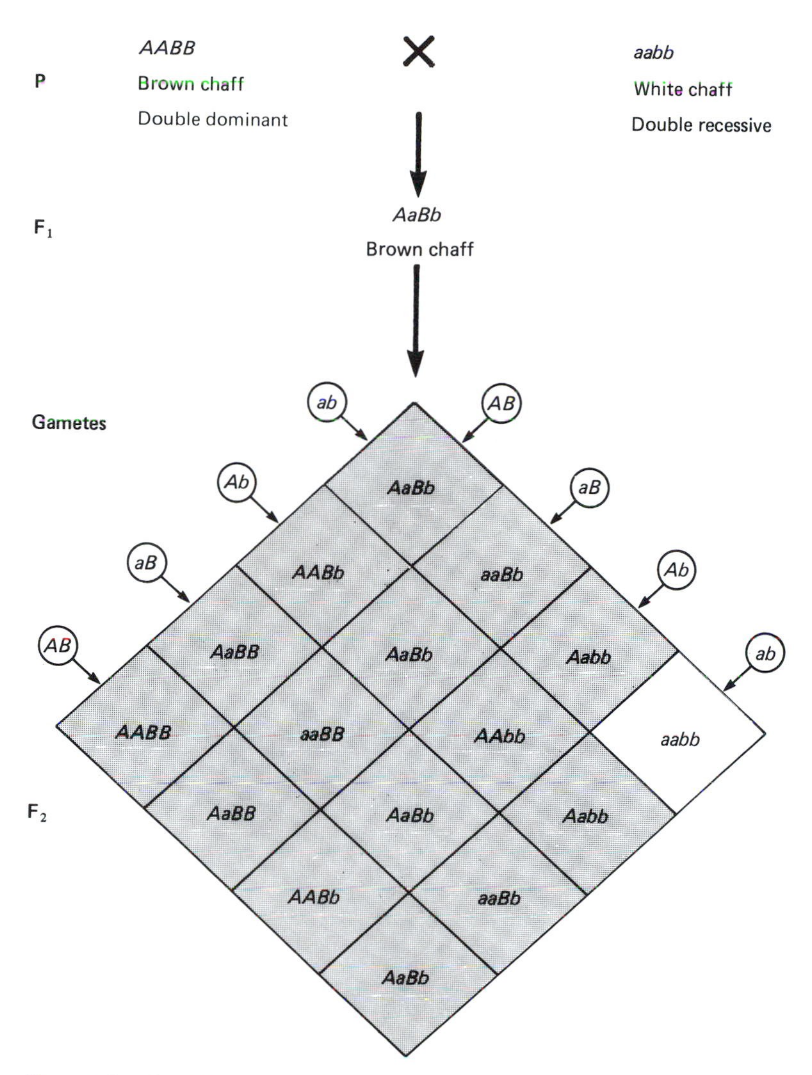

Fig. 4.4. Chaff colour in Wheat (*Triticum*). (From Nilsson-Ehle, 1909.)

the genetics of continuous variation. To demonstrate the principles we will examine a hypothetical case of flower colour (Fig. 4.5). In this model we postulate that two different genes are involved: *A* and *B* being the dominant alleles determining red flower colour, alleles *a* and *b* determining white flower colour. In the example, we assume, however, that the effects of *A* and

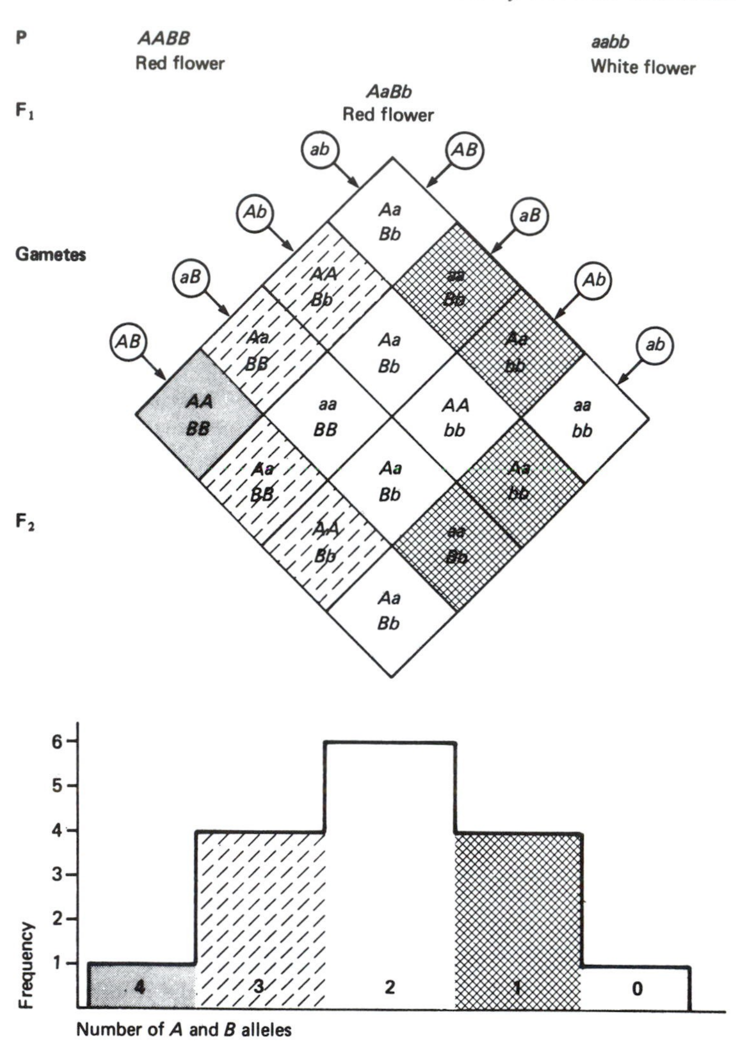

Fig. 4.5. Flower colour: a hypothetical case.

B are additive, the degree of red colour in the flower depending upon the number of *A* and *B* alleles present in an individual. Examination of the F_2 'chequer-board' shows that one-sixteenth of the progeny has four red alleles, four-sixteenths have three, six-sixteenths have two, four-sixteenths have one, and one-sixteenth has none. It should be noted that our example still shows Mendelian segregation of 15 red: 1 white on a broad classification, in detail with four different categories of red. Expressing the frequencies as a histogram, we obtain a distribution which bears a striking resemblance to a normal curve.

Consider now what might happen if a larger number of genes was involved. With six dominant genes, all additive in their effect for red colour, the F_2 would show very many categories of individuals and a closer fit to a normal curve. Of great importance, too, the parental genotypes *AAB-BCCDDEEFF* (red) and *aabbccddeeff* (white) would be very infrequently segregated in the F_2. In fact only 1/4096 of the progeny would be *AABBCCDDEEFF* and, even more important, there would be a similar proportion of *aabbccddeeff* which would be the only white phenotype. In actual practice, if the cross were made, even though Mendelian segregation had taken place at gamete formation in the F_1 plants, it is quite likely (especially if the F_2 is represented by a small number of plants) that no *aabbccddeeff* plants would be recovered at all. The F_2 progeny would then all be red-flowered, in different degrees, giving a normal distribution curve.

Turning now to an actual experiment, the *Nicotiana* crosses of East which we referred to earlier (Table 4.3), far from demonstrating blending inheritance, may more satisfactorily be interpreted on the basis of multiple factors affecting corolla length. The two variants of *Nicotiana* used differ in corolla length, and the F_1 from the cross is intermediate in length, indicating the absence or incompleteness of dominance. In the F_2 a wide array of corolla sizes is found, the frequency distribution approximating to a normal curve. Note that the extreme 'parental' corolla sizes are not represented in the data. East considered that there were probably four genes involved in the determination of corolla length.

Many investigations of continuous variation patterns in nature have given similar results to those of East, and elaborate genetic and statistical experiments since that time have demonstrated the general validity of the multiple-factor hypothesis. Such systems, in which the character is determined by several genes, are usually called *polygenic*.

Physical basis of Mendelian inheritance

So far we have not discussed the physical nature of Mendel's factors. In Mendel's day, little or nothing was known about the physico-chemical basis of heredity but there was plenty of theoretical speculation. Nägeli, for instance, postulated a genetically active 'idioplasm'. By the time Mendel's work was rediscovered in 1900 the situation, however, was very different. The latter half of the nineteenth century had seen an enormous increase in interest in the microscopic study of plant and animal cells. Certain technical innovations such as the use of stained material (carmine was introduced in the 1850s, and haematoxylin and aniline dyes in the 1860s) and the perfecting of apochromatic lenses (by Abbé in 1886) enabled biologists to make a close study of all aspects of cell division and development (Hughes, 1959). It is impossible in the space available to review the results of these studies in any detail, but the main conclusion was that the chromosomes discovered in cell division were clearly very important in heredity.

It was found that each species has a characteristic number, the diploid number, of chromosomes, visible in stained preparations of meristematic cells. The account that follows applies to diploid organisms, i.e. those whose nuclei contain two like sets of chromosomes, one set from pollen and one from the egg. There are, however, haploid organisms, e.g. certain fungi, whose nuclei contain only one set of chromosomes. In higher plants there is generally consistency of number, size and form of chromosomes of the meristematic cells of root-tip and shoot apex where chromosomes divide by *mitosis*. Essentially each chromosome divides into two daughter chromosomes and at the end of the process the two groups of daughter chromosomes are separated from each other by a new cell wall (Fig. 4.6).

Studies of the division of chromosomes in young anthers and in ovules revealed a different kind of nuclear division, the so-called *reduction division* or *meiosis* (Fig. 4.7). In this process the chromosome number is halved, the four derivatives having the haploid chromosome number. This halving compensates for the doubling in chromosome number following fertilisation of egg by sperm. Thus in a diploid plant a haploid complement of chromosomes has come from each parent. Microscopic examination of favourable material establishes a most interesting fact; if maternal and paternal haploid complements are examined they are normally found to be exactly alike in appearance (except in the case of certain sex chromosomes). In the early stages of meiosis, homologous chromosomes, from maternal and paternal sources, pair together. Studies as early as those of Rückert (1892) suggested that in this paired state exchanges of chromosome material occurred.

This very brief outline of mitosis and meiosis gives some idea of the sort of knowledge about chromosomes which was available at the beginning of the century. It was only a short time after the discovery of Mendel's work that various biologists, Boveri, Strasburger and Correns among them, saw a possible connection between Mendelian segregation and chromosome disjunction. It was probably Sutton (1902, 1903), however, who first set out with clarity a cytological explanation of Mendel's findings. In his view, the separation of maternal and paternal chromosomes of a homologous pair at

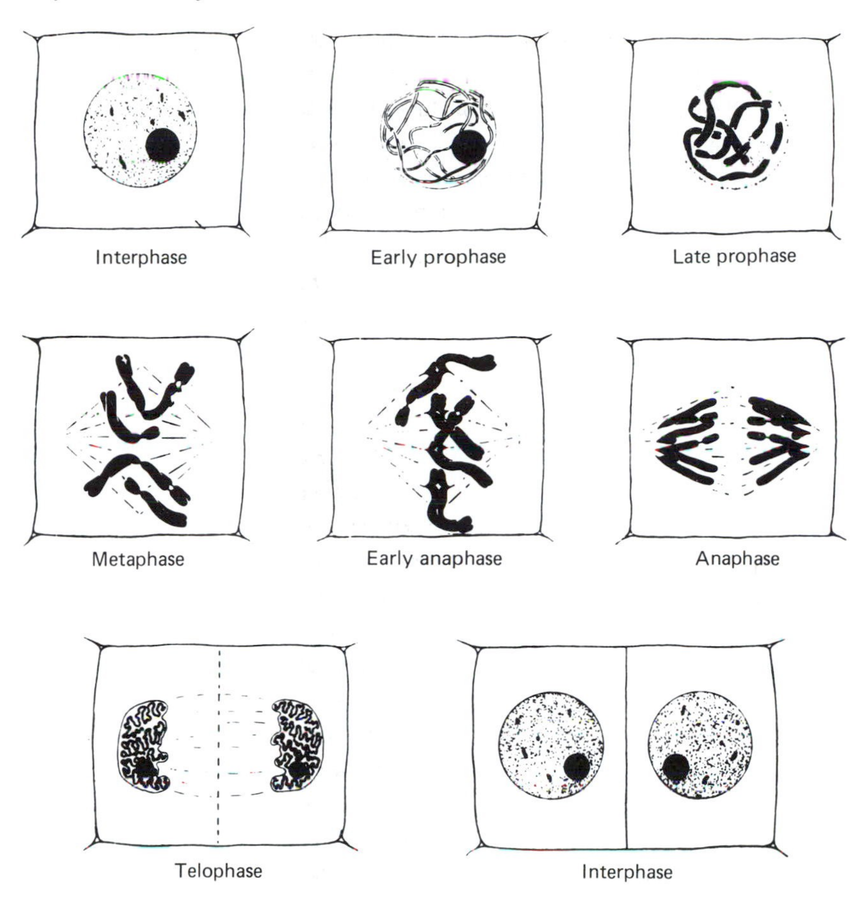

Fig. 4.6. The stages in mitosis in an organism with two pairs of chromosomes. (From McLeish & Snoad, 1962.)

the end of the first stage of meiosis resembled the postulated separation of factors which Mendel suggested occurred at gamete formation. Further, if the orientation of pairs on the spindle was at random, a number of combinations of maternal and paternal chromosomes would be obtained in the gametes. If the chromosome number was very small the number of combinations would also be relatively small; on the other hand, a diploid chromosome number as low as 16 would give 65 536 possible zygotic combinations (Table 4.4, Fig. 4.8) (see Sutton, 1903). As many plants have chromosome numbers higher than this, a huge number of combinations is possible. We have here the beginnings of the chromosome theory of heredity, which is the basis of all modern genetics.

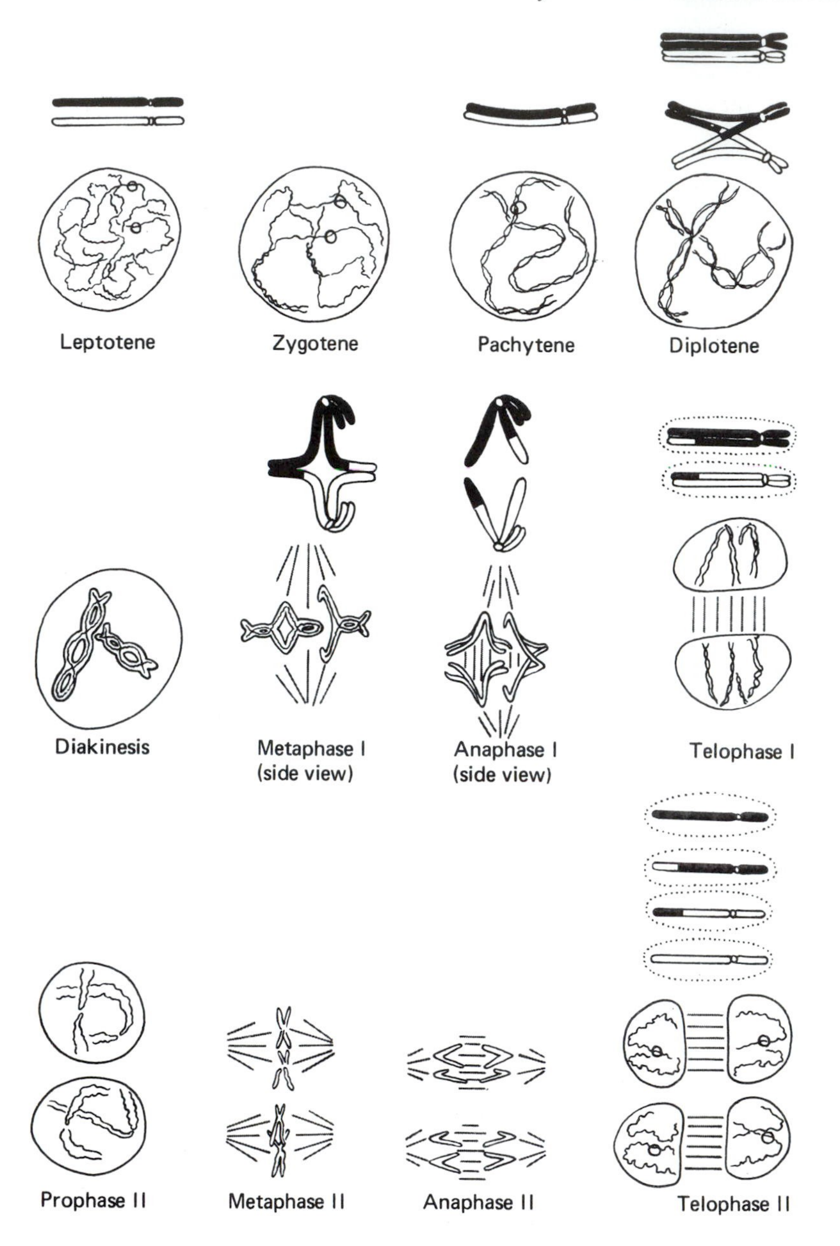

Fig. 4.7. The stages in meiosis in an organism with two pairs of chromosomes. The formation of one bivalent and its subsequent behaviour are shown diagrammatically above appropriate stages. (From Whitehouse, 1965.)

Table 4.4. *Possible zygotic combinations (After Sutton, 1903)*

Chromosome number			
Diploid	Haploid	Combinations in gametes	Combinations in eventual zygotes
2	1	2	4
4	2	4	16
6	3	8	64
8	4	16	256
10	5	32	1024
12	6	64	4096
14	7	128	16384
16	8	256	65 536
18	9	512	262 144
20	10	1024	1 048 576
22	11	2048	4 194 304
24	12	4096	16 777 216
26	13	8192	67 108 864
28	14	16 384	268 435 456
30	15	32 768	1 073 741 824
32	16	65 536	4 294 967 296
34	17	131 072	17 179 869 184
36	18	262 144	68 719 476 736

Mendel postulated in his experiments the independent segregation of factors and this view received support from the early geneticists. There were, however, increasing signs in the first decade of this century that not all genes segregate independently. Bateson, Saunders & Punnett in 1905, working with two-factor crosses in Sweet Peas (*Lathyrus odoratus*), did not get 9:3:3:1 ratios in F_2 families. Similar aberrant results were obtained from many organisms, amongst them the Fruit Fly (*Drosophila*) and Peas. Many biologists followed Mendel in experimenting with Peas, and up to 1917 an additional 25 character-pairs were examined (White, 1917). A very interesting series of crosses was made by de Vilmorin (1910, 1911), and subsequently by de Vilmorin & Bateson (1911) and Pellew (1913) working with 'Acacia' Peas, a variant characterised by the absence of the normal leaf tendrils. The absence of tendrils was associated with wrinkled seed. The cross 'Acacia' × round seed and tendrilled leaf gave an F_1 with round seed and tendrilled leaves. The F_2, instead of segregating to give 9:3:3:1, gave the results in Table 4.5.

It is quite clear that the two factors are not segregating independently: the grandparental combinations of wrinkled seed/no tendril and round

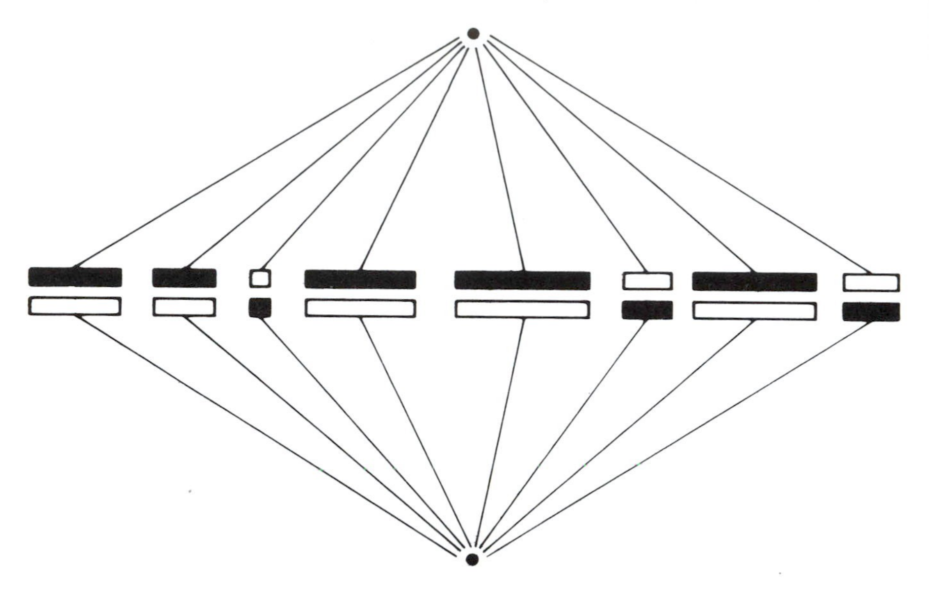

Fig. 4.8. The random orientation of bivalents at meiosis. At fertilisation a male and female gamete fuse, and each has a haploid set of chromosomes. Meiosis takes place in the diploid phase of the life cycle. Homologous chromosomes form bivalents (maternal chromosomes, dark; paternal chromosomes, white), which orientate themselves at random about the equatorial region of the cell. The diagram represents one of the possible meiotic metaphase arrangements of the chromosome complement in a diploid cell with 16 chromosomes (8 pairs). As Table 4.4. shows, in a plant with $2n = 16$ there are 256 possible patterns of arrangement of paternal and maternal chromosomes. As homologous chromosomes may carry different alleles, independent orientation means that many different combinations of maternal and paternal genes may be obtained in the gametes.

seed/tendril are being recovered with too high frequencies. Various explanations were offered for this phenomenon, which came to be called 'partial linkage', later abbreviated to 'linkage'. Bateson & Punnett (1911) favoured an obscure 'reduplication' hypothesis; as time went by, however, the views of Morgan prevailed. He suggested that partially linked groups of factors were together on the same chromosome. In *Drosophila melanogaster*, where $n = 4$, the extensive researches of Morgan and his colleagues established beyond doubt the existence of four such linkage groups. In Pea, where $n = 7$, there are seven linkage groups. It is of considerable interest that two of the seven traits studied by Mendel are now known to be linked (Lamprecht, 1961).

Table 4.5. *'Acacia' Peas, results of various experiments as reported in White (1917)*

Source	Wrinkled seed, no tendril	Wrinkled seed, tendril	Round seed, no tendril	Round seed, tendril
de Vilmorin	70	5	2	113
de Vilmorin	99	4	1	170
Bateson	64	1	4	210
Pellew	564	15	20	1466

In the formation of a diploid organism the two gametes each carry one set of linkage groups – the haploid chromosome number. The appearance of occasional recombinants in small numbers in a cross such as that in Table 4.4 was accounted for by postulating an exchange of parts by homologous pairs in the first stage of maturation division. Evidence of such an exchange was seen in the chiasmata of prophase (Fig. 4.9).

There is insufficient space to examine further the progress and implications of the chromosome theory. A number of excellent books exist which discuss chromosome mapping, and give full and detailed evidence for all the postulates in the chromosome theory of heredity (for example, Whitehouse, 1973).

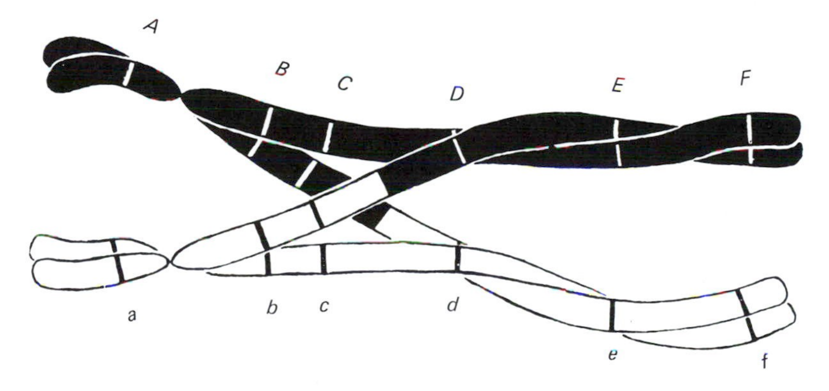

Fig. 4.9. A bivalent at diplotene with a single chiasma. The position of some genes (represented by letters) is indicated. Crossing-over has occurred between one chromatid of the maternal chromosome (white) and one chromatid of the paternal chromosome (dark). (From McLeish & Snoad, 1962.)

5

Post-Darwinian ideas about evolution

For 40 years after the publication of the *Origin*, Darwin's ideas were a source of tremendous public controversy. For this reason he never received any awards from the state, although he was awarded honorary degrees and decorations in plenty. Despite thousands of sermons attacking the idea of evolution by natural selection, more and more biologists became convinced of the truth of Darwin's view. The biological literature of the period is full of papers speculating about the adaptive significance of various structures, the probable course of evolution in the plant and animal kingdoms, and so-called 'evolutionary trees' showing phylogenetic relationships (Fig. 5.1). Some of this work is of lasting interest; but there was a depressing tendency in the later years of the period for armchair biologists to produce highly speculative theories and there was a lack of critical experiment with living material. Towards the end of the century, however, there were signs of an increasing interest in the possibility of using experiments for the investigation of evolutionary problems.

Experimental investigation of evolution

A good example of this change in climate is provided by the controversy which enlivened the pages of *Nature* and the editorials of *The Gardeners' Chronicle* in 1895. At a discussion meeting of the Royal Society, the Director of Kew, W. T. Thiselton-Dyer, had shown specimens of the 'feral' type and cultivated variants of what he called *Cineraria cruenta* – the gardener's Cineraria – and an extended account of his ideas was printed in *Nature* (1895). He suggested, as befitted an ardent and orthodox disciple of Darwin, that as far as was known the Garden Cineraria was derived from *Senecio* (*Cineraria*) *cruentus* from the Canaries 'by the accumulation of small differences'. Bateson (1895*a,b*) responded to this view in a lengthy

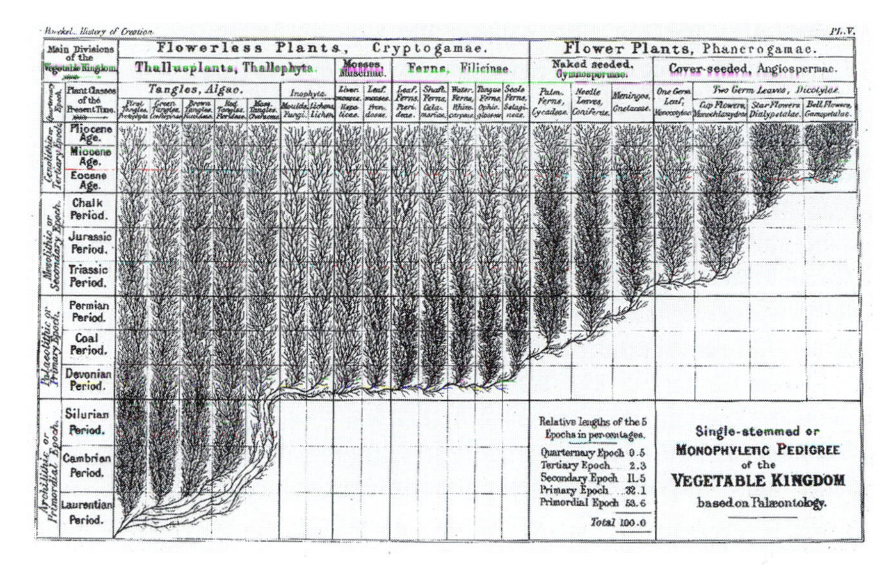

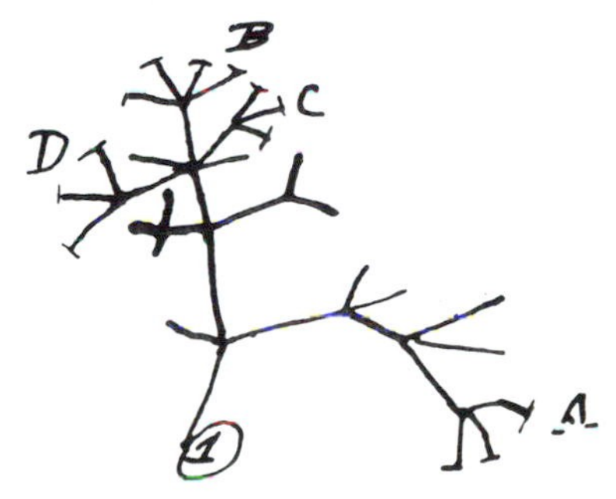

Fig. 5.1. Tentative sketch of a phylogenetic tree from Darwin's notebook (1837) (see De Beer, 1960–61) which contrasts with the baroque splendour of Haeckel's highly speculative 'monophyletic pedigree of the vegetable kingdom' of 1876.

letter to the Editor, questioning the assertions of Thiselton-Dyer. Bateson concluded, after a study of the literature, that modern Garden Cinerarias arose from hybridisation between several distinct species, that selection was practised on variable hybrid progeny, and that 'sports' may have been important, as well as subsequent improvements as a consequence of the selection of small-scale variation. The arguments in *Nature* continued back and forth with four letters from Thiselton-Dyer, three from Bateson and three from the biometrician Weldon. It became clear that argument could not settle the issue of the origin of the Garden Cineraria. The possibility that experimental hybridisation might shed light on the variation patterns occurred to Bateson, who enlisted the help of Lynch, Curator of the University Botanic Garden in Cambridge. Lynch raised stocks and made a number of artificial crosses, some of which were exhibited at a meeting of the Cambridge Philosophical Society in 1897. The report of the meeting (Bateson, 1897) says that the experiments 'were entirely consistent with the view that Cineraria was a hybrid between several species'. Lynch's experiments were published in detail in 1900. Here we have a clear case of speculation about evolution leading directly to experiment. It is interesting that a more recent review of the Cineraria problem (Barkley, 1966) reveals that it has received little attention since these early experiments.

During the period 1892–1910 some of the first experiments investigating natural selection were carried out. Darwin had written in the *Origin* (Chapter 4): 'Can we doubt (remembering that many more individuals are born than can possibly survive) that individuals having any advantage, however slight, over others, would have the best chance of surviving and of procreating their kind?' In 1895 Weldon wrote: 'The questions raised by the Darwinian hypothesis are purely statistical, and the statistical method is the only one at present obvious by which that hypothesis can be experimentally checked. In order to estimate the effect of small variations upon the chance of survival, in a given species, it is necessary to measure first, the percentage of young animals exhibiting this variation; secondly, the percentage of adults in which it is present' (Weldon, 1895*b*). If the percentage of adults exhibiting the variation proved to be less than that in young animals, then some of the young animals must have been lost before reaching adulthood and a measure of the advantage or disadvantage of the variation could be obtained. In putting these novel ideas to the test, Weldon (1898) investigated the variation in the Crab, *Carcinus maenas*, at a site on Plymouth Sound. While it is inappropriate to give details of his results, we may note that his findings offered some support for the initial hypothesis. After several years' investigation (1892–8) he also concluded that the population was unstable. He deduced that changes were caused by the increasing amounts of china clay and sewage in the waters of Plymouth Sound, and carried out experiments investigating the death rate of captive Crabs subjected to foul water. Crabs survived captivity in clean water but only a portion of the variable population – those with small frontal breadth relative to their carapace size – survived in foul water. While it is true that these experiments may be criticised on grounds of sampling technique (and, indeed, in other ways), they do represent a major step forward in the design of investigations into natural selection and, as we shall see, are a model for some more recent botanical studies.

Weldon was also instrumental in encouraging some of the first field studies of selection. Di Cesnola (1904), a student of Weldon's at Oxford, noticed that in Italy there were green and brown variants of Praying Mantis (*Mantis religiosa*). The green variant was found in grasses and the brown on vegetation burnt by the sun. In an experiment which lasted several days, individuals were tethered by silk threads as follows:

1. In a green grassy area – 20 green and 45 brown individuals.
2. In a brown area – 25 green and 20 brown individuals.

The 25 greens in the brown area and 35 of the browns in the green area were taken by birds or ants. It is significant that the individuals which matched the background vegetation were untouched. These studies were forerunners of many experiments by zoologists to test the supposed adaptive significance of protective coloration (Cott, 1940). As in the case of the investigations on Crabs, it is obvious that the experiment with *Mantis* is not beyond criticism, but it is based, nevertheless, on a novel approach to the study of the force of natural selection.

In plants too, studies of variation led to insights into natural selection. For example, in 1895, von Wettstein described the phenomenon of seasonal dimorphism in a number of hemiparasitic genera including *Euphrasia*, *Odontites* and *Rhinanthus*. As a result of careful investigation, many species appeared to have two subspecies, viz. an early summer flowering variant (aestival) and a later variant (autumnal). He considered that the practice of haymaking in Central Europe was important in the origin of the two types of subspecies. The maintenance of the annual habit was only possible if plants either fruited before midsummer grass-cutting or elongated and matured after the hay crop had been taken. Plants flowering or with immature fruit at the time of haymaking would fail to reproduce, and selection would therefore favour both early and late flowering. This work provoked a good deal of controversy, especially about which subspecies was ancestral and whether patterns were as simple as was suggested by von Wettstein. For full details of the historical studies in this area see Borg (1972).

Interesting results were also obtained from agricultural species. Brand & Waldron (1910) and Waldron (1912), working at Dickinson, North Dakota, USA, cultivated 68 samples of the important legume forage crop Lucerne or Alfalfa (*Medicago sativa*) collected from different parts of the world. Plants from Mongolia proved to be cold-resistant, whilst those from Arabia and Peru, on the other hand, were frost-sensitive. During the severe winter of 1908–9, the pattern of losses due to frost damage provided interesting information. For instance, in a strain from Utah, many plants died of frost damage, but out of the progeny of three especially resistant plants originating from this stock only 3.5% died. Brand & Waldron deduced that a frost-hardy strain could originate from extreme individuals of an otherwise frost-sensitive stock.

There was also interesting research on 'races' in hemiparasitic plants. Thus, three variants of Mistletoe (*Viscum album*) were described by von Tubeuf (1923) from broadleaved trees, Fir and Pine. Each was morphologically distinct in such characters as size and shape of leaves, and colour of berries. Some attempt was made (only partially successful) to test, by transfer of seed, whether there were three different physiological races of Mistletoe, each adapted to a different host.

The mutation theory of evolution

The experimental approaches employed by biologists at the turn of the century not only provided insights into biometry and natural selection, but also provoked Bateson, De Vries and others to propose a rival theory to that of Darwin – the mutation theory of evolution. The theory stressed the importance of 'sports' and various other abruptly occurring new variants. While the theory was claimed to be new, it is clear, as we saw in Chapter 2, that it grew out of a longstanding interest in the subject of 'sports'.

Darwin argued that species were ever-changing entities, the products of natural selection; his thesis was descent with modification, involving continual and gradual change. De Vries and Bateson did not deny the existence of natural selection; in fact it still played a key role in their ideas of evolution. What was different was their view that new species arose abruptly by 'mutation'. They confined the significant changeability of species to distinct and probably short periods. They accepted the theory of descent with modification, but thought that the changes occurred abruptly, interspersed with periods of stability.

What evidence could the 'mutationalists' find in support of their theory? First, they examined cases of the apparent abrupt evolution of new persistent variants. Most famous of these were plants of the genus *Oenothera* studied by De Vries (1905). Secondly, they discussed problems of heredity. For the 'mutation theory' it was *discontinuous* patterns which were significant and, with the re-finding of Mendel's work in 1900, Bateson did not fail to point out that this provided a mechanism explaining discontinuous variation. In his view, continuous variation was the product of environmental factors.

A third piece of evidence was also forthcoming. In Johannsen's experiment, which we introduced in Chapter 4, selection had no effect upon mean seed weight. Try as he might, from a particular line he could not select a strain with larger or smaller beans. Some variations did occur, but Johannsen ascribed this to the effect of the environment. Bateson and others went further and argued that all 'fluctuating variations' found in nature were environmentally based. For the 'mutationalists' natural selection occurred only when the products of mutation were being sorted out.

By the beginning of the century a curious situation had developed. In opposing camps were the 'Mendelian-mutationalists' and the 'Darwinian-biometricians' (Waddington, 1966; Crew, 1966; Provine, 1971, 1987). The Darwinian-biometricians, for the most part, remained loyal to Darwin's theory of gradual change. It is remarkable that this group of mathematically minded scientists opposed Mendel's views, preferring instead Galton's law of ancestral inheritance. Indeed, the animosity between the forceful personalities involved contributed to a delay in the development of the subject. Furthermore, it has been pointed out by Fisher (1958), the lack of mathematical understanding in biologists possibly contributed to the neglect of Mendel's work at the time of its publication, yet, paradoxically, on its re-finding it was the mathematical biologists who opposed it.

Neo-Darwinism

In the early decades of the twentieth century, many biologists thought that Darwinism had been eclipsed and was indeed dead (Huxley, 1942). Then in the 1930s, through the work of Fisher (1929), Haldane (1932), Sewall Wright (Provine,1986) and many others, an integration of Mendelian genetics and Darwin's evolutionary theory took place. 'Neo-Darwinism' was born.

There was an acceptance that inheritance is particulate and that genetic material is borne on chromosomes. Segregation and recombination were widely demonstrated. Gradually, more and more evidence against the 'mutation theory of evolution' view was discovered. First, intensive genetic and cytological studies of many species, including species of *Oenothera*, were carried out and it became obvious that the new persistent variants found in *Oenothera* were of several different sorts. Some were simple mutants; others were polyploid derivatives; and a further group was the result of complex interchanges of chromosome segments (see Chapter 6). Other species did not give the same results and the *Oenothera* situation was seen to be unique in its complexity.

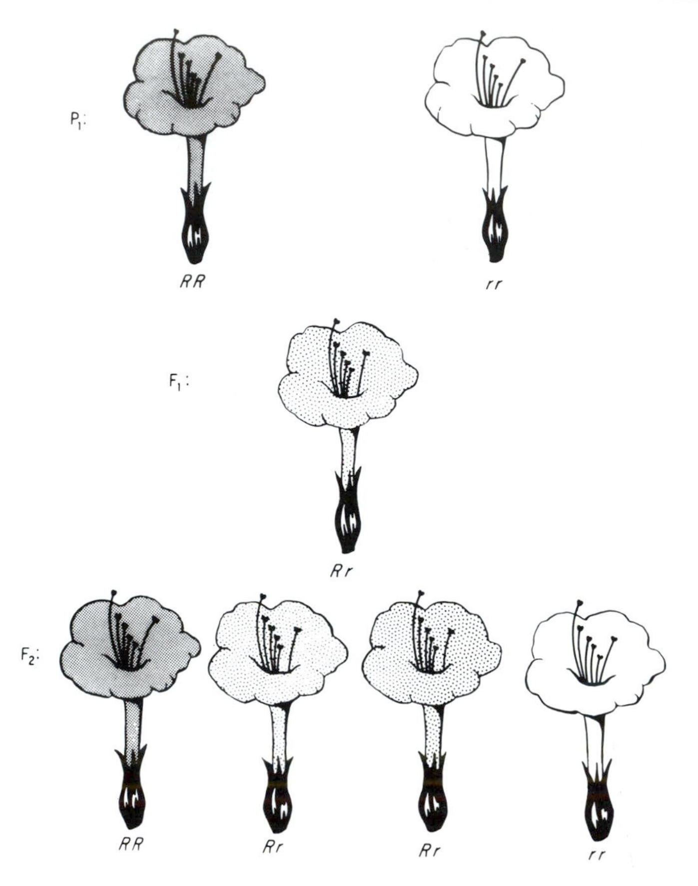

Fig. 5.2. *Mirabilis jalapa.* At first sight the production of pink-flowered offspring from red- and white-flowered parents looks like a case of blending inheritance, but, as the diagram shows, the situation can be explained in simple Mendelian terms if we assume the absence of dominance. Reprinted with permission of Macmillan Publishing Co. Inc. from Strickberger (1976) after Correns. © 1976 Monroe Strickberger.

Secondly, further sets of data became available to compare with Johannsen's results. Habitual self-fertilisation is characteristic of French Beans and, as we shall see later, this leads to genetic homozygosity. The ineffectiveness of selection in his experiment is understandable considering the breeding system. However, it is only in the absence of genetic variability that selection is ineffective, a fact attested by many successful selection experiments with outbreeding organisms.

The idea of blending inheritance was finally demolished by work carried out at this time. Darwin's idea of the persistence of favoured variants under a regime of blending inheritance necessitated a high mutation rate. Mathematical calculations, based on the natural mutation rate, showed that its incidence was much lower than that required to support the idea of blending (Fisher, 1929). Further, it became clear that cases of inheritance that were at first explained by blending were explicable in terms of Mendelian genetics – some as instances of systems with no dominance (Fig. 5.2), and others, as we saw in Chapter 4, as examples of inheritance controlled by many factors. Thus, with the disappearance of the grounds for believing that speciation only occurred by 'mutation' and the demise of the theory of blending inheritance, a more unified science emerged and, as we shall see in the following chapters, it is now accepted that both gradual change and abrupt events play a part in speciation.

6

Modern views on the basis of variation

Molecular basis of heredity

The early geneticists visualised genes as 'beads threaded on a string'. In 1944, the physicist Schrödinger, in his fascinating book *What is life?*, suggested that genes could be complex organic molecules in which endless possible variations in detailed atomic structure could be responsible for codes specifying the stages of ontogenetic development. More recently, as a result of the application of sophisticated biochemical and physical techniques to biological material, spectacular progress has been made in our understanding of the nature of the hereditary material. In the past it was considered that proteins might be the carriers of genetic information, but now it has been established that deoxyribonucleic acid (DNA) is the primary hereditary material. The achievements of molecular biology have been so great and the progress so rapid that an enormous body of information now exists. A revolution in biology is taking place – no less of a revolution than that caused by Darwin's theory of evolution by natural selection. Obviously in this short book we cannot survey even the main findings of these new approaches; instead we must be content to stress some of the ways in which this new knowledge of molecular biology illuminates various aspects of plant variation and microevolution. The long historical development of research into the molecular basis of heredity has been reviewed by Olby (1974, 1985) and Portugal & Cohen (1977). For comprehensive accounts of recent developments in molecular biology, Lawrence (1989), Lewin (1997) and Watson *et al.* (1992) may be consulted.

The helical structure of DNA was established by Watson & Crick (1953). (For recollections of the circumstances leading up to its discovery, see Watson, 1968, and Crick, 1988). Figure 6.1 shows diagrammatically the sugar–phosphate 'backbone' of the two complementary chains, which are held together by hydrogen bonding of nitrogenous bases. As indicated in

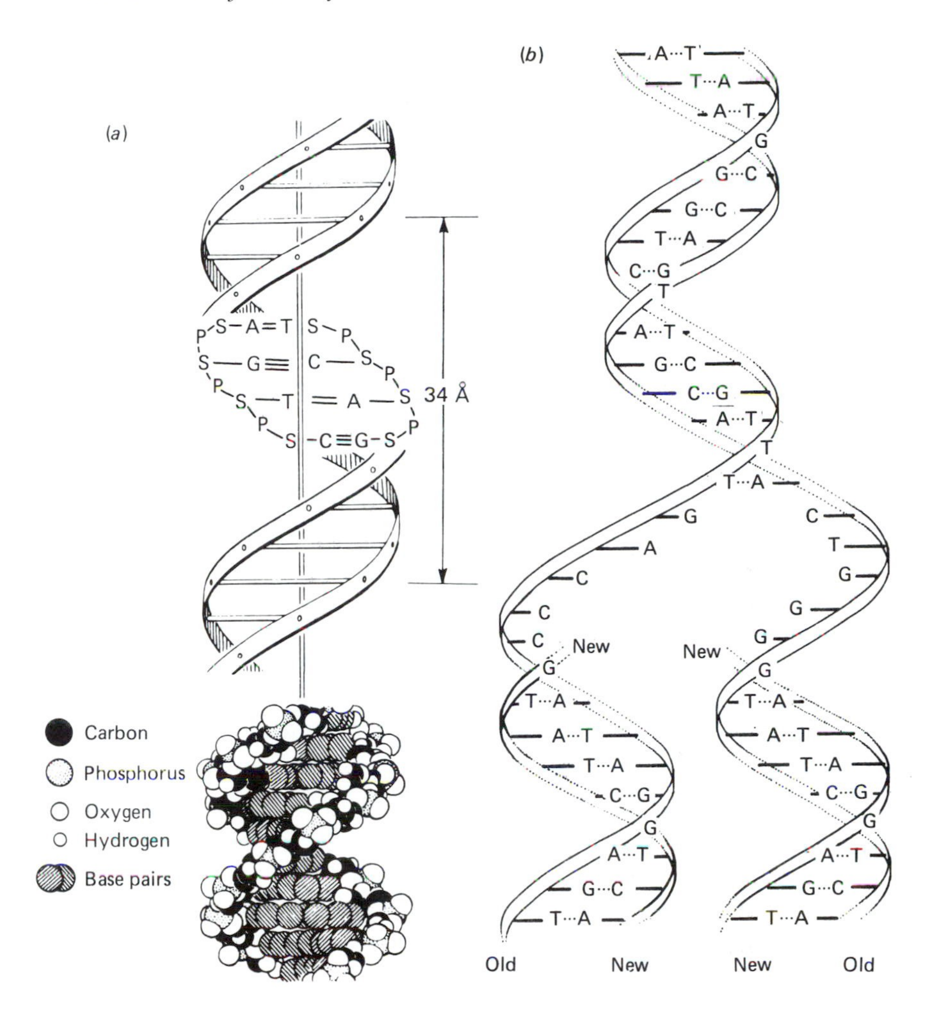

Fig. 6.1. The structure and mode of replication of DNA. (*a*) Structure of DNA, showing the complementary paired strands linked together by hydrogen bonds. Adenine (A) and thymine (T) always pair together, as do guanine (G) and cytosine (C). The backbone of the DNA molecule consists of the phosphate (P) and deoxyribose sugar (S) groups. Thus, the DNA molecule is built up from many basic units called nucleotides, each of which consists of a base linked to a molecule of deoxyribose plus a phosphate group. (From Hayes, 1964, and Berry, 1977.) (*b*) Replication of DNA as suggested by Watson & Crick. The complementary strands are separated to produce a replication fork and each forms the template for the synthesis of a complementary daughter strand. Reprinted by permission of Watson, J. D., *Molecular biology of the gene*, Menlo Park, California. The Benjamin Cummings Publishing Company, Inc. 1965.

the diagram, the pairing of these bases is highly specific. Four different bases are found in DNA: adenine (A) always pairs with thymine (T) and cytosine (C) with guanine (G).

DNA has a number of important properties. First, by enzymic activity, it may replicate without change, in a semi-conservative fashion, i.e. the bonds between the base pairs are broken at a replication fork (Fig. 6.1*b*). Each half of the double helix acts as a template for synthesis. There is evidence, in higher organisms, that replication begins at several points (replicons) in the long DNA molecule and that, from each point of initiation, replication proceeds in opposite directions along the two separated strands. There may be as many as 35000 such points in the DNA of *Vicia faba* (Lewin, 1990).

Secondly, DNA, in its sequence of bases, provides a code of genetic information. Most of the DNA in plants resides in the nucleus, but, as we shall see, there is also DNA in organelles, elsewhere in the cell. The coded information in the nuclear DNA, transcribed in the nucleus, is then translated into action outside the nuclear membrane in the production of a range of proteins by a process that assembles 20 amino acids in different combinations. In the transcription of information, DNA serves as a template for ribonucleic acid (RNA: which differs from DNA in being usually single-stranded, having ribose sugar in its nucleotides and the base uracil (U) instead of T).

Primary protein structure is specified by genes. At the molecular level a gene is a length of sequences of nucleotides of DNA. Starting from a fixed point, the code of the gene, represented by the base sequences, is read as a series of non-overlapping triplets of nucleotides (codons). Almost all the amino acids are specified by more than one triplet. Thus, there is 'redundancy' in the code (Fig. 6.2).

Some genes have coding and non-coding (intron) regions and the initial RNA molecule is reduced in length by the removal of the introns. Thus, the coded information in the DNA molecule is accurately transcribed in the complementary messenger RNA (mRNA) molecule, which moves out of the nucleus to the ribosomes (Fig. 6.3). Here it acts as a template on which the amino acids appropriate to the genetic code are joined together to form polypeptide chains and, ultimately, functioning proteins. The 20 different amino acids found in proteins are brought to the ribosomes for assembly by 20 different transfer RNA (t RNA) molecules (Fig. 6.4).

A third property of DNA, of great biological significance, is its molecular stability, a stability which nevertheless permits some change. In the growth and development of the cells of a plant, each cell usually receives an exact

First base	*Second base*							
	U		*C*		*A*		*G*	
U	UUU UUC UUA UUG	Phe Leu	UCU UCC UCA UCG	Ser	UAU UAC UAA UAG	Tyr Stop Stop	UGU UGC UGA UGG	Cys Stop Trp
C	CUU CUC CUA CUG	Leu	CCU CCC CCA CCG	Pro	CAU CAC CAA CAG	His Gln	CGU CGC CGA CGG	Arg
A	AUU AUC AUA AUG	Ile Met	ACU ACC ACA ACG	Thr	AAU AAC AAA AAG	Asn	AGU AGC AGA AGG	Ser Arg
G	GUU GUC GUA GUG	Val	GCU GCC GCA GCG	Ala	GAU GAC GAA GAG	Asp Glu	GGU GGC GGA GGG	Gly

Fig. 6.2. The genetic code. The triplet codons given are those of the RNA which provides the template for the production of the polypeptides from amino acids. The amino acids are given the standard three letter abbrevations: Met (Methionine); Val (Valine), etc. Note that almost all the amino acids are specified by more than one triplet. Stop codons mark the end of a sequence specifying a particular protein. (Reprinted by permission of Addison Wesley Longman Ltd. from Lawrence, 1989.)

copy of the hereditary material. Indeed, there is evidence that, as the process of replication of DNA occurs, a 'proof-reading' mechanism operates, correcting errors. However, despite the mechanisms to perpetuate the DNA in an unchanged form, permanent change in base sequence may occur. Such mutations may be spontaneous – or artificially induced by chemicals or radiation. With regard to the frequency of mutation in all the loci of an organism, an estimate is available from studies of *Drosophila*. Some 5% of the gametes produced in the current generation are likely to contain a mutation (see Rothwell, 1993).

A number of different types of mutational change have been recognised

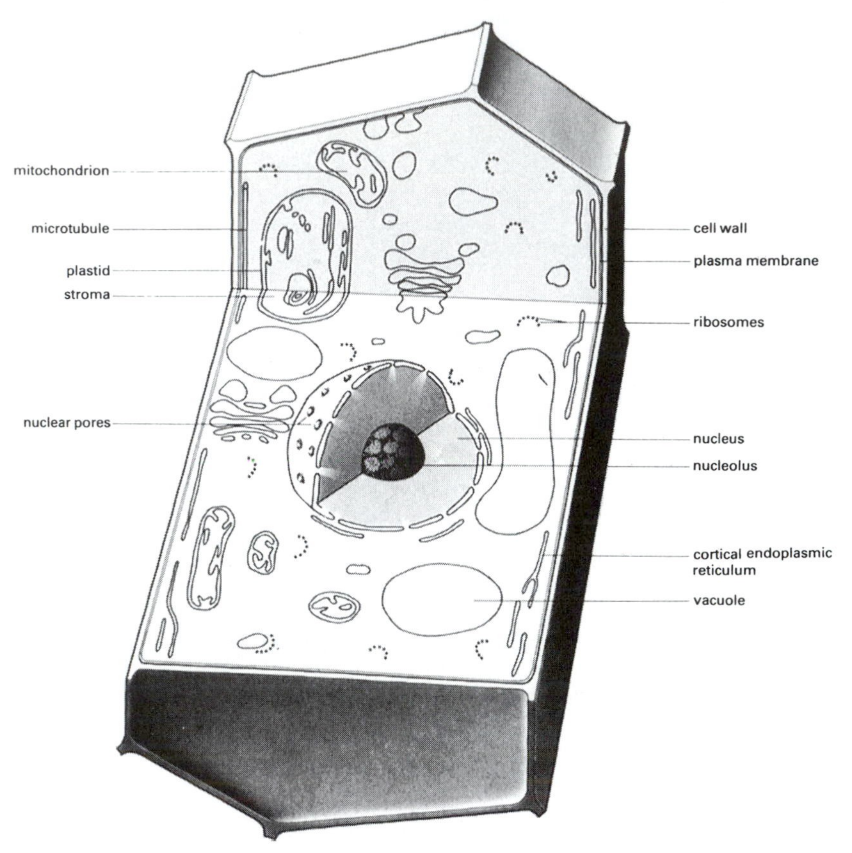

Fig. 6.3. Diagram of a generalised plant cell, showing its component parts, and their distribution in the cell. (From Friday & Ingram, 1985.)

(Fig. 6.5). With reference to an original sequence of bases in the DNA, a point mutation may occur, resulting in an altered codon and possibly the insertion of a different amino acid into a polypeptide chain (Fig. 6.5*a*). As the coding bases are read as non-overlapping triplets, frameshift, insertion and deletion of bases will cause a mutation by altering the codons (Fig. 6.5*b–d*). Furthermore, given the redundancy of the code, the change of a nucleotide may sometimes produce a 'silent' mutation as no change to the amino acid sequence results. (Fig. 6.5*e*).

A fourth property of DNA, of crucial importance in evolution, is the existence of mechanisms to enable reciprocal exchanges of segments of nuclear DNA. There is evidence that the unreplicated chromosome of higher organisms contains a single very long DNA molecule, forming a complex structure with nuclear proteins, in particular histones. This

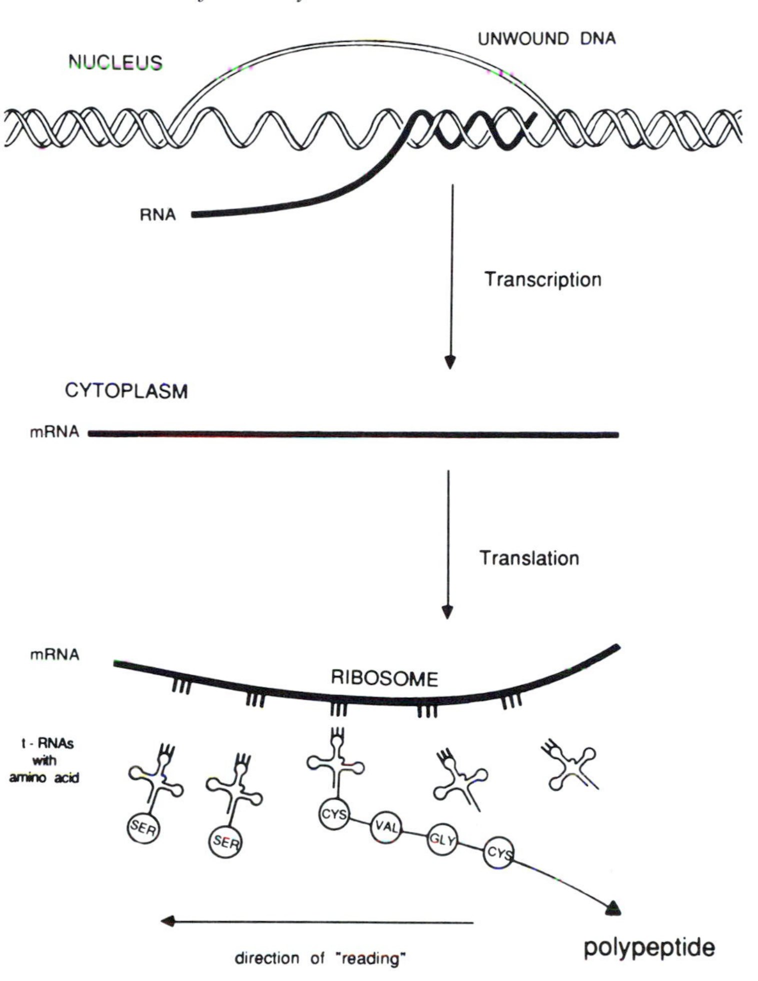

Fig. 6.4. The transcription and translation of genetic information in the production of polypeptides. The information in the DNA is transcribed in the mRNA. The translation of this information occurs outside the nuclear membrane. The assembly of amino acids into polypeptides is achieved as each three base codon in the mRNA in turn forms a complementary 'pair' with the anti-codon of an appropriate tRNA. There are 20 such tRNAs each carrying a different amino acid. Reprinted by permission of John Wiley & Sons, Inc. from Crawford (1990*b*), *Plant molecular systematics*. © 1990.

complex, in its contracted state, is much folded and supercoiled. Mitosis is a process which allows orderly assignment of the identical products of DNA replication to daughter nuclei. In contrast, the process of meiosis permits the reciprocal exchange of DNA segments, as crossing over takes place

...ATG GTG CTC TCT ATA GCT...
Met Val Leu Ser Ile Ala

a Point mutation

...ATG GTG **G**TC TCT ATA GCT...
Met Val Phe Ser Ile Ala...

b Frameshift

...ATG GTG CTC **AT**T CTA TAG...
Met Val Leu Ile Leu Stop

c Insertion (leading to frameshift)

...ATG GTT **CGA TAT CTC TGT G**CT CTC TAT AGC T...
Met Val Arg Tyr Leu Cys Ala Leu Tyr Ser...

d Deletion (– 11 nucleotides)

...ATG GTG CCT...
Met Val Pro... (the rest of the sequence will be frameshifted)

e Silent mutation

...ATG GTG CT**A** TCT ATA GCT...
Met Val Leu Ser Ile Ala... (no change)

Fig. 6.5. The diagram shows the effect of different types of mutation on an original sequence of triplets of DNA (top line) each coding for a different amino acid (second line). The amino acids are given the standard three letter abbreviations: Met (Methionine); Val (Valine), etc. (*a*) Point mutation resulting from the substitution of one base by another. (*b–d*) Mutation in the reading frame of the triplets caused by the insertion or deletion of bases. (*e*) Because of the redundancy of the code, a base change may lead to a silent mutation as it does not result in a change in the amino acid sequence. (Reprinted by permission of Addison Wesley Longman Ltd. from Lawrence, 1989.)

between the homologous pairs of chromosomes. Evidence of such exchanges is provided by the chiasmata seen at the diplotene stage (Fig. 4.9).

In previous chapters we discussed the development of ideas about the underlying causes of variation. Having considered the basic properties of DNA, we are now in a position to examine, in more detail, how molecular studies have revolutionised our understanding of a number of important questions.

Mutation

The term mutation has had an odd history (Mayr, 1963). In the seventeenth century it was used to describe changes in the life cycles of insects, and in the nineteenth century it was used by palaeontologists to refer to markedly new variants in a line of fossils. As we saw in Chapter 5, De Vries (1905) used the term for new phenotypes that arose abruptly in stocks of various plant species. The term mutation was used by the early geneticists, as now, to describe the spontaneous origin of new genetic variation, much of it Mendelian and allelic in nature. For instance, in *Drosophila* it was established that the alleles of particular genes (for example, eye colour) arose as rare, spontaneous changes from the normal or 'wild type', and it was eventually demonstrated that any gene has a low but measurable 'mutation rate' by which the particular wild type allele can change to a mutant allele. Another important principle was also established when it was shown that mutations are reversible. One means of estimating mutation rate is given in Fig. 6.6 and some figures for spontaneous mutation rate in Maize are set out in Table 6.1.

As we have seen, molecular studies have transformed our understanding of mutation. The gene may consist of several thousand nucleotide pairs (Lewin, 1990) and in a molecular structure of this size and complexity there are obviously many different structural changes possible at the level of the DNA. This fact illuminates an early finding of genetics, that sometimes in genetic studies many different alleles of the same gene are found. In the cases that Mendel examined there were only two alternative states of the gene but, as we have seen, it was later found that 'multiple alleles' are common.

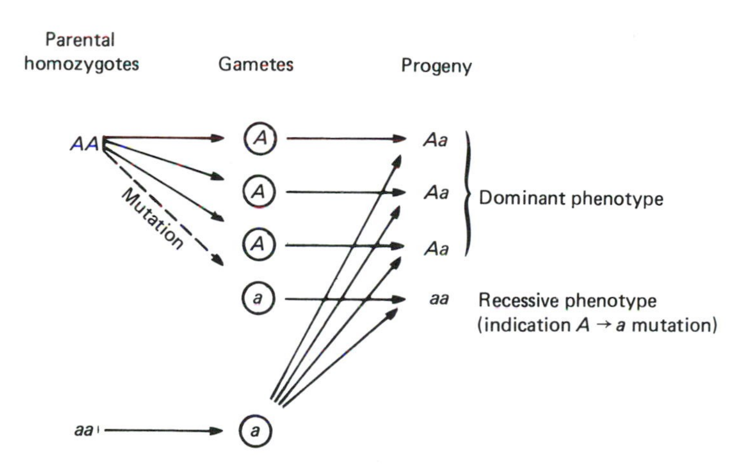

Fig. 6.6. The spontaneous mutation rate ($A \rightarrow a$) may be obtained from the proportion of *a* gametes produced by an *AA* individual, and detected by backcrosses as shown in the diagram. (From Edwards, 1977.) Other methods of estimating mutation rates are given in Strickberger (1976).

Table 6.1. *Spontaneous mutation rates in Maize (From Stadler, 1942)*

Gene	Number of gametes tested	Number of mutations	Frequency per million gametes
R	554 786	273	492
I	265 391	28	106
Pr	647 102	7	11
Su	1 678 736	4	2
C	426 923	1	2
Y	1 745 280	4	2
Sh	2 469 285	3	1
Wx	1 503 744	0	0

These multiple alleles can now be visualised as different variants in base sequence at the level of the DNA.

A large body of experimental work has demonstrated that the natural mutation rate may be increased by a wide range of 'mutagens'. These include radiation of a wave-length shorter than visible light, especially ultraviolet light and X-rays, as well as many chemical substances. In the case of chemical mutagens, some only cause mutation in certain organisms and then only at a certain stage in the life cycle. Spontaneous and induced mutants are often 'deleterious' in the sense that the phenotype is defective, and may not survive in competition with the normal, wild type genotypes.

The phenomenon of spontaneous mutation in plants is little understood. Studies of *Crepis capillaris* have shown a remarkable increase in spontaneous mutation rate with increasing altitude, with the highest values at 3000 m above sea-level (Bruhin, 1950). As Bruhin (1951) was able to increase the mutation rate in *C. capillaris* achenes treated at 45°C for 16–20 days, temperature – especially in the strongly insolated soils of alpine areas – was thought to be an important factor in nature. There is also evidence that, as seeds age, the spontaneous mutation rate increases. Such increases may be due to metabolic changes in the seeds, and D'Amato & Hoffmann-Ostenhof (1956) have discussed the role of naturally occurring mutagenic substances (or precursors which under certain circumstances may be converted to mutagens) in producing mutation in seeds and other plant structures.

Mutation may also occur as a consequence of man's activities. From more than 900 chemicals, of the order of 100 000 pesticide formulations have been devised as fungicides, herbicides and insecticides. In bioassay tests of some of these formulations, genetic or chromosomal damage has been detected (Sharma & Panneerselvam, 1990). Genetic abnormalities caused by radioactive substances have long been known or suspected and it is not surprising that genetic damage has been discovered in plants growing near the Chernobyl Atomic Station in the Ukraine, where the major accident in April 1986 caused a huge release of radiation (Syomov, Ptitsyna & Sergeeva, 1992).

Mutation often leads to small changes in phenotype. However, gross teratological changes, involving malformations and abnormal growths are sometimes reported (Heslop-Harrison, 1952). In some cases it has been discovered that mutational changes of large phenotypic effect are controlled in simple Mendelian fashion (see review by Hilu, 1983). We have already introduced the vexed question of the role of 'sports' in evolution and will return to the subject again in later chapters. It is important to note, in the present context, that simply inherited mutations may cause not only small changes but also very large effects on the phenotype. For instance, Hilu notes that 'evolutionary trends in the corolla (petals) include loss or reduction in the number of petals, fusion of parts, and tendency towards asymmetry. Various mutations that are generally simply inherited have been found to reconstruct these trends'.

It is of great interest that these characteristics, which represent major pathways in the evolution of the genera and families of flowering plants, are inherited in such a simple Mendelian fashion. By employing mutant variants, particularly in *Antirrhinum majus* (Scrophulariaceae) and *Arabidopsis thaliana* (Cruciferae/Brassicaceae) considerable progress is being made on the developmental genetics of the flower and the inflorescence (Coen, 1991). Investigations of the peloric mutants of *Antirrhinum* have proved informative (see Chapter 2 for descriptions and early studies of *Peloria* in *Linaria vulgaris* , which is in the same family). It has been concluded that 'the basic mechanisms that define organ identity in the developing flowers appear to be the same in both species [*Antirrhinum majus* and *Arabidopsis thaliana*]... This shows that the basic processes of floral organ specification are evolutionarily old, and that the very different forms taken by flowers of different plant families, at least amongst the dicotyledonous plants, result from recent modifications of the developmental processes' (Coen & Meyerwitz, 1991).

As part of a general interest in floral structure, geneticists have also recently studied meristic variation in the number of equivalent parts in the capitula of Compositae (Asteraceae). As we saw in Chapter 3, the number of ray-florets in a sample is often, but not always, found to be one of the Fibonacci series (1, 2, 3, 5, 8, 13, 21, 34, etc.). Attempts are still being made to develop a general model to account for such 'canalisation' of development (Battjes, Bachmann & Bouman, 1992; Battjes, Vischer & Bachmann, 1993 *a*, *b*).

Cytological differences

Early cytologists demonstrated that a particular diploid number of chromosomes was normally characteristic for a species. For example, the French cytologist Guignard (1891) made a clear drawing of the stages of mitosis in very young anthers of *Lilium martagon*, establishing the diploid number for that species as 24. Descriptive cytology has made great strides in the present century and it is instructive to compare Guignard's drawings shown in Fig. 6.7*a* with a diagrammatic representation of the haploid set of chromosomes (the idiogram) of *L. martagon* prepared by Stewart (1947) shown in (Fig. 6.7*b*). A close study of these shows that individual chromosomes of the set are distinct in such features as total length, length of arms and the presence of secondary constrictions. Fig. 6.7*b* also shows that there are differences between idiograms of *Lilium* species. The term karyotype is used for a more faithful representation of the chromosomes in drawings or photographs (see, for example, Fig. 6.8.)

An examination of *L. martagon* root-tip mitosis demonstrates that the diploid chromosome number is 24 and that this is the characteristic number for the species. Studies on many Linnaean species have revealed, however, that different individuals may have different chromosome numbers. In diploid organisms, homologous pairs of chromosomes are found. Occasionally, however, misdivision of the paired chromosomes at meiosis may give gametes with more or fewer chromosomes than the haploid number. Chromosomes may be missing or represented more than once. Gametes with an incomplete haploid complement are usually defective, but those with one or more additional chromosomes may be fertilised and may develop into adult plants. Thus, in a large sample of plants there may occur individuals, called aneuploids, which have the different chromosomes of the set present in different numbers. For example, in a sample of 4000 plants of *Crepis tectorum* ($2n = 8$) Navashin (1926) found:

10 plants with $2n = 2x + 1 = 9$
4 plants with $2n = 2x + 2 = 10$
4 plants with $2n = 2x + 3 = 11$

Gametes may also be produced which contain the unreduced number of chromosomes. Fusions between unreduced and 'normal' gametes (and between unreduced gametes) will produce in many cases viable plants with elevated chromosome numbers. For example, *Campanula rotundifolia* plants are divisible into three main groups, $2n = 34$, $2n = 68$ and $2n = 102$. In many genera, individual species form a polyploid series, in which high

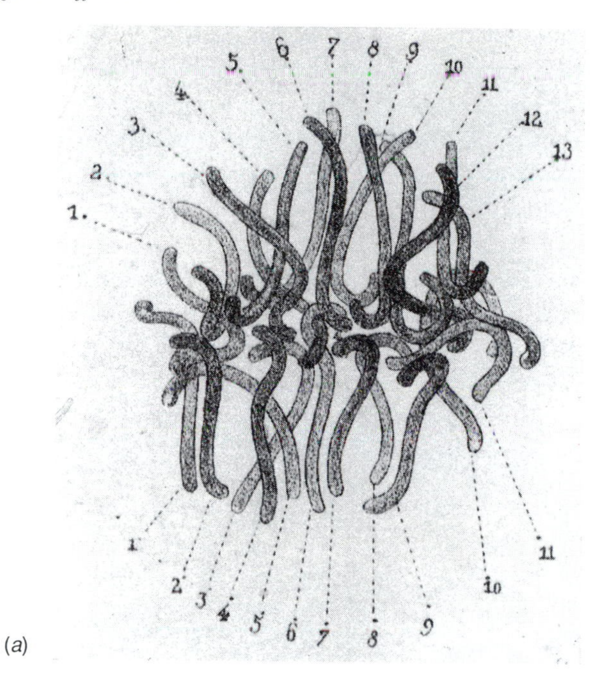

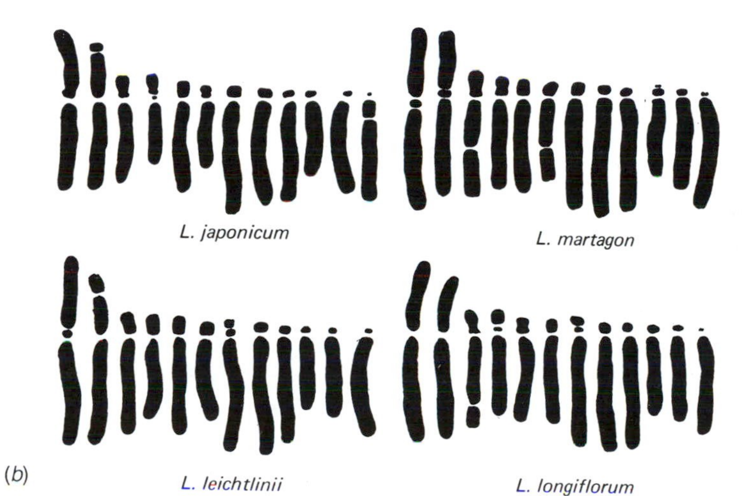

Fig. 6.7. Chromosomes of *Lilium* (*a*) Mitosis in very young anthers of *Lilium martagon* ($2n = 24$) ($\times$ 1500). (From Guignard, 1891). (*b*) Idiograms of species of *Lilium* ($\times$ 660). (From Stewart, 1947.)

numbers are simple multiples of the lowest haploid number. The widespread occurrence of polyploidy in higher plants emphasises the extreme importance of the phenomenon, to which we return in later chapters.

In cytological investigations of individuals belonging to the same species, differences in karyotype may be found. In species which have separate male and female individuals, the sex-determining mechanism may involve distinct sex chromosomes (Grant, 1975; Richards, 1986). Also, chromosomes additional to the normal complement may be found (Fig. 6.8). Such accessory, supernumerary or 'B' chromosomes as they are often called, occur in at least 1372 flowering plant species (Jones, 1995). They are restricted to outbreeding species. Evidence suggests that they do not

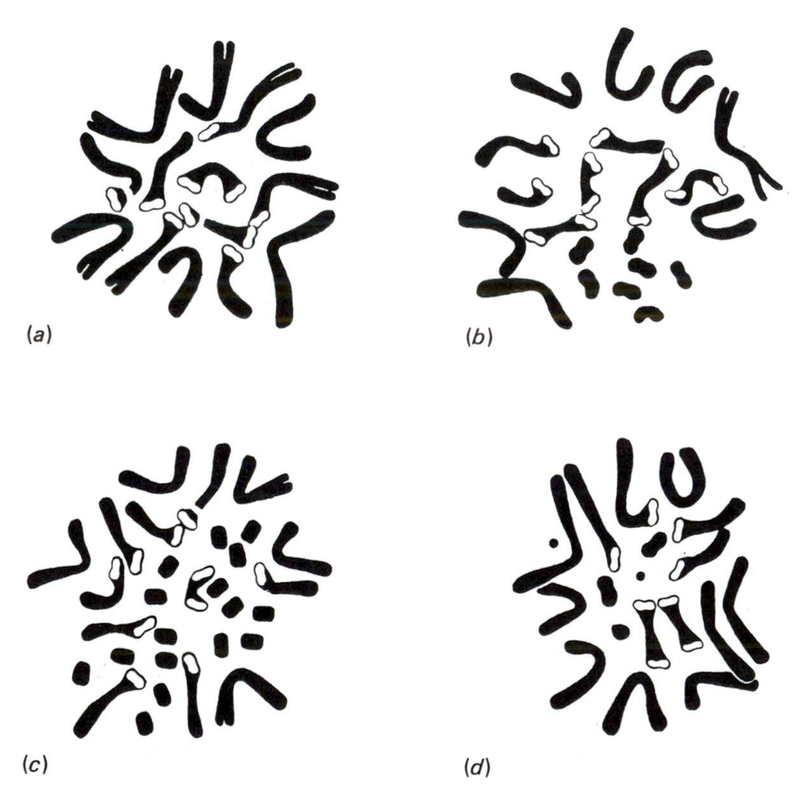

Fig. 6.8. Metaphase plates from root-tips of different plants of *Festuca pratensis*. In some plants B chromosomes are absent, as in (*a*) ($2n = 14$). In others B chromosomes are present, as in (*b*) ($2n = 14 + 7B$) and (*c*) ($2n = 14 + 16B$). B chromosomes may be the same size, as in (*c*), or different sizes as in (*d*) ($2n = 14 + 5B$). (All × 2700.) (From Bosemark, 1954.)

usually pair with the larger members of the set (the 'A' chromosomes) and have irregular non-Mendelian modes of inheritance. Often they do not appear to produce major phenotypic effects in plants which have them, leading to the suggestion that they are essentially 'selfish' structures maintained without function. However, in some studies functional genes have been located on B chromosomes. How Bs are maintained in populations has been the subject of much research, for example in *Crepis capillaris* (Parker *et al.*, 1991). In this species, which is polymorphic for B chromosomes in southern but not northern Britain, the behaviour of the Bs in mitosis leads to their accumulation in the floral parts. However, the frequency of Bs is not constant between populations and it is suggested that selection operates to regulate numbers of Bs, which are more frequent in populations growing under optimal conditions than in those found in sub-optimal situations.

In addition to gross features of chromosome morphology, different parts of the chromosomes may show differential staining. A combination of cytological and genetic studies indicates regions of the chromosomes, called heterochromatic segments, different from 'normal' euchromatic regions. Heterochromatin has several distinctive cytological properties, being condensed when euchromatin is extended (Fincham, 1983). In some taxa, the presence of heterochromatin may be accentuated by cold treatments before staining (Darlington, 1956). There may be a consistent pattern of distribution of heterochromatin, so-called constitutive heterochromatin, e.g. around the centromeres, but in many taxa there is variation in the pattern and a class of 'facultative' heterochromatin has been distinguished (Rees & Jones, 1977). B chromosomes are often, but not always, heterochromatic.

Further aspects of the karyotype may be revealed by different stains that give different patterns of banding. Treatment with quinacrine, for example, produces 'Q-bands' visible under ultraviolet light in fluorescence microscopy. Preparations treated with Giemsa dye give 'G-bands'. 'R-bands' may be revealed in heat-treated preparations, while 'C-bands' may become evident in chromosomes stained with Giemsa followed by alkaline extraction. Molecular studies have revealed that banding may indicate concentrations of particular bases in the DNA molecule, e.g. 'Q-bands' are rich in A–T bases, while 'R-bands' are thought to occur in areas with high levels of G–C bases.

Studies of meiosis may also reveal cytological differences between plants. Figure 6.9 shows diagrammatically the main kinds of chromosomal change which result in genetic effects and visible cytological peculiarities at meiosis

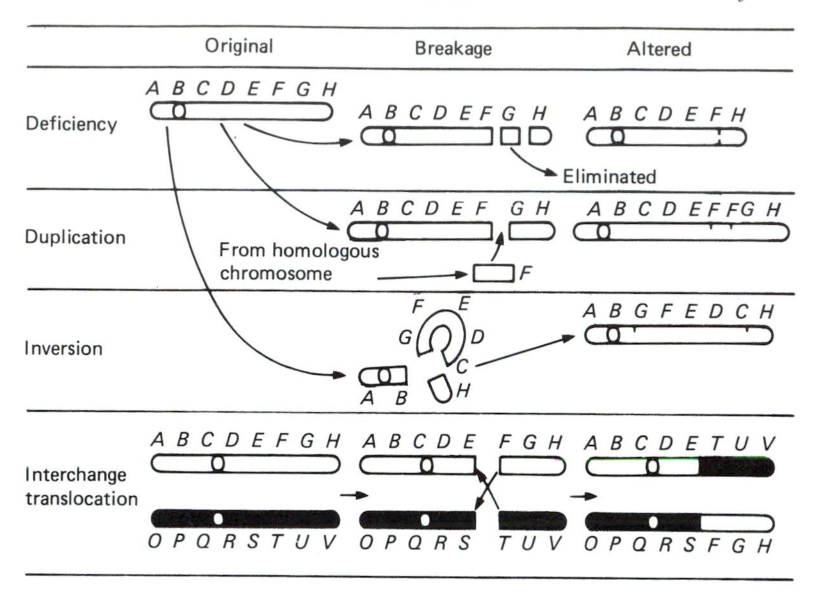

Fig. 6.9. Diagrams to show how chromosome breakage and reunion can give rise to the four principal changes which chromosomes undergo. (After Stebbins, 1966.)

in the heterozygote. Such inversions, translocations, deletions and duplications of segments of chromosomes, which represent gross changes at the level of the DNA, are to be found in naturally occurring individuals as well as in experimental material. Clearly they provide a further important source of variation of great evolutionary significance. We may note at this point that the 'mutations' of De Vries mentioned in Chapter 5 were of different kinds: many arose from the peculiar cytological situation in *Oenothera*, where the species is a complex heterozygote in which all the chromosomes exhibit interchanges, but occasionally more or less homozygous 'mutant' individuals are produced.

Non-Mendelian inheritance

Molecular studies have transformed our understanding of many puzzling experiments carried out by pioneer geneticists. For example, Correns, one of the rediscoverers of Mendel's work, showed as early as 1909 that in the familiar American garden flower Four-O'Clock or Marvel of Peru (*Mirabilis jalapa*), some variants with yellowish-green or variegated leaves showed normal Mendelian segregation, whilst a particular variant, *albomaculata*, with yellowish-white variegation, did not. Plants of *albomaculata* produced occasional shoots which were wholly green and others which were white. Flowers on green shoots gave only green progeny whether pollinated from flowers on green, variegated or white shoots, and, conversely, flowers on white shoots, whatever the pollen parent, gave only white progeny (which died in the seedling stage). Variegated shoots gave all three kinds of progeny. In two respects this inheritance was clearly non-Mendelian: first, because the offspring resembled closely the female parent and the reciprocal crosses gave entirely different results; and secondly, because there was no regularity in the proportions of the phenotypes in the segregating families.

Over the years many other cases of non-Mendelian inheritance were discovered, e.g. leaf variegation in the Garden Geranium (*Pelargonium* species), the Evening Primrose (*Oenothera* species) and many other genera (Kirk & Tilney-Basset, 1978; Evenari, 1989).

Molecular studies have greatly increased our understanding of non-Mendelian inheritance, for it has been discovered that DNA is present in the chloroplasts and mitochondria of plant cells and that this DNA expresses its genes. In reporting the behaviour of this DNA, geneticists now use the term 'extranuclear inheritance' rather than the older term 'cytoplasmic inheritance'. Self-replicating closed circles of DNA occur in the chloroplast DNA (cpDNA) and in the mitochondrial DNA (mtDNA). The genetics of these DNAs will depend upon how replicas are distributed as proto-organelles. In a cross, are they inherited both from the male and the female side? Investigations have revealed different behaviour. For instance, in conifers inheritance of plastid transmission is paternal (Schaal, O'Kane & Rogstad, 1991) and in flowering plants inheritance is often maternal, but, in some cases, it is paternal or biparental (Harris & Ingram, 1991).

Modern techniques used in studying genetic variation

The biologist interested in plant variation and evolution is presented with an ever-increasing choice of techniques: examining the DNA itself, studying the metabolic products of the activity of the cell, or making deductions about relationships by studying the appearance and/or behaviour of the phenotype – the final manifestation of the interaction of genotype and environment. As we have seen in earlier chapters, biologists first studied phenotypes. Then, when better microscopes became available in the latter part of the nineteenth century, they examined the chromosomes. Later in the twentieth century, as the necessary techniques became available, they made genetic and phylogenetic inferences based on the secondary metabolites found in samples of plants. The historical development of this area has been reviewed by Harborne & Turner (1984) in a comprehensive survey of research on plant scents, odours, bitter and toxic defence agents, pigments, and storage metabolites, etc.

Recently, in making judgements about heredity and relationship it has proved possible to study the DNA directly. As there is redundancy in the genetic code, not all changes at the molecular level may produce changes in the secondary chemicals or the phenotype. Thus, two genotypes with different DNA may be indistinguishable in studies of downstream gene products (Crawford, 1990). For definitive studies many biologists choose to examine the DNA itself. While in the present context it is not necessary to review all the methods available for the study of genetic variation, it is instructive to examine the techniques used in the study of variation in enzymes and in the examination of DNA, for these approaches have contributed enormously to our understanding of variation and evolution and have the potential to widen our understanding even further.

Electrophoretic studies of enzymes

In research first developed in the 1950s and 1960s, it proved possible to study proteins, particularly enzymes, by their physical and chemical properties, using the technique known as gel electrophoresis. Such studies have provided the means of testing many hypotheses in plant systematics, etc. (see Gottlieb, 1984; Soltis & Soltis, 1990; Hillis, Moritz & Mable, 1996).

Synthetic polymer, starch or agar gels are prepared. A homogenate of plant material (roots, flowers, but most often leaves (in buffer)) is inserted into a sample well in the gel and an electric field is generated across the gel (Fig. 6.10*a*). Depending on their electrical charge and size, molecules of protein separate in migrating to the positive or negative pole. Enzymes, which are the most commonly studied proteins, are present in minute quantities in the gel. To locate them, for they are invisible, a substrate is added together with a dye which produces a colour reaction at the site of enzymic breakdown of the substrate. This combination of electrophoresis, followed by staining, yields a zymogram (Fig. 6.10*b*).

When homogenates from different individuals of the same taxon are run in parallel across the gel, different patterns of banding may be found for each class of enzyme. Biochemists have concluded, as a result of critical studies of selected enzymes, that such enzymes are polymorphic, that is, they consists of a family of related structures which catalyse the same

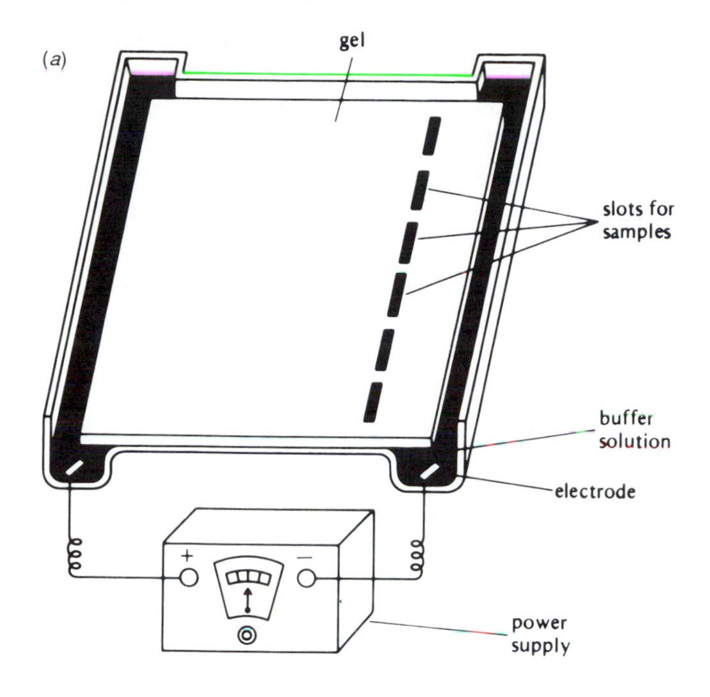

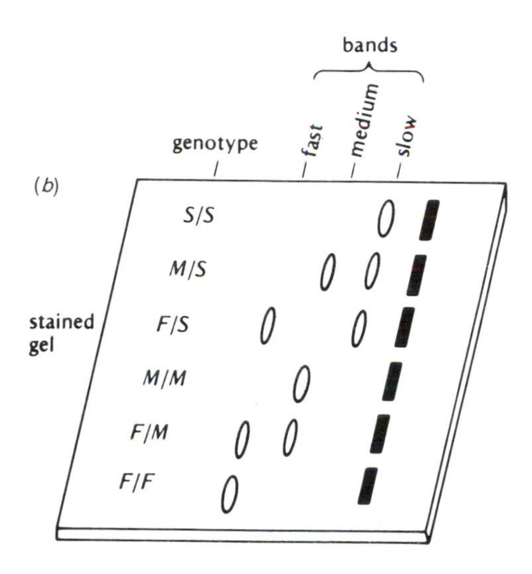

Fig. 6.10. (*a*) Apparatus for gel electrophoresis. (*b*) A zymogram produced by staining a gel (see text for a discussion of the interpretation of banding patterns). (From Strickberger, 1985.)

general reaction. There is evidence, in some closely studied examples, that the members of each family differ in amino acid composition, as a consequence of mutation at the DNA level at the locus (or loci) encoding the enzyme. Thus, mutations in DNA may result in changes in the detailed pattern of amino acids in the enzyme molecule and the substitution of one amino acid for another may yield an enzyme molecule of different mobility under electrophoresis.

If the different enzyme variants are encoded by different alleles at a locus, they are referred to as allozymes. Where more than one locus is involved they are called isozymes. The term isozyme is also used in cases where the genetic basis of the enzyme variation is unknown (Crawford, 1990). For simplicity, when we refer to different studies of electrophoretic variants of enzymes in this book, we use the word isozyme(s). For details of the known or presumed genetic basis of the enzyme polymorphism in particular cases, the reader is referred to the original papers cited in the references.

Banding patterns on zymograms may be simple or complex, depending on the genetic context and whether the enzyme has one, two or more polypeptide subunits (see Gottlieb, 1981*a*; Crawford, 1990). To illustrate the principles behind the determination of the genetic basis of banding, we may consider the simplest situation, where the enzyme has a single polypeptide band (i.e. it is monomeric) and the enzyme is encoded by allozymes (Fig. 6.10*b*). Dominance and recessiveness do not apply, alleles are co-dominant, that is to say in the heterozygote both alleles are expressed. Thus, homozygotes give single bands on the gel and, as there may be a range of such homozygotes with bands of different mobility (say slow S/S; medium M/M; fast F/F), there will be a range of different heterozygotes (M/S; F/S; F/M) and each will produce two bands of appropriate mobility. Where the enzyme is made up of two or more polypeptides, heterozygotes of allozymes will produce several bands on the gel. More complex patterns occur in situations where enzymes are encoded by more than one locus and in polyploid plants with several genomes.

In interpreting zymograms three points must be borne in mind:

1. In the extraction of leaves, etc. care must be taken to exclude secondary plant products (gums, latexes, etc.), as these might seriously distort the banding patterns.
2. A knowledge of the genetics of different banding patterns is essential. Returning to the example set out above, the supposition that a plant is heterozygous for two alleles encoding an enzyme with a single polypeptide may be put to the test. First, the detection of apparent heterozygosity

may be made on a single leaf of a plant. Selfing this plant (say of genotype F/M) may readily be accomplished if it is self-compatible and segregation patterns appropriate to a heterozygote may be looked for in the seedling progenies (in this case 1 F/F: 2 F/M: 1 M/M). Secondly, as an aid to interpreting banding patterns in a particular diploid, investigations may be made to see if enzymes are expressed in the haploid structures of the same plant (pollen in flowering plants (Weeden & Gottlieb, 1979), the megasporophytic tissues surrounding the embryo in conifers, and gametophytes of ferns and bryophytes (Wendel & Weeden, 1990)).

3. Only 30% of the base changes in DNA will lead to changes in amino acids yielding a difference in net charge and thereby changing the electrophoretic mobility of the enzymes (Gottlieb, 1981*a*; Crawford, 1990). Furthermore, sometimes null alleles are encountered in studies of plant enzymes. These are 'alleles that are no longer transcribed or that code for defective polypeptides lacking enzymic activity' (Weeden & Wendel, 1990).

Analysis of DNA

As an alternative to studying variation in enzymes, which are the downstream products of the DNA, recent advances in technology have made it possible to examine the genetic and evolutionary relationships of groups of plants by determining the sequence of bases in the DNA itself (Lewin, 1990). Indeed, automatic DNA sequencing machines have been developed (Rothwell, 1993), but they are expensive and not yet widely available. Two ways of studying variation in DNA have commonly been used (Weising *et al.*, 1995; Hillis, Moritz & Mable, 1996):

1. Hybridisation-based fingerprinting

Specific breakage is achieved by treating extracted DNA with special enzymes, the so-called type II restriction endonucleases, which are more commonly referred to as restriction enzymes. These enzymes are produced by bacteria and degrade 'foreign' DNA. The DNA of a given bacterium is not attacked by its own restriction enzymes, as it is slightly different chemically. More than 100 restriction enzymes have been extracted from different species of bacteria. Commonly used enzymes recognise and cut at pallindromic sequences. When the base sequence on one DNA strand at the point of breakage is compared with the sequence on the other strand, the same base sequence is found but read in the reverse direction (Fig. 6.11). The points of

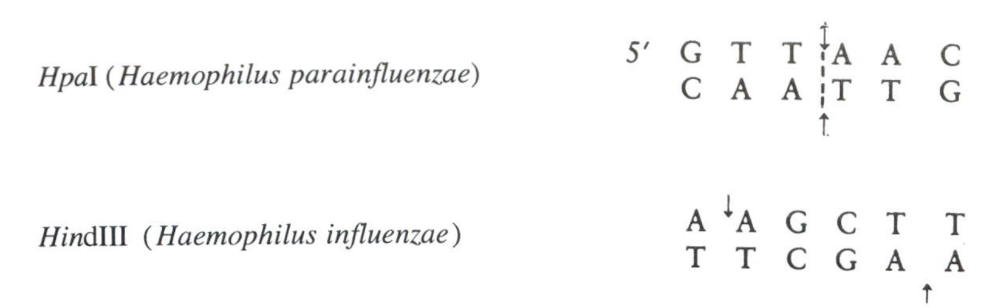

Fig. 6.11. Examples of restriction enzymes and the way they cut the DNA molecule. Standard abbreviations are used for the different enzymes, e.g. *Hpa*I (from the bacterium *Haemophilus parainfluenzae*) and *Hin*dIII (from *Haemophilus influenzae*). *Hpa*I makes a transverse cut in the DNA; in contrast *Hin*dIII produces a staggered cut. Note the palindromic nature of the base sequence at the restriction sites. (Reprinted by permission of Addison Wesley Longman Ltd. from Lawrence, 1989.)

cleavage are, therefore, so precisely determined that restriction sites may be regarded as specific genetic markers. There is a distinct probability that any sizeable length of DNA will have sites which can be cleaved by a particular restriction enzyme. Variation between individuals in restriction fragment length of a region of the genome arise by point mutations altering the restriction sites (either generating or deleting a site), or by chromosomal rearrangements such as insertion or deletion of a section of DNA.

After treatment with one or a number of restriction enzymes the DNA fragments formed may then be separated by gel electrophoresis, the principles of which we have discussed above. DNA is uniformly negatively charged and fragments travel from anode to cathode, separation being entirely dependent on size. Thus, small fragments of DNA travel further from the origin than large fragments. The result of treatment with restriction enzymes followed by electrophoresis is to produce a smear of DNA fragments from a given sample.

The precision with which pairing occurs in DNA has allowed the development of methods of identification of specific fragments in the smear. First, the DNA duplex is dissociated by sodium hydroxide treatment in the gel and then transferred to a sheet of nitro-cellulose or nylon by blotting – the Southern blotting technique, so-called after its inventor. Thus, the DNA is transferred from a gel to a solid support matrix, the relative position of the DNA fragments having been maintained. Then, the membrane is incubated to fix the DNA, or treated with ultraviolet light to cross-link the DNA to the matrix. A single-stranded probe is then added. This may be a specific gene, or segments of DNA or RNA from a known source labelled with radioactivity or with chemiluminescent chemicals. The probe and the DNA of the sample will hybridise if the base sequences of the two are complementary. The nitro-cellulose membrane is then washed; only the probe DNA that has hybridised will remain on the membrane. Single pieces of the sample DNA may 'match' the sequences in the probe. However, if the restriction enzymes have broken into several pieces the sample sequence matching that of the probe, then several fragments may be labelled. In the case of radioactively labelled probes, the location of the labelled fragments is revealed by placing an appropriately sensitive X-ray film in contact with the nitro-cellulose sheet (Fig. 6.12) (Watson *et al.*, 1992).

2. *Fingerprints produced using the polymerase chain reaction (PCR)*

This method involves the *in vitro* amplification of particular segments of DNA (Fig. 6.13). The reaction requires cycles of precisely controlled heating

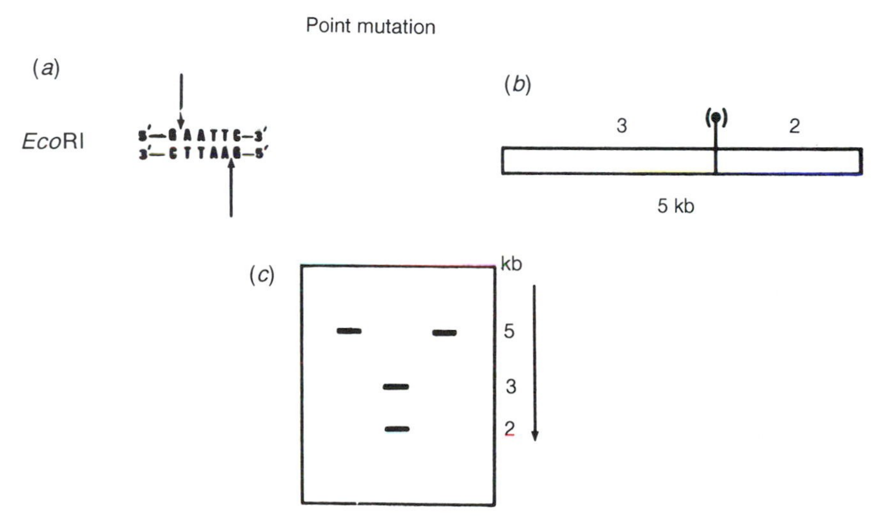

Fig. 6.12. The use of restriction enzymes, electrophoresis and probes in the study of variation. In a study of three 5 kilobase (kb) samples I, II and III. (*a*) The DNA of sample II has a restriction site cleaved by *Eco*RI (an enzyme from *Escherichia coli*) which is not present in samples I and III. (*b*) Such a site could arise by point mutation. (*c*) After electrophoresis and the use of a radioactive probe (see text for further details), the intact 5 kb fragment is detected in samples I and III. However, in sample II the 5 kb fragment is missing. Two fragments (2 and 3 kb long) are present instead. This type of variation – usually explored with many more than three samples – is called restriction fragment length polymorphism (RFLP). (From Schaal, Leverich & Rogstad, 1991).

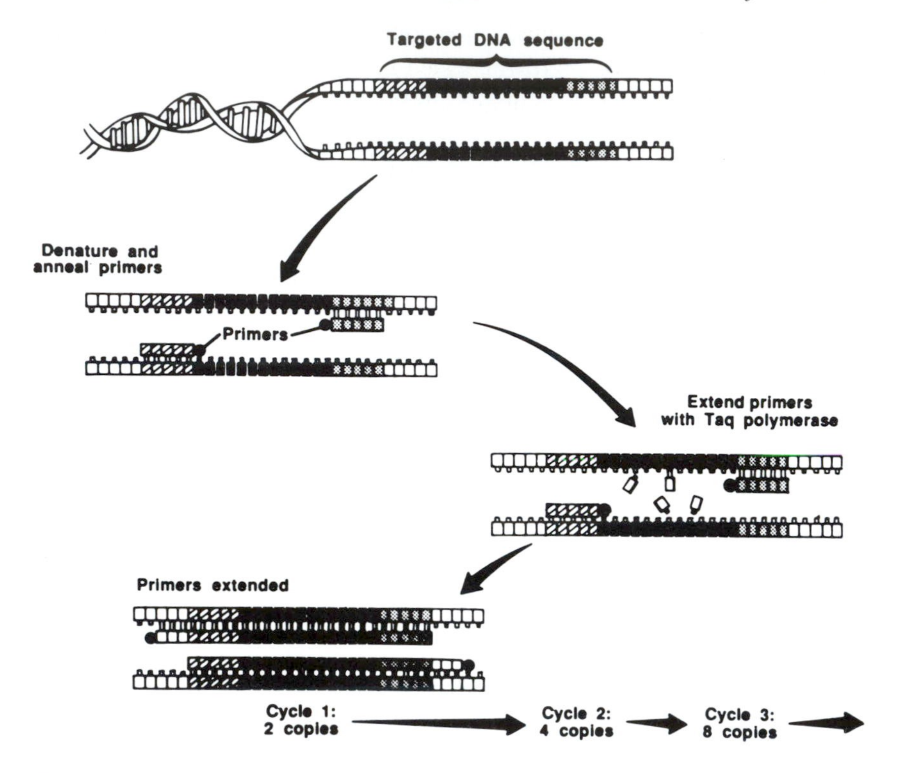

Fig. 6.13. Amplification of DNA using the polymerase chain reaction (PCR). Each cycle consists of heat denaturation of the target molecules, annealing of an oligonucleotide primer to each target complementary DNA strand, and extension of the primers using Taq polymerase. The newly synthesized molecules are denatured to provide double the number of templates for the next cycle. (From Kirby, 1990.)

and cooling. The figures given below indicate the sort of temperature needed at each stage of the sequence.

A sample of DNA is denatured by heat treatment at *c.* 95 °C, to produce single stranded DNA. On cooling, these strands are used as templates for the generation of the complementary DNA strand mediated by the DNA polymerase. It has been discovered that, for the reaction to work, small double-stranded portions of DNA are necessary to act as starting blocks for synthesis of the new strand. This is provided by annealing a short specific segment of DNA (called a primer) at a lower temperature (35–55 °C) depending on the length and base sequence. Then, at a temperature of *c.* 72 °C, in the presence of DNA polymerase and the necessary nucleotide bases, the synthesis of new strands complementary to the original DNA is completed. Two copies of the DNA have now been produced. If the procedure just outlined is repeated many times, then an ever-increasing number of copies of a fragment of the original DNA may be made. The PCR method has been automated, an advance made possible by the isolation of a heat-stable form of DNA polymerase (Taq) from *Thermophilus aquaticus*, a bacterium living in hot springs (Tindall & Kunkel, 1988). Unlike the normal form of the enzyme, Taq is able to withstand the high temperatures at the denaturing part of the cycle without loss of activity. PCR amplification may be carried out using primers specific to known sequences, which amplify a particular section of the DNA. As an alternative, arbitrarily chosen groups of nucleotides may be used as primers in the PCR reaction, the so-called random amplified polymorphic DNA (RAPD) method. Recently, another approach has been devised that combines restriction fragment length polymorphism (RFLP) analysis and PCR; this is known as amplified fragment length polymorphism (AFLP). Specifically devised radioactive or fluorescence-labelled primers are used in conjunction with restriction fragments of DNA and a subset of these fragments is amplified (for details of the method see Weising *et al*., 1995). In all these cases, the primer 'pairs' with exactly complementary base sequences and, as several such sequences may be present in a sample under test, a number of sections of DNA of variable length are amplified, which can then be separated by electrophoresis and visualised by autoradiography, fluorescence or by silver staining as appropriate to provide a DNA fingerprint. Different fingerprint patterns may be found in different taxa (Fig. 6.14).

It is of great interest that PCR techniques can be used not only for DNA extracted from living plants but also from other sources. Thus, DNA has been extracted from the Quagga, an extinct fossil member of the Horse family, (Higuchi *et al*., 1984); from a fossil Magnolia of the Miocene age (Golenberg *et al*., 1990); and from archaeological remains such as Egyptian mummies (Cherfas, 1991). DNA has also been recovered from museum specimens of birds (Ellegren, 1991) and from herbarium material of plants (Arnheim, White & Rainey, 1990; Rogers, 1994).

Use of DNA in studies of variation

At appropriate places in this book we will be discussing the results of molecular studies. At this point in our review of modern ideas about the basis of variation between plants, it is important to consider some of the properties of DNA, which has been isolated and studied from many

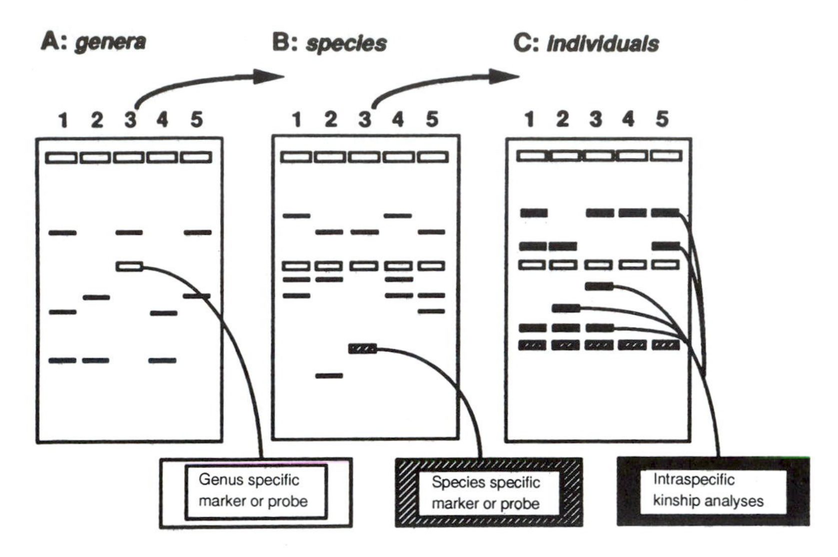

Fig. 6.14. Use of polymorphic and non-polymorphic RAPD fragments for genome scanning. Schematic diagrams of RAPD fingerprints of genera, species and individuals illustrate the potential use of polymorphic and non-polymorphic RAPD fragments as diagnostic markers. In gel A, a genus 3 specific band is identified by its presence in all species of that genus (and consequently all individuals of all species of genus 3 tested); this genus-specific fragment can serve as a diagnostic DNA marker in RAPD analyses or a genus-specific probe in RFLP analyses. Similarly, a species-specific fragment (gel B) is identified by its presence in all individuals of species 3, that can serve as a species-specific RAPD marker or probe. Finally, gel C shows five polymorphic RAPD fragments between different individuals of species 3; these may be used for intraspecific analyses, such as paternity determinations by means of RAPD fingerprinting. (From Hadrys, Balick & Schierwater, 1992.)

different sources (Rothwell, 1993). In DNA, some sequences of bases are present as single copies – the so-called unique sequences- whilst several copies of other sequences may be present. Indeed, some may be present in very large numbers, either interspersed amongst other sequences or found as tandem repeats of a basic sequence with or without variants. Repeats may be of long motifs of 10–60 base pairs (minisatellites) or, in a spectrum of possibilities, reiterations of 1–5 base pairs (microsatellites). Weising *et al.* (1995) may be consulted for a review of molecular studies of fingerprints of tandem repeats in plant genomes.

The amount of DNA in the haploid genome (the C-value) is very variable between groups, even between related species. For example, *Vicia faba* has 1.2×10^{-7} kilobase pairs, while *V. lathyroides* has only 2.4×10^{-6}. The largest genomes are found in the monocotyledons (Bennett & Smith, 1991), however, recent studies reveal that large genomes are restricted to certain lineages in the Liliiflorae, Commelinales, Alismatales and Araceae, and most monocotyledons have small genomes (Bharathan, Lambert & Galbraith, 1994). A very interesting hypothesis has come from the study of C-values and other molecular studies. Only a very small percentage of DNA present in the genome is needed for the coding of proteins. There would seem to be an absolute excess of DNA. Terms such as 'selfish', 'parasitic' or 'junk' DNA have been coined to specify such apparently useless material (Lewin, 1990). We do not yet know why there is so much DNA in the nuclei of cells.

As we have seen, modern methods have made direct analysis of the genetic material possible. There are different levels of variation in DNA: some regions are constant within and between related species; and others are hypervariable so that each individual in a population may be different, allowing the production of genetic fingerprints in humans (Lewin, 1990) and plants (Tzuri *et al.*, 1991). Thus, different parts of the genome are appropriate for different studies. The problem in hand will dictate the type of DNA chosen for study. In most higher plants, the plastids in the pollen do not enter the zygote and therefore, chloroplast DNA is maternally inherited and, as a consequence, not subject to the recombination processes of meiosis. The study of such relatively conserved regions of the DNA have been very informative in the investigation of hybridisation and phylogenetic relationships. As we shall see in later chapters, the use of molecular tools has allowed progress to be made in the study of population variation, breeding systems, gene flow, hybridisation, introgression, ancestry of polyploids, etc. (Bachmann, 1994). Moreover, it has also proved possible to examine evolutionary relationships between groups of plants so distantly related that crossing is impossible.

However, as in all forms of biological investigation, there are a number of problems to be faced. Sometimes it is very difficult to extract DNA for analysis, but many different extraction protocols have been devised and progress has been made in extracting DNA from recalcitrant plants (Sytsma, 1994). There are also problems of sampling. Sometimes only a few samples are studied and only small segments of the genome are examined. Eminent molecular biologists have pointed out that it is possible to overextend the analysis (Schaal, Leverich & Rogstad, 1991). It could easily be assumed that there is no variation in a particular DNA sequence within a taxon. Such an assumption may not be warranted. For instance, it has sometimes been assumed that chloroplast DNA is invariable within species, but intraspecific variation has been detected (Harris & Ingram, 1991). There is also a concern about the reproducibility of the results. As Bachmann (1994) points out, in the RAPD method, primers will sometimes bind to sites containing one or two base mismatches. Rigorous standardisation of the conditions may reduce the number of unreliable amplification products. Returning to sampling strategies in molecular taxonomic studies, it is legitimate to ask the question; are the samples, selected for phylogenetic and other studies, truly representative of the variation in the taxa from which they are drawn? Finally, contamination of samples, either at source or during processing, is a serious concern, especially with PCR methods. The technique is so powerful that even minute traces of contaminant DNA may be amplified. Very high standards of laboratory technique are needed, with the routine use of replicate samples.

Patterns of variation

Having discussed recent advances in our understanding of the molecular basis of heredity, we may now examine several other major components of variation in plants. In previous chapters, we have discussed the variation patterns found in samples taken from plant populations and suggested that for descriptive purposes it may be useful to distinguish three broad types of individual variation – viz. genetically determined, environmentally induced and developmental variation. This notion of three types of variation, however, gives an oversimplified picture of the nature of individual variation. There is strong evidence for the proposition that the phenotype and behaviour of the plant are determined by interactions between genotype and environment. As we shall see, different genotypes react differently to a given set of environmental conditions and plants of identical genotype produce different phenotypes under contrasting environmental conditions. Moreover, the *interactions* between genotype and environment are further complicated by the complex sequence of changes which occurs as a plant develops from an embryo to the mature fruiting state.

Phenotypic variation

In the growth of an organism from fertilised zygote, a minute quantity of DNA in the cell plays a vital role in determining the characteristics of the mature phenotype (perhaps a tree 30 m high) that is organised from raw materials drawn from outside the plant. There are complex close-knit interactions between genotype and environment at the level of the cell and of the whole plant. Concerning the genetic control of cellular processes, it seems likely that they are of such interlocking complexity that it is no longer possible to claim that one gene alone determines a particular character (Strickberger, 1985). A gene provides information which, in an appropriate environment, will contribute to a particular phenotype. There is no certainty that a particular gene will even manifest itself in all environments. For example, in *Lotus corniculatus* certain plants known to possess alleles appropriate to the production of hydrogen cyanide (HCN) when their foliage is crushed do not in fact produce the cyanogenic reaction in every circumstance (Dawson, 1941). Also, it has been found that a particular Barley variant produces *albino* phenotypes out of doors, yet the same genotype grown at higher temperatures in a glasshouse has normal foliage (Collins, 1927).

In general, plants show greater phenotypic variability than is found in higher animals. In animals, individual variation is apparently held within very tight bounds by the early precision of mechanisms determining irreversibly the form and relationship of the main organs. To some extent this is true of plant structures, but there is a very important difference between the plant and animal. This relates to the fact that there is a persistent meristem or region of undifferentiated growing tissue in even the most short-lived of ephemeral plants, on which a succession of organs of limited growth is initiated. The result of this difference is that the individual plant is open to much more environmentally induced variation over a much greater part of its life than is the animal.

Here it may be helpful to consider the causes of phenotypic variation in plants. If material of a given genotype is divided into separate pieces (ramets) and grown in two or more different environments, different genotype–environment interactions may be produced. While the plant responds to the environment as a whole, it is possible, by appropriate experiments involving changes in particular factors, to deduce that certain elements of the environment are of particular importance. Test environments may differ in soil properties (e.g. water table, base status, level of toxic ions) and aspects of climate or microclimate (e.g. temperature extremes, wind exposure, rainfall). In their competition with one another in experiment or in natural communities, plants show many diverse interactions, for example to the effects of shading. Also of importance may be the influence of chemical substances (at present often hypothetical) produced by one taxon in suppressing the growth of others, so-called allelopathic effects (Rice, 1984; Putnam & Tang, 1986; Rizvi & Rizvi, 1992). The availability of organisms to form the natural symbioses characteristic of many plants is a critical factor in natural and experimental situations (Harley & Harley, 1987), as are the presence and severity of grazing, pest attack and disease. Given the variety of different environments, a great diversity of genotype–environment interactions is possible in the growth and reproduction of individual plants. However, we should note that the stresses of the environment may kill the plant, restrict its growth or perhaps prevent reproduction.

Developmental variation

Phenotypic variation should be viewed within the context of developmental variation. In most flowering plants the cotyledon stage of the seedling is very different from the adult – a fact which has, of course, excited the interests of botanists from early times, and which provided the basis for John Ray's inspired division of flowering plants into the monocotyledons and dicotyledons, a division which we still use as a primary grouping. The developmental transition between the simple cotyledon and the often lobed or dissected mature leaf is generally rather abrupt, and provides the most familiar example of a phenomenon which Goebel (1897) called 'heteroblastic development', or the change from a juvenile to an adult phase accompanied by more or less abrupt changes in morphology. Strictly speaking, the cases which Goebel and others have mainly called 'heteroblastic' are those in which a juvenile leaf other than the cotyledons contrasts more or less clearly with an adult one, as, for example, in the case of Gorse (*Ulex europaeus*), in which the seedlings produce trifoliate leaves (of a type which is normal in related genera) before the simple ones (Fig. 6.15). There seems, however, much to be said for extending the term to cover all ontogenetic phase changes (in leaf-shape, etc.), including the cotyledons, and including changes between early and late phases which are more gradual.

If we investigate closely the detailed development of a plant with an adult leaf clearly different from the juvenile, the commonest situation is likely to be as shown in Fig. 6.16 for Morning Glory (*Ipomoea purpurea* (*I. caerulea*)), a common tropical climbing herb. In this illustration, taken from the work of Njoku (1956), the top line shows the shape of the first ten leaves of a plant grown in the shade and the second line shows the same series from a plant in full daylight. Here two things are evident: the development of the adult three-lobed leaf-shape is gradual; and the onset of the three-lobed leaf-shape in the developmental series is greatly modified by the environment, in this case by light. Figure 6.16 also illustrates the effect of transferring the

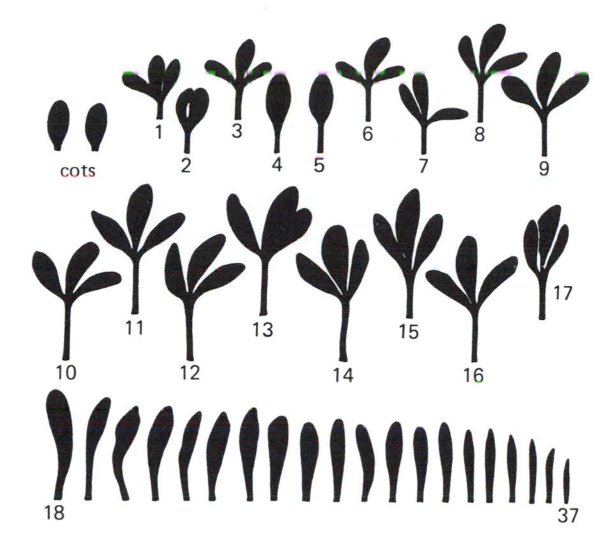

Fig. 6.15. Juvenile (top two rows) and adult (bottom row) foliage in *Ulex europaeus*. The leaves are numbered in sequence. (From Millener, 1961.)

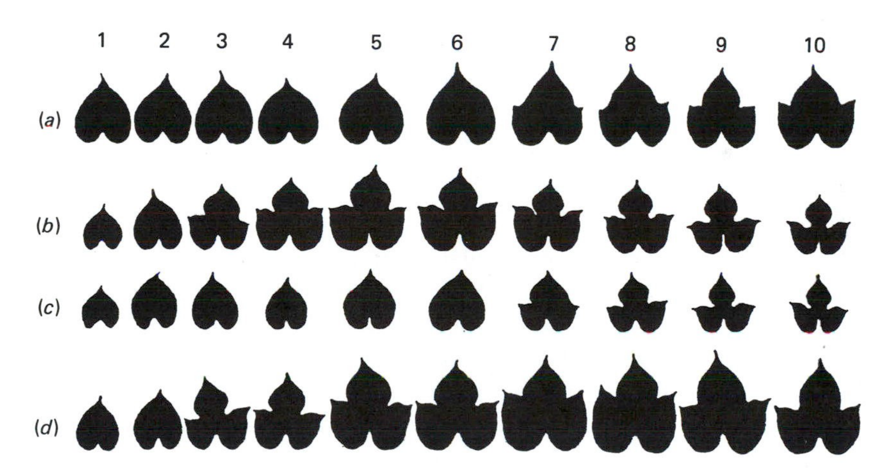

Fig. 6.16. Njoku's experiment with *Ipomoea purpurea* (*I. caerulea*) (× 0.125). The first ten leaves are shown of: (*a*) a plant grown in the shade; (*b*) a plant grown in the light; (*c*) a plant transferred from shade to light as the second leaf unfolded; and (*d*) a plant transferred from light to shade as the second leaf unfolded. (From Njoku, 1956.)

plant from shade to light at the stage of the unfolding of the second leaf and of transferring it from light to shade. In both cases there is a 'time-lag' in reaction which lasts until the sixth or seventh leaf, suggesting that the developmental processes which determine the form of the mature leaf are operating at an early stage in the differentiation of leaf primordia at the growing point and that once these have reached a certain stage the effect of environmental factors is no longer operative. The important point to note is the irreversibility of change at a certain stage in the development of mature structures and its relation to developmental variation.

Another example of a situation where changes are generally irreversible is found in Ivy (*Hedera helix*). The wild plant, common in Britain and Atlantic Europe, shows a very marked heterophylly, the familiar lobed 'ivy-leaf' being produced exclusively on non-flowering shoots which are normally flattened and adapted for growth attached to trees or on the woodland floor. The flowering shoots, in contrast, are erect, branch more or less radially and bear simple leaves. Intermediate shoots and leaves are rare (Fig. 6.17). Seedlings, as would be expected, produce lobed leaves and quickly assume the vegetative phase. If, however, portions of *either* vegetative or reproductive shoots are detached and rooted separately, the plants so produced normally continue to grow in the manner characteristic of the particular phase. This is apparently true even of intermediate shoots. In this way, whole plants of *H. helix* with a more or less erect habit and simple leaves can be propagated, apparently indefinitely, though 'reversion' to the juvenile vegetative phase can be induced, for example, by repeated cutting, by grafting on to the juvenile stock or by spraying with the growth substance gibberellic acid (GA_3). As treatment with abscisic acid (ABA) reverses the effect of GA_3, it would appear that the balance of gibberellins and ABA is involved in the transition from juvenile to adult state (Salisbury & Ross, 1992).

We must now look briefly at the phenomenon of flowering. Here there is a great deal of information, derived in part from studies of cultivated plants in which flower and fruit production are economically important (Ridge, 1991). The reproductive phase in the development of the individual plant is usually marked by an abrupt change in pattern of growth at the apex. The floral parts originate, like the leaves, as lateral outgrowths at the growing point, but the internodes, which were very obvious between the successive leaves, are suddenly greatly reduced or suppressed, so that whorls or spirals of tightly packed floral parts are produced. There is great variation between different plants as to the influence of the environment upon the initiation of flowering, but there is usually some detectable effect, and in the cases of

Fig. 6.17. Heterophylly in *Hedera helix* (× *c.* 0.33). Note the simple leaves on flowering shoots. (From Ross-Craig, 1959.)

photoperiodic response the effect can be very great indeed. Certain species require exposure, often for very brief periods only, to a particular daylength before they will pass into the reproductive phase which, once initiated, can continue whether the particular day length conditions are still present or not. This kind of adaptation to daylength has, of course, obvious importance in terms of wild populations; it may mean (as, for example, with certain plants of American origin in Europe) that the plant may be unable to flower and fruit when introduced into a country where the particular daylength conditions it requires are not present, and in this way the spread of a species may be restricted. Other species may require cold treatment before flowering is initiated, a subject discussed by Taiz & Zeiger (1991), Salisbury & Ross (1992) and Hopkins (1995), who provide a useful survey of the present state of knowledge on flowering and its physiological control.

Phenotypic plasticity

We have discussed various general aspects of the development of the phenotype and suggested the possibility of different genotype–environment interactions. While the study of these interactions has not always received the attention it deserves, a steady flow of important work has emerged (Bradshaw, 1965; Schlichting, 1986; Sultan, 1987; Thompson, 1991) and there is now good evidence for the following hypotheses:

1. Different characters of the phenotype show different degrees of plasticity.
2. The extent of phenotypic plasticity differs in different taxa.
3. Phenotypic plasticity is under genetic control.

The following examples serve to illustrate these ideas, but it is important to note that very rarely have all aspects of the control of plasticity been examined in any particular case.

Many investigations have studied the relative plasticity of different characters. Indeed, we have already referred to the question in Chapter 3. The classic studies of *Potentilla glandulosa* in California by Clausen and associates provide a further example (Bradshaw, 1965). As we shall see in Chapter 8, they discovered, in experiments involving clone transplants into a number of different environments, that there was considerable plasticity in vegetative parts, including plant height, and the number of shoots, leaves and flowers. In contrast, there was less plasticity in shape and marginal serration of the leaves, in the shape of the inflorescence and in the size of parts of the flower (Clausen, Keck & Hiesey, 1940).

The variable extent of plasticity in related taxa is beautifully illustrated by the group of species of *Ranunculus* subgenus *Batrachium* which exhibit heterophylly (Fig. 6.18 and Table 6.2). Species growing on mud or in very shallow water produce only floating leaves, while taxa from deep or swiftly flowing water develop only finely divided submerged leaves. In contrast, species inhabiting shallow water produce both types of leaves (Cook, 1968; Zander & Wiegleb, 1987; Webster, 1988).

Experiments with *Persicaria amphibia* (*Polygonum amphibium*) have also yielded valuable information on phenotypic plasticity. Plants growing on land and in water have very different phenotypes (Fig. 6.19). If ramets of cloned material are separately grown in conditions simulating land, waterlogged and submerged conditions, then the degree to which a particular individual may produce both the 'land' and 'water' phenotype may be investigated. Studies by Turesson (1961) and Mitchell (1968) suggest

Table 6.2. *The occurrence of heterophylly in British species of* Ranunculus *subgenus* Batrachium *(From Bradshaw, 1965)*

		Leaves	
Species	Habitat	Floating	Submerged
R. hederaceus	Mud or shallow water	Many	—
R. omiophyllus	Streams and muddy places	Many	—
R. tripartitus	Muddy ditches and shallow ponds	Some	Many
R. fluitans	Rapidly flowing rivers and streams	—	Many
R. circinatus	Ditches, streams, ponds and lakes	—	Many
R. trichophyllus	Ponds, ditches, slow streams	—	Many
R. aquatilis	Ponds, streams, ditches, rivers	Some	Many
R. peltatus	Lakes, ponds, slow streams	Many	Many
R. baudotii	Brackish streams, ditches, ponds	Some	Many

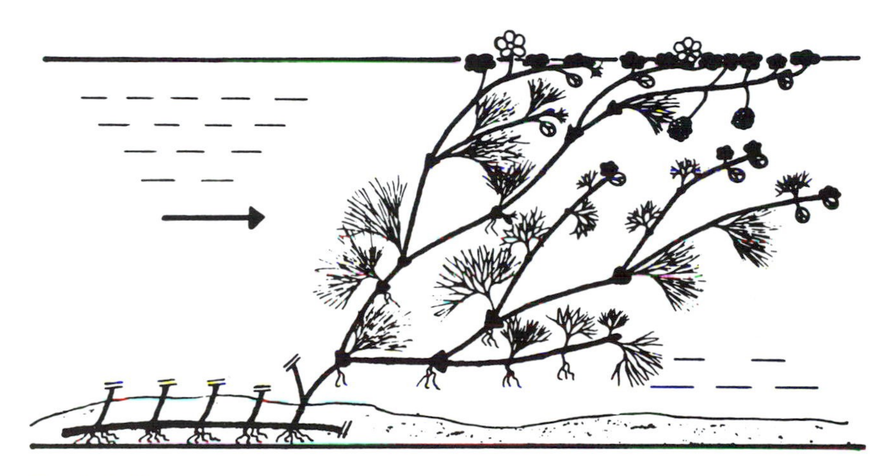

Fig. 6.18. Generalised diagram showing heterophylly in *Ranunculus* subgenus *Batrachium.* For convenience only one shoot is illustrated. Close examination has revealed seasonal variation in leaf form, especially in the submerged leaves. (From Zander & Wiegleb, 1987.)

that different individuals (presumably different genotypes) show different degrees of plasticity.

Considering a general model of plasticity, it has been discovered that different genotypes, faced with the same range of different test conditions, produce a different repertoire of responses and some are able to produce a

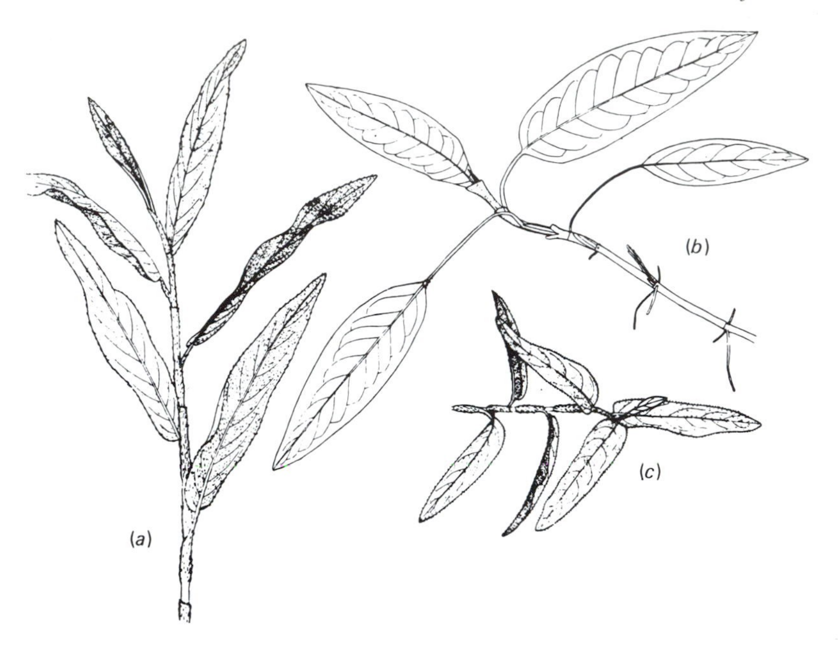

Fig. 6.19. *Persicaria amphibia* (*Polygonum amphibium*) (× 0.5). (*a*) Terrestrial form, growing on land near water. (*b*) Aquatic form, with submerged stems and floating leaves. (*c*) Form growing in damp sand dunes. (From Massart, 1902.)

wider range of phenotypes than others. Using the terminology of Johannsen (1911) and Schmalhausen (1949), different taxa show different 'norms of reaction'. Morisset & Boutin (1984) consider that there is a clear tendency for pioneer species, in early successional habitats, to show greater plasticity than related taxa in later successional vegetation types. However, Schlichting & Levin (1984) point out that there is insufficient information to permit critical tests of hypotheses and, hence, difficulties in making generalisations. Clearly, many interacting factors must be considered including levels of heterozygosity, degree of relatedness of the taxa, as well as ecological factors. Plasticity remains a comparatively neglected area in the study of plant variation. For a critical discussion of the design and interpretation of experiments on plasticity, Sultan (1987) and. Thompson (1991) may be consulted.

Our analysis of phenotypic responses has stressed morphological aspects of variation, the sort of variation which has fascinated and frustrated many taxonomists. However, we must also see phenotypic plasticity through the eyes of the plant physiologist and ecologist. Differences in underlying physiological processes associated with morphological plasticity must not be overlooked and the success or otherwise of reproduction in different conditions must be considered. Finally, many botanists consider that genotype–environment interactions are adaptively significant and that such adaptive responses have evolved under natural selection (Bradshaw, 1965; Schlichting, 1986; Sultan, 1987; Thompson, 1991). We will examine this hypothesis in Chapter 9.

7
Breeding systems

The study of a range of plants in the wild or on display in a Botanic Garden reveals a bewildering array of floral types. Many books describe in some detail a selection of pollination mechanisms, often discussing in highly technical botanical language the variety of floral structure. Special terms have quite properly been devised by botanists to enable them to write concise, accurate plant descriptions. Although the botanical literature reports extensively on the structures involved in reproduction, in our opinion it does not pay sufficient attention to the variety of breeding systems in plants, systems of which complex structures and pollination mechanisms are only a part.

We have noted in earlier chapters the role of the internal sources of variation, namely mutation and recombination, and have seen that a vast number of gametic types is theoretically possible as a consequence of these factors. Which gametes are actually brought together to form the zygotes, however, depends, to a great extent, upon the breeding system of the plant concerned. Other factors such as pollen and seed dispersal are also important, and we discuss these later in the book.

In this chapter, as a prelude to our discussions of variation within and between populations, we will show how studies of breeding behaviour have developed and make it clear how a knowledge of breeding systems provides an indispensable framework for understanding the complexities of patterns found in nature.

There are three basic breeding mechanisms: outbreeding, self-fertilisation and apomixis, which we examine in turn.

Outbreeding

In many animal groups outbreeding – crossing between different individuals – is rendered likely by sexual differentiation. In higher plants, however, separation of the sexes is the exception. According to Richards (1986) only about 4% of the flowering plants are dioecious. The incidence of dioecy varies in different floras. While only *c*. 3% of British flowering plants are dioecious, much larger percentages are reported, for example, from groups of distant oceanic islands such as Hawaii (28%) and New Zealand (13%). Tropical forests may also have high numbers of dioecious species (e.g. 23% of dioecious species were reported in a sample of woody plants from a lowland rain forest in Costa Rica (Bawa, Perry & Beach, 1985)).

Most higher plants are hermaphrodite, and the typical angiosperm flower has a zone of pollen-bearing stamens (androecium) surrounding a gynoecium containing one or more ovules. Even in the case of simple unisexual flowers of the catkin-bearing woody plants (Amentiflorae), male and female flowers are usually found on the same individual. Such juxtaposition of stamens and ovules suggests that self-fertilisation would be the most likely mode of reproduction, and it is interesting to trace the historical development of ideas which force us to conclude that many plants are adapted to facilitate cross-fertilisation (i.e. crossing between different individuals) and to minimise or prevent self-fertilisation (Whitehouse, 1959; Richards, 1986; Barrett, 1988).

The sexual function of flowers was established in the seventeenth and eighteenth centuries (Proctor & Yeo, 1973; Proctor, Yeo & Lack, 1996). In 1793, Sprengel published his classic book in which he produced excellent descriptions of the wind and insect pollination of plants, but he was apparently unaware that flowers are primarily adapted for cross-fertilisation. It was Darwin who concluded, after a detailed study of many Orchid species, that the Orchid flower was 'constructed so as to permit of, or to favour, or to necessitate cross-fertilisation'. Reflecting further on the Orchid studies, Darwin wrote (1876):

> It often occurred to me that it would be advisable to try whether seedlings from cross-fertilised flowers were in any way superior to those from self-fertilised flowers. But as no instance was known with animals of any evil appearing in a single generation from the closest possible interbreeding, that is between brothers and sisters, I thought that the same rule would hold good with plants...

However, in 1866 Darwin was already studying inheritance in *Linaria vulgaris*, and raised two large beds of seedlings of self-fertilised and cross-fertilised individuals. To his surprise the 'crossed' plants, when fully grown, were taller and more vigorous than the 'selfed' progeny. Darwin's interest was thoroughly aroused and he investigated, over many years, the effect of cross- and self-fertilisation in a number of species, e.g. *Ipomoea purpurea, Mimulus luteus, Digitalis purpurea, Zea mays.* He gave great attention to experimental design; for example, his basic comparative test was devised as follows. Seed from cross- and self-fertilised plants was germinated and a 'crossed' seedling was matched against a 'selfed' seedling, several such comparisons being made in each of a number of pots. A partition was placed between the two sets of seedlings but in such a way as to make sure that both sets of plants were equally illuminated. Other types of pot experiment and garden trial were attempted and the effect of crossing and selfing in some cases was studied for a number of generations. He examined one or a number of measures of performance such as height, weight or fertility. With some exceptions, his results revealed that the progeny of cross-fertilised plants were superior in performance when compared with the progeny from self-fertilisations. An example of his results with Maize is set out in Table 7.1.

As a result of his experiments, which were a landmark in the study of breeding behaviour, Darwin was able to see how the enormous range of floral types and physiological differences in behaviour, such as different times of maturity of stamens and stigma on the same flower, could be viewed as adaptations to ensure cross-fertilisation. Why it might be beneficial for progeny to be crossbred rather than the product of self-fertilisation, Darwin was unable to decide.

In his experiments on 'cross' and 'self' progenies, Darwin discovered that in some cases the attempt to produce progeny by self-fertilisation failed – certain plants were self-sterile. Examples of self-sterility had also been noted by other botanists. Again Darwin was unable to account satisfactorily for this phenomenon.

Great progress has been made this century in our understanding of self-sterility, which may arise from a number of causes. In most cases self-incompatibility is involved. A fertile hermaphrodite plant is incapable of producing zygotes following self-pollination. There is a mechanism operating at the pre-zygotic stage – involving both pollen and style or stigma tissue – preventing self-fertilisation.

The first, and inconclusive, studies of self-incompatibility were made by Correns (1913) in experiments with *Cardamine pratensis*. He suggested that a genetic mechanism was involved, but it was not until the experiments of East & Mangelsdorf (1925) working with *Nicotiana* species that the genetics of self-incompatibility became clearer. The elements of the scheme they proposed – a gametophytic incompatibility system – have now been found to apply to very many, but not all, incompatible species (De Nettancourt, 1977; Sims, 1993).

Table 7.1. *Darwin's experiments with Maize* (Zea mays); *height (in inches) of young plants raised from seeds obtained by cross- and self-fertilisation*

	Cross-fertilisation	Self-fertilisation
Pot I	$23^4/_8$	$17^3/_8$
	12	$20^3/_8$
	21	20
Pot II	22	20
	$19^1/_8$	$18^3/_8$
	$21^4/_8$	$18^5/_8$
Pot III	$22^1/_8$	$18^5/_8$
	$20^3/_8$	$15^2/_8$
	$18^2/_8$	$16^4/_8$
	$21^5/_8$	18
	$23^2/_8$	$16^2/_8$
Pot IV	21	18
	$22^1/_8$	$12^6/_8$
	23	$15^4/_8$
	12	18

While we might accept Darwin's (1876) general conclusions from his experiments that progeny from cross-fertilised plants were generally taller, etc. than progeny from self-fertilised plants, it is interesting to discover, as Darwin himself realised, whether the results of a particular experiment were statistically significantly different. Darwin consulted his cousin Galton who, employing the crude statistical techniques available at the end of the nineteenth century, had to conclude that the experiment was based on too few plants. It was not until 1935 that Fisher made a close analysis of Darwin's experiments. He concluded that the experimental design was fundamentally sound – comparing the growth of an outcrossed with that of a selfed seedling. (However, planting several plants in each pot must have led to competition and perhaps a better planting arrangement could have been devised.) Fisher used statistical tests first devised in the early years of the twentieth century for studying small samples, and discovered that the Maize result above was just statistically significant.

The stages are as follows:

1. Each plant is heterozygous for the gene *S*; e.g. S_1 S_2. (Many, perhaps scores, of alleles occur: say S_3, S_4, S_5, etc.) The style and stigma in the flowers of a plant are maternal tissue containing nuclei with the diploid chromosome number and in consequence style and stigma contain S_1S_2 in a plant of genotype S_1S_2.
2. At meiosis prior to pollen formation, segregation occurs and pollen, which contains nuclei with a haploid chromosome number, receives one of the two *S* alleles. The pollen hereafter behaves in pollination in accordance with its *S* allele genotype.
3. Pollen arrives at the stigma, which may be especially adapted for pollination by wind, insect or other means. Often stigmas are 'wet' from the secretion of exudates, but in some cases they are 'dry', for example in the grasses.
4. The pollen grain hydrates and germinates to give a pollen tube which grows intracellularly through a special tract of transmitting tissue. In some species the tract is solid, while in others it lines the hollow style.
5. If the *S* allele present in the pollen is also found in the style, the growth of the pollen tube in the style is progressively slowed over several hours, the tip of the incompatible tube growing abnormally and becoming occluded with callose. The accumulation of callose in incompatible tubes provides a means of exploring the behaviour of pollen from different sources, as callose 'plugs' fluoresce in ultraviolet light when stained with decolorised aniline blue. If pollen and stylar tissues have dissimilar *S* alleles, growth of the pollen tube proceeds normally and fertilisation may occur. (In some plants, however, for example the grasses, the incompatibility reaction appears to occur at or near the stigmatic surface rather than in the style.) The identity of the substances involved in incompatible reactions is reviewed by Sims (1993) and Hughes (1996).

Given such a genetic mechanism, it is now clear why certain self-fertilisations fail to produce offspring. The *S* alleles segregating in the pollen (say S_1S_2) are both represented in the stylar tissue of the same plant and all pollen is arrested in its growth down the style. The consequences of self-pollination and pollination by individuals of various '*S*' genotypes is displayed in Fig. 7.1. It is important to note that a given plant will not necessarily be fully fertile with all other *individuals* of the same species, even if they differ substantially genetically. The *S* alleles represented in pollen and style are decisive. The self-incompatibility reaction arises from an interaction between the gene products of pollen and style, preventing free

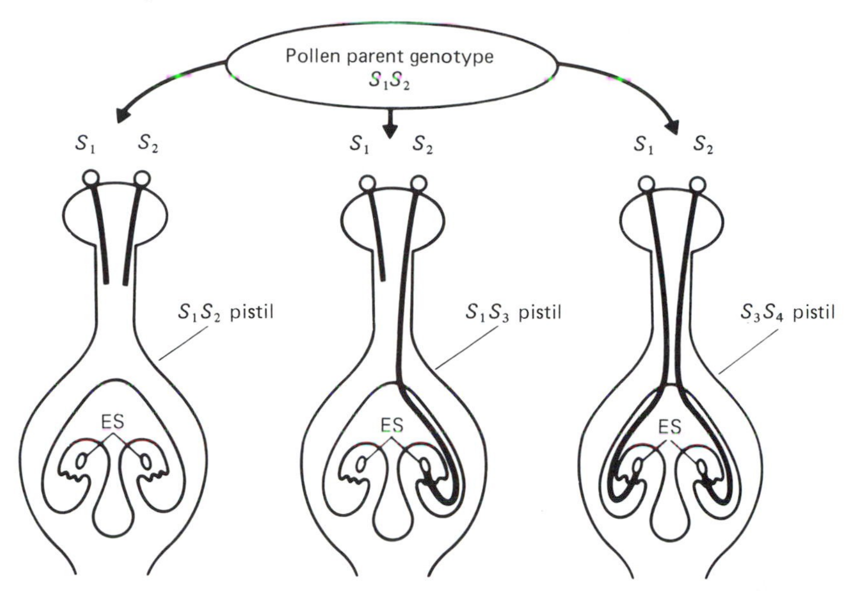

Fig. 7.1. Gametophytic self-incompatibility. A pollen parent of genotype S_1S_2 will be infertile, semi-fertile or fully fertile according to the genotype of the female plant. In most species with this system, incompatible pollen tubes are inhibited in the style. ES = embryo-sac. (From Heslop-Harrison, 1978). Note that in some families gametophytic self-incompatibility is more complex genetically, being controlled by two or more loci (De Nettancourt, 1977).

access of pollen tubes to the ovule, permitting pollen tube growth in certain cross-fertilisations and preventing pollen tube growth on self-pollination (Sims, 1993).

In 1950, American botanists studying *Parthenium* (Gerstel, 1950) and *Crepis* (Hughes & Babcock, 1950) discovered what has come to be known as the sporophytic incompatibility system. The following are the important differences from the gametophytic system just outlined:

1. Individuals are heterozygous for the *S* gene, say S_1S_2, and, as before, segregation occurs at meiosis preceding pollen formation. The pollen, notwithstanding the 1:1 ratio of segregation of *S* alleles, behaves as if it had the genotype of the plant which produced it. Evidence suggests that proteins, amongst which are recognition components responsible for the incompatibility reaction, are produced by the tapetal cells of the anther pollen sacs. These substances are exported to the pollen wall during its development (Sims, 1993).

2. In general, pollen is rejected if its pollen parent shares the same *S* allele as the stigma and style of the female parent. (However, in some species there are more complex reactions, including dominance of certain alleles.) The results of certain cross-pollinations are set out in Fig. 7.2.
3. The site of pollen tube failure is often different in sporophytic systems, occurring on or close to the stigmatic surface. As in gametophytic systems, the accumulation of plugs of callose indicate incompatible pollinations.

There are a number of important differences between the two systems. 'Sporophytic' plants generally have a dry stigmatic surface, a rapid incompatibility reaction (occurring in minutes rather than in hours) and tricellular pollen, whereas, plants with gametophytic incompatibility usually have bicellular pollen.

Darwin's list of self-sterile plants was quite short, but self-incompatibility is now thought to be widely distributed in both the monocotyledons and dicotyledons. It has been reported in 71 families (Barrett, 1988).

Another finding of considerable interest is that plants with gametophytic incompatibility systems are all homomorphic, that is, there are no structural differences associated with different *S* alleles. In contrast, some plants

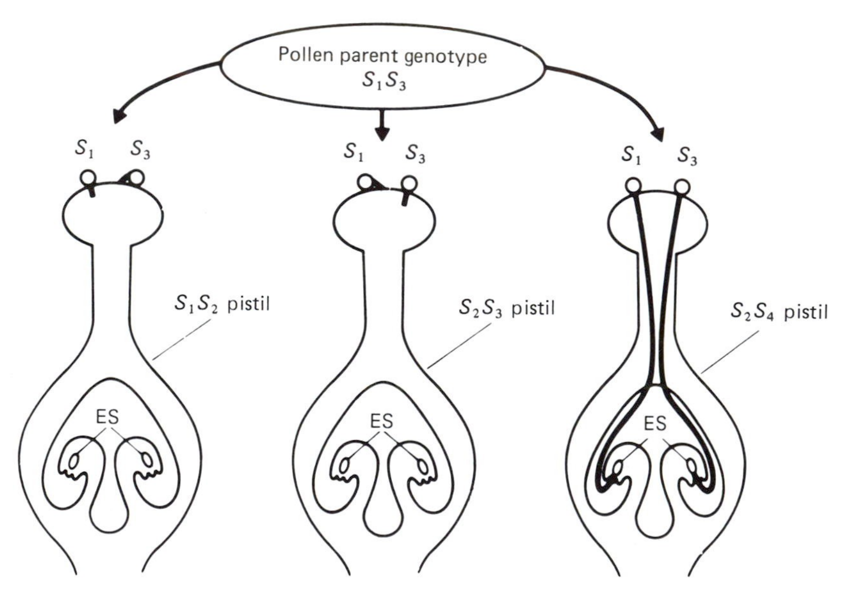

Fig. 7.2. Sporophytic incompatibility. In this diagram the *S* alleles are presumed to act independently. Other relations are known, including dominance and mutual weakening. ES = embryo-sac. (From Heslop-Harrison, 1978.)

with sporophytic incompatibility systems are heteromorphic. (Heterostyly is reported from approximately 25 angiosperm families (Barrett, 1992).) The existence of different forms of flowers in the same species was the subject of some of the most famous studies made by Darwin. He carried out crossing experiments with heterostylous plants of both the distylic and tristylic type (Fig. 7.3). In the case of the Cowslip (*Primula veris*), he showed that there are two forms of the flower, 'pin' and 'thrum', each with a characteristic syndrome of characters (Fig. 7.4). The pollinations pin × thrum and thrum × pin yielded good seed set; selfing pins or thrums or crossing pin × pin and thrum × thrum yielded very much less seed. In view of the comparative self-sterility of pin and thrum plants, Darwin concluded that distyly was a device favouring outcrossing. Pollinators visiting *P. veris* were likely to pick up pollen in a pattern related to anther position. Subsequent visits would transfer pollen to stigmas at 'an equivalent height'.

Subsequent study has revealed the genetics of the situation in *Primula*.

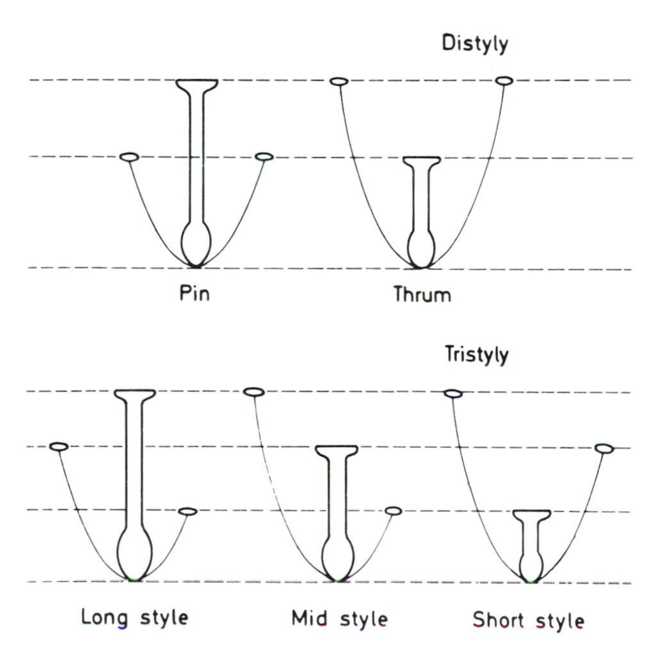

Fig. 7.3. Symbolic representation of distyly and tristyly. In each system, the compatible pollinations only involve anthers and styles at the same level, and therefore the following are incompatible combinations: pin × pin, thrum × thrum, long × long, mid × mid, and short × short. (From De Nettancourt, 1977.) Darwin (1877) listed 14 families in which heterostyly had been confirmed. This list has now been extended to *c.* 25 families (Barrett, 1992).

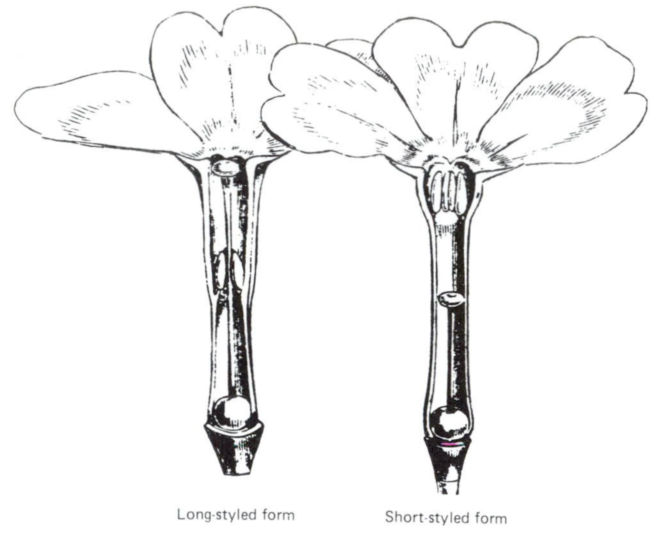

Fig. 7.4. Heterostyly in *Primula veris* (× 2). Darwin (1877*a*) discovered that the long-styled (pin) form always had a much larger pistil with a globular, rough stigma standing high above the anthers. Pin pollen was oblong in shape. In contrast the short-styled (thrum) plants always had a short pistil about half the length of the corolla, with a smooth stigma (depressed on the summit). Pollen in the anthers, which stood above the stigma, was spherical and larger than in the pin.

Dimorphy and the incompatibility reaction are controlled by blocks of tightly linked genes, sometimes called 'supergenes'. Thrum plants are heterozygous *GPA*/*gpa*, while pin plants are homozygous *gpa*/*gpa*, where:

Controlling female characters. *G*: short style, short stigmatic papillae and thrum female incompatibility versus *g*: long style, long papillae and pin female incompatibility.

Controlling male characters. *P*: large pollen, thrum male incompatibility versus *p*: small pollen and pin male incompatibility.

Controlling anther position. *A*: high anthers (as in thrum) versus *a*: low anthers (as in pin).

It has been found that other heteromorphic sporophytic plants are di-allelic (Lewis, 1979). The genetic information in the 'supergene' is very tightly linked, but occasionally crossing over occurs to give various types of homostyle (Richards, 1986). We will consider the breeding behaviour of these variants below. Ganders (1979) has provided a valuable review of the characters in which heterostylous plants vary (style length/colour; stigma size/shape/papilla shape; stamen position/anther size/pollen size and number/pollen shape, size, sculpturing and food reserves; corolla size/pubescence).

Darwin also studied Purple Loosestrife (*Lythrum salicaria*), in which three kinds of flowers are found corresponding to different levels of anthers and stigmas in the flowers; 18 different interpollinations were needed to investigate the effects of six kinds of anther and three kinds of style, and these Darwin carried out. The number of seeds in six pollinations between anthers and stigmas at the same height was much greater than those produced in other pollinations.

Late-acting self-incompatibility systems

Studies of self-incompatibility have usually concerned themselves with the pre-zygotic mechanisms of differential growth of pollen tubes of different genotypes on the stigma and in the style. However, studies have revealed various pre- and post-zygotic reactions in the ovary. Barrett (1988) points out that the possibility of post-zygotic rejection of selfs is often 'excluded from definitions of self-incompatibility', but the review of Seavey & Bawa (1986); makes it plain that such possibilities must be considered, notwithstanding the difficulties of distinguishing between late-acting and inbreeding effects (see below).

Self-fertilisation

The second main breeding method involves self-fertilisation. In his account of cross- and self-fertilisation of plants, Darwin (1876) reported a list of species which yielded seed when covered by a net to exclude insect visitors. In 1877, he discussed the presence in some species of specialised flowers that never open, pollination occurring automatically in the closed bud, which gives rise directly to a fruiting structure. Such flowers, which are of course self-fertilising, are referred to as cleistogamous. Following a literature search and as a result of his own researches, Darwin listed 55 genera in which cleistogamous flowers had been found and investigations by botanists over the years have added to the list (Uphof, 1938). In some cases cleistogamy appears to be more or less obligatory, as in several grass species (McLean & Ivimey-Cook, 1956), but, as we shall see below, cleistogamous flowers occur seasonally in some species.

Following observations and experiments, botanists have concluded that some species are predominantly self-fertilising, but in many cases this is a presumption based on the following types of evidence. It may be readily demonstrated that some species are self-compatible, and it can also be observed that some plants, visited by insects in summer, can flower and fruit in autumn, winter or spring, when insects are absent. Further, progenies of some species are remarkably uniform in appearance and it is assumed that such uniformity results from persistent inbreeding. As we shall see later such evidence is circumstantial and for definitive studies the use of genetic markers is essential in the study of breeding behaviour.

If some plants are predominantly or obligately self-fertilising, how is this to be equated with the demonstration by Darwin of general superiority of outbred plants? This paradoxical situation will be examined below.

Apomixis

In this third reproductive mode, reproduction is achieved without fertilisation, the sexual process being wholly or partly lost (the term and its definition are according to Winkler, 1908). Two types of system are found: vegetative apomixis and agamospermy.

Vegetative apomixis

Some plants, lacking the means of sexual reproduction, reproduce entirely vegetatively. For instance, female plants of the dioecious waterweed, *Elodea canadensis*, were introduced into Europe in the nineteenth century and these introductions, even in the absence of male plants, spread widely by vegetative growth and fragmentation (Sauer, 1988). In other cases, specialised structures such as small, aerial, readily detachable bulbils may be produced (Fig. 7. 5). Plants arising from vegetative spread or specialised propagules will, unless somatic mutations occur, have the same genotype as the parent plant. Many cultivars have arisen by somatic mutation (Hartman & Kester, 1975) but until molecular tools became available we had no idea of the frequency of such mutation. There is now a good deal of evidence for somatic changes in the genome but the significance of this for population variation has yet to be determined (Schaal, 1988; Gill *et al.*, 1995).

Agamospermy

In certain plants, normal seed is set but no sexual fusion has occurred in its production. Offspring have the genetic constitution of the plant which

Fig. 7.5. Bulbils. (*a*) *Saxifraga cernua*: 1. (× 1); 2. A cluster of bulbils; 3. Bulbils in various stages of development. (From Kerner, 1895.) (*b*) *Poa alpina*: 1. *P. alpina* with bulbils replacing its flowers (× 1); 2. A portion of the inflorescence; 3. A miniature grass-plant developed between the glumes of a spikelet of *P. alpina*.

produces them. A plant reproducing by seed apomixis or agamospermy has all the advantages of the seed habit without the risks which may be associated with pollination. As there is no essential genetic difference between simple agamospermy and asexual reproduction, Winkler grouped these two types of reproduction under the common term of apomixis. Some botanists do not regard these vegetative systems as apomictic and would restrict the use of the term apomixis to agamospermy (see below). Others regard vegetative spread in the absence of sexual reproduction as a variant of apomixis (Stebbins, 1950; Stace, 1989), while excluding those cases where vegetative propagation – by means of rhizomes, stolons, runners, etc. – occurs in sexually reproducing plants.

Agamospermy was first described in 1841 by J. Smith in plants of the Australian species *Alchornea ilicifolia* growing at Kew Gardens. The deduction that seed development had occurred without fertilisation could safely be made, as the Kew collections consisted entirely of female plants. Embryological studies, which revealed some of the underlying mechanisms of seed apomixis, were carried out by Murbeck and Strasburger on *Alchemilla* and *Antennaria* species at the turn of the century. Since those classical studies, a wealth of detailed examples has accumulated, and several very important generalisations can now be made. First, apomictic species occur very widely in higher plants, both in the ferns and the angiosperms; but no apomictic gymnosperms are known. Secondly, there are certain flowering plant families which show a great deal of apomixis affecting several genera: the outstanding familiar examples in the Northern Temperate flora are in the Rosaceae and the Compositae (Asteraceae). Thirdly, there is in these families a rather obvious correlation between taxonomic difficulty and the occurrence of apomixis; many of the so-called 'critical' genera or species-groups of the nineteenth century, in which the taxonomists found extreme difficulty in reconciling the points of view of the 'splitters' and the 'lumpers' as to specific delimitation, turn out to be agamospermic. Familiar examples are the genera *Rubus* and *Sorbus* in the Rosaceae, and *Hieracium* and *Taraxacum* in the Compositae.

Complete agamospermy is not difficult to detect. For example, in most plants of the Common Dandelion (*Taraxacum officinale*), emasculation of all the florets, if performed carefully, followed by 'bagging' of the capitulum, will still result in a perfect head of fruit (Fig. 7. 6). Partial apomixis, on the other hand, may be very difficult to detect merely from emasculation experiments, for a low proportion of seed set could so easily be due to chance contamination with pollen. Even more difficult are the cases of *pseudogamy*, in which pollination is necessary for seed formation but nevertheless the embryo is not formed by sexual fusion. Indeed, the detection of pseudogamous situations is so difficult that we may well suspect them to be more common than we know at present. Maternal (matroclinous) inheritance is usually an indication of pseudogamy. If an apparent cross between two plants differing obviously in easily scored characters produces a rather uniform F_1 resembling very closely the female parent, one should look for pseudogamy in the details of embryo formation. This is how the phenomenon was first suspected in the case of *Ranunculus auricomus*. It has been demonstrated in 'crosses' between species of

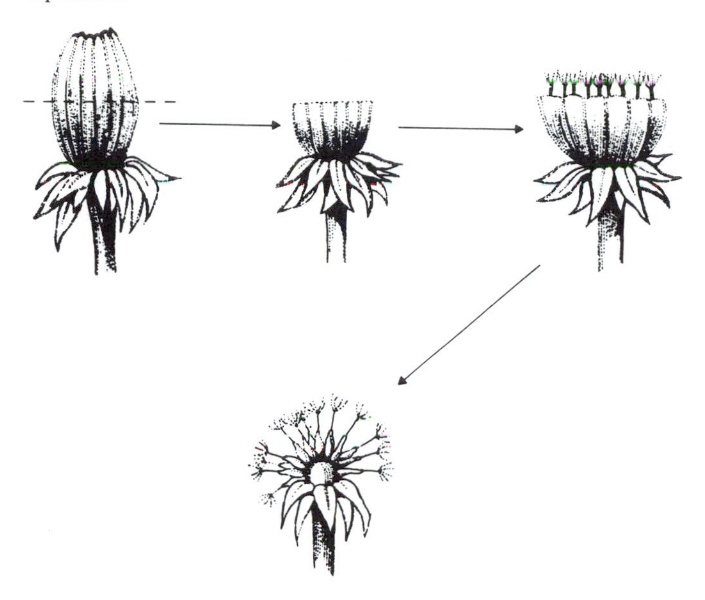

Fig. 7.6. Emasculation of an agamospermous *Taraxacum*. The top portion of the bud, containing anthers and stigmas, are sliced off with a razor, and, notwithstanding the loss of these structures, fertile seed is set apomictically (× 1). (From Richards, 1986.)

Potentilla (Fig. 7.7). More recently, the use of molecular markers is proving effective in the study of apomixis. If 'crosses' are made between plants with different genetic markers, then the hypothesis that progenies are maternal in their genotype can be critically tested (Asker & Jerling, 1992).

Embryology of apomixis

It is not possible to give within the restricted space available here any detailed account of the embryological and cytological complexity of apomictic groups. For this, reference must be made to the standard works on apomixis by Gustafsson (1946, 1947*a*, *b*), Battaglia (1963), Nogler (1984), Asker & Jerling (1992) and Naumova (1993). Nevertheless, the subject is interesting and, as apomictic plants are found in all parts of the world, a general account is necessary.

When we come to look at the causes of apomixis at the embryological level, we find that we are not dealing with a single, standard pattern of development, but with a whole range of possible situations having in common only one feature, namely that they involve abandonment of the

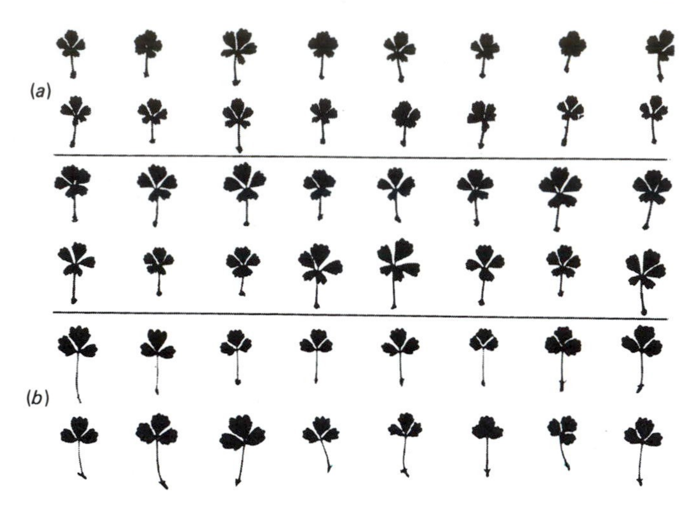

Fig. 7.7. Leaves of *Potentilla* species, showing maternal (matroclinous) inheritance – evidence that the attempted 'cross' between two species has produced only pseudogamous offspring (× 0.2). Top two rows: *P. neumanniana* (*P. tabernaemontani*) (*a*) Middle two rows: offspring of attempted cross between (*a*) and (*b*), showing leaf shape of (*a*). Bottom two rows: *P. arenaria* (*b*), a variant with only three leaflets, used as pollen parent.

fusion of gametes in the normal sexual process as a necessary preliminary to embryo and seed development. Like many biological topics where our knowledge has accumulated rapidly, we find terminological difficulties and must do the best we can in this situation. The terms here used generally follow Gustafsson; where Battaglia differs substantially, we have indicated the alternative term in brackets.

Before apomictic situations can be understood, we must know the normal pattern for a sexually reproducing flowering plant. (In ferns, apomixis is necessarily somewhat different because the gametophyte is a free-living plant separate from the sporophyte; most fern apomixis is technically *apogamy*, in which vegetative cells of the gametophyte give rise to a new embryo directly, thus omitting the stage of gamete production.)

Fig. 7.8*a* illustrates in diagrammatic form the essential features of the development of the ovule and pollen grains of a typical angiosperm. Note the following points:

1. A mature ovule ready for fertilisation contains a single embryo-sac, which corresponds to the free-living gametophyte generation in the ferns (where it is called a 'prothallus').

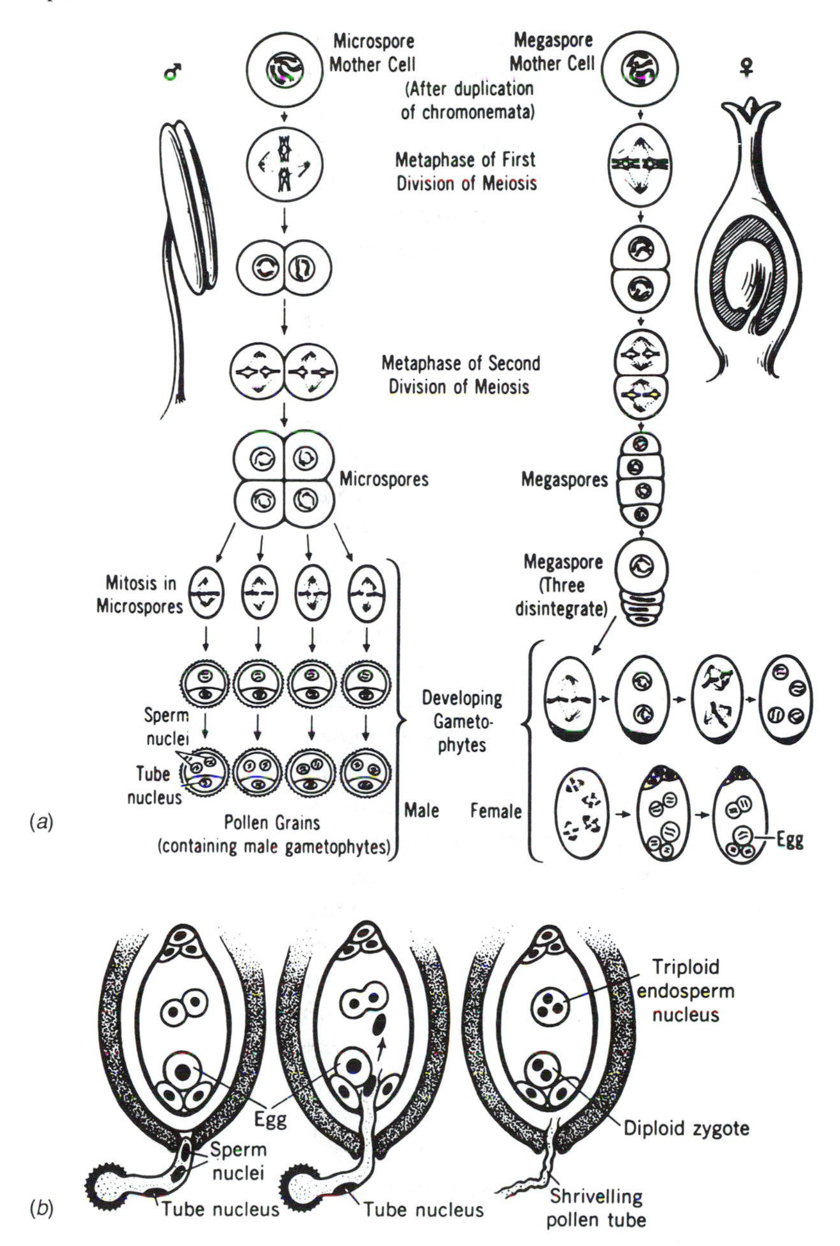

Fig. 7.8. (*a*) Development of pollen and ovules in sexual angiosperms. (*b*) Fertilisation in angiosperms. Two sperm nuclei unite with the egg and the central nucleus, respectively. (From Winchester, 1966.)

2. This embryo-sac contains eight nuclei, one of which is an egg-cell or female gamete; the embryo-sac has developed by means of three ordinary mitotic nuclear divisions from a single cell, the megaspore.
3. The megaspore originated by meiotic division as one of four products of an initial megaspore mother-cell; the other three nuclei degenerated early.

The great majority of apomictic deviations from this pattern involve the production of an apparently normal embryo-sac, from which, however, an egg-cell develops directly, without fertilisation by a male nucleus from a pollen tube.

In the normal sexual process, the meiotic division which occurs in the formation of the megaspore (and also a similar meiotic division in the formation of the microspore or pollen grain) results in the production of a cell with a single set of chromosomes – the so-called haploid state. Subsequent mitotic divisions replicate this haploid set, so that the gametophyte generation, and the male and female gametes produced, are all haploid. The sexual fusion of egg-cell (female gamete) with pollen tube generative nucleus (male gamete) shown in Fig. 7.8*b* restores the diploid state in the zygote, which then divides mitotically to give the embryo sporophyte; this grows eventually, after the germination of the seed, into a mature diploid plant. This cycle of haploid gametophyte generation succeeded by diploid sporophyte generation is of fundamental significance in the plant kingdom and can be traced from the more complex algae right through to the flowering plants. The cytological differences between the generations are accompanied in all land plants by very obvious morphological differences. The apparent simplicity of the life cycle of the flowering plant, involving pollination, seed setting and dispersal, disguises a complex evolutionary history of suppression of the free-living gametophyte generation and the free-swimming gametes, which are still present in the more primitive members of the land flora such as the ferns. Returning to apomictic flowering plants, we find it is in the production of an embryo-sac with the unreduced number of chromosomes that their deviation normally shows. Such an embryo-sac has naturally a diploid egg-cell, which requires no complement of chromosomes from a male gamete to restore the normal sporophyte number. In this way, the commonest kinds of apomixis cut out the meiotic stages from the life cycle, so that the possibilities of variation generated by sexual reproduction are lost. It is for this reason that apomictic reproduction is genetically equivalent to vegetative propagation.

If the origin of the unreduced gametophyte is investigated, it is generally possible to distinguish between two situations. In the first, which is called *diplospory* (*gonial apospory*), the gametophyte arises from an unreduced megaspore; whereas in *apospory* (*somatic apospory*) it arises from an ordinary somatic cell of the sporophyte, which is, of course, also unreduced. Obviously the diplosporous condition involves a less radical departure from the normal sexual pattern than does the aposporous. In plants where the megaspore mother-cell is clearly differentiated in an early stage in the ovule (that is, the archesporium is unicellular), there is no difficulty in deciding between these two possibilities if apomixis is present; this is the case, for example, in the Compositae (Asteraceae) (Fig. 7.9). However, there are some groups (for example, Rosaceae) in which there is a multicellular archesporium and in such cases a decision as to whether a particular cell which has undergone apomictic development is or is not part of the archesporial tissue may be difficult or even almost arbitrary. Thus, the fact that both diplospory and apospory occur in *Potentilla neumanniana* (*P. tabernaemontani*) is not indicative of any fundamental difference in this case (Smith, 1963*a*, *b*). Asker & Jerling (1992) may be consulted for details of the different variants of diplospory and apospory that have been detected in the embryology of the angiosperms.

Diplospory and apospory are the commonest apomictic situations, but we must briefly mention the range of possibilities which Battaglia calls aneuspory. In such cases the megaspore mother-cell undergoes a more or less irregular meiosis to form the megaspore. In apomictic *Taraxacum* at the first division of meiosis, there is usually no chromosome pairing and, instead of producing two nuclei, a single restitution nucleus is produced. The second division then produces a dyad of unreduced cells (instead of the normal tetrad) and the lower one of these functions as a gametophyte initial, giving the normal eight-nucleate embryo-sac. In some cases chromosome pairing does occur in *Taraxacum*, and this may allow some crossing-over and reassortment of genetic material which is not possible in the simple cases of diplospory and apospory. 'Sub-sexual' complexities of this kind may be more widespread, and more important in their effect on variation patterns, than has yet been established.

One final question concerns the function of pollination in pseudogamous species. In most cases, it seems likely that the characteristic fusion of one of the male nuclei with the polar nuclei of the embryo-sac to form the endosperm does take place in pseudogamous apomicts, although it is understandably difficult to demonstrate the actual fusion process. This would explain why pollination remains necessary for proper seed formation in spite of the apomictic origin of the embryo, for we could assume that, in the absence of nutritive endosperm tissue, the normal embryo development could not take place.

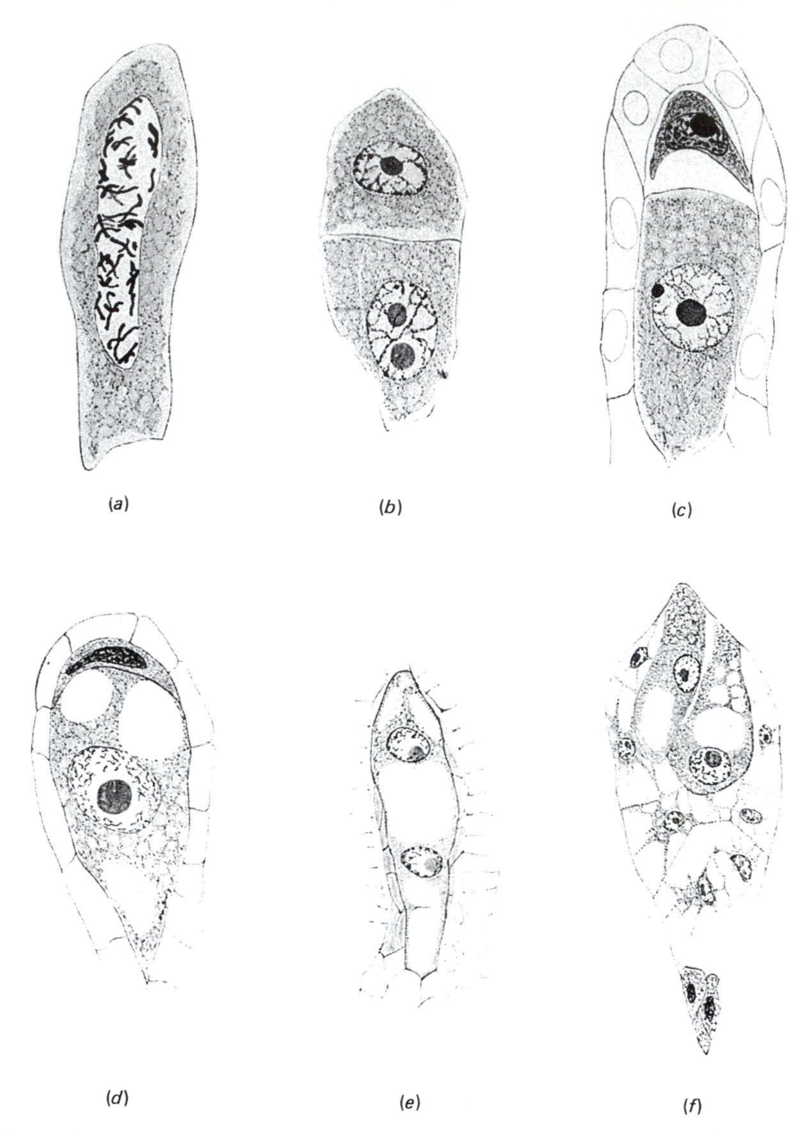

Fig. 7.9. Diplospory in *Taraxacum*. (*a*) Restitution nucleus. (*b*) Dyad of unreduced cells (megaspores) formed after division of the restitution nucleus. (*c*) Degeneration of the upper cell, and development of the lower one to a functional megaspore. (*d*) Functional megaspore. (*e*) Binucleate embryo-sac formed from megaspore. (*f*) Mature embryo-sac, with egg-cell (ovum) and synergidae. In the formation of restitution nuclei, irregular chromosome behaviour may give rise to nuclei, and ultimately to seeds, with chromosome numbers different from that of the parent plant. For example, Sørensen & Guđjónsson (1946) produced from triploid *Taraxacum* ($2n = 3x = 24$) some plants with $2n = 3x - 2 = 22$, $2n = 3x - 1 = 23$ and $2n = 3x + 2 = 26$. Aberrants were also found with unaltered chromosome numbers. In these cases it seems likely that some pairing of chromosomes occurs in the embryo-sac mother-cells and crossing-over takes place, giving rise to plants of different genotype from that of the parent. (Drawings from Osawa, 1913.) ((*a*) × 1100; (*b*), (*c*), (*d*) × 860; (*e*) × 400; (*f*) × 270.)

Consequences of different reproductive modes

Having discussed the three main modes of reproduction, we may now examine the consequences of reproduction in each mode.

What happens as a result of repeated self-fertilisation is highly important in our understanding of breeding systems. (For an interesting review of the history of this subject, see Wright, 1977.) Studies of inbreeding in Maize by East and Shull at the beginning of the century provided an important model, which has been confirmed in many other studies of crop plants. In the heterozygous diploid, the dominant allele often shelters recessive alleles that are deleterious in the homozygous state. Self-fertilisation quickly results in the segregation of lethal or sublethal types as homozygous recessives are produced. (Unless especially looked for in the seedling stage these types, which may die at a very early stage of growth, may be undetected even in garden or glasshouse culture.) Further selfings produce rapid separation of the material into uniform lines, often called pure lines, differing from each other in various vegetative and reproductive characteristics (Fig. 7.10). The continued selfing of uniform lines may be rendered impossible as some plants may become weak or sterile. Surviving lines may be characterised by plants of reduced vigour and fertility. If plants of pure lines originating from different parental stocks are crossed together, hybrid vigour (so-called heterosis) may be demonstrated. Such hybrid plants are characteristically of great vegetative vigour and high fertility (Table 7.2). It is important to note that crossing genetically closely related plants from lines derived by repeated self-fertilisation from the same original parental stock will not give heterotic plants. It is clear from the model that repeated

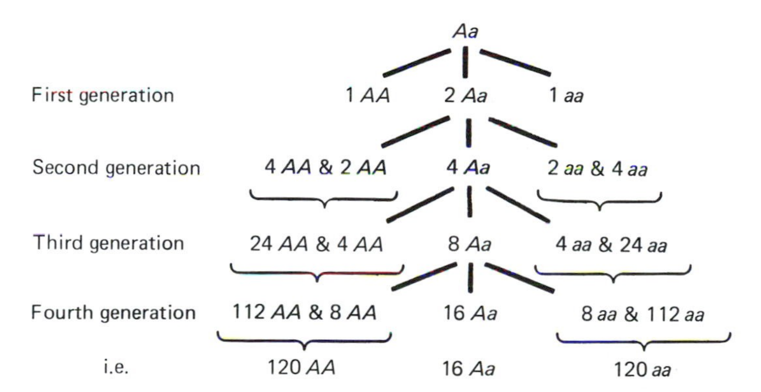

Fig. 7.10. Diagram showing the effect of selfing on a heterozygote. The proportion of heterozygous individuals rapidly declines in successive generations, and 'pure lines' are established. (From Wilmott, 1949.)

Table 7.2. *Hybrid vigour in Maize* (Zea). *Crossing inbred lines of Maize P_1 and P_2 yields F_1 plants showing hybrid vigour. Repeated self-fertilisation through several generations results in diminution in height and loss of yield (From Jones, 1924)*

	Parents		Successive generations							
	P_1	P_2	F_1	F_2	F_3	F_4	F_5	F_6	F_7	F_8
Number of generations selfed	17	16	0	1	2	3	4	5	6	7
Mean height (inches)	67.9	58.3	94.6	82.0	77.6	76.8	67.4	63.1	59.6	58.8
Mean ear length (cm)	8.4	10.7	16.2	14.1	14.7	12.1	9.4	9.9	11.0	10.7
Mean yield (bushels per acre)	19.5	19.6	101.2	69.1	42.7	44.1	22.5	27.3	24.5	27.2

self-fertilisation will yield complete homozygosity in a few generations unless the heterozygous state is favoured by selection (Fig. 7.10). However, many plants are polyploids, having more than two representatives of a gene, and, as we shall see in Chapter 12, a greater number of generations will be required to produce complete homozygosity in such plants.

The phenomenon of heterosis, so pronounced in experimental crosses with inbred lines, was not a new discovery, being often reported in the studies of early plant hybridists (Roberts, 1929). The underlying causes of loss of vigour or fertility on repeated selfing and the heterotic effects in products of crossing inbred lines have been the subject of intense study. For accounts of this controversial subject the reader is referred to Charlesworth & Charlesworth (1987) and Strickberger (1985).

In considering generalisations which might be made on the effects of inbreeding, it is important to note that our ideas are based largely on results with crop plant species. However, a few studies of wild plants have also been made (Charlesworth & Charlesworth, 1987). Judged by seed set, germination rate, plant size, fertility or survival, there is clear evidence of inbreeding depression, even for those species that are regularly self-fertilised. However, it should be noted that most of the data sets do not provide estimates of all components of fitness.

Using isozyme markers, it has proved possible to test some of the predictions of models of breeding behaviour. Gottlieb (1981*a*) compared the variation in outbreeding species with selfers. As predicted, selfers have statistically significantly less genetic variation in the mean proportion of loci that are polymorphic, the mean number of alleles at polymorphic loci and the mean level of heterozygosity.

DNA fingerprint methods have also been employed to examine variation in species with different breeding systems. For example, Wolff, Rogstad & Schaal (1994) investigated population variation in three *Plantago* species. *P. major*, which is regularly self-fertilised, showed relatively little variation within populations. In contrast, the obligately outcrossing *P. lanceolata* exhibited high variation within populations.

Advantages and disadvantages of different breeding systems

One possible advantage of repeated self-fertilisation might be that well-adapted genotypes could be replicated with little change. A further advantage, especially in extreme or marginal habitats, where crossing between plants might be hazardous or fail altogether, is that self-fertilisation is an assured method of producing offspring (Lloyd, 1979*a*, *b*). An appreciation of the long-term disadvantages of inbreeding enable us to recognise the advantages of the outbreeding mode of reproduction. As we have seen, structural features or physiological mechanisms prevent, or discourage, self-fertilisation and lead to crossing between different individuals. The role of incompatibility mechanisms is very important in considering breeding within populations. While some fruits (or seeds) from a given parent may be dispersed some considerable distance, many fall close to the parent, developing and flowering as a family group. This group, which may include parents and other relatives in plants with a long life cycle, is made up of genetically related individuals. Crossing between close relatives leads to inbreeding depression, although complete homozygosity is not achieved so swiftly by such matings. Lewis (1979) has shown that incompatibility systems will restrict crossing between close relatives, the degree of restriction depending on the genetic mechanism of the incompatibility system. Moreover, the outbreeding enforced by self-incompatibility allows new mutations arising in different individuals to be 'brought together' and (perhaps of equal importance) a few exceptionally favourable progeny may be produced (Richards, 1986).

In general terms, obligate outbreeding would appear to have advantages, but there are costs to be borne. Compared with regularly self-fertilising species, a greater amount of biomass has to be employed in producing flowers, nectar, etc. If only one genotype is present in an area, it may not be able to reproduce sexually. Reproduction may be rendered uncertain or unlikely by environmental factors influencing, for example, pollinating insects. Moreover, given that plants which survive to reproduce successfully in a habitat are well adapted, outbreeding might seem to offer only the possibility of loss of such variants as each generation produces new variability. While this variability may include individuals of high fitness, there may be a considerable 'genetic load' of less fit progeny.

The third mode of reproduction – apomixis (either by vegetative (asexual) means or by agamospermy) – facilitates the reproduction in quantity of well-adapted plants of maternal genotype, with little or no genetic load. The agamospermic development of large quantities of identical progeny has been likened to the production line of the Model-T Ford (Marshall & Brown, 1981). In some apomictic plants the pollen is defective and it has been suggested that the 'male costs' may be lower (Richards, 1986). Apomixis may also offer the possibility of reproduction by seed in plants with 'odd' or unbalanced chromosome numbers, such plants being unable to produce viable products at meiosis and likely to be totally or partially seed-sterile in sexual reproduction. Seed apomixis provides all the advantages of the seed habit (dispersal of propagules and a potential means of survival through unfavourable seasons). As we shall see later, apomictic plants are often of polyploid and hybrid origin. Therefore, this type of breeding system may be viewed as a means of conserving high heterozygosity (Asker & Jerling, 1992). Apomixis would also appear to be important at the edge of the range of many species, allowing populations to persist in territory in which various factors – including lack of insects – limit the extent, or exclude the possibility, of sexual reproduction (e.g. at high altitude or latitude).

While we might postulate various advantages of the apomictic mode of reproduction, it is clear that there are also some theoretical disadvantages. Mutations arising in different lineages cannot be brought together. The generation of 'new' variability would seem, at first sight, to be restricted or prevented in apomictic plants. We return to this question later in the chapter.

Vegetative apomixis would appear to have limitations related to senescence and diseases. Whilst higher animals usually have a clearly defined lifespan, it is not clear whether there are ageing processes in perennial plants which would restrict or prevent natural asexual reproduction in the long-term. Observations on cultivated *Citrus* plants suggest that repeated vegetative propagation leads to senescence (Frost, 1938). Moreover, reproduction by vegetative means carries with it the possibility that virus or disease might build up in the plant. A very good example (Richards, 1986) is provided by the sterile hybrid (*Primula* × *scapeosa*) produced in 1949 by crossing two Himalayan species. The plant was subsequently propagated by vegetative means and widely planted in gardens in different parts of the world. By 1982, it had ceased to be grown, as the plant had became debilitated by infection with cucumber mosaic cucumovirus. There is evidence that the sexual cycle provides a means of 'purging' the plant system of some disease organisms by providing a 'clean egg' (Richards, 1986). However, some viruses (Matthews, 1991) and other disease organisms are seed-borne.

An analysis of the three modes of reproduction reveals, therefore, that each has its advantages and disadvantages. While it might be important in certain open, early successional habitats to reproduce unchanged, well-adapted genotypes, by self- fertilisation or apomixis, in other habitats, with high spatial, temporal and biotic heterogeneity, plants capable of producing variable progeny would appear to be at a selective advantage. Two commonly employed verbal models help us to see the force of this argument. 'The Tangled Bank' model – drawn from a famous quotation in Darwin's *Origin* about the interactions of plants – emphasises the heterogeneity of plant communities (Bell, 1982), whereas a second model considers the predator/prey and host/pathogen cycles at work in plant communities. Plants must produce the 'new' genetic variation necessary to 'stay in the game'. They have been likened to the Red Queen in *Alice in Wonderland* who ran to stay in the same place.

In addition to spatial, temporal and biotic heterogeneity, we must also consider vegetational and climate change over the longer time scale. Intuitively, we can appreciate that a lineage lacking the capacity to produce variation might be at a serious selective disadvantage in competition with lineages capable of change. A lack of variation might prevent a lineage from withstanding the selection pressures associated with, say, migration during a period of global climate change. Important too, and perhaps less well known, are systematic changes, for example in climate, in the short term. Thus, there is evidence from documents, plant remains, works of art, etc., for fluctuations in climate in recent historical times (Lamb, 1970).

Given the spatial and temporal variability of habitats, and the fluctuations in climate, as well as the marked differences in weather in successive years, it is not surprising that detailed studies of the breeding behaviour of particular species reveal that, instead of reproducing entirely in one of the three modes outlined above, many plants reproduce by several methods. Thus, a species may produce invariant progeny by one mode of reproduction, while at the same, or a different time, generate variation by another mode of breeding. A different balance of variance and invariance may be seen in the products of different species or lineages within species (Mather, 1966).

Breeding systems in wild populations

In order to understand and discover the actual breeding system in the field, detailed studies at the population level are necessary, investigating natural lineages of plants and their production. Inferring the breeding behaviour from flower structure or pollinator activities would seem to be fraught with difficulty. For instance, the flowers of certain taxa of *Calyptridium* (Portulacaceae), which are regularly visited by insects, and therefore on logical grounds likely to be cross-pollinated, are in fact regularly self-pollinated by the insects which visit them (Hinton, 1976).

We may now examine, in outline, a number of different situations. showing how many plants have a capacity to produce both variance and invariance, and discuss a number of experiments which have shown how environmental factors provide a 'trigger', switching the plant from one mode of reproduction to another (Heslop-Harrison, 1964; Asker & Jerling, 1992).

Outbreeding combined with vegetative reproduction

Individuals of many self-incompatible species, producing variable progeny by outbreeding, are capable of considerable lateral spread. Decay of plant connections yields clonal patches of well-adapted genotypes often of considerable size and age, e.g. *Trifolium repens* (Harberd, 1963) and *Lysimachia nummularia* (Dahlgren, 1922).

Outbreeding in association with vivipary

Some species of genera such as *Agrostis*, *Allium*, *Deschampsia*, *Festuca*, *Poa* and *Saxifraga* have the capacity to reproduce not only by the sexual processes but also by vivipary, a condition in which tiny plantlets are produced in the inflorescence instead of, or mixed with, ordinary florets. In normal sexual reproduction, the generation of variation is possible, whilst the viviparous propagules reproduce the genotype of the plant which produces them (Fig. 7.5).

In experiments with *Poa bulbosa* (Youngner, 1960) it was shown that conditions of long daylength and a short cold period followed by high temperatures yielded sexual inflorescences. In contrast, short daylength and low temperatures yielded viviparous inflorescences, and mixed panicles of sexual and viviparous products resulted from long day/low temperature and short day/high temperature combinations.

Outbreeding combined with occasional self-fertilisation

Often it is not clear whether self-incompatibility mechanisms totally prevent selfing in nature. However, there are many cases known where largely self-incompatible species are capable of producing seed on selfing, for example the *Primula veris* stocks studied by Darwin. There is some experimental evidence which suggests that self-fertilisation may occur in 'self-incompatible' species under certain conditions, for example, in material subjected to high temperatures, in situations where pollination of ripe stigmas is long delayed, or at the end of the flowering season (De Nettancourt, 1977). The rigidity, or otherwise, of incompatibility systems under field conditions requires further study.

Outbreeding combined with regular self-fertilisation

In some species (e.g. *Viola*) the spring-formed insect-pollinated flowers allow the possibility of outbreeding, but in the summer, cleistogamous flowers are produced in which self-fertilisation is automatic (Fig. 7.11). Daylength is critical in the regulation of flower type. Borgström (1939) has shown that plants grown under 13–15 hours light per day produce normal flowers, whilst longer days (> 17 hours), typical of early summer, induce the formation of cleistogamous flowers.

Cleistogamy has been regarded as a rather rare phenomenon, but the capacity to produce cleistogamous flowers may be more widespread than hitherto realised (Richards, 1979). Lord (1981) reported that cleistogamous flowers have been discovered in 56 families and 287 species. In some species, such as *Cardamine chenopodifolia*, there are normal flowers above ground, while cleistogamous flowers are produced below ground. The production of normal flowers is 'expensive' and the production of cleistogamous flowers under adverse conditions of shade, temperature and light may be seen as a 'fail-safe' strategy. In *Impatiens capensis* the energetic costs of producing seeds from cleistogamous flowers appears to be only two-thirds that of outcrossed seeds (Waller, 1979). Moreover, seed from normal flowers – produced with the possibility of outcrossing – had better germination, survivorship and competitive success when compared with the obligatorily selfed progenies of cleistogamous flowers (Waller, 1984).

In many plant species, male sterile individuals occur as rare mutants but in some species, perhaps in a greater number than previously acknowledged, populations have a high proportion of female plants together with hermaphrodite individuals. Such species, which are particularly frequent in the Labiatae (Lamiaceae), are referred to as gynodioecious, a term coined by Darwin (1877*a*). Darwin's early studies of *Thymus* (Fig. 7.12) are particularly interesting. Gynodioecy permits the generation of variation in the crossing of hermaphrodite and female plants, whilst allowing the possibility of selfing in hermaphrodites, although it should be noted that some gynodioecious plants are self-incompatible, e.g. *Plantago lanceolata* (Hooglander, Lumaret & Bos, 1993).

While the gynodioecious breeding system is quite common, there is only one confirmed case of androdioecy. The breeding behaviour of *Datisca glomerata*, with populations composed of hermaphrodites and males, has recently been investigated by Fritsch & Rieseberg (1992), using molecular markers.

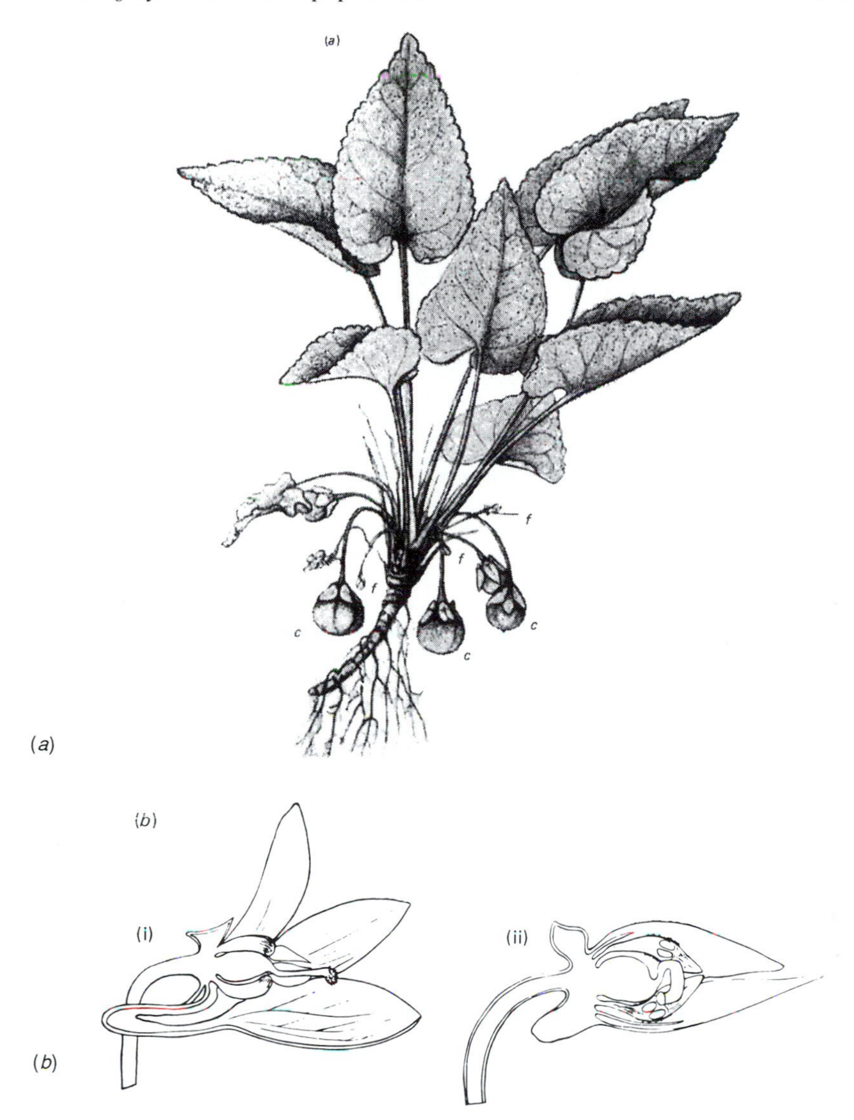

Fig. 7.11. Cleistogamy in *Viola*. (*a*) *Viola hirta*: plant with cleistogamous flowers, *f*, and developing capsules, *c* (× 0.66). (*b*) *Viola riviniana*: open (i) and cleistogamous (ii) flowers in longitudinal section, showing in the latter the crumpled style in contact with the developing anthers (× 2.0). (From McLean & Ivimey-Cook, 1956.)

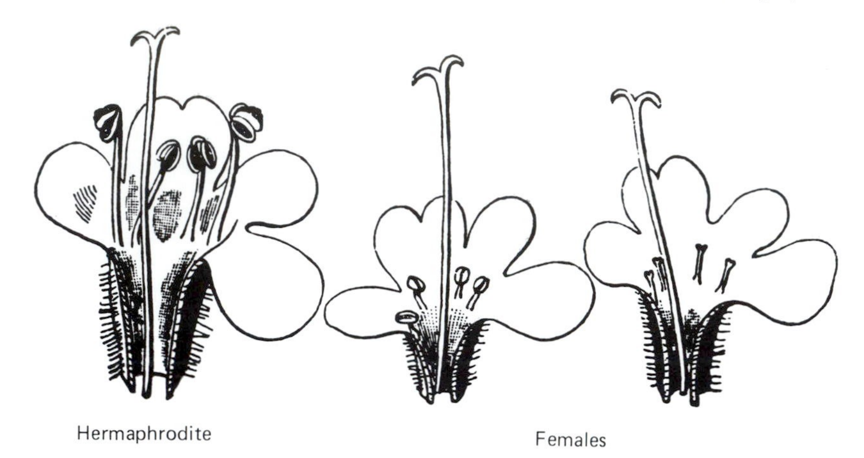

Fig. 7.12. Gynodioecy in *Thymus vulgaris* from sketches sent to Darwin from Mentone. The larger flowered hermaphrodite is figured left and the smaller flowered male sterile 'female' flower is shown right. (From Darwin, 1877*a*.)

Darwin raised plants from purchased seed and discovered that the male sterile plants produced a greater quantity of seed than the hermaphrodites. He suggested that this extra seed production was 'compensation' resulting from the reallocation of resources in the reproduction of the male sterile plants. Studies by Assouad *et al.* (1978) confirm that male sterile plants produce more seed than hermaphrodites in this species. However, seed production is not the only factor to consider. Male sterile plants outcross with the hermaphrodite plants. Hermaphrodite plants, on the other hand, may reproduce by self-fertilisation with the possibility of inbreeding depression (Charlesworth & Charlesworth, 1978).

Dommée, Assouad & Valdeyron (1978) discovered an interesting pattern of distribution of the two flower-types in *Thymus vulgaris* in southern France. In unstable conditions – grasslands and old fields – the percentage of male steriles was high, allowing maximum seed production and outcrossing. In contrast, in more stable areas of rocky outcrops, the frequency of male steriles was much lower and there may be a greater level of autogamy. It is suggested that the frequency of male steriles in populations is subject to natural selection.

The genetics of gynodioecy has proved to be complex. In those species most extensively studied, sex determination has been found to be determined by nuclear and cytoplasmic genes. For a review of present knowledge of the biology and genetics of gynodioecious plants, Lewis & Crowe (1956), Lloyd (1975), Widen (1992) and Maurice *et al.* (1993) may be consulted, together with references cited therein. For a full review of flower size dimorphism in plants see Delph (1996).

Table 7.3. *Estimates of the percentage outcrossing in different self-compatible species. (Data from different authors who used a variety of techniques, from Richards, 1986)*

Species	Estimated outcrossing (%)
Festuca microstachys	0–0.01
Spergula arvensis	0–3
Hordeum vulgare	1–2
H. jubatum	1–3
H. spontaneum	0–10
Avena barbata	1–8
Galeopsis tetrahit	0–16
Avena fatua	1–12
Senecio vulgaris (unrayed)	0–5
Senecio vulgaris (rayed)	12–26
Thlaspi alpestre	5.25
Trifolium hirtum	1–10
Lupinus affinis	0–29
L. bicolor	13–50
L. nanus	0–100
Gilia achilleifolia	15–96
Clarkia temborlensis	8–83
C. exilis	43–89
Lycopersicon pimpinellifolium	0–84
Limnanthes alba	43–97
Plectritis congesta	48–80
Helianthus annuus	60–91
Eucalyptus obliqua	64–84
E. pauciflora	62–84
Vicia faba	70
Mimulus guttatus	87.6
Cheiranthus cheiri	92
Clarkia unguiculata	96
Pinus ponderosa	96

Mixed reproduction: selfing and outcrossing in different proportions

Recent research emphasises the wide range of behaviour and floral morphology in self-compatible species. Thus, in Table 7.3 we can see that the percentage outcrossing in some species is very small, whilst, in others it may be more than 50%. Many factors may influence the rate of outcrossing, including floral structure, the behaviour of pollinators and environmental conditions.

In an important reappraisal, Lloyd (1980) makes the very important point that self-pollination is not a single unvarying process. He distinguishes between models of:

1. 'Prior' self-fertilisation, where the ovules are spontaneously fertilised without any opportunities for crossing.
2. 'Competition' between self and outcrossed products following the visits of pollinating insects.
3. 'Delayed' self-fertilisation where there is an opportunity for cross-fertilisation and, should this fail, selfing will occur.

This analysis raises some very important issues concerning the relative competitive abilities of self- and cross-pollen. Darwin (1876) studied the progenies from self-compatible plants grown close to plants of a different variety or amongst plants of the same variety. It might be expected that, under such conditions, only selfed progenies would be produced. However, on the basis of character combinations, or vigour in the progeny, Darwin identified outcrossed plants in the progenies of seed produced by particular mother plants. This led him to postulate that the outcrossed pollen was 'prepotent' on the stigmas of the plants – in competition with 'self' pollen 'outcrossing' pollen 'won'. Support for this hypothesis has come from a few but not all experimental tests (see Lloyd, 1992 for details). However, given Lloyd's reappraisal of self-compatible plants, it would seem important to re-examine pollen competition very carefully for 'prepotency can cause outcrossing to prevail whenever it is possible and allow self-fertilisation by default whenever outcrossing fails... Altogether, the best-of-both-worlds mechanisms may be the cause of intermediate selfing frequencies in the angiosperms as a whole, but the extent of their occurrence is largely unproven at present' (Lloyd, 1992).

Another mixed breeding system has been described. In some self-compatible species the flowering sequence is 'synchronised so there is little or no overlap between staminate and pistillate phases of an individual plant' (Cruden & Hermann-Parker, 1977). Thus there is, in effect, temporal dioecism and, as different plants are likely to be at different phases, outcrossing is encouraged. However, if flowers are unpollinated selfing can take place.

Facultative and obligatory apomixis

In some genera, apomixis seems to have replaced completely the sexual processes in the great majority of species. In the Lady's Mantles (*Al-*

chemilla), for example, plants of the common northern European species-group to which Linnaeus gave the general name *A. vulgaris* show defective pollen, often degenerating in the tetrad stage, and precociously ripening fruit – sure indications that pollination is not necessary for seed formation. Indeed, with the exception of a very dwarf alpine species *A. pentaphyllea* and a very few alpine taxa belonging to another subsection of the genus, so far as is known, all *Alchemilla* species in Europe are apomictic. It is interesting that Robert Buser, the Swiss expert on *Alchemilla* who achieved an unrivalled knowledge of plants in the field, herbarium and cultivation, had rightly suspected from field evidence that certain puzzling intermediate populations in the Alps were hybrids of *A. pentaphyllea* and other species before anything was known of their genetical complexity. Obligatory apomixis, as is shown by all the 'vulgaris' Alchemillas, is accompanied by a relatively straightforward pattern of variation; the collective Linnaean species *A. vulgaris* and *A. alpina* consist (in Europe) of some 300 taxonomically distinguishable microspecies, many of which are wide-ranging and no more difficult to identify than many sexual species in other genera. It is the number of microspecies involved and the relative complexity of the detailed morphological differences between them, which make such critical groups the concern of so few botanists (Fig. 7.13).

Studies of apomixis are increasingly confirming an extremely important conclusion; total or obligatory apomixis is much less common than partial or facultative apomixis. Indeed, since we can never be certain that sexual reproduction is quite ruled out, even in cases such as *Alchemilla*, it may be that strict obligate agamospermy does not occur. In facultatively apomictic plants, amongst them the taxonomically difficult genera of *Rubus* and *Potentilla* in the Rosaceae and *Pilosella* in the Compositae (Asteraceae), apomictic embryos develop automatically without sexual fusion, but in addition some egg-cells are produced with the reduced number of chromosomes. Such eggs may be fertilised giving rise to sexually produced embryos. It is predicted that obligate apomixis may yield little variation. In contrast, facultative apomixis combines a capacity to reproduce successful genotypes unchanged, with a mechanism allowing the generation of variation on an occasional or regular basis.

Facultative apomixis is well illustrated by the adventitious embryony in the genus *Citrus*. According to Asker & Jerling (1992) 'usually, pollination is followed by the double fertilisation of a reduced sexual embryo sac. The embryo and endosperm start to develop, and the stimulus of embryo development ... results in the growth of further adventive [apomictic] embryos in the nucellus... Typically one of the apomictic embryos invades the embryo sac, out-competing the other apomictic and sexual embryos, and utilizes the sexually produced endosperm. In other cases, the sexual embryo and one (or more) of the apomictic embryos coexist, sharing the same endosperm.' Thus sexual and apomictic progenies can be produced.

A good example of how occasional sexual crossing might yield a complex pattern is provided by the British representatives of the Series Aureae of *Potentilla*, which are, so far as is known, all pseudogamous apomicts. Most British plants can be classified as either *Potentilla neumanniana* (*P. tabernaemontani*), a rhizomatous, mat-forming perennial of chalk and limestone mainly in the lowlands, or *P. crantzii*, a non-rhizomatous

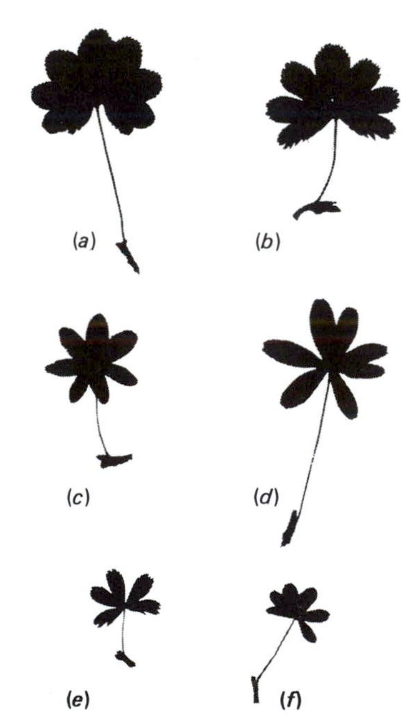

Fig. 7.13. Leaves of apomictic *Alchemilla* species, showing a range of form from the simple, lobed 'vulgaris' shape, (*a*), to the compound 'alpina' shape, (*f*). (*a*) *A. fulgens* (Pyrenees). (*b*) *A. faeroensis* (Faeroes, East Iceland). (*c*) *A. conjuncta* (Jura, West Alps). (*d*) *A. plicatula* (Pyrenees, Alps, Balkan Mountains). (*e*) *A. subsericea* (Alps). (*f*) *A. alpina* (Arctic, North European Mountains, Alps). *A. faeroensis*, with about 220 chromosomes, is almost certainly of ancient allopolyploid origin from a 'vulgaris' and an 'alpina' species. Nearly all traces of sexual reproduction are now lost in the present-day representatives of these groups ($\times$ 1/3).

perennial with an unbranched woody stock, typically found on calcareous cliffs in upland or mountain areas in Britain. In some parts of upland Britain, especially in northern England, however, puzzling intermediate plants occur, which obscure the otherwise fairly clear distinction between the two species. These plants mostly have higher chromosome numbers than normal specimens of either species. During a detailed study, Smith (1963*a*, *b*, 1971) showed that, if large numbers of progeny were raised from 'crosses', occasional aberrant individuals could be detected because they differed from the normal offspring which resembled closely the female parent. One such individual, raised from the 'cross' between a *P. neumanniana* plant with $2n = 49$ from Fleam Dyke, Cambridgeshire, as a female parent, and a *P. crantzii* plant with $2n = 42$ from Ben Lawers, Scotland, as pollen parent, was found to have 70 chromosomes. There can be little doubt that, in this case, an unreduced egg-cell with 49 chromosomes had been fertilised by a normal reduced pollen grain with 21 chromosomes, to give a zygote, and eventually a mature sporophyte plant, with $2n = 70$. A hybrid apomict had been synthesised. Such a plant can reproduce both vegetatively (to a limited extent) and agamospermously, and might have established a more or less uniform population in nature.

Moving to a more general model, sexual reproduction may not be rare in facultative apomictic plants, and hybridisation between different variants could take place many times, giving populations differing subtly from each other according to the exact genetic constitution of their parents. Such events are particularly likely in pseudogamous species, where the pollen is largely normal. Furthermore, pollen from an obligate apomict may take part in sexual crosses in which the female parent is a sexual or facultatively apomictic plant (Asker & Jerling, 1992). Even the very limited sexuality of *Alchemilla* may for this reason be of far greater significance than we think.

Environmental control of facultative apomixis

It has been found in experimental studies of *Dichanthium aristatum*, for example, that environmental factors are important in determining the pattern of facultative apomixis. Thus, daylength is important in determining the balance between apomictic and sexual reproduction. Under continuous short days, up to 79% of embryos produced were apomictic; under long days, after floral induction in short days, only about 47% of embryos were aposporously produced (Knox & Heslop-Harrison, 1963).

The use of molecular markers in the study of the reproductive behaviour of apomictic plants

Studies of population variation in reproductive behaviour have been revolutionised by the used of genetic markers, and provide the potential for the resolution of many questions about apomixis. The study of apomictic plants by modern methods is in its infancy, but published accounts shed important light on a number of issues.

1. *Are apomictic plants less variable than those reproducing by sexual means*? Antonius & Nybom (1994) studied restriction fragment length polymorphism in the genus *Rubus*, using the M13 repeat sequence as a hybridisation probe. They discovered, as might be predicted, that every one of 24 different plants of the sexual species *R. idaeus* had a different genetic profile. In contrast, all the samples of an apomictic species *R. nessensis* were completely identical for the part of the genome examined.

2. *Do apomictic plants have a higher heterozygosity than their sexual relatives*? In a study of isozyme markers in sexual and agamospermous *Eupatorium altissimum* by Yahara *et al.* (1991), low heterozygosity was discovered in both sexual and agamospermic races. This approach provides the means to make critical studies of population variation in other apomictic species.

3. *Are populations of apomictic plants invariable*? The variation in a population of *Arabis holboellii* var. *retrofracta* was studied using isozymes. Progeny testing revealed no variation within the progenies of different seed-parents, but there was a good deal of variation between seed-parents (Roy & Rieseberg, 1989).

Studies of DNA profiles indicated that there were at least 15 different genotypes in a sample of 20 plants of the sexual species *Rubus occidentalis*. Such variation in a sexually reproducing species is not unexpected. However, it is of considerable interest that five different genotypes were detected in a sample of presumed apomictic *R. pensilvanicus* (Nybom & Schaal, 1990). Further studies are needed to determine the degree of variability in this facultatively apomictic genus.

4. *How far do progenies of obligate apomictic plants vary*? Studying restriction site variation in DNA, King & Schaal (1990) surveyed 714 offspring within 31 lineages of obligately apomictic *Taraxacum officinale*.

While the progeny of many of the lines proved to be invariable, in two lines 'mutants' were detected. 'Mutant, aberrant and deviant individuals' have often been identified in studies of apomictic plants, but this pioneering study provides critical identification of the mutational change.

Mogie (1985) studying isozymes of species of the section Hamata of *Taraxacum*, discovered both inter- and intraspecific variation. Using electrophoretic markers, Hughes & Richards (1988) compared populations of species of the same genus with different breeding systems. Sexual outbreeders were genetically highly diverse, polymorphic and heterozygous. Inbreeders were more genetically uniform and homozygous, while apomictic taxa proved to be polymorphic, highly heterozygous with variable populations.

5. *What variation is detectable in facultatively apomictic plants*? In the past such analyses were often based on laborious cytological and embryological studies. For instance, four different types of behaviour were detected in studies of *Agropyron scabrum* (Hair, 1956):

1. Plants completely and normally sexual.
2. Plants facultatively apomictic.
3. Plants predominantly apomictic, meiosis suppressed in 'female', not in 'male'.
4. Plants obligatorily apomictic, suppression of meiosis in 'female' and 'male'.

Reproductive behaviour is more quickly and decisively determined by studies using appropriate molecular markers. For instance, Bayer, Ritland & Purdy (1990) analysed the behaviour of populations of *Antennaria media*. On the basis of the patterns of segregation of four polymorphic enzyme loci studied by electrophoresis, the population was found to consist of sexual, partially apomictic and totally apomictic plants. Another example is provided by the studies of isozymes in *Allium tuberosum* by Kojima, Nagato & Hinata (1991). They discovered the precise mixture of sexual reproduction (10%) and apomixis (90%) in progenies.

We predict that major advances in our understanding of apomixis will come with the use of molecular tools, but it is important that the methods are used in conjunction with embryological and cytological investigations. Thus, it should, for example, be possible to study the genetic composition of the endosperm and increase our understanding of the extent and nature of genetic variation within and between taxa.

Evolution of breeding systems

Having discussed different breeding systems found in higher plants, it is appropriate at this point to consider the evolution of the different systems. At the outset to is important to stress that we do not know the breeding system of the early flowering plants. However, a suggestion has been made by Whitehouse (1950), based on the widespread occurrence of self-incompatibility mechanisms in many flowering plant families. He proposed that this condition is ancestral and postulated that the significance of the closed carpel 'lies in the protection of the ovules, not from desiccation or the attack of animals, but from fertilisation by the individual's own pollen, without appreciably restricting cross-fertilisation'. De Nettancourt (1977) supports Whitehouse's view, and discusses the implications of the corollary proposition, namely that self-compatibility is a secondary, derived condition in modern angiosperms. He agrees that the relative advantages of outbreeding and inbreeding, as we have just seen, are likely to be very different under different environmental conditions. It is possible to see both outbreeding and inbreeding – and indeed apomixis – as potentially adaptive under different conditions. This would imply that the initial diversification through the Cretaceous period was possible because of the efficiency of the outbreeding mechanism, but in later periods other factors were selectively more important, which favoured other breeding systems.

A great deal has been published on the evolution of flowering plant breeding systems (see, for example, Barrett & Harder, 1996). Here we note a number of the hypotheses concerning major trends. Molecular methods of investigating the genetic relationships of groups may provide the means of testing some of these phylogenetic hypotheses.

First, many botanists consider that gametophytic self-incompatibility (multi-allelic) is ancestral and that sporophytic systems are derived from it. For instance, Richards (1986) considers that 'sporophytic systems are probably secondary in origin, having evolved polyphyletically from self-compatible, partially inbred populations, encouraged by renewed selection pressures for outbreeding'. With regard to the underlying control of self-incompatibility, Beach & Kess (1980) and Muenchow (1982) take the view that homomorphic and heteromorphic sporophytic systems have essentially the same mechanism. However, Gibbs (1986) considers that such an assumption is premature, and that molecular studies of the *S* locus will be needed to establish the relationship between the two systems. So far molecular studies have attempted to identify and characterise the gene products involved in the self-incompatibility reaction, rather than consider evolutionary issues. In providing an important review of recent studies,

Sims (1993) writes that the 'self-incompatibility response is likely to be far more complex than suggested by historical models'.

With regard to the evolution of heterostyly, it seems possible that the ancestral stocks from which they arose were homomorphic (Charlesworth & Charlesworth, 1979) and that heterostyly arose as a mechanism favouring precision of pollination, thus increasing the chance of a stigma receiving compatible pollen (Yeo, 1975; Barrettt, 1992). Lloyd & Webb (1992), however, have examined the case for considering that at least some of the features associated with heterostyly could have evolved prior to self-incompatibility.

A second major evolutionary trend has been identified by many botanists. Self-compatibility has evolved from obligate self-incompatibility. There have been many reviews of the subject (see Wyatt, 1988; Barrett, 1989 *a*, *b*). Evidence suggests that such a change has occurred independently many times in plant evolution. For instance, diversification in the genus *Phlox* in relation to pollen vectors has been studied by Grant & Grant (1965) (Fig. 7.14). They conclude that the group co-evolved with different pollinators – bats, bees, hawk moths, humming-birds, etc. Six lines of specialised radiation were postulated, each with its independently evolved small-flowered self-fertilising autogamous species. Self-compatible species may also show other character changes (Ornduff, 1969). Some 21 such changes have been detected in detailed studies of a number of genera (Table 7.4).

However, self-compatible species are not always small-flowered. Adaptive radiation – in relation to such vectors as humming-bird, butterfly and moth – has been studied in the Orchidaceae (Pilj & Dodson, 1966) and, again, there are many instances of reversions to autogamy, but the flowers may not be reduced in size. It is now clear that self-compatible species are not all evolving towards complete autogamy. As we have seen above, many 'mixed' breeding systems occur (see Table 7.4). Such systems can provide the advantages of outcrossing, while allowing the reproductive assurance of self-fertilisation.

Concerning the steps in the development of self-compatible variants, a number of intensive studies of distylous plants have been very revealing, e.g. *Primula* (Ernst, 1955), *Armeria* (Baker, 1966) and the *Turnera ulmifolia* complex (Barrett, 1985). These have shown how crossing over in the supergene which controls the distylous breeding system has given rise to self-fertile homostyle plants. The evidence that self-incompatible diploids have sometimes given rise to self-compatible polyploid derivatives will be considered in Chapter 12.

A second example of the evolution of self-compatibility is provided by the breakdown of tristyly in *Eichhornia* (Barrett, 1980*a*, *b*, 1985, 1989*b*). For

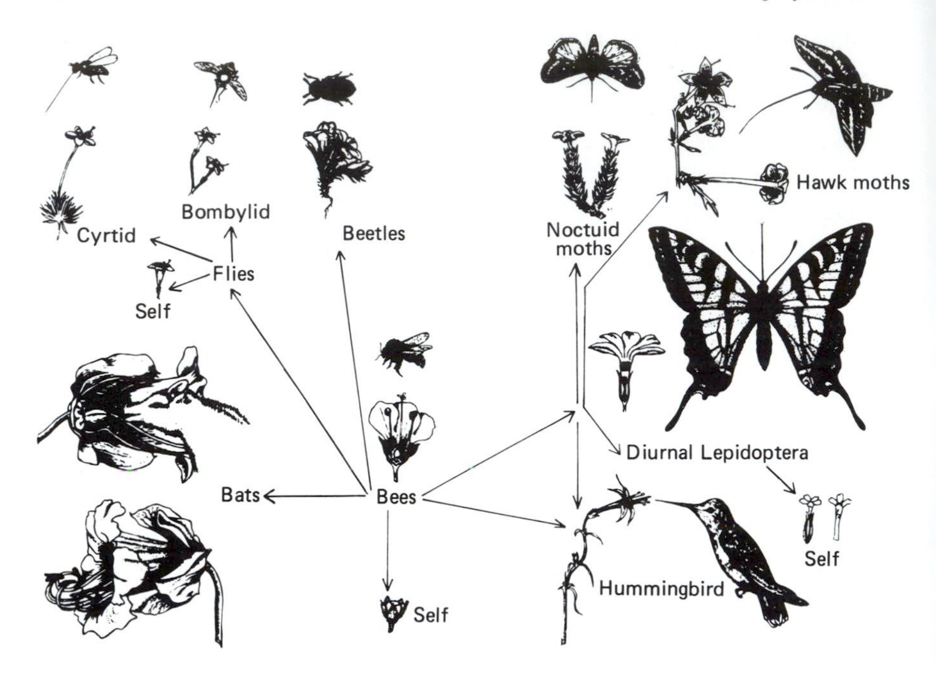

Fig. 7.14. It seems probable that, as a consequence of natural selection, the ancestors of a single group may through evolutionary divergence become adapted to a range of environments. A splendid example of presumed adaptive radiation for pollination by different pollen vectors is illustrated by present-day members of the Phlox family (Polemoniaceae). The following species of pollinators and flowers are illustrated as representative of each group: bees, *Polemonium reptans* and *Bombus americanorum*; bats, *Cobaea scandens* and *Leptonycteris nivalis*; cyrtid flies, *Linanthus androsaceus croceus* and *Eulonchus smaragdinus*; bombylid flies, *Gilia tenuiflora* and *Bombylius lancifer*; beetles, *Linanthus parryae* and *Trichochrous* sp. (Melyridae); noctuid moths, *Phlox caespitosa* and *Euxoa messoria*; hawk moths (Sphingidae), *Ipomopsis tenuituba* and *Celerio lineata*; diurnal Lepidoptera, *Leptodactylon californicum* and *Papilio philenor*; birds, *Ipomopsis aggregata* and *Stellula calliope*; self (autogamous), *Polemonium micranthum* (bottom), *Gilia splendens*, desert form (left), *Phlox gracilis* (lower right). (From Stebbins, 1974, redrawn from Grant & Grant, 1965.)

instance, in Brazil and the Caribbean large populations of *E. paniculata* which is a tristylous species are self-incompatible. In contrast, small populations are often dimorphic (some plants show self-compatibility) or monomorphic (all plants fully self-compatible). It seems possible that such self-compatible variants might be at a selective advantage in areas of drought and human disturbance. However, the loss of the different morphs could be the result of chance effects following reduction in population size or resulting from random dispersal.

Table 7.4. *List of 21 morphological and phenological character changes that are often associated with the evolution of autogamy (Modified from Ornduff, 1969, by Wyatt, 1988)*

Outcrossing progenitors	Autogamous derivatives
Flowers many	Flowers fewer
Pedicels or peduncles long	Pedicels or peduncles shorter
Sepals large	Sepals smaller
Corollas rotate	Corollas funnelform, cylindric, or closed
Petals large	Petals smaller
Petals emarginate	Petals less emarginate
Floral colour pattern contrasting	Floral colour pattern less contrasting
Nectaries present	Nectaries reduced or absent
Flowers scented	Flower scentless
Nectar guides present	Nectar guides absent
Anthers long	Anthers shorter
Anthers extrorse	Anthers introrse
Anthers distant from stigma	Anthers adjacent to stigma
Pollen grains many	Pollen grains fewer
Pollen presented	Pollen not presented
Pistil long	Pistil shorter
Stamens longer or shorter than pistil	Stamens equal to pistil
Style exserted	Style included
Stigmatic area well defined, pubescent	Stigmatic area poorly defined, less pubescent
Stigma receptivity and anther dehiscence asynchronous	Stigma receptivity and anther dehiscence synchronous
Many ovules per flower	Fewer ovules per flower

A third set of evolutionary trends links the evolution of self-compatibility and dioecy. Baker (1955, 1967) has suggested that, with self-compatible individuals, a single propagule is sufficient to start a sexually reproducing colony on an island or in a new habitat. In contrast, a single individual of a self-incompatible species would be unable to found a new colony reproducing by seed. There is support for this theory (elevated to a law by Stebbins, 1957): for example, in a study of plants of the Galapagos Islands, out of the 52 species studied, all those for which conclusive evidence was available proved to be self-compatible (McMullen, 1987). Support for Baker's view has also come from studies of *Nigella* in the eastern Mediterranean (Strid, 1970) and *Papaver* (Fig. 7.15) on Scottish islands (Richards, 1986). However, as we noted at the beginning of this chapter, dioecious plants are disproportionately common on certain islands, e.g. New Zealand and Hawaii. Maybe, long-range dispersal brought both sexes to the islands in two or more

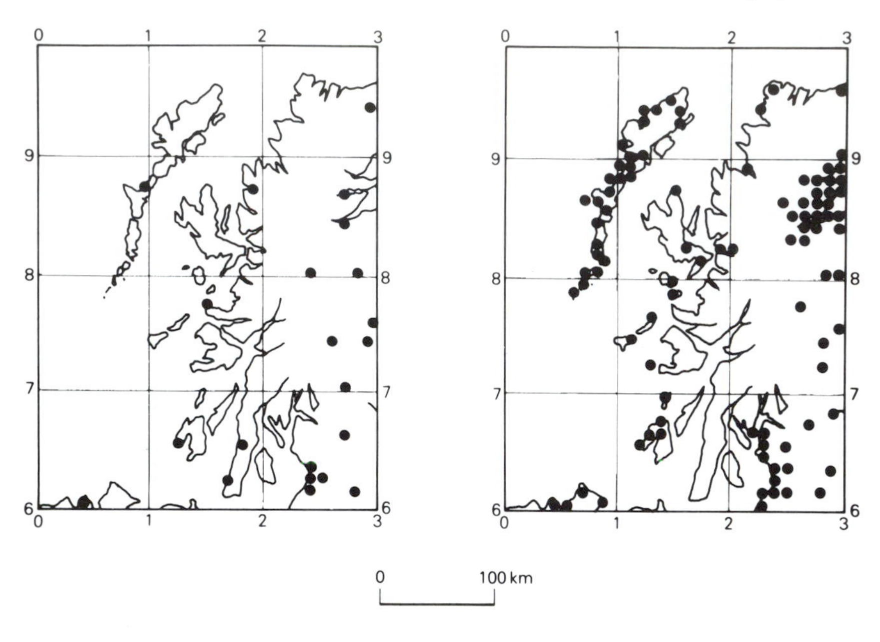

Fig. 7.15. Distribution in western Scotland of the self-incompatible Poppy, *Papaver rhoeas* (left), and the self-compatible *P. dubium* (right). *P. dubium* has colonised many more of the Scottish Isles. (From Richards, 1986. Maps reproduced with permission from *Atlas of the British flora*, Perring & Walters, 1976.)

separate colonising events or perhaps on a single occasion a many seeded propagule arrived. However, another possibility must be considered. Self-compatible colonising populations may suffer inbreeding depression and it has been suggested that selection may favour mutations in sex expression to yield, perhaps via gynodioecy, a fully dioecious system with its obligate outbreeding system. Cox (1988) and Richards (1986) may be consulted for a full discussion of the possibilities. This is not the only possible way that dioecy might arise; Ornduff (1966) provides evidence that, in *Nymphoides*, dioecy arose directly from heterostyly. (See also Lloyd, 1979*b*, for an important review of this area.)

Finally, other major trends may be briefly considered. After a full review of all the recent and historic evidence, Asker & Jerling (1992) consider the commonly held belief that sexual relatives of most agamospermous plants are self-incompatible and, therefore, apomictic variants are likely to have arisen from such plants. In many genera this appears to be the case, for instance in *Limonium*, the self-incompatible sexual relatives are heterostylous (Ingrouille & Stace, 1985). Berry, Tobe & Gómez (1991) suggest that in the genus *Erythroxylum* agamospermy is derived from distyly. However, the tendency for near relatives to be self-incompatible is not an exclusive trend. In *Antennaria media* (*A. alpina* group), apomixis is found in a dioecious plant, the female being the only variant present over much of its range (Bayer, Ritland & Purdy, 1990). Furthermore, the relatives of apomictic *Citrus*, *Potentilla* and *Aphanes* are self-compatible. Finally, we note the very important observation, considered in detail in Chapter 12, that apomixis is found in the polyploid species of a group (Stebbins, 1950).

Concluding remarks

To conclude our brief survey of the breeding systems in higher plants, we stress two important points.

First, a major theme of this chapter is quite rightly the way 'reproductive features regulate the degree of inbreeding' (Charnov, 1988). For completeness, however, it is essential to introduce, briefly, another perspective on breeding behaviour and its evolution. Charnov stresses that both 'pollen and ovules contribute equal numbers of chromosomal genes to seeds'. Therefore, the 'male' (pollen) side of reproductive biology must be considered. It may be postulated that natural selection acts on both 'male' and 'female' reproduction. Does success in the 'female' mode reduce success in the 'male'? Does evolution by natural selection result in compromises in 'male' and 'female' success? How is this translated into use of resources in the flower in producing various structures in the diverse groups of higher plants? All these fascinating questions are addressed in Lovett Doust & Lovett Doust (1988). A key question is whether it is possible to measure 'female' and 'male' reproductive investment and success? At present many important questions have not been properly examined experimentally, as there are problems of deciding which parts of a hermaphrodite flower are acting as part of 'male' function and which are properly 'female'.

A second major point concerns the many hypotheses put forward regarding the evolution of different breeding systems. We stress that such evolution should be viewed in a population context. In studying both short- and long-term changes in breeding behaviour, models must consider the relative fitness of mutants. In the past, morphological and genetical studies, crossing experiments, etc. have provided many insights. However, with the development of molecular tools, we face the exciting prospect that a range of hypotheses may be more critically tested, leading to an increased understanding of evolutionary changes in breeding systems in higher plants, including the the evolution and breakdown of heterostyly (Barrett, 1992).

Our discussions of variation have brought us to the point where we have explored the *potential* variation generated by the variety of plant breeding systems. Plants grow in populations, and we may now turn our attention to a review of our knowledge of variation between and within populations. A discussion of the taxonomic treatment of different groups exhibiting diverse variation patterns is deferred until some of the evidence about patterns of variation and processes of change has been examined.

8

Infraspecific variation and the ecotype concept

We saw in Chapter 7 how different breeding systems can be expected to produce different patterns of variation. If we are to understand the variation patterns actually found in nature and the processes which give rise to these patterns, we must discover how the potential variation in seeds relates to the variation of reproductively mature plants. Historically, the first advances in this field were made by means of comparisons between plants belonging to the same species but from different populations. Taxonomists, biometricians and, later, geneticists became interested in genetic variation in the wild and many of their studies converged at one point, namely the controversy over the 'reality' of the infraspecific groups, which could be distinguished in nature, whether they were the subspecies or varieties of the taxonomist or the 'local races' of the biometrician. A new look at this old question was provided by the famous research of Turesson published in the early 1920s.

Turesson's pioneer studies and other experiments

At the time of Turesson's experiments, the question of the reality of 'local races' was combined with another controversial issue, namely how much of the observed variation in natural populations was the result of the direct modification of plants subjected to severe environmental stresses? By the end of the nineteenth century, many botanists reasoned that distinctive infraspecific variants were merely 'habitat modifications'. Turesson, however, pointed out that in all previous cases known to him, only a partial test of the 'habitat modification' hypothesis had been carried out. For example, he considered the studies of *Lathyrus japonicus* undertaken by Schmidt (1899). Baltic populations of this plant have dorsiventral leaves, whilst on the North Sea coast of Denmark the plant has isolateral leaves. Schmidt showed, by experiment, that watering the Baltic variant with sodium chloride solutions induced a leaf structure typical of Danish plants. Given that the North Sea has a higher percentage content of salt than Baltic waters, Schmidt deduced that the leaf structure of the plants on the North Sea coast of Denmark was merely a habitat modification.

The logic of this type of deduction did not satisfy Turesson. His approach to the problem was to grow samples of several variants of a species in a standard garden, to see if 'distinctiveness' was retained or lost. He collected living plants (and in certain cases seeds) of many common species from a variety of natural habitats in southern Sweden and grew them in experimental gardens first at Malmö (1916–18) and subsequently at the Institute of Genetics at Åkarp. In this way he studied, for example, shade variants, dwarf lowland plants from coastal habitats and succulent variants, in most cases growing these plants alongside collections of the same species collected from ordinary inland habitats (Turesson, 1922*a*, *b*).

In some cases the distinctness of the variants was lost in cultivation in an inland garden, but usually the distinctive plants originating from extreme habitats retained their characteristics in cultivation even in the absence of shading, salting, etc. These observations were clearly at odds with the notion that extreme variants were nothing more than habitat modifications and the persistence of distinct variants under standard conditions suggested to Turesson that the variation had a genetic basis.

Many of Turesson's early experiments were carried out on the Composite *Hieracium umbellatum*. This plant is common in southern Sweden where its principal habitats – woodland, sandy fields, dunes and cliff tops – may all be found. In each of these habitats a distinctive plant was discovered in the field. By careful sampling and cultivation, Turesson found that, with few exceptions (for example, certain prostrate plants from sandy fields) distinctive variants retained their characteristics in cultivation. The results of these experiments were consistent with those obtained in studies of other species and again Turesson considered that patterns of residual difference had a genetic basis.

H. umbellatum is a common plant in southern Sweden, and Turesson was able to collect many samples from each habitat type. A close study of his extensive collections after a number of years of cultivation suggested to him the possibility that habitat-correlated patterns of genetic variation were present, that is to say, in a particular habitat of *H. umbellatum* a certain race of characteristic morphology was invariably present. In the appropriate habitat, there was to be found a dune race, a woodland race, etc. Turesson called these local races 'ecotypes' and described five, as follows (note that in these descriptions he considered anatomical and physiological traits (e.g. flowering times) as well as morphological features):

1. *An ecotype from shifting dunes.*
   Narrow leaves and slender, less erect, sometimes more or less prostrate stems. Marked power of shoot regeneration in autumn. Leaves tough and thick with three to four layers of palisade cells. Fruiting in early September.
2. *An ecotype from sandy fields and stationary dunes.*
   As 1, but power of shoot regeneration in autumn weak or lacking. Extremely prostrate in growth habit.
3. *An ecotype from western sea cliffs.*
   Broad leaves and more or less prostrate stems. Growth form contracted and bushy. Cells of leaves more or less distended. Fruiting late September to early October.
4. *An ecotype from eastern sea cliffs.*
   As 3, but plants tall and almost as erect as in 5.
5. *An ecotype from open woodland.*
   Stout, erect plants with lanceolate leaves of intermediate width. Leaves thinner with two or, at most, three palisade layers. Fruiting in September.

Turesson notes that additional ecotypes might be discovered in future studies.

H. umbellatum is a member of a genus famed for its apomictic reproduction. In considering Turesson's results it seems essential, therefore, to take into account the breeding behaviour of the plant. In a partial examination of the breeding system of his material, Turesson performed castration experiments, removing the upper half of unopened flower-heads with a razor. No fruits developed. This evidence supports the view that reproduction is sexual and not obligately apomictic. Plants of *H. umbellatum* proved in fact to be self-incompatible, and artificial crosses between plants of the dune ecotype and between plants of the cliff ecotype produced progenies in which the ecotypic characteristics of each were perpetuated, confirming the genetic basis of the discovered differences. Lövkvist (1962) has re-sampled at many of Turesson's *H. umbellatum* sites, and found broadly similar patterns of variation in cultivation trials. He also re-examined the breeding system of southern Swedish material of *H. umbellatum* and found no evidence of apomixis. However, apomixis *has* been reported in this species (e.g. Bergman, 1935, 1941, and references cited therein) and may influence patterns of variation elsewhere.

Considering the origin of ecotypes, Turesson made two important deductions. He concluded, first, that the finding of widespread habitat-correlated genetic variation does not support the view that the variation patterns are largely governed by chance; rather the evidence suggests that natural selection operates in natural populations, well-adapted genotypes being selected in each habitat. This idea is expressed many times in Turesson's writings, for example, he says (1925): 'Ecotypes ... do not originate through sporadic variation preserved by chance isolation; they are, on the contrary, to be considered as products arising through the sorting and controlling effect of habitat factors upon the heterogeneous species-population'. Turesson further concluded that a close study of the variation within and between ecotypes of *H. umbellatum* revealed patterns of leaf morphology which suggested a 'local' origin for coastal ecotypes from the widespread inland populations. It was possible that an appropriate ecotype could be produced many times, that is to say polytopically, and it was not necessary to postulate the invasion of Sweden by fully formed standard ecotypes after the last glaciation.

In a series of long papers published from 1922 onwards, Turesson eventually described ecotypes in more than 50 common European species. His first papers were about the plants of southern Sweden, but later (1925, 1930) he experimented with material collected from distant localities in all parts of Europe and also showed physiological differences between some of his stocks (1927*a*, *b*). Analysis of the behaviour of his extensive collections in cultivation enabled him eventually to distinguish two kinds of ecotypes, namely edaphic and climatic ecotypes, where the most important environmental effects were soil type (as in the case of *H. umbellatum* in southern Sweden) and the climatic influences, respectively.

As early as the beginning of the eighteenth century, there was a considerable amount of observational evidence that common species did not flower at the same time in different localities. For example, Linnaeus (1737) noted the different flowering times of Marsh Marigold (*Caltha palustris*) (March in the Netherlands, April to May in different parts of Sweden, June in Lapland). Quetelet (1846), having studied the dates of first flowering of Lilac (*Syringa vulgaris*) in different parts of Europe, came to the conclusion that there was a retardation of 34 days for each advance of 10° northwards in latitude. He also compared flowering at different altitudes above sea-level, and discovered a retardation of 5 days for every 100 m increase in elevation. The important environmental factor controlling flowering was thought to be temperature. Turesson, studying the behaviour in cultivation of a large number of spring-flowering species, clearly demonstrated the importance of persisting genetic differences between plants originating from different climatic regions. Southern plants of such species flowered earlier in Turesson's experimental garden than plants of the same species collected from northern latitudes. He suggested that this group of plants is adapted to flower in the period immediately preceding the leafing of trees, a phenomenon which occurs earlier in the year in southern latitudes than in northern Europe.

In the botanical literature of the nineteenth century, there are scattered reports that alpine plants flower earlier than lowland ones when both are cultivated in lowland gardens. Turesson's extensive experiments with species such as *Campanula rotundifolia* (Table 8.1) and *Geum rivale* enabled him to demonstrate that alpine ecotypes were smaller and retained their early flowering habit in cultivation. He also carried out research upon summer-flowering plants, showing that northern ecotypes were early flowering and of moderate height, while southern plants were late-flowering and tall. Western Europe was characterised by late-flowering plants of low growth; from eastern Europe, on the other hand, came taller early-flowering ecotypes.

Turesson's contribution to our understanding of the patterns of variation within species is of very great importance; he demonstrated clearly the widespread occurrence of infraspecific habitat-correlated genetic variation. Adaptation to the environment was sometimes by plastic responses, but more frequently it had a genetic basis. Such studies were grouped together under the name of 'genecology' and the work was the model for many studies by other botanists. The work of Stapledon (1928) is of special interest. Using the common pasture grass *Dactylis glomerata*, he studied the influence of hay cutting and animal grazing, and described a third class of ecotype, namely the 'biotic ecotype'. His work is summarised in Table 8.2.

Scandinavian botanists have made many notable contributions to genecology and it is appropriate at this point to give an example of the important experiments of Bøcher. He used the Turessonian technique of cultivation in a standard garden to examine the variation and flowering behaviour of collections of many European plants, and carried the analysis of variation into an important new area, namely the study of the timing of flowering in relation to the life history of the plant. For example, he discovered in cultivation experiments with *Prunella vulgaris* (1949) that there were two main growth types in Europe, namely plants with a short vegetative phase, flowering in their first year, and plants with a longer vegetative phase, flowering in their second year. This latter group was further subdivided into plants which were short-lived and perennial types. The distribution of the two main types – first- and second-year flowerers –

Table 8.1. *Geographic variation in* Campanula rotundifolia

(a) Results of transplant experiments from Turesson (1925) (Means of five measurements given)

		Field no.	Transplanted from	Length of stems (mm)	Width of middle-stem-leaves (mm)	Number of flowers on stems	Length of corolla (mm)	Width of corolla in the middle (mm)	Width of corolla at mouth (mm)	Length of corolla lobes (mm)	Length of calyx lobes (mm)	Power of regeneration of basal rosette-leaves	Year of collec-tion	No. of plants
Norway		99	Vitemölla	547.75	2.18	23.25	18.63	16.50	22.45	7.33	6.33	none-weak	1920	8
and		206	Åhus	650.54	2.16	27.49	19.93	16.45	22.82	7.65	7.29	none-weak	1922	13
Sweden		270	Ulriksdal	334.30	1.86	20.33	17.13	14.99	20.2	7.02	5.63	none-weak	1921	14
	M	298	Åre	308.43	2.97	11.5	22.12	21.0	25.91	9.54	5.88	mostly strong	1921	17
		349	Bergen	378.67	2.73	9.19	20.56	20.53	25.67	8.41	6.36	weak-strong	1922	7
		240	Trondhjem	336	2.03	15.97	21.0	18.34	25.06	8.39	5.80	weak-strong	1922	14
Central	M	19–25	Abisko (seeds)	250.10	1.99	13.97	24.47	20.48	27.68	9.32	7.90	strong	1921	seeds
Europe		770	Freiburg	278.56	2.12	19.86	20.44	18.89	24.54	8.5	6.56	none-weak	1923	16
	M	796	Feldberg	224.66	4.29	6.88	23.45	21.82	25.32	8.76	7.89	strong	1923	14

(b) Progeny trial, from Turesson (1930)

Field no.		Source	No. of plants	Height (cm)			Earliness of flowering[a]		
				Mean	σ	m$\pm$	Mean	σ	m$\pm$
770		Freiburg	20	68.9	5.89	1.32	1.60	0.35	0.29
796	M	Feldberg	20	29.5	2.41	0.54	5.00	0.00	0.00
270		Ulriksdal	20	47.1	5.47	1.22	2.80	0.44	0.10
298	M	Åre	20	33.4	3.75	0.84	5.00	0.00	0.00

[a]note that a large mean corresponds to earlier flowering

M = montane localities

Table 8.2. *Biotic ecotypes in* Dactylis glomerata. *Stapledon (1928) discovered that grassland use determined the type of* Dactylis *present in a particular area*

		Per cent growth type				Per cent over 100 cm	Per cent flowering behaviour			
		Hay	'Cup'	Tussock	Pasture		Early 1	2	3	Late 4
Commercial hay stocks	A	59	36	2	3	78	40	50	9	1
	B	66	31	1	2	78	61	32	6	1
Old pastures		15	23	6	56	15	11	35	38	16
Hedgerows and thickets		26	35	25	14	31	17	35	34	14

Hay types with their taller early-flowering plants were distinct from the shorter, later-flowering plants characteristic of grazed pasture. Pasture types had many more tillers than hay types and a smaller percentage of tillers produced inflorescences. Plants from hedgerows had a wide range of variants. Even though this experiment did not reveal a discontinuous pattern of variation, Stapledon was content to interpret his results in terms of 'biotic ecotypes'. (See Warwick & Briggs, 1978*a*, *b*, for a partial review of recent work on 'hay' and 'pasture' ecotypes.)

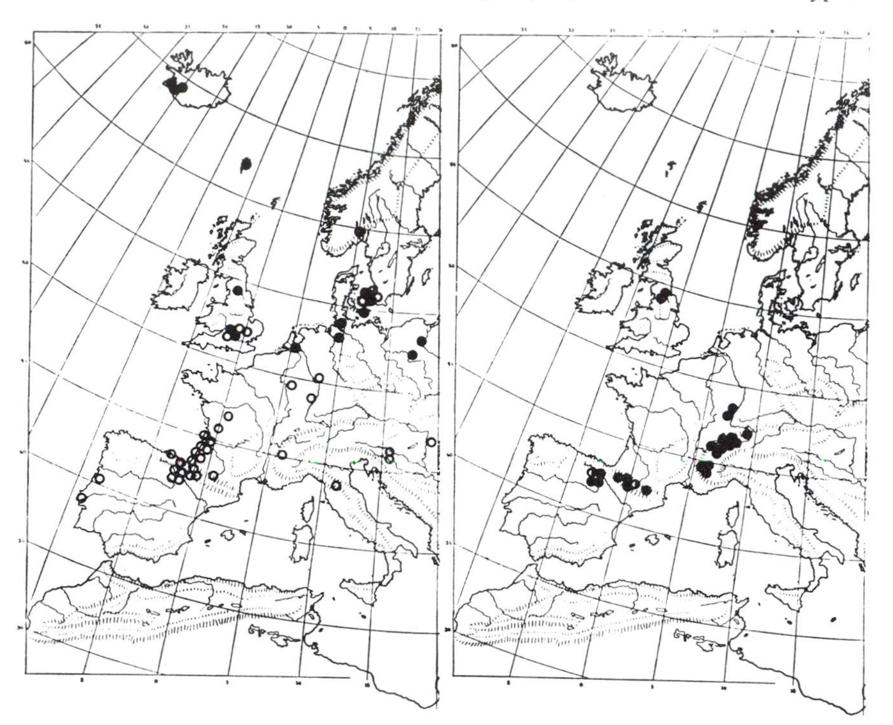

Fig. 8.1. Distribution in Europe of first-year flowering and second-year flowering types of *Prunella vulgaris*: the first (with short rosette stage) indicated by open rings, the second (with long-lasting rosette stage) by filled circles. All 75 samples were sown and cultivated simultaneously during 1950–51. On the map on the left are 51 lowland samples. On the map on the right are 24 samples from montane stations. The tendency towards second-year flowering in the northerly direction and from the lowland to the highland areas is evident. (From Bøcher, 1963.)

proved most interesting (Fig. 8.1); for example, in Mediterranean regions subject to summer drought, only short-lived annual plants were found, whilst in areas with different climatic conditions biennial or perennial types were characteristic. Such patterns are likely to be the result of natural selection: only those plants whose life history 'fits' the growing season of a particular area will survive in the long-term.

Experiments by American botanists

Some of the most famous experiments on ecotypes were carried out by Clausen, Keck & Hiesey (1940) on different species of plants collected on a

200-mile transect across Central California, from a 'Mediterranean' climate in the west to an 'alpine' climate in the east. Turesson's method of studying ecotypes was to grow all his collections in a lowland garden. Such a method has the limitation that it may not allow certain traits to be revealed (e.g. tolerance or sensitivity to frost or drought). In an attempt to overcome this difficulty, Clausen and his co-workers carried out experiments with many gardens, and finally used three: at Stanford (30 m above sea-level), Mather (1400 m) and Timberline (3050 m). To illustrate the very different conditions in the gardens, Fig. 8.2 gives climatic details for sites near Stanford and Timberline. Of special importance are the extremes of temperature and the differences in the length of the growing season. In each garden, plants were grown spaced out in weed-free plots protected from grazing. The experimental plantings consisted, in the main, of clonally propagated stocks, each individual being grown and divided, and a ramet of each planted in each garden. Thus the growth and performance of each individual from samples collected from a range of different sites could be studied in a 'Mediterranean', an intermediate and an 'alpine' garden. Climatic ecotypes were studied in many species, particular attention being paid to *Potentilla glandulosa*, a species found from the coastal hills near the west coast of California to high altitudes in the Sierra Nevada. Their experiments made it possible to test the behaviour of diverse stocks in very different standard gardens. For example, they discovered that most lowland stocks died in the harsh climate of the alpine garden, and at the Stanford garden plants originating from high altitude remained winter-dormant under conditions which stimulated growth of lowland samples. Clausen and his associates (1940) decided that there were four distinct climatic ecotypes in *P. glandulosa*, corresponding to the following taxa: subsp. *typica* (lowland); subsp. *reflexa* and subsp. *hanseni* (intermediate altitudes); and subsp. *nevadensis* (alpine) (Table 8.3). Clausen & Hiesey (1958) suggested that each subspecies was in fact made up of two or more ecotypes. Their hypothesis that ecotypic variants of *P. glandulosa* differed genetically received support from a comprehensive series of crossing experiments.

Other American botanists made studies of ecotypes using the transplant stations at Stanford, Mather and Timberline. Lawrence (1945), for example, studied ecotypes of the grass *Deschampsia cespitosa*, discovering differences in survival in different stations (Fig. 8.3). Of special interest were his studies of reproduction in the different transplants; although all individuals survived at Timberline, only the stocks native to that area were able to produce seeds in the short growing season. Such a finding, which is of crucial importance in understanding the genecology of the species, could

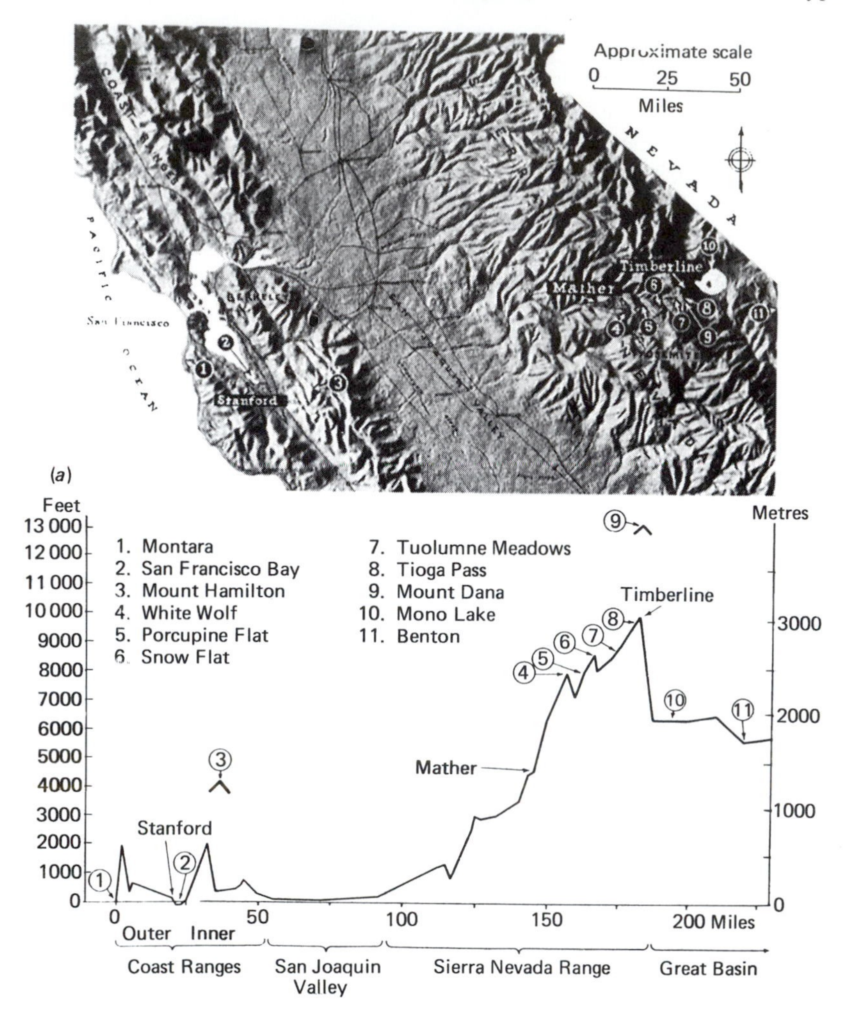

Fig. 8.2. Map and climatic details for Stanford and Timberline sites in Central California, used for the famous transplant experiments of Clausen, Keck & Hiesey. (*a*) Diagrammatic transect showing heights above sea-level. (*b*) Graphs showing annual variation in temperature and precipitation (US Weather Bureau data for 1925–35 inclusive) near Stanford and Timberline. In the lowland site with 'Mediterranean-type' climate (Stanford), active growth is possible throughout the year, whereas at Timberline (*c.* 3000 m) the active growth period is restricted to July and August. At Stanford, average annual precipitation was 31.7 cm; there was no snowfall except for traces in 1931 and 1932. At Timberline, average annual precipitation was 74.1 cm. Maximum temperature = average of the highest monthly temperatures. Mean temperature = average of the mean monthly temperatures obtained from daily readings. Precipitation = average monthly precipitation. Minimum temperature = average of the lowest monthly temperatures. (From Clausen, Keck & Hiesey, 1940.)

(*b*)

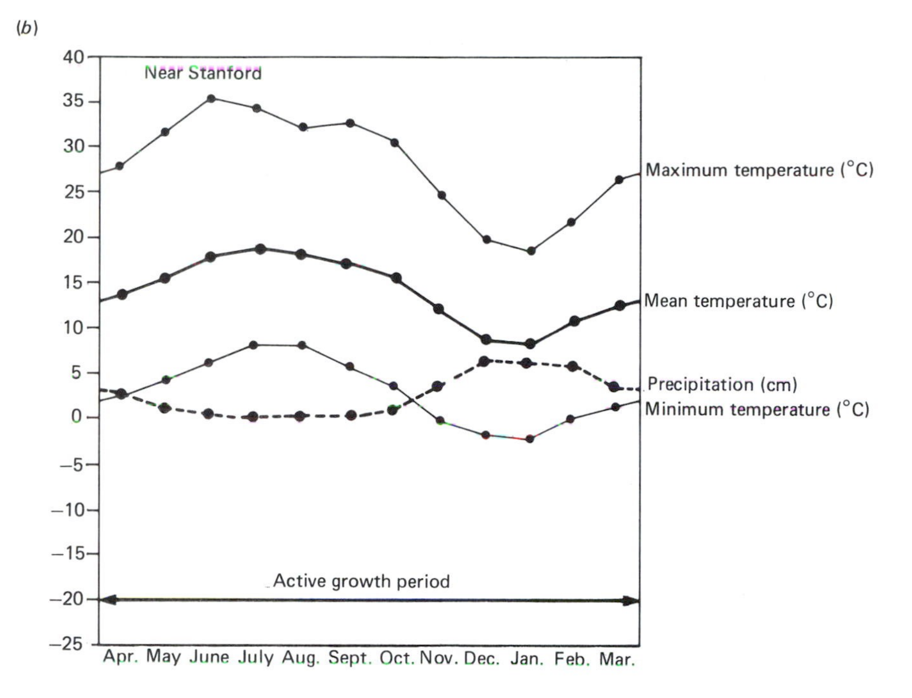
Near Stanford
40
35
30
25
20
15
10
5
0
−5
−10
−15
−20
−25
Maximum temperature (°C)
Mean temperature (°C)
Precipitation (cm)
Minimum temperature (°C)
Active growth period
Apr. May June July Aug. Sept. Oct. Nov. Dec. Jan. Feb. Mar.

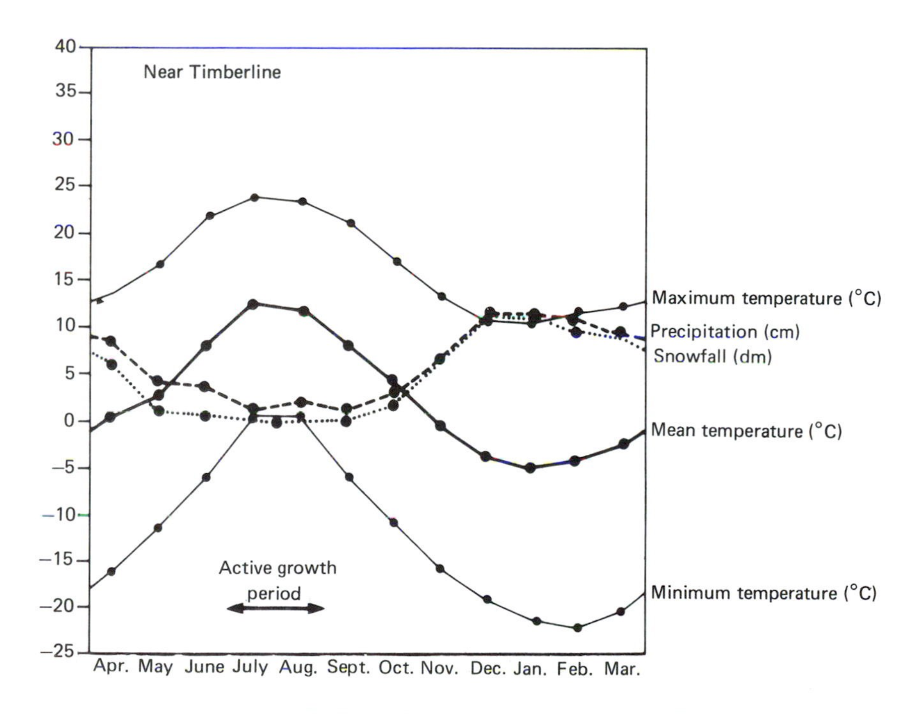
Near Timberline
40
35
30
25
20
15
10
5
0
−5
−10
−15
−20
−25
Maximum temperature (°C)
Precipitation (cm)
Snowfall (dm)
Mean temperature (°C)
Minimum temperature (°C)
Active growth period
Apr. May June July Aug. Sept. Oct. Nov. Dec. Jan. Feb. Mar.

Table 8.3. *A summary of the characteristics of the ecotypic subspecies of* Potentilla glandulosa *along the Central Californian transect (Data from Clausen & Hiesey, 1958, as summarised by Heslop-Harrison, 1964)*

	typica	*reflexa*	*hanseni*	*nevadensis*
Distribution	Coast ranges and lower Sierra Nevada	Low and middle altitudes of Sierra Nevada	Meadows, midaltitudes of Sierra Nevada	High altitudes of Sierra Nevada
Habitat	Soft chaparral and open woods	Dryish, open timbered slopes	Moist meadows	Moist, sunny slopes
Climatic tolerances as experimentally determined	Coastal to middle altitudes	Coastal to middle altitudes	Middle and high altitudes (poor survival near coast)	Middle and high altitudes (poor survival near coast)
Seasonal periodicity at Stanford (alt. 30 m)	Winter- and summer-active	Winter-active or -dormant; summer-active	Winter-dormant, summer-active	Winter-dormant, summer-active
Internal variation	Wide, probably several 'ecotypes'	Wide, probably several 'ecotypes'	Wide, at least two 'ecotypes'	Moderate, at least two 'ecotypes'
Self-compatibility	Self-fertile	Self-fertile	Undetermined	Self-sterile

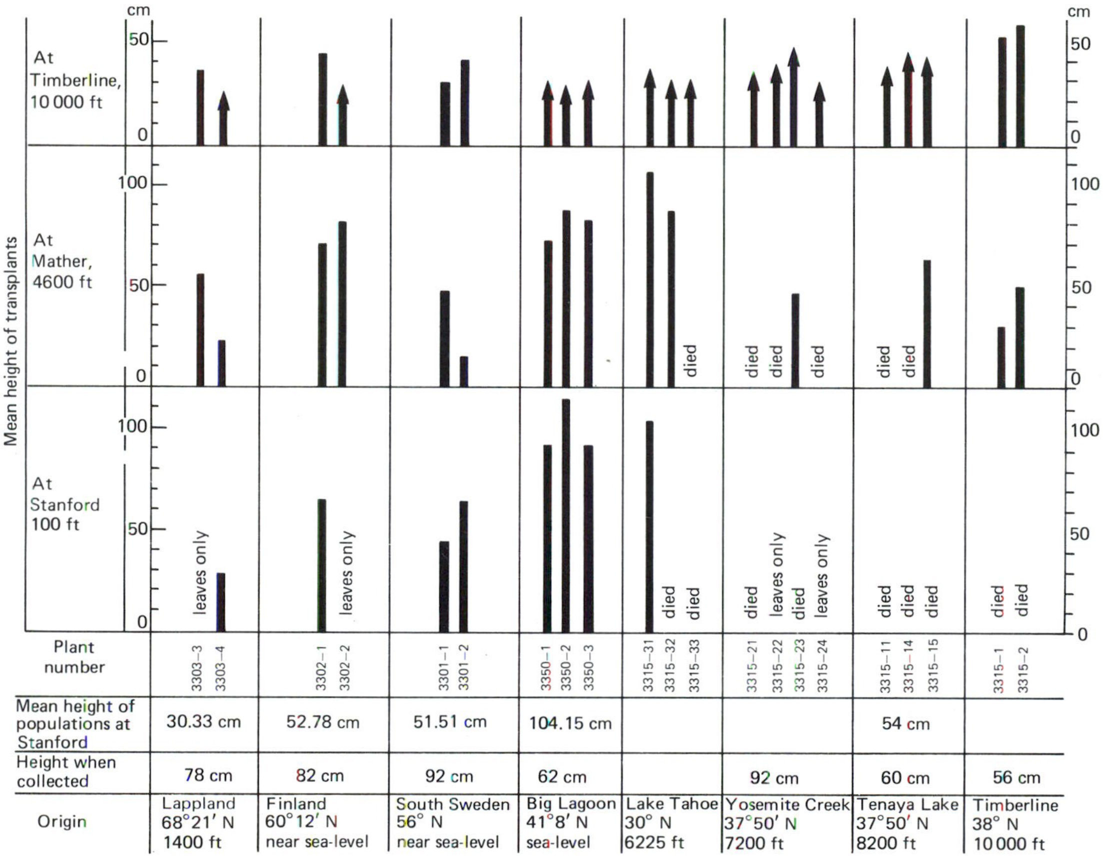

Mean height of transplants

At Timberline, 10 000 ft

At Mather, 4600 ft

At Stanford 100 ft

Plant number	3303–3, 3303–4	3302–1, 3302–2	3301–1, 3301–2	3350–1, 3350–2, 3350–3	3315–31, 3315–32, 3315–33	3315–21, 3315–22, 3315–23, 3315–24	3315–11, 3315–14, 3315–15	3315–1, 3315–2
At Mather, 4600 ft					3315–33 died	3315–21 died, 3315–22 died, 3315–24 died	3315–11 died, 3315–14 died	
At Stanford 100 ft	3303–3 leaves only	3302–2 leaves only			3315–32 died, 3315–33 died	3315–21 died, 3315–22 leaves only, 3315–23 died, 3315–24 leaves only	3315–11 died, 3315–14 died, 3315–15 died	3315–1 died, 3315–2 died
Mean height of populations at Stanford	30.33 cm	52.78 cm	51.51 cm	104.15 cm			54 cm	
Height when collected	78 cm	82 cm	92 cm	62 cm		92 cm	60 cm	56 cm
Origin	Lappland 68°21′ N 1400 ft	Finland 60°12′ N near sea-level	South Sweden 56° N near sea-level	Big Lagoon 41°8′ N sea-level	Lake Tahoe 30° N 6225 ft	Yosemite Creek 37°50′ N 7200 ft	Tenaya Lake 37°50′ N 8200 ft	Timberline 38° N 10 000 ft

Fig. 8.3. Mean heights of plants of *Deschampsia cespitosa* (*D. caespitosa*) of diverse origin grown at the transplant stations Stanford, Mather and Timberline, in Central California. (The arrowheads on certain columns at Timberline signify that these individuals did not reach maturity in any year.) (From Lawrence, 1945.)

not have been revealed in a lowland garden. A further point of general interest is revealed by their results with plants of *D. cespitosa* from Finland (latitude 60° N) and South Sweden (latitude 56° N). When these plants were grown at low altitudes at Stanford (38° N), many of them became viviparous, a character not expressed in their native habitats. Growth in a garden with very different climatic characteristics may provoke an unusual response from plants.

Experiments with several gardens separated by great distances are expensive to maintain, and botanists have devised ways of investigating ecotypes by varying the conditions in a single garden or laboratory. Turesson's experiments were carried out in a lowland garden on fertile soil and in describing edaphic ecotypes he inferred the importance of soil differences in the wild. A more direct approach to the study of patterns of variation in relation to edaphic factors was made by Kruckeberg (1951, 1954). In one experiment, fruits of *Achillea borealis* were collected from serpentine and non-serpentine sites in California. (Serpentine is a rock type which gives rise to soil with high levels of magnesium and low levels of calcium.) Two tons each, of a serpentine and a fertile soil, were collected and transported to the University of California Botanical Gardens, and stocks were grown from seed in soil bins, or pots, of the two soil types. Stocks raised from seed of plants native to serpentine soils grew well on the serpentine test soil, but, in contrast, plants from other soil types (shales, basalt, etc.) generally (though not always) grew badly or died (Fig. 8.4). Kruckeberg's results on *A. borealis* and other species are consistent with the idea that a common species found on different soil types may be made up of a number of edaphic ecotypes.

A second example of the way in which diverse stocks may be presented with different environments in one garden or laboratory is provided by the use of glasshouses, growth chambers, etc., in which daylength, temperature and other factors may be varied. Samples may be tested in a variety of artificially controlled environments, in which, for instance, the responses of different stocks may be monitored under different daylengths. In the first experiments studying the effect of different daylengths, plants were grown on movable trucks. After a period of natural daylight plants were moved into light-proof structures where they could be either in total darkness or given supplementary light from artificial sources. A good example of this type of experiment is provided by Larsen (1947), who studied *Andropogon scoparius*, a widespread and important forage grass in North America. Plants were collected from 12 localities from 28°15′N in Texas to 47°10′N in North Dakota. The grasses were given constant daylengths of 13, 14 and 15

Fig. 8.4. Experiments with *Achillea borealis*, grown on serpentine soil (above) and non-serpentine soil (below). All eight samples grew well on the fertile, non-serpentine soil, whilst three of the four samples from the non-serpentine soil (161, 125, 198) grew badly on serpentine soil. The fourth sample from non-serpentine soil (206) grew unexpectedly well on serpentine soil, however. (From Kruckeberg, 1951.)

hours of light. None of the 12 samples flowered at 13 hours. Plants from the southern USA required a 14-hour photoperiod for floral induction, but a photoperiod of 15 hours was necessary for flowering in many northern plants. Figure 8.5 illustrates the relation between latitude and daylength at different times of year. *A. scoparius* plants growing in the southern USA naturally come into flower after receiving a photoperiod of 14 hours. Northern plants, with longer summer days, need a 15-hour day to come into flower.

As more sophisticated equipment became available, growth chambers were constructed in which many environmental factors (e.g. temperature, daylength) could be controlled. Adjacent chambers could be used to subject plants to different conditions. A splendid example of such studies is provided by the experiments of Mooney & Billings (1961) who studied *Oxyria digyna* collected from sites between 38° N and 76° N in North America. Other botanists have continued to be fascinated by the different photoperiodic responses of plants from different geographic areas. The work of McMillan (1970, 1971) on *Xanthium strumarium* provides an impressive example. Physiological studies are advancing our understanding of ecotypes, and the reviews of Heslop-Harrison (1964), Hiesey & Milner (1965) and Bannister (1976) may be consulted for details of early studies. In recent years, the field of physiological ecology has expanded greatly. Specialist reviews give details of research in different fields: e.g. frost survival (Sakai & Larcher, 1987); photosynthesis (Evans, Caemmerer & Adams, 1988); multiple stresses (Mooney, Winner & Pell, 1991); resource

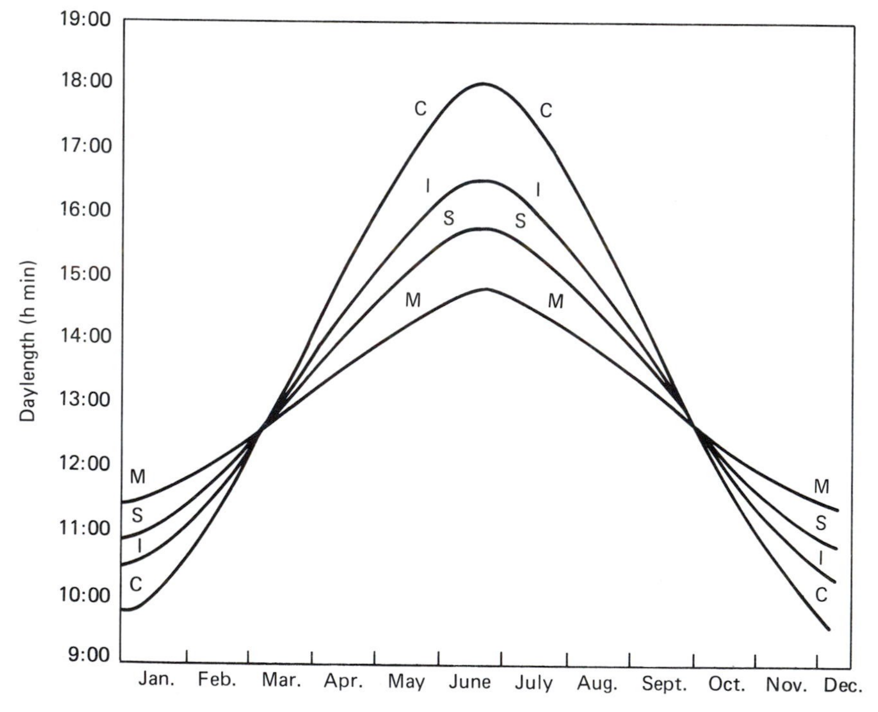

Fig. 8.5. Relation between latitude and daylength at different times of the year. Daylength includes twilight of that intensity receivable when the sun is 6° or less below the horizon, thus adding about 1 hour to the daylength between sunrise and sunset. M = Miami, FL, latitude *c.* 26°N; S = San Francisco, CA, *c.* 37°N; I = Ithaca, NY, *c.* 42°N; C = southern Canada, 50°N. (From Curtis & Clark, 1950.)

use (Townsend & Calow, 1981); and physiological ecology of woody plants (Kozlowski, Kramer & Pallardy, 1991).

The widespread occurrence of ecotypes

As a result of experiments in which plants have been grown in gardens or under controlled conditions, ecotypes have been described in hundreds of species. There is evidence that ecotypes occur not only in outbreeding species but also in species apparently predominantly inbreeding. There are also numerous studies of facultatively apomictic plants in which ecotypic patterns have been described, for example the grass *Poa pratensis* (Smith, Nielsen & Ahlgren, 1946) and *Potentilla gracilis* (Clausen, Keck & Hiesey, 1940).

Of special interest is the finding of genetic heterogeneity in plants which are apparently obligately apomictic. Turesson (1943) discovered, within collections of European *Alchemilla glabra*, *A. monticola* (*A. pastoralis*) and *A. filicaulis*, that plants from Lapland and montane areas were earlier flowering in cultivation than lowland stocks. The patterns of variation appeared to be ecotypic, but Turesson called the variants 'agamotypes' in recognition of the breeding system of *Alchemilla*. Bradshaw (1963*a*, *b*, 1964) and Walters (1970, 1986*a*) have described dwarf variants of an ecotypic nature in *Alchemilla*, the origin of which is plausibly due to selection in response to grazing by sheep (Fig. 8.6).

Clines

In the experiments outlined above, the researchers were content to describe their material in terms of distinct local races, often using the term 'ecotype'. However, the ecotype concept was not without its critics. Langlet (1934), for example, pointed out that the most important habitat factors, such as temperature and rainfall, commonly varied in a continuous fashion, and thus one would expect graded variation in many widespread species rather than discontinuous variation.

Support for this view was provided by Gregor (1930, 1938) who made an intensive study of *Plantago maritima* in northern Britain. Representative seed collections were made and plants were grown in an experimental garden of the Scottish Society for Research in Plant Breeding. Table 8.4 gives results obtained by Gregor (1946) in similar studies. In this case all three sample zones are from the Forth estuary in eastern Scotland. If collections of *P. maritima* taken from different sites along a gradient from

Fig. 8.6. Dwarf variants of *Alchemilla* from the North Pennine Hills, England. Four transplants from grazed mountain localities, grown under standard conditions for nine months at Durham University experimental grounds. Top row: two separate transplants of *A. minima* from Ingleborough. Bottom row: *A. filicaulis*; left, transplant from Mickle Fell; right, transplant from Moor House National Nature Reserve. *A. minima* retains its very dwarf habit in cultivation, in contrast to *A. filicaulis*. (From Bradshaw, 1964.)

high to low salt concentration are compared, a progressive increase in scape height is found. In a similar fashion there are increases in: scape volume and thickness; leaf length, breadth and spread; and seed length. Figure 8.7 illustrates the different growth-habit types found in *P. maritima*. As Table 8.4 shows, it is only in the upper marsh that erect plants predominate.

In 1938, Huxley, after surveying the literature, coined the useful term 'cline' for character variations in relation to environmental gradients. Thus, a graded pattern associated with ecological gradients is referred to as an ecocline (a good example of this is Gregor's *P. maritima* result). If the pattern is correlated with geographical factors, the term topocline can be employed. Clinal variation has been described in a large number of species and a small selection of examples is given in Table 8.5 and Fig. 8.8.

Table 8.4. *Results of soil analyses (air-dried samples) and cultivation experiments with* Plantago maritima *(Gregor, 1946)*

Habitat	Mean scape length (cm)	Habit grades (Percentage of sample in each grade)				
		1	2	3	4	5
Waterlogged mud zone (salt concentration 2.5%)	23.0 ± 0.58	74.5	21.6	3.9	—	—
Intermediate habitats with intermediate salt concentrations	38.6 ± 0.57	10.8	20.6	66.7	2.0	—
Fertile coastal meadow above high tide mark (salt concentration 0.25%)	48.9 ± 0.54	—	2.0	61.6	35.4	1.0

How far are intraspecific patterns of variation explicable in terms of ecotypes and clines? Experiments, for example, by Bradshaw (1959*a*, *b*, *c*, 1960) on the grass *Agrostis capillaris* (*A. tenuis*), have shown that much more complex patterns may be found in nature. Careful collections of living specimens of this grass were made mostly from localities in Wales. The stocks were grown, and then cloned material was planted into a number of experimental plots in North and Mid-Wales, with an altitudinal range from sea-level to about 800 m. A wide range of different responses was demonstrated by these experiments. Not only were plants different morphologically but there were also physiological differences. For example, certain plants grew well on soils containing lead and other heavy metal residues; others, indistinguishable from them morphologically, died on this type of soil (we shall return to this interesting phenomenon of tolerance of heavy metal ions in Chapter 9). At this point it is important to note that Bradshaw could not delimit ecotypes in *A. capillaris*. This was not because extreme variants were not found in extreme habitats. On the contrary, many very distinctive plants were discovered: for instance, dense cushion plants from the exposed Atlantic cliffs at West Dale, South Wales. The problem was that, even though habitat-correlated variation could be demonstrated, the fact that all kinds of intermediate plants were discovered made it utterly impossible to decide where one ecotype ended and another began.

Does the concept of clines help in this situation? Bradshaw studied his material closely with this idea in mind. In many areas, even though clines

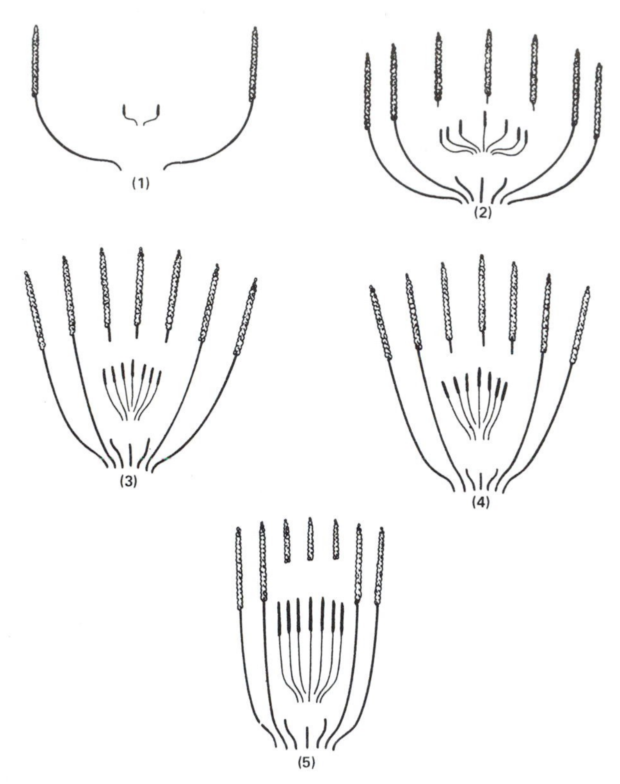

Fig. 8.7. Variation in *Plantago maritima*. For purposes of classification, Gregor divided his material into five grades, illustrated diagrammatically here. There was, however, no sharp line of demarcation between one grade and the next. (From Gregor, 1930, 1938.)

might be described, he decided that the environmental gradients and the associated variation were too complex.

What then, determines the patterns of intraspecific variation found in the wild? How can one reconcile the distinct ecotypes of Turesson and Clausen with the complex variation found by Bradshaw and many other researchers?

Factors influencing the variation pattern

Of first importance is the type of sampling technique used. Turesson and many other botanists collected widely spaced samples, whereas Gregor and Bradshaw carried out intensive sampling in small areas. Widely spaced samples taken from extreme habitats may exhibit a pattern of distinct

Table 8.5. *Some examples of clinal variation*

Species	Variation	Reference
Allium schoenoprasum	Longitudinal cline in chromosome banding pattern in eastern North America	Tardif & Morisset (1991)
Anthoxanthum odoratum	Clines for various characters at mine/pasture boundary	Antonovics & Bradshaw (1970)
Asclepias tuberosa	Clines for flower colour and leaf-shape in North America	Woodson (1964)
Blandfordia grandiflora	Morphological & reproductive characters in Australia	Ramsey, Cairns & Vaughton (1994)
Dactylis glomerata complex	Clinal variation in European populations in glutamate oxaloacetate transaminase gene frequencies (GOT I Locus)	Lumaret (1984)
Eschscholzia californica	Clines in California for various features	Cook (1962)
Eucalyptus spp.	Graded patterns of leaf glaucousness with extreme 'waxy' types in exposed habitats	Thomas & Barber (1974)
Geranium robertianum	Clines for hairiness	Baker (1954)
Geranium sanguineum	Decrease in leaf-lobe breadth west to east in Europe	Bøcher & Lewis (1962)
Holcus lanatus	First-year flowering in Southeast Europe. Second-year flowering in northern Europe	Bøcher & Larsen (1958)
Juniperus virginiana	Clines in terpenoid content northeast Texas to Washington DC	Flake, von Rudloff & Turner (1969)
Lotus corniculatus	Flower colour variation in North England. Dark-keeled variant rare in West, increasing in frequency eastwards.	Crawford & Jones (1986)
Pinus strobus	Decrease in leaf length and number of stomata, increase in number of resin ducts, with increasing latitude in North America	Mergen (1963)
Silene latifolia	Clinal variation in seed morphology	Prentice (1986)
Viola riviniana	Clines in plant size	Valentine (1941)

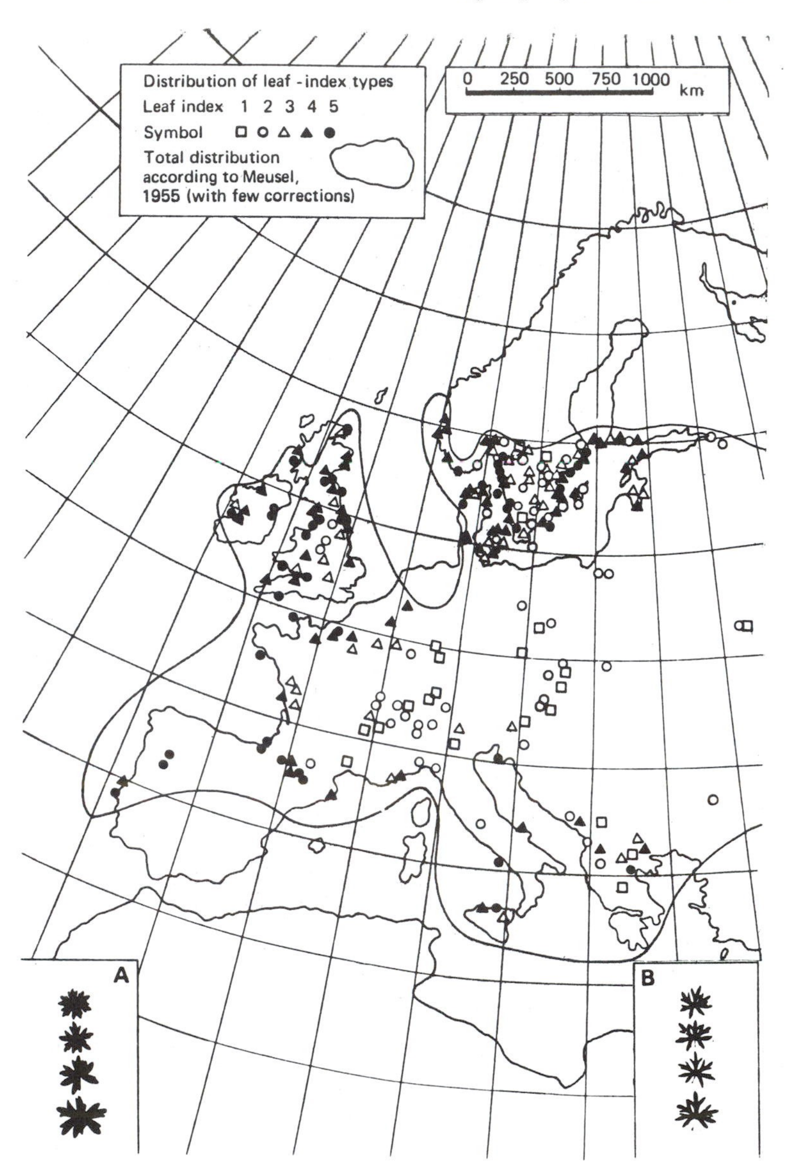

Fig. 8.8. Clinal variation in *Geranium sanguineum* (Bøcher & Lewis, 1962). At first sight there seems to be a more or less simple topocline for leaf-lobe width across Europe, plants from North and West Europe usually having broad-lobed leaves (leaf index 4 and 5), as in inset A. On the other hand, material from continental Europe often has narrow leaf-lobes (leaf index 1 and 2), as in inset B. The distribution map of leaf index values for herbarium material suggests, however, that the variation is more complex. It seems likely, in view of the occurrence of broad-lobed plants on the east coast of Sweden and in the Mediterranean area, that this leaf type is associated with coastal climatic conditions. Narrow-lobed plants, found in dry limestones of inland Britain and Sweden, seem to be found wherever continental climatic conditions occur.

ecotypes. In contrast, samples taken from along smooth, regular gradients of soil or altitude may well give a pattern of clinal variation in the experimental garden. If, however, sampling is carried out in small areas, the plants being collected at random rather than along particular gradients, then experiment might reveal very complex patterns. Thus, in a very real sense, the mode of sampling largely determines the patterns 'discovered' in cultivation experiments.

Another aspect of sampling is important. An experimenter can choose either to collect a representative seed sample or to dig up mature plants. If both types of sampling are carried out on a single population, different patterns of variation might well be found. This is because mature plants have survived the rigours of natural selection. Seed collections, on the other hand, give an estimate of potential rather than actual variation. If several adjacent populations in different environments are examined, in a case where pollen can be transported from one population to another, sampling of mature individuals might well reveal a pattern of more or less distinct ecotypes. On the other hand, because of gene flow between populations, seed samples will seem to reveal a more complex pattern in the same case.

Ecological, historical and geographical factors also influence the patterns discovered in experiments. If a species is found as small, non-contiguous populations, or if it has populations inhabiting two or more very different types of habitat, then the pattern of variation in the wild is more likely to be that of distinct ecotypes. In contrast, common species, which throughout their geographical range are more or less continuously distributed over many habitats, will in all probability exhibit complex patterns of continuous variation. Also, the mode of pollination is important. Small populations of insect-pollinated species often exhibit ecotypic discontinuities, but these are less likely to occur in widespread wind-pollinated species.

Since Turesson's time there has clearly been a change of outlook. Ecotypes are now regarded as nothing more than prominent reference points in an array of less distinct ecotypic populations (Gregor, 1944). Some experimenters have been reluctant to designate ecotypes; they have instead carefully recorded the patterns of 'ecotypic differentiation' found in particular experiments (see, for example, Quinn, 1978). However, despite the difficulties of defining the word 'ecotype', on-line searches of the Institute for Scientific Information database of recent scientific publications (via BIDS; Bath Information and Data Services, England) reveal that it is still being used for local and regional variants. Also, some of the regional, biochemical and developmental variants of *Arabidopsis thaliana* are referred to as 'ecotypes' in British, European and US stock lists and in publications.

With hindsight one can see in Turesson's own results the possibility that, in common species, variation patterns were more complex than the ecotype concept implied. For instance, where sandy fields and dunes were found as adjacent habitats, a considerable number of intermediate *Hieracium umbellatum* plants were found linking the two ecotypes. Similarly, in *Leontodon autumnalis*, Turesson (1922*b*) found a complex situation where meadows and pastureland ran down to the sea.

The refining of genecological experiments

Early cultivation experiments were often very crude; a few plants were dug up in the wild and planted in a garden. As we have just seen, the pattern of sampling will to a very large extent determine the outcome of an experiment. Furthermore, while a simple garden technique may serve to study major differences between population samples, the study of fine scale variation has resulted in the devising of improved cultivation and other experiments. Thus, as genecology has developed, the methods of sampling and cultivation have been refined to enable statistical analysis of finer and finer differences between samples of plants.

Sampling populations

Much time and effort may be spent in growing and measuring plants and analysing results, but very little attention may have been given to sampling strategies; indeed, the word 'strategy' may be entirely inappropriate for samples of seed snatched at brief roadside stops on car journeys or obtained from Botanic Garden seed lists.

If statistical analysis is to be performed on the results, then ideally a random sample of plants must be collected. Ward (1974) has described a simple way in which two people may collect such a sample. Having decided on the area to be sampled, the recorders count the number of individuals in the area (or subsection of an area if the plant population is very large). A decision is then made on the size of the sample, say 25 plants out of 250. Using a table of random numbers (as found, for instance, in Fisher & Yates, 1963) or numbers 'drawn out of a hat', 25 numbers within the range 1–250 are 'selected' and placed in ascending order: say 5, 8, 14, 27, etc. On traversing the sample area again, one person calls out the number of each individual, 1–250, while the second person labels the individuals to be sampled, the 5th, 8th, etc., as determined by the random numbers. This random sample is then used for experimental investigation. Other methods of random sampling are discussed by Yates (1960), Cochran (1963), Greig-Smith (1964) and Green (1979). There are some theoretical and practical difficulties to be faced in undertaking such a sampling procedure, which will now be considered.

If the experimenter is studying apparent hybridisation, a random sample might not include all the 'interesting' plants of an area. A deliberate sampling of the plants of the area might be more appropriate in such circumstances. Should the study involve the investigation of variation across a vegetational discontinuity, e.g. woodland to grassland, it might be more informative to collect plants from a transect (sampling at, say, metre intervals) across the ecotone rather than collect a random sample. All will depend on the hypothesis being tested. There are many habitats where the collection of random samples is very difficult (e.g. tropical rain forests, aquatic and wetland habitats, cliffs). However, where the collection of such samples is a practical possibility it should be seriously considered.

Since populations often contain individuals at all stages of growth and development from seeds and seedlings to adult plants, a truly random sample should perhaps contain individuals in several different age classes. In practice, a subset of the population is often sampled. The following might usefully be distinguished:

1. 'Individuals' present as ungerminated seed in the soil ('seed bank').
2. Seedlings, a transitory stage in many habitats, but more important in some plant communities. For instance, in tropical rain forests many tree species growing in deep shade have a long seedling stage; only if disturbance in the canopy causes greater illumination of the ground flora do the seedling trees develop into adults.
3. Immature individuals.
4. Mature individuals.
5. Seeds attached to 4.
6. Diseased and damaged plants. Sometimes, as in the case of the 'choke' disease of grasses caused by fungus, the plants are vegetatively vigorous but the fungal infestation suppresses the formation of inflorescences (Bradshaw, 1959*c*).

Subsets 4 and 5 are most commonly sampled by experimenters. Different subsets may reveal quite different spectra of variation; we shall see examples in Chapter 9 when we consider attempts to study the effects of natural selection by comparing the variation of different subsets in cultivation.

One of the biggest difficulties in sampling populations concerns the definition of an individual. In open vegetation it is usually possible to define individuals in annual plants and to see patches of individual perennial plants. In closed swards, however, the problem is more difficult. Sometimes the presence of 'marker' genes (e.g. leaf marks in *Trifolium repens*: Davies, 1963) might reveal the extent of particular individuals; but such markers are rare. Theoretically it might be possible to trace root systems in an attempt to establish the extent of individuals, but the practical difficulties are enormous. Furthermore, in some plants, e.g. certain forest trees, root-grafts occur which unite the root systems of several different individuals (Graham & Bormann, 1966; Böhm, 1979). Recent studies of patterns of allozymes in Strangler Figs (*Ficus* species), which form a woody sheath around many tropical trees, provide evidence that apparent individuals are in reality genetic mosaics, caused by root fusions of a number of plants (Thomson *et al.*, 1991).

The problem of defining the individual is further complicated by clone formation, in which the vegetative continuity of an individual breaks down, producing a clonal patch of several individuals of identical genotype (Harper, 1978). Evidence for clonal populations is usually circumstantial, but direct evidence is available in the case of certain self-incompatible species which are very variable morphologically. Variability has been studied in garden trials of population samples, and the material classified into different individuals on the basis of morphology, phenology, susceptibility to pests and diseases, etc. The behaviour of different plants in crossing experiments is then studied. Crosses between dissimilar-looking plants may yield a 'full seed-set', from which we may infer that the plants have different *S* alleles and are different genotypically. Conversely, crosses between plants which are morphologically indistinguishable may yield little (or no) seed, and can be thought to share the same *S* alleles and to be of the same genotype or 'isoclonal'. This method was used to study variation in populations of the grass *Festuca rubra* (see Table 8.6). Some caution is necessary in interpreting experiments of this type, as the method depends upon a thorough knowledge of the type of incompatibility mechanism involved – a requirement almost never satisfied with wild species.

Recently, investigations using molecular methods have greatly increased our understanding of populations of clonally propagating species. For instance, in a study of a population of Bracken (*Pteridium aquilinum*) in Virginia using 6 polymorphic isozyme loci, as many as 45 genotypes were detected in the study area (Parks & Werth, 1993). Moreover, some of these clones were very extensive. In an investigation of a population of the same

Table 8.6. *Some examples of studies of clones (Lines of evidence: F. = field observations; C. = cultivation trials; H. = hybridisations; I. = electrophoretic studies of isozymes; M. = DNA fingerprinting)*

C.	*Anemone nemorosa* (von Bothmer *et al*., 1971): large number of clonal patches, of limited size, in Swedish habitats.
C.F.I.	*Betula glandulosa* (Hermanutz, Innes & Weis, 1989): clones mapped on Baffin Island, at northern limit of species.
F.I.	*Decondon verticillatus* (Eckert & Barrett, 1993): a survey of this tristylous species reveals that at the northern margin of its range in eastern North America, populations may consist of only one of the three style variants and reproduction is exclusively by clonal propagation.
C.H.	*Festuca rubra* (Harberd, 1961): evidence of many genetically different individuals in a study of an area of South Scotland. One particular variant occurred at points *c*. 220 m apart. If this area was achieved by radial growth then the clone must be *c*. 400–1000 years old. However, perhaps the present distribution has been achieved by dispersal of fragments by animals or other causes, or as a consequence of vivipary, which has been recorded in this species (Smith, 1965). Widespread clones also found in *Festuca ovina* (Harberd, 1962).
F.I.	*Larrea tridentata* (Sternberg, 1976; Vasek, 1980): extensive clonal patches, visible on aerial photographs, in the Mojave Desert, California. By radiocarbon dating oldest clone may be 11 700 years old. Isozyme studies reveal that parts of apparent clones are indeed isoclonal.
C.H.	*Lysimachia nummularia* (Dahlgren, 1922; Bittrich & Kadereit, 1988): self-sterile clones found in many parts of North and Central Europe; presumably sexual reproduction only takes place in populations where individuals with different *S*-alleles occur together.
F.M.	*Phragmites australis* (Neuhaus *et al*., 1993): large and small clones 'mapped' in Berlin and Northeast Germany using DNA fingerprinting (*a*. by digestion using restriction enzymes *Alu*I or *Dra*I, with the oligonucleotide $[GATA]_4$ used as a probe in hybridisation; or *b*. by RAPD reactions followed by separation of the amplification products on agarose gels and staining with ethidium bromide).
F.M.	*Populus tremuloides* (Rogstad, Nybom & Schaal, 1991): clones of various sizes mapped using DNA fingerprinting; Fig. 8.9 (digestion with restriction enzymes *Dra*I, *Hae*III or *Hinf*I and hybridization with the M13 probe).
F.I.	*Solidago altissima* (Maddox *et al*.,1989): using isozyme markers, clones were mapped in sites at different stages of old field succession near Ithaca, New York.
F.	*Ulmus* spp. (Rackham, 1975): by studying in British woodlands patterns of morphological variation together with incidence of fungal diseases and timing of coming into leaf and leaf fall, evidence of very extensive clonal patches was discovered.

species in North Wales, an extensive triploid clone was detected and this was mapped using isozyme markers (Sheffield *et al.*, 1993). On the basis of various lines of evidence it would seem that extensive clones, probably of great age in some instances, occur in some habitats (see Table 8.6 and Fig. 8.9).

Why is knowledge of the extent of individual genotypes important in sampling? Suppose we collect two population samples, A and B. Fortuitously, sample A could consist of 25 pieces of a widespread clone, whilst sample B could consist of material of 25 genetically different individuals. A comparison of the two 'populations' in a cultivation trial is likely to show that they are different, but interpreting this difference as a real population difference could be misleading. Perhaps population A *is* largely composed of the clonally propagated individuals of one genotype, while B is variable; on the other hand, populations A and B might both be variable, and the multiple sampling of one clone in population A might be merely the

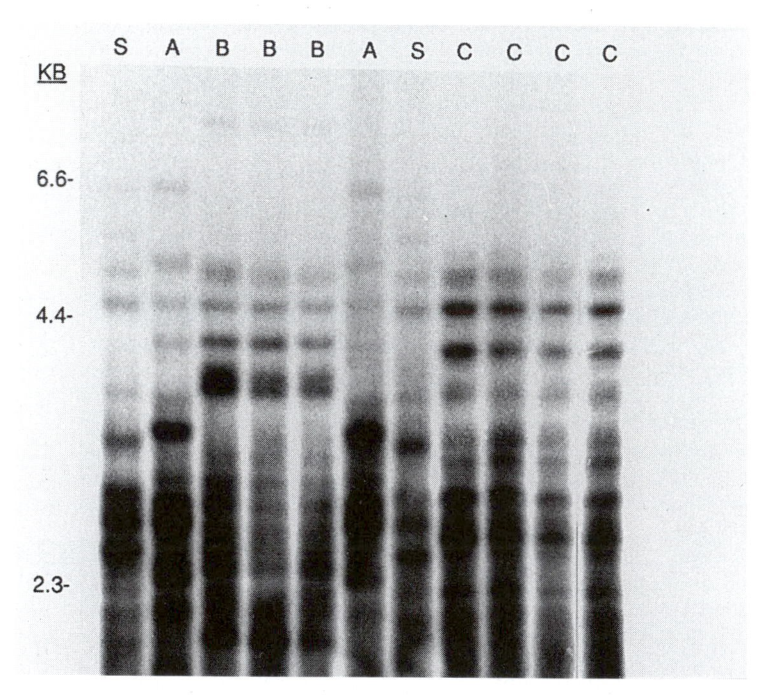

Fig. 8.9. The use of DNA fingerprinting to examine genetic diversity and clonal growth in *Populus tremuloides*. (From Rogstad, Nybom & Schaal, 1991). Initially, on morphological grounds, it was postulated that material from a site in Colorado was of two clones. DNA fingerprinting revealed that three clonal genotypes (A, B, C) occurred in the area. S = standards.

consequence of poor sampling technique. Harberd and others, who have made a special study of the problem (Harberd, 1957, 1958; Wilkins, 1959, 1960; Ward, 1974), recommend that spaced samples be collected from populations. From all the evidence available the probable maximum extent of clonal patches is estimated. Sampling at points separated by distances greater than this estimated clonal patch size is then carried out. Widely spaced samples are to be recommended to counteract another problem which arises in studying plant populations. Fruits and seeds are often shed very close to the plant which produced them and 'family groups of close relatives', perhaps involving several generations, may be found (see, for example, Linhart *et al.*, 1981). Distorted comparisons can arise if a sample containing a group of closely related plants is matched against a set whose members are totally unrelated.

It must be noted, however, that wide spacing of samples is somewhat at odds with the present trend of studying small systems in detail. Such studies as those of Smith (1965, 1972), and Harper and associates (Harper, 1983) reveal enormous variation within sites. There seems to be no easy solution to the problems raised by clonal populations; the experimentalist must make the best judgement possible in each situation in relation to the hypothesis under consideration.

Another question to be resolved before sampling is undertaken concerns the number of sites and samples within sites. Suppose we study a single site with two different soil types, A and B. Patterns of variation may be revealed in samples drawn from the two subsites A and B, and at the end of an experiment some differences related to soil type may be found in plants originating from the two subsites. The experimenter must then decide whether the differences are ecotypic or whether they owe their origin to random variation. With one A/B comparison it is difficult to rule out random events (Wilkins, 1959). A more penetrating study of the patterns of variation might be made by studying several areas, where subsites of type A and B are juxtaposed. Furthermore, in collecting from the wild, a bulk seed sample may be made to represent each of the subsites A and B, or the seeds from a random collection of mature individuals may be separately collected and packeted at each subsite. Family lines may then be grown, patterns of variation within lines offering some insights into the breeding system of the plants under study. This type of sampling – a hierarchical or nested pattern – has much to recommend it, allowing not only a number of A/B comparisons to be made, but also providing some information on variation within subsites. For instance, the plants under study might be obligate apomicts; while the progenies of different 'seed parents' might differ, there might be little or no variation within progenies. In this circumstance, the cultivation of plants from bulked seed samples would fail to reveal an important strand in the variation pattern.

Cultivation experiments

A study of variation usually requires cultivation of plants. This is true not only of field collections brought into a common environment to investigate the nature of variation patterns, but also of many sophisticated genetic and physiological studies. In many cases, the experimenter wishes to grow material from diverse sources under the same conditions. Thus, if population samples are collected in the wild and if there are interesting phenotypic differences between populations, a Turessonian cultivation experiment might be carried out, to see if differences between populations persist in cultivation.

At first sight a requirement to grow material 'under the same conditions' appears to present little difficulty. A moment's reflection, however, is sufficient to remind the reader of the variation in soil fertility, drainage, pests and diseases within even the most uniform experimental plot in the garden or field. The notion that glasshouses provide a uniform environment is quickly dispelled by studying investigations of yields of vegetable crops on benches in different parts of experimental glasshouses (see, for example, the little-known experiments of Lawrence, 1950).

In designing genecological experiments, the botanist has had much to learn from the agricultural scientist. Farmers wish to grow high-yielding varieties of crop plants and, since the middle of the last century, research workers have struggled to perfect experiments designed to study yield. In this short book we cannot provide a complete review of this interesting subject and will confine our attentions to a few important general issues. Notable advances in the design of field experiments came with the work of Fisher, who studied the famous long-term Broadbalk Wheat experiment at Rothamsted Research Station in South England (Fig. 8.10). A book on the life of Fisher (Box, 1978) provides a useful historical review of field experimentation and explores in detail Fisher's many contributions to the subject.

The basic ideas behind the design of cultivation trials are as follows:

1. Experiments must be designed with sufficient replications of the varieties, populations, treatments, etc. Thus, in a simple experiment on yield in, say, Spring Wheat, several plots of each variety must be grown.

Fig. 8.10. Layout of the famous Broadbalk field experiment at Rothamsted, Herts, England, studied by R. A. Fisher. Experimental crops of Winter Wheat have been grown continuously in these plots since 1843. Photograph taken in 1954. © Rothamsted Experimental Station.

2. Soil fertility and other edaphic factors often vary across garden plots and fields, but it is commonly found that adjacent sites have similar fertility, etc. Thus, Fisher (1935) recommended that the ground available be divided into uniform blocks (not necessarily square). Each block should contain a full complement of the material under study. Within blocks the small plots of each variety should be *randomly* arranged. In early experiments in agriculture and forestry it was hoped that, by careful husbandry, varieties could be given the same conditions. But a critical approach to experimentation suggests that this is a forlorn hope; it is impossible to ignore the variability induced by environmental factors. With a proper layout of experiments, differences between blocks can be *measured* to give an estimate of the random element of variation introduced into the experiment.
3. Another important factor in the design of field experiments is the effect of position. If plants are growing in blocks, those in the centre of the block will be surrounded by neighbouring individuals; in contrast, plants on the margins of blocks are likely to be adjacent to bare soil and subject to very different amounts of root and shoot competition. Thus, it is

recommended that 'guard rows' of similar plants be planted around the blocks, to provide uniform conditions for the experimental material. Guard rows, usually of the same species as the plants under study, are discarded at final harvesting of the experiment.

It is clear that these ideas can with profit be incorporated into the design of genecological experiments, and indeed advanced field trial techniques were employed in the famous genecological experiments of Gregor and his associates in studies of variation in *Plantago maritima*, to which we have already referred (Gregor, 1930, 1939; Gregor, Davey & Lang, 1936; Gregor & Lang, 1950).

In a simple genecological experiment each individual, say of plants A, B, C and D, may be clonally propagated, the experimental garden may be divided into small blocks and a ramet of each individual of A, B, C and D planted in a weed-free plot surrounded by guard rows of the same species. The position of each ramet within blocks is determined by random numbers.

The fundamental ideas influencing the layout of simple field trials may also be incorporated into the design of more complex genecological experiments such as population trials, family lines and experiments involving populations given various treatments. Several excellent books with fully worked examples of various designs are now available for the biologist. Especially suitable for beginners are: Salmon & Hanson (1964); Bishop (1971); Parker (1973); and Clarke (1980). More advanced treatment will be found in: Campbell (1974); Ridgman (1975); Snedecor & Cochran (1980); Sokal & Rohlf (1981); Yates (1981); Stuart (1984); and Mead (1988).

Studies of agricultural crops have resulted in other important insights into the design of field experiments. At first sight it would seem reasonable to suppose that repeated experiments with the same varieties (or genotypes) would 'give the same results'. In practice there are considerable differences from year to year in the results of experiments estimating yield in cultivated stocks. The principal causes of variability are differences in weather, and changes in the incidence and severity of various pests and diseases (which are themselves probably correlated with past or present weather conditions). An experiment by Nelson (1967) emphasises the importance of year-by-year differences in a genecological experiment. He studied variation in *Prunella vulgaris* collected from many sites in the USA, by growing material at Berkeley, California. Usually there is no winter frost in this area, but, exceptionally, a very cold period occurred from 20–24 January 1962, providing him with a unique experiment, which revealed that some of his plants were frost-sensitive.

In designing genecological experiments other factors must be taken into account:

1. In experiments begun with samples of seeds, Roach & Wulff (1987) point out that there may be maternal effects, i.e. there may be a 'contribution of the maternal parent to the phenotype of its offspring beyond the equal chromosomal contribution expected from each parent'. Such contributions may be: (a) cytoplasmic; (b) flow from the greater contribution of maternal genes (2*n*) than male genes (*n*) to the (3*n*) endosperm; or (c) result from the fact that maternal tissues contribute to developing fruits and seeds. Thus, maternal effects are often manifest in differences in seed size and mineral composition. Roach & Wulff (1987) review the various techniques available to measure maternal effects in cultivation trials. Evidence for paternal effects – via 'male' cytoplasm – is also reviewed in the same paper. Maternal and paternal effects are examples of what are sometimes called 'carry-over' effects.
2. 'Carry-over' effects are also possible in experiments begun with vegetative material, such as clone transplants. Thus, the length of an experiment may be crucial if the investigation involves material dug up from the wild and transplanted into a garden for, as Turesson (1961) discovered, an extended period of adjustment may be necessary before plants may be said to have outgrown the effects of their original habitats. Indeed, it may be difficult to convince a sceptic that a complete adjustment is ever made, especially in the case of woody plants. Experiments with herbaceous plants have also been revealing. For instance, from a 43-year-old pasture in Canada, Evans & Turkington (1988) grew samples of *Trifolium repens* collected from patches dominated by different species of grasses. At the end of a field trial, lasting for 4 months, significant differences were detected between the samples for a number of characters. Then, a second trial was begun with the same samples, using material produced by vegetative propagation. It is of very great interest that there were no significant differences between the samples after 27 months. This experiment is a clear indication of the importance of 'carry-over' effects in relation to the duration of experiments. A second example makes some further important points. 'Carry-over' effects could arise from the use of unequal-sized pieces of material used to begin clone experiments. In a study of many facets of garden trials, Davies & Snaydon (1989) examined the effect of tiller size – small versus large – in the grass *Anthoxanthum odoratum* on a number of measures of performance in a garden trial. They discovered no evidence

for a major problem with 'carry-over' effects in this case, as there were no differences in survival, height or date of flowering. However, large tillers produced slightly larger plants.

3. The pretreatment of seeds and seedlings prior to the experiment is very important. There will be differences in the speed of development of plants between those sown as seed and those set out in the field as young plants. The timing of the experiment in relation to such seasonal factors as cold periods may also be crucial. Thus, some plants will not flower unless subjected to cold treatments, and spring and autumn sowing will yield different results.
4. The treatment of plants during the experiment has a profound effect upon their growth and performance. The experimenter must decide whether to water plants in dry weather, apply fertilisers, etc.
5. The incidence of pests and diseases causes considerable problems. In particular, experiments in glasshouses often turn into a struggle to control various insect and fungal pests, and the liberal use of pesticides may be the only means of 'preserving' the experiment. It is important to realise that 'spot-treatments' of badly infected individual plants may seriously affect the randomised design of the experiment. In the design of garden and field trials, on the other hand, the decision is often taken to allow non-catastrophic invasions of pests and diseases to take their toll on the experimental material. In this way it may be possible to see if any individuals or populations are resistant to fungal or insect attack. Studies of the effects of non-fatal pests and diseases may add a further dimension to our knowledge of population variation.
6. Agricultural experiments are often designed to be left until the final harvest when estimates of yield are made on fruiting material, and in other cases the experiment is so constructed as to permit regular intermediate harvests at selected periods between sowing and final harvest. Such experiments may be poor models for experiments in the ecological genetics of plants, in which a great deal of information may be gathered by 'non-destructive scoring' of the plants over weeks or months. For instance, given adequate spacing between plants, plant height at different times could be measured, and the timing of flowering and fruiting could be studied. Also, samples of leaves could be removed for study, provided that all the material in the experiment is treated alike. Thus, a good deal of quantitative information might be obtained by repeated scoring of an experiment. Sometimes it is unnecessary to make measurements; the stages of development or incidence of damage by pests may be recorded by classifying the material into a small number of 'character states'.

The designed experiment

So far we have discussed a number of important factors in the design of genecological experiments. For both the experimenter and the botanist who wishes to interpret the scientific literature, it is crucial to take proper account of the problems, and possibilities of sampling and cultivation. These are elements in a larger canvas, however. Many authors have stressed that genecologists should aim at a *designed experiment* in which hypothesis, sampling, cultivation, analysis and interpretation all take their proper places.

The generation of germinal ideas is a mysterious process. Armed with a knowledge of the literature, provoked by the observations and comments of others, the botanist notices something of interest in the patterns of variation. From this initial interest an idea emerges for an experiment. The process by which ideas occur to experimenters is not to be seen as a mechanical process, but as a creative act much as is required for practice of the arts. Next, the experimenter formulates a hypothesis leading to an experimental investigation, the results of which are used to consider whether the hypothesis is confirmed or rejected. As part of the investigation the results may be subjected to statistical tests.

The best way to appreciate the different elements in the designed experiment is to study an example. We have chosen to present the results of a simple study on *Plantago major* (Warwick & Briggs, 1979). Our account should be seen as a simplified introduction to a central concern of science, namely how to devise, execute and interpret experiments. We hope that biologists reading our account will be encouraged to study the many excellent introductory books (which we have noted above) on the design and statistical analysis of experiments.

An experiment to study the variation in **Plantago major** *growing on droves (grassy tracks) at Wicken Fen Nature Reserve, Cambridgeshire, England*

Many thousands of visitors visit the famous Wicken Fen Nature Reserve each year and the droves (grassy tracks) which cross the Fen are subject to severe trampling pressure. *P. major* occurs in the heavily trampled areas (as a small, prostrate plant) and also in the adjacent grassy sward (in which it is a larger, erect plant).

Ecotypic differentiation has been reported in *P. major* (Turesson, 1925; Groot & Boschuizen, 1970; Mølgaard, 1976) and, as we shall see in Chapter 9 there is evidence from a number of genecological studies which suggests that differentiation might occur over short distances, despite gene flow. Therefore, the possibility exists that dwarf, prostrate variants might be selected on the pathway, while taller plants would be at a premium in the adjacent grassy sward. Thus, we could formulate the hypothesis that samples taken from the wild might retain their distinctness in cultivation. As the differences involved are those of size, our hypothesis is not very precise in its present form. We cannot make any definite prediction as to the degree of difference to be retained; indeed as we are dealing with quantitative differences it is not at first sight clear how one can make a prediction as to the degree of difference which 'needs' to be retained in order to accept the hypothesis. So far our hypothesis is too vague. However, a precise hypothesis is possible in this case, namely that on cultivation we expect *no* difference between groups of plants after cultivation. Such a hypothesis is known as a 'null hypothesis'. The concept of the null hypothesis is widely used in biology and such a hypothesis, that zero difference is expected between two sample groups, should always be formulated as part of a designed experiment, for a precise initial hypothesis is likely to lead to a well-designed investigation.

Unbiased samples, 10 from the trampled area and 10 from adjacent grassy swards, were collected in the autumn of 1974. *P. major* is not a clonally propagating species (although it may be cloned in gardens: Marsden-Jones & Turrill, 1945), but spaced samples were taken at least 10 m apart. Plant material was potted up in John Innes No. 1 compost and the pots, which were randomly arranged, were plunged to the rims in the sand of an outdoor plunge bed. Spacing between pots was very generous and guard rows were not necessary.

In order to allow us to examine the null hypothesis, a statistical test is necessary to enable us to compare the two groups of samples. The test should allow us to compare the variation between and within groups. Clearly variation between groups (from trampled path versus adjacent grassy sward) is only likely to be significant if it can be shown to be significantly greater than variation within groups. We shall use for our test the analysis of variance technique, which works by estimating the significance of variation between groups by comparing it with variation within groups. The variation in some measurable trait of 20 plants of *P. major* is, by this test, partitioned in such a way as to enable us to see the variation due to subsites at Wicken, while at the same time giving us an estimate of the variation within groups.

The steps in the analysis of variance are a simple extension of those used in Chapter 3. To recapitulate, we showed that:

$$\text{variance } (s^2) = \frac{\sum(x - \bar{x})^2}{n - 1}$$

The sum of the squares of the deviations from the mean could be calculated by subtraction of each value from the mean, squaring the difference and summing the resulting squared deviations. Alternatively, we suggested that, if a calculating machine is available, the sum of (deviations from mean)^2 (sum of squares) could more readily be calculated by employing the formula:

$$\text{sum of squares} = \sum x^2 - \frac{\left(\sum x\right)^2}{n}$$

Where $\frac{\left(\sum x\right)^2}{n}$ is known as the Correction Factor or Term, C.

We may now examine (Table 8.7) the steps in the calculation of simple analysis of variance on the *P. major* experiment. The *null hypothesis* is that there is no difference in leaf length between plants grown from trampled drove and from grassy sward.

If this null hypothesis is to be confirmed then there should be little or no difference between the variances between and within groups. To estimate the relative size of these two variances we calculate the variance ratio (the F value – in honour of R.A. Fisher who developed analysis of variance). If, however, there is a real difference between groups, we would expect variation between groups to exceed that of the variance within groups. Tables of probabilities appropriate to different values of F are available. In the case of the *P. major* experiment, it is clear that a good deal of the variation is *within* groups and that the difference between groups is small. The mean values are very similar. Indeed, leaf length of plants grown from the small plants of the trampled area slightly exceeds that for the samples from the tall sward. The null hypothesis, that there is no statistically significant difference between the two groups of plants, is supported by our results. On the strength of present evidence, we have no reason to suppose that ecotypic differentiation has occurred in the trampled and tall sward subsites.

The *P. major* investigation was part of a more extensive study of this species in various grasslands (Warwick & Briggs, 1979, 1980*b*). Table 8.8 sets out another comparison. Small phenotypes were found not only in trampled areas (as on the droves at Wicken), but also in closely mown lawns. Samples of plants from the Botanic Garden lawn in Cambridge and

Table 8.7. Plantago major: *length of longest leaf (cm) in plants after* c. 10 *months cultivation in the Botanic Garden, Cambridge*

	Wicken Fen: trampled areas on droves	Wicken Fen: grassy swards adjacent to droves
	30.5	33.3
	33.4	28.0
	25.5	21.9
	34.2	26.0
	27.4	24.0
	26.5	28.4
	31.5	32.2
	29.3	27.0
	24.8	26.3
	28.0	26.0
Mean	29.110	27.310
Total	291.100	273.100

Grand total = 564.200

$$\text{Correction factor} = \frac{564.200^2}{20} = 15916.082$$

$$\text{Sum of squares (total)} = (30.5^2 + 33.4^2 + 25.5^2 \ldots 26.0^2) - C$$
$$= 16133.080 - 15916.082 = 216.998$$

Having calculated the total sum of squares we now calculate the variations between and within groups.
Between groups is estimated by

$$\frac{291.100^2}{10} + \frac{273.100^2}{10} - C = 15932.282 - 15916.082 = 16.200$$

Within groups is estimated by subtracting 16.200 from the total sum of squares.
Within groups sum of squares = 216.998 − 16.200 = 200.798
Subdivision of the sum of squares into its two parts has been accomplished and the degrees of freedom (19 in all: one less than the number of observations) may now be determined for each component. Between groups: 2 groups, therefore 1 degree of freedom. Within groups: 10 observations per group, each loses 1 degree of freedom, total 18.
The analysis of variance may now be set out in a table showing the sources of variation, the divisions of degrees of freedom and sum of squares. Mean squares (variances) are now calculated. The between-groups mean square gives the variance of the two groups about the grand mean, while the within-groups variance gives the variance of individual values about the two sample means.

Source of variation	Degrees of freedom	Sum of squares	Mean square (variance)	Variance ratio (F)	Probability
Between groups	1	16.200	16.200	1.452	> 0.05
Within groups	18	200.798	11.155		
Total	19	216.998			

Table 8.8. Plantago major: *the effect of* c. *10 months cultivation on samples of small phenotype from Wicken droves and Botanic Garden lawns; length of longest leaf (cm)*

	Wicken: trampled areas on droves	Botanic Garden lawns
	30.5	11.8
	33.4	20.7
	25.5	8.9
	34.2	22.6
	27.4	24.0
	26.5	14.1
	31.5	13.1
	29.3	16.0
	24.8	12.5
	28.0	12.0
Mean	29.110	15.570
Total	291.100	155.700

Grand total = 446.800

$$\text{Correction factor} = \frac{446.800^2}{20} = 9981.512$$

$$\text{Sum of squares (total)} = 30.5^2 + 33.4^2 + 25.5^2 \ldots 12.0^2 - C$$
$$= 11228.860 - 9981.512 = 1247.348$$

$$\text{Between groups sum of squares} = \frac{291.100^2}{10} + \frac{155.700^2}{10} - C$$
$$= 916.658$$

Within groups sum of squares = 1247.348 − 916.658 = 330.690

Source of variation	Degrees of freedom	Sum of squares	Mean square (variance)	Variance ratio (F)	Probability
Between groups	1	916.658	916.658	49.895	< 0.001
Within groups	18	330.690	18.372		
Total	19	1247.348			

from Wicken droves (trampled areas) were compared. The variation between groups in this case is statistically significantly greater than the variation within groups. Therefore, the null hypothesis, namely that samples do not differ in leaf length, receives no support from the experiment. There would appear to be a real difference in leaf length between the two samples. In Warwick & Briggs (1979, 1980*b*) details are given of the highly distinctive plants of *P. major* discovered in the lawns of Cambridge colleges and gardens.

Our examples of analysis of variance are of a very simple kind, with division of the variation into two parts. Much more elaborate experiments may be devised and the 'overall variation' discovered in experiments may be divided into many parts estimating, where appropriate, the variation due to blocks, population differences, family lines within populations, interacting factors, random events, etc. By looking at the relative magnitude of different segments of the variation, very considerable insights into population variation may be obtained.

Analysis of variance is a most elegant technique, which must, however, be used with care. It should only be employed in analyses where the results are 'normally distributed' and in which the variances of the contributing population samples, treatment values, etc., are equal or approximately so. Various tests have been devised to study the 'properties' of arrays of figures to see if they are appropriate for analysis of variance (for details of Bartlett's test see Salmon & Hanson, 1964; Sokal & Rohlf, 1981). Sometimes it is possible to 'transform' the results to produce equality of variances. For instance, the unsatisfactory raw data may be converted to square roots or to logarithms. If the results cannot be satisfactorily transformed, then other statistical tests – so called non-parametric tests – may be applied (Sokal & Rohlf, 1981). Such tests do not make any assumptions that the figures from the experiment are normally distributed or have equal variances. Non-parametric tests should be more widely used in biology, for the results of many experiments and observations show enormous departures from normality.

The interpretation of experiments

Whatever the results of particular experiments, there are usually grounds for a cautious interpretation of genecological studies.

However many plants are grown, or studied in experiments, the size of samples that can conveniently be grown is often minute relative to the size of wild or semi-natural populations. For instance, according to the estimates of Barling (1955), populations of *Ranunculus bulbosus* may reach 257 000 per acre in the English Cotswolds, and continuous populations in adjacent fields of pasture were estimated to contain 14 000 000 plants. Such figures are by no means exceptional.

Many experiments are carried out in conditions remote from those in nature. For example, studies of metal tolerance in plants involve measurements of root growth in very simple culture solutions (see Chapter 9). In the wild, plants grow in soils where conditions are quite different. The attempt to simplify situations in order to study individual factors is clearly justified, but the investigator must not make too facile an extrapolation from simple laboratory tests to the natural situation.

As in the case of metal tolerance, many experimentalists isolate individual factors of presumed importance and make special studies of the tolerances of population samples. A fascination with the study of critical or limiting factors should not blind the student of evolution to the fact that the concept of a factor is an abstraction. Often particular factors are chosen for study, largely because the means to control or vary them in precise ways are available in laboratories. It is often forgotten that plants respond to their environments as a functioning whole. Realising this difficulty, a number of botanists are becoming interested in experimental studies of the adaptive significance of variation in plants by carrying out experiments in the field. The garden trial with its weed-free, spaced plants is not entirely satisfactory as a means of studying 'adaptation', for the competitive interaction between plants is absent. Thus, there has been a revival of interest in the reciprocal clone-transplant experiment, in which cloned material of diverse origin is transplanted into swards subject to different treatments. By close mapping and labelling of plants, the survival and growth of transplants may be studied (see Chapter 9). Care in the layout and recording of such experiments may overcome the difficulties which, as we saw in Chapter 6, cast doubt on the historic studies of Bonnier and Clements. By studying the way plants behave in such experiments, the experimentalist may have a very direct insight into the responses of plants to 'whole' environments. There is obviously a place for both tolerance tests and reciprocal transplant investigations in the repertoire of techniques available to the genecologist.

A final problem facing the experimentalist is that of deciding the causes of the underlying patterns of variation under study. Even after long and complex experiments, it is not possible to conclude with certainty that residual variation in, say, a garden trial, is 'genetic' in origin; breeding experiments are necessary to see if characteristics are transmitted by seed. As we shall see in the next chapter, which reviews recent studies in genecology, the advent of molecular methods has transformed the study of population variation.

9

Recent advances in genecology

In attempting a short review of contemporary developments in genecology, we note that not only are botanists still fascinated by patterns of variation between and within populations, but also that a number of other issues have emerged to enrich genecological research.

1. Turessonian genecology is based on the premise that patterns of ecotypic variation are the result of natural selection. In recent years, special efforts have been made to study the amount of genetic variation in populations, the speed of action of natural selection, and factors limiting the responses of populations of plants. Experimentalists have also studied other population phenomena, such as gene flow and chance effects. Such studies have become more precise with the development of molecular tools, including isozymes and DNA profiles.
2. Early studies in genecology often, but not always, involved the study of widely spaced sites and the examination of gross differences in phenotype, say as large as those found in Mendel's stocks of Peas – tall (*c*. 2 m or more) versus dwarf (*c*. 0.5 m). Almost any garden trial, however crude, would hardly fail to 'reveal' such differences in height. However, with the recent surge of interest in studying small sites as systems has come a fascination with small-scale variation. As we saw in Chapter 8, the study of this type of variation requires good sampling techniques, combined with proper attention to design, layout, scoring and analysis of experiment. In the repertoire of techniques available to the genecologist, the transplant experiment is proving a powerful tool in analysing population variation. In such experiments, plants are set out in natural or artificial swards.
3. Evidence for patterns and processes within populations studied in detail is likely to be more complete if some notion of the history of study sites is

obtained. Thus, many genecologists studying evidence from different sources try to see present-day vegetation in its historical context. As we shall see later in this chapter, a knowledge of the history of study areas has provided a framework in which the speed of evolutionary change may be judged.

4. In times past, many genecologists chose to study more or less natural communities (Heslop-Harrison, 1964). However, many insights have come from agricultural experiments, and it has recently been realised that many of the most fascinating areas for the study of selection and other population processes are to be found in man-disturbed habitats (Bishop & Cook, 1981; Bradshaw & McNeilly, 1981; Taylor, Pitelka & Clegg, 1991; Davies, 1993). In such areas, the botanist often finds the juxtaposition of extreme habitats, for instance: areas with heavy metal pollution can occur as islands in a sea of pasture land; grasslands managed for grazing may be found close to hayfields or arable land; lawns are surrounded by flowerbeds; and improved grasslands, subject to fertiliser and pesticide treatments, are located next to untreated grasslands.
5. Another major development in ecological research has influenced the thinking of genecologists. Ecologists now study populations in detail, recording the lifespans of individuals, the effects of migration and competition, and details of resource allocation and reproductive success (see, for example, Harper, 1977; Willson, 1983; Silvertown & Lovett Doust, 1993). The close study of plants in the field, coupled with imaginative experiments, has begun to offer considerable insight into the factors affecting population size and evolutionary aspects of plant–animal interactions (Shorrocks, 1984; Cockburn, 1991). Population biologists are also considering the implications of genetic variation between and within populations, and are often interested in genecological issues as ecologists: in contrast, in the early years of the subject many genecologists had an initial interest in genetics or taxonomy.

Variation in populations

Isozymes are being increasingly used to estimate the level of genetic variation in populations. (See reviews of a growing body of information by Hamrick, Linhart & Mitton, 1979; Gottlieb, 1981*a*; Hamrick & Loveless, 1989; Hamrick & Godt, 1989.) A sufficient number of surveys have now been carried out to permit some generalisations to be made. Populations of species with different life-form, breeding system, pollination type, seed dispersal, etc. have different levels of variation (Table 9.1). Also, as we

Table 9.1. *The amount of allozyme variation for different groups of higher plants. (From Brown & Schoen, 1992, using information in Hamrick & Godt, 1989.) The measures of variation are: A = allelic diversity, as number of alleles per locus; h = a measure of gene diversity;* G_{ST} *= the extent of population divergence, as the proportion of the total diversity found among populations* ($G_{ST} = 1 - h_p/h_s$)*; and N = number of species. Subscripts* s *and* p *refer to measures within species and populations, respectively.*

Many comparisons may be made between different groups of species. For instance, note that, in comparison with outcrossing species, selfers show less diversity within species and populations but greater divergence between populations. Also, that endemic and narrowly distributed species are less variable than widespread species

Categories	N	Species A^a_S	h_S	Population A_P	h_P	Divergence G_{ST}
Breeding System – Pollination						
Selfing	78	1.69	0.12	1.31	0.07	0.51
Mixed						
– animal	60	1.68	0.12	1.43	0.09	0.22
– wind	9	2.18	0.19	1.99	0.20	0.10
Outcrossing						
– animal	124	1.99	0.17	1.54	0.12	0.20
– wind	102	2.40	0.16	1.80	0.15	0.10
Geographic Range						
Endemic	52	1.80	0.10	1.39	0.06	0.25
Narrow	82	1.83	0.14	1.45	0.11	0.24
Regional	180	1.94	0.15	1.55	0.12	0.22
Widespread	85	2.29	0.20	1.72	0.16	0.21
Life Form						
Annual	146	2.07	0.16	1.48	0.11	0.36
Perennial herbs	119	1.70	0.12	1.40	0.10	0.23
Long-lived woody	110	2.19	0.18	1.79	0.15	0.08
Seed Dispersal						
Gravity	164	1.81	0.14	1.45	0.10	0.28
Attached	52	2.96	0.20	1.68	0.14	0.26
Explosive	23	1.48	0.09	1.25	0.06	0.24
Ingested	39	1.69	0.18	1.48	0.13	0.22
Wind	105	2.10	0.14	1.70	0.12	0.14

saw in Chapter 7, DNA fingerprint techniques are providing valuable insights into population variation.

Plant populations

The word 'population', like so many other familiar terms, seems to present little difficulty until we have to define it. In statistics, the concept of a population is an abstraction signifying a theoretically large assemblage of individuals from which a particular group under consideration is a sample. Most biological uses of the term, however, imply the total of organisms belonging to a particular taxonomic group (or 'taxon'), which are found in a particular place at a particular time. The population unit of outstanding significance in the study of variation is quite different, being the local 'interbreeding group of individuals sharing a common gene pool' (Dobzhansky, 1935, in Rieger, Michaelis & Green, 1976). In earlier editions of this book, the term 'gamodeme' has been used for such populations. It is with regret that we abandon the use of this term, but must face the fact that the 'deme' terminology has not been accepted by genecologists. It was devised by Gilmour and associates (Gilmour & Gregor, 1939; Gilmour & Heslop-Harrison, 1954; Gilmour & Walters, 1963; Walters, 1989*a*, *b*). Details of the system are given in the Glossary. Population geneticists have coined other names – Mendelian or panmictic populations – for groups essentially similar to gamodemes. All these units represent idealised 'model systems' (Maynard Smith, 1989; Silvertown & Lovett Doust, 1993). As we shall see, it is not possible to determine the exact limits of Mendelian populations. The best that can be achieved is the use of statistical models to determine the probable limits of gene flow.

The model underlying the idealised population is the Hardy-Weinberg Law. In a population with two alleles A and a, which are selectively neutral and in which mating is at random, the expected frequencies of A and a (p and q, respectively) are:

$$(p + q)^2 = p^2 + 2pq + q^2 = 1$$

This represents the 'null' model (Cockburn, 1991) for a very large population of a diploid organism, with sexual reproduction, non-overlapping generations, the same allele frequencies in male and female, no gene flow or migration, and no selection operating for that locus. In such a situation, the relative frequencies of the alleles will not change from generation to generation.

However, several factors will cause deviations from the Hardy-Weinberg equilibrium and many have been modelled by population geneticists (Silvertown & Lovett Doust, 1993). These are: mutation; 'linkage' of genes of various kinds; non-random mating; gene flow; chance events; and natural selection. Such model systems provide a very valuable framework for the genecologist but, as we are concerned in the main with field studies, we will use mostly verbal models, at the same time giving details of published sources containing mathematical treatments.

We may now look at the factors involved in deviations from the Hardy-Weinberg equilibrium. Mutation rates are usually very low and do not influence populations in the short term. Concerning the 'linkage' of genes, we have already considered, for example, the breeding system in *Primula* where, a 'supergene' of closely linked genes controls distyly. Our account of the various breeding systems found in plants (Chapter 7) provides many excellent examples of situations, including self-incompatibility, autogamy and apomixis, where the Hardy-Weinberg 'null' position is vitiated by various kinds of non-random mating. Pollinator behaviour may also produce non-random (so-called assortative) mating. For instance, in *Raphanus raphanistrum*, which is sometimes polymorphic for flower colour (white/yellow), pollinators may show strong preferences for one flower colour variant (Kay, 1978).

Gene flow

In order to determine which plants are interbreeding it is necessary to study patterns of gene flow, i.e. the source and ultimate destination of fertilising pollen, and the pattern of dispersal of fruits and seeds. Until recently, direct measures of gene flow had been made only in crop plants, where genetic markers were available. For wild plants only circumstantial evidence was available, based on a study of the movement of pollen and seed. However, as we shall see, molecular tools are now providing, within certain limits, direct evidence of actual gene flow and paternity in wild plants.

Gene flow: early ideas

Until the 1970s, models of gene flow in natural populations were not clearly formulated but, according to Ehrlich & Raven (1969), many biologists thought that such flow was extensive (see, for example, Merrell, 1962; Mayr, 1963). This view was supported by the following observations of naturalists and others:

1. Pollen, present in enormous quantities in the air at appropriate times of

the year, is widely dispersed (even being detected, for example, 50 kilometres (km) out to sea in the Gulf of Bothnia: Hesselman, 1919).
2. The various mechanisms for ensuring insect and wind pollination were assumed to be widely effective.
3. The assumption was made that fruits and seeds were well adapted to wide dispersal in nature.

These general impressions led naturally to the view that the members of a widespread species may be united in the same gene pool by gene flow. The species was viewed as the important evolutionary unit, gene flow amongst its members making it a breeding unit. Speciation was seen as the breakdown of the cohesive gene pool of the species by the evolution of isolating mechanisms.

This view that gene flow is widespread and 'effective', and that interbreeding populations are therefore large, has been subject to radical reappraisal. It is now contended that gene flow in nature may be much more restricted than previously thought, in the distribution of both pollen grains and seed. Thus, species may now be visualised as comprising a multitude of small sub-populations rather than single or a few large ones (Ehrlich & Raven, 1969; Levin, 1978*a*). Evidence of gene flow comes from various lines of study (Levin & Kerster, 1974; Richards & Ibrahim, 1978; Levin, 1988; Falk & Holsinger, 1991).

Gene flow: agricultural experiments

Some of the most interesting information on gene flow has come from studies of crop plants (Levin & Kerster, 1974), forest trees (Wright, 1953) and weeds (Cousens & Mortimer, 1995). The development of superior cultivars by plant breeding requires the production of large quantities of 'pure' seed, and there has, therefore, been a good deal of interest in estimating the minimum distance required to prevent crossing between two different cultivars. Table 9.2 shows the isolation requirement for a number of crop species. It is clear that only a relatively small distance is required in many cases to prevent contamination. Experiments with mixtures of cultivars differing in genetic markers have also provided important insights. Thus, the incidence of hybridisation between mixtures of two crops depends upon planting arrangements in the field (Fig. 9.1). The degree of crossing also depends on the presence and disposition of taller plants. For example, in an experiment studying gene flow in Cotton (*Gossypium*), which is insect pollinated, the degree of intercrossing of two cultivars of Cotton was very

Table 9.2. *Isolation requirements for seed crops (After Kernick, 1961, from Levin & Kerster, 1974)*

Species	Breeding system[a]	Pollination agent[b]	Isolation requirement (m)
Gossypium spp.	S	I	200
Linum usitatissimum	S	I	100–300
Camellia sinensis	S	I	800–3000
Lactuca sativa	S	I	30–60
Avena sativa	S	W	180
Hordeum vulgare	S	W	180
Oryza sativa	S	W	15–30
Sorghum vulgare	S	W	190–270
Triticum aestivum	S	W	1.5–3.0
Cajanus cajan	SC	I	180–360
Citrullus vulgaris	SC	I	900
Apium graveolens	SC	I	1100
Carthamus tinctorius	SC	I	180–270
Papaver somniferum	SC	I	360
Vicia faba	SC	I	90–180
Pastinaca sativa	SC	I	500
Voandzeia subterranea	SC	I	180–360
Nicotiana tabacum	SC	I	400
Coffea arabica	SC	IW	500
Hevea brasiliensis	C	I	2000
Helianthus annuus	C	I	800
Brassica campestris	C	I	900
Daucus carota	C	I	900
Lycopersicum esculentum	C	I	30–60
Anethum graveolens	C	I	300
Brassica oleracea	C	I	600
Allium cepa	C	I	900
Raphanus sativus	C	I	270–300
Brassica oleracea	C	I	970
Carica papaya	C	I	100–1600
Zea mays	C	W	180
Chenopodium ambrosoides	C	W	180–360
Cannabis sativa	C	W	500
Secale cereale	C	W	180
Beta vulgaris	C	IW	3200

[a]S, species which are predominantly self-fertilising; SC, species in which self- and cross-fertilisation are of similar importance; C, species in which there is predominant or exclusive cross-fertilisation.
[b]I, insect pollination; W, wind pollination.

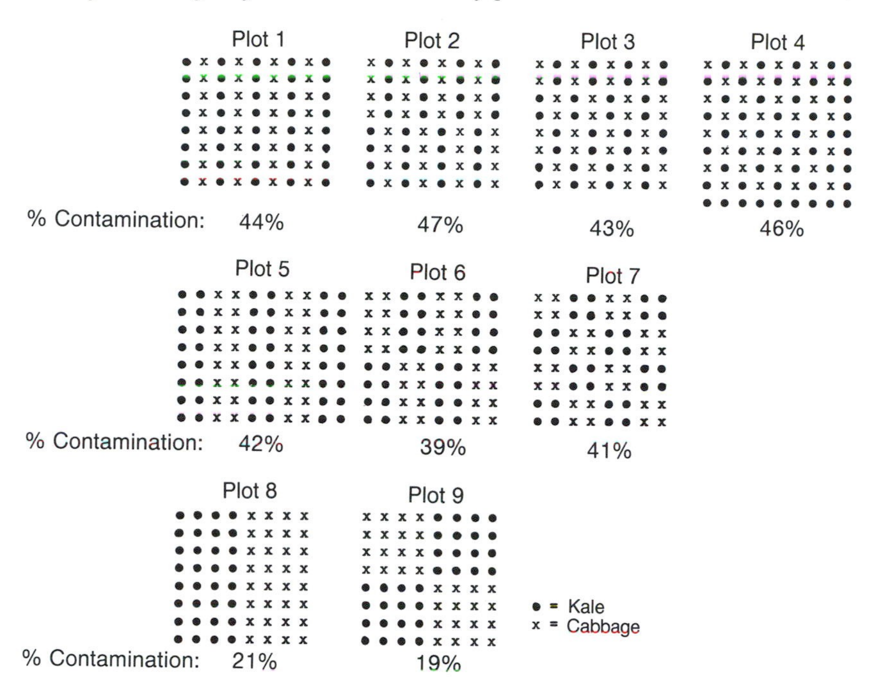

Fig. 9.1. Block planting layout to show the relation between the pattern of planting and the production of hybrid seed. • = Kale; x = Cabbage (both cultivars of *Brassica oleracea*). Distance between plants = 50 cm. (After Nieuwhof, 1963, from Levin & Kerster, 1974.)

much influenced by growing barriers of Maize (*Zea*), *c*. 2–3 m high, between the rows of plants (Pope, Simpson & Duncan, 1944). In the agricultural landscape, the presence of hedgerows and plantations has been known to act as a barrier to crossing (Jensen & Bogh, 1941; Jones & Brooks, 1952).

Gene flow: insights from the movement of pollen

Govindaraju (1988) has reviewed the available evidence on pollen movement in 115 species. On average, gene flow in wind-pollinated species is greater than that of those pollinated by animals.

Wind pollination.

By sampling the pollen content of the air, using sticky slides or pollen traps, it has frequently been discovered that from an isolated source pollen is distributed in a leptokurtic fashion, much pollen falling relatively close to the parent plant and only a small portion travelling some distance from the parent. In interpreting the various studies, it is important to note the marked effects of wind speed and direction, as well as the effect of the height of pollen presentation, pollen grain sizes, and the influence of environmental heterogeneity in biotic and topographical factors. It is also important to study the actual pollination mechanism in the field. Figure 9.2 shows that a plant thought on grounds of morphology to be wind pollinated is in fact visited by insect pollinators.

Animal pollination.

In recent years the study of plant–animal relationships has proved a major growth area in biology. While many plants are apparently well adapted for insect (or bird) pollination, it does not follow that those plants are all

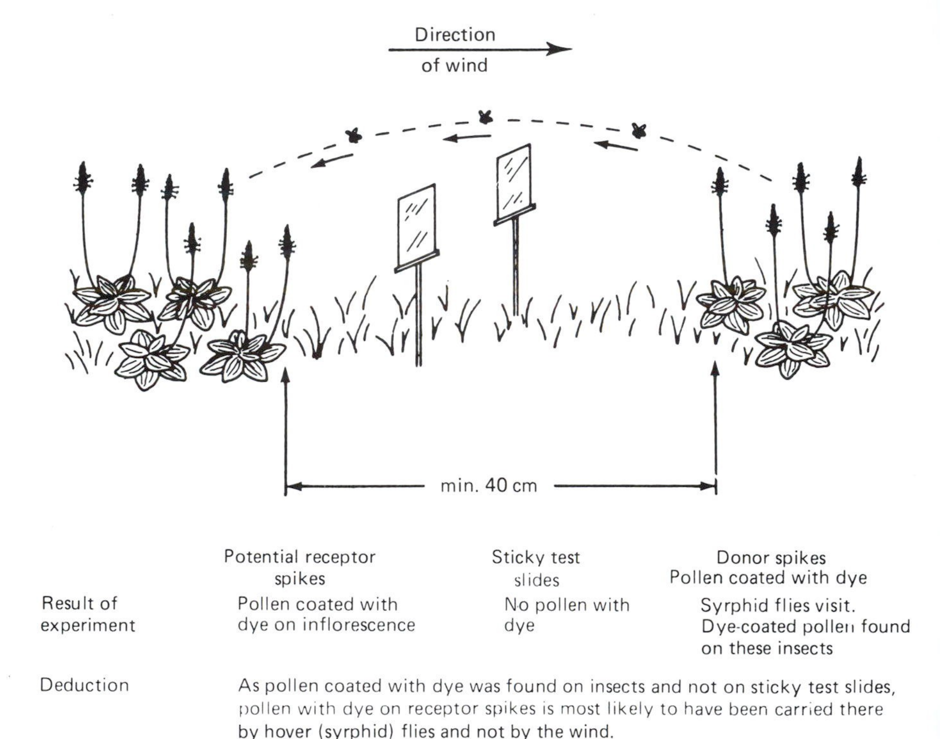

Fig. 9.2. Insect pollination of supposedly wind-pollinated plants: sketch of a field experiment with *Plantago lanceolata*. (From Stelleman, 1978.)

effectively pollinated. Competition for pollinators may occur (Dafni, 1992) or there may be too few plants in the population for all to be effectively pollinated (Jennersten, 1988). Entomologists are now inclined to study insect behaviour in terms of cost-benefit analysis. Foraging for food (pollen or nectar) is an energy-consuming activity, and 'optimal foraging' may be undertaken by insects, that is to say, they visit the flowers offering the best return for the energy expended in searching. The flowers of a particular species under study must now be viewed as an array of 'floral offerings' presented to a wide variety of insects, and many factors such as the degree of faithfulness of an insect to particular plant species, as well as patch size, spatial pattern and density effects, site heterogeneity, and seasonal changes must be taken into account. Given such complexities, the only course of action open to the student of evolution is to study populations individually. Insects and birds must be observed to see which plants are actually visited. As an aid to such studies, pollen may be stained (Simpson, 1954; Sindu & Singh, 1961) or made radioactive (Schlising & Turpin, 1971), so that patterns of dispersal may be discovered. Sometimes intraspecific differences in size of pollen grains have enabled observations of gene flow to be undertaken (e.g. Richards & Ibrahim, 1978). Dafni (1992) provides a critical review of the methods available for studying pollen flow. In summary, as with wind dispersal of pollen, leptokurtic distribution of pollen is likely as a consequence of the activities of pollen vectors.

Gene flow: studies of seed dispersal

Wind dispersal.

The distribution of propagules may be studied by catching seeds or fruits on sticky tapes or in traps, or by looking for seedlings around individual plants. Sometimes the distribution of marker genes from a carrier parent may be examined (see, for example, Bannister, 1965, for a study of the distribution of various markers in plants of progeny found in a study of *Pinus radiata*). Dispersal of medium to large seeds probably conforms to a leptokurtic distribution, though the 'tail' may be very long. Risking a generalisation, Levin & Kerster (1974) consider that even in species with fruits and seeds apparently well adapted with wings or plumes, most fruits or seeds travel relatively short distances from the parent plant. Much remains to be discovered about wind dispersal; experimental approaches should be encouraged to displace the simple notion that, if a plant has a plumed fruit or seed, *ipso facto* its progeny must be widely scattered over large areas at each generation. Detailed studies are needed to determine what happens in particular species. For instance, can it simply be assumed that all the seeds produced by a plant behave in the same way? Cheplick & Quinn (1982, 1983) have discovered a very interesting reproductive strategy in the grass *Amphicarpum purshii*. Plants produce not only normal spikelets of 'far dispersed' aerial fruits, but also larger and heavier subterranean 'near dispersed' cleistogamous fruits. In many Composites, dimorphic or polymorphic achenes are found. For example, in *Heterotheca latifolia* the disc-florets produce a thin-walled short-lived fruit with a pappus, facilitating dispersal by the wind. In contrast, the ray-florets give rise to thick, fibrous achenes with no pappus, and these achenes form a seed bank in the soil (Venable & Levin, 1985). The multiple dispersal strategies of plant species have been reviewed by Schoen & Lloyd (1984), and seed heteromorphism and germination behaviour by Silvertown (1984).

Animal dispersal.

Current interest in population dynamics and food resources has stimulated research into the relationships of animals to plant propagules. Very little is known, however, about the primary distribution of propagules by animals (Harper, 1977). Cousens & Mortimer (1995) provide some excellent examples of animal dispersal of weeds, both by adhesion to fur and by the ingestion of seeds, subsequently defecated in a viable condition. However, undoubtedly much remains to be discovered about primary dispersal. Also, secondary dispersal may complicate seed flow in some species. For instance, in some species of the genus *Viola*, primary dispersal is by means of explosive release and wind, but each seed has a protein-rich elaiosome on its surface. This is a food source for ants, which carry them back to their nests, giving a different ultimate pattern of seed dispersal (Beattie, 1978; Huxley & Cutler, 1991).

Gene flow: studies using molecular tools

The estimation of gene flow in wild populations, using observations of pollen and seed dispersal, is problematic, but many of the difficulties are avoided if unique genetic markers are employed. For example, Muller (1977) examined the distribution of a unique allele (of the enzyme leucine aminopeptidase; LAP) in a population of *Pinus sylvestris* and discovered the allele in progenies of trees 80 m from the source. If unique alleles are not found in the wild, researchers have often studied gene flow in artificial populations using allozymes (Smyth & Hamrick, 1987). For example,

Schaal (1980) used an allozyme marker to study the movement of pollen in populations of *Lupinus texensis*. She found that the estimate of gene flow obtained from the distribution of allozymes in progenies was greater than that obtained by the study of pollen flow. A number of indirect methods have been employed in estimating gene flow. If there are some unique alleles in a population, then gene flow between it and other populations can be estimated (for details, see Hamrick, 1990).

Polymorphic allozyme markers have also been used in the studies of paternity. In a study of the dioecious species *Chamaelirium luteum* in a forest in North Carolina, the genotypes of many plants were established using 11 electrophoretic markers (Meagher & Thompson, 1987). From this detailed knowledge it was possible to determine the realised gene flow patterns and 'most-likely' pollen parent of progenies of various plants (Fig. 9.3). Studies of paternity have also been made in populations of *Raphanus* (Ellstrand & Marshall, 1986). Typically, 1–4 (mode 2) pollen parents have fertilised the different ovules on each maternal parent.

Recently, a further advance in the study of gene flow has been made, in studying populations of the dioecious species *Silene latifolia* (*S. alba*). McCauley (1994) writes: 'With maternal inheritance the genetic structure of the chloroplast DNA should reflect seed movement, whereas the genetic structure of the nuclear-encoded allozyme loci should reflect the movement of both seeds and pollen'. The results of the study revealed that both seeds and pollen contribute significantly to gene flow.

Studies using polymorphic allozymes have led to higher estimates of gene flow than those discovered by studying the movement of pollen and seeds, e.g. Ellstrand & Marshall (1985) discovered that 8–18% of the gene flow in *Raphanus sativus* may occur between populations separated by as much as a kilometre. Furthermore, Friedman & Adams (1985) discovered that 40% of the pollen fertilising a stand of *Pinus taeda* came from wild populations at least 400 m away.

Studies of isozymes have also provided important new information on long-distance dispersal to isolated oceanic islands. For example, the fern, *Asplenium adiantum-nigrum*, is the putative allopolyploid derivative of two strictly European diploid species. The tetraploid occurs on the extremely isolated islands of the Hawaiian archipelago. Samples of the species were collected and isozymes were studied (Ranker, Floyd & Trapp, 1994). There is evidence that the species is highly inbred and that different collections were genetically different. The pattern of difference could not simply be accounted for by mutation. Since the parental diploids are not known from the islands, the differences between samples is likely to be the result of

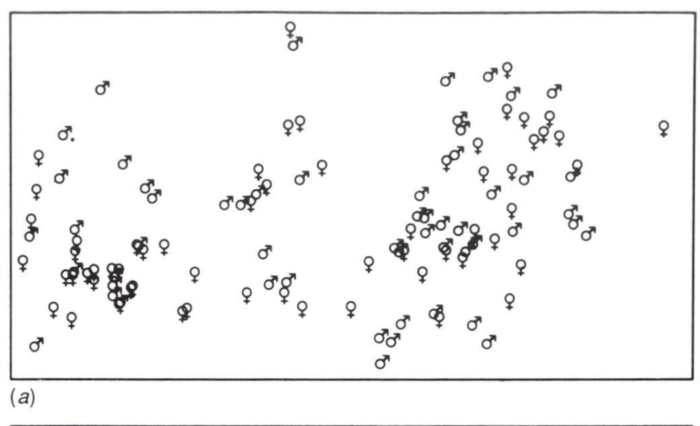

(*a*)

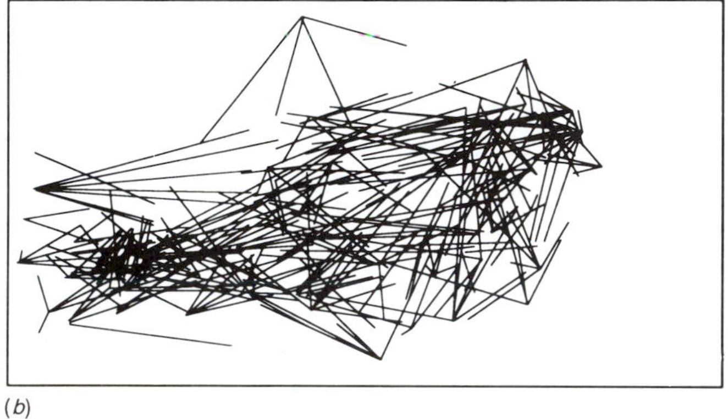

(*b*)

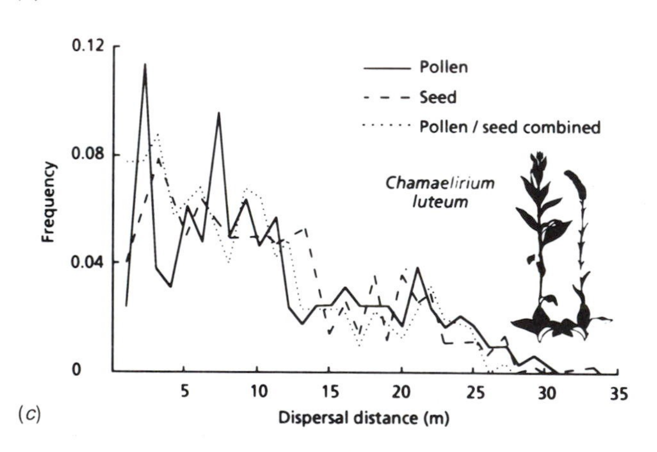

(*c*)

Fig. 9.3. Map of a population of *Chamaelirium luteum* showing: (*a*) the distribution of males and females; (*b*) lines connecting females with their mates in 381 natural crosses; and (*c*) the dispersal of pollen and seed within the population (Meagher, 1986, and Meagher & Thompson, 1987, from Silvertown & Lovett Doust, 1993).

multiple colonisation events (a minimum of three and a possible maximum of 17) by natural long-distance dispersal of the dust-like spores of this fern.

Some multiple introductions have resulted from human activities. For example, the medicinal and ornamental plant *Bryonia alba* was introduced to the western USA from Europe in the nineteenth century. Studies of the allozymes suggest that there were two or possibly three separate introductions (Novak & Mack,1995).

'Neighbourhoods' in wild populations

To determine the limits of Mendelian populations it would be necessary to have exact knowledge of gene flow. As we have already pointed out, population biologists have had to settle for *estimating* the size of 'breeding groups'. To estimate the number of individuals randomly mating and the area they occupy, Wright (1943, 1946) devised a 'neighbourhood' model, applicable to animal populations. 'A neighbourhood is defined as an area from which about 86% of the parents of some central individual may be treated as if drawn at random' (Levin, 1988). Discussion of the equations used is beyond the scope of this book, but full details are presented in Levin & Kerster (1974) and Crawford (1984). It is clear that Wright's equations, first formulated for bisexual mobile animal populations, are not ideally suited to the study of hermaphrodite, sessile and often clonal plants. Modified equations have been produced to take account of clonal spread, rate of outbreeding, pollen flow and seed dispersal (Crawford, 1984). Calculations of neighbourhood sizes have been made for a number of species (Table 9.3).

Several difficulties arise in the calculation of neighbourhoods in plants.

1. Pollinators. In estimates based upon pollinator flight distances, it is unclear whether pollen carry-over occurs to a second flower, and to a third, etc. (See Harder & Barrett, 1996, for a detailed discussion of pollen dispersal and mating patterns.) It is also difficult to take full account of insect behaviour, as there may be some directionality in their flight paths.
2. Seed dispersal. It often proves difficult to study seed dispersal in the wild and estimates involving spaced plants have had to be used (see, for example, Gliddon & Saleem, 1985, who studied *Trifolium repens*).
3. Dispersal distances. In the determination of both pollen and seed dispersal it is impossible to make an accurate assessment of the 'tail' of the distribution in natural vegetation.

Table 9.3. *Neighbourhoods in various herbs (Published estimates, details in Levin, 1988)*

Species	N_a (m^2)	N_a diameter (m)	N_e
Phlox pilosa	41	3.61	533
Liatris cylindracea	66	4.58	633
Liatris aspera	35	3.34	175
Viola pedata	48	3.91	432
Viola rostrata‡	25	2.82	167
Primula veris	30	3.09	7.4
Plantago lanceolata	10	1.78	17
Avena barbata	4	1.13	140

‡Estimates do not take into account selfing by cleistogamic flowers.
N_a = Neighbourhood Area
N_e = Neighbourhood Size

4. Choice of study area. Technical difficulties of watching pollinators and studying seed dispersal might have encouraged the experimenter to choose a relatively 'simple' open area with a low number of plants. It is unclear whether the results in Table 9.3 are typical of the species in question. Clearly, the neighbourhood size is greatly influenced by the density of the species being studied and the height of the vegetation in which it is growing.
5. Estimates of neighbourhoods do not represent constants for particular populations. Year-by-year differences in neighbourhoods are likely in pollinator activity, etc. Populations of plants with overlapping generations also present a complicating factor. For instance, biennial plants often have first-year non-flowering rosette individuals along with flowering and fruiting second-year plants. At first sight it would appear that there are two separate gene pools in biennial plants. However, studies (e.g. of *Senecio jacobaea*: Harper, 1977) have shown that second-year plants, prevented from flowering by insect or other damage, may flower in their third or later years.

It is clear that there are many problems in estimating neighbourhoods. Those who have studied the subject in detail are most aware of the difficulties. Thus, Levin (1988) states 'the neighbourhood values that I and others have calculated are only rough approximations'. However, the effort to try to estimate neighbourhoods has been clearly justified, as it has helped to transform our view of populations of plants in the field. Evidence from the few species studied in detail suggests that the dispersal of fertilising pollen and seeds is restricted, that neighbouring plants are likely to mate together, and that adjacent plants in a population are often likely to be genetically related. However, many more studies are required, especially in view of recent studies using molecular markers, which suggest the possibility of wider gene flow. Thus, we should avoid dogmatic generalisation about neighbourhoods on the basis of a few studies of plants mostly of herbaceous habit. Clearly, the 'neighbourhoods' of plants will depend upon growth form and the dispersal mechanism of the species in question. Thus, neighbourhoods in tree species with wind dispersed pollen and species with dust-like seed (e.g. Orchids) or small spores (e.g. ferns, fungi) may be larger.

Effects of chance

Chance may have a profound effect in restricting the variation in populations. This is particularly true if the population is reduced by a bottleneck effect to a very small size (Fig. 9.4). Wright (1931) was the first to point out that, in small populations, irregular random fluctuations in gene frequency occur which may result in the fixation or loss of one or more alleles 'without regard to their adaptive value' (Rieger, Michaelis & Green, 1976). Such random changes in gene frequency are known as 'genetic drift' or the 'Sewall Wright' effect. The effects of genetic drift may be very important in populations of endangered species, and we will be considering such species in Chapter 15. While models and computer simulations suggest that genetic drift may be crucial in small populations, it is exceedingly difficult, if not impossible, to separate the effects of chance and selection. Nevertheless, it is sometimes claimed that the effects of chance may be revealed in cases where selectively neutral traits are being considered. For instance, Rafinski (1979) investigated stigma colour polymorphism (white versus orange) in *Crocus scepusiensis* found growing in the Gorce mountains of Poland. Populations differed widely in frequency of the different morphs and Rafinski considered that the variation could be the result of chance events. The assumption that stigma colour is a neutral trait may not be justified, however, for the polymorphism may be adaptively significant in pollination, in ways we do not understand.

Chance also operates through the so-called 'founder effect'. A new

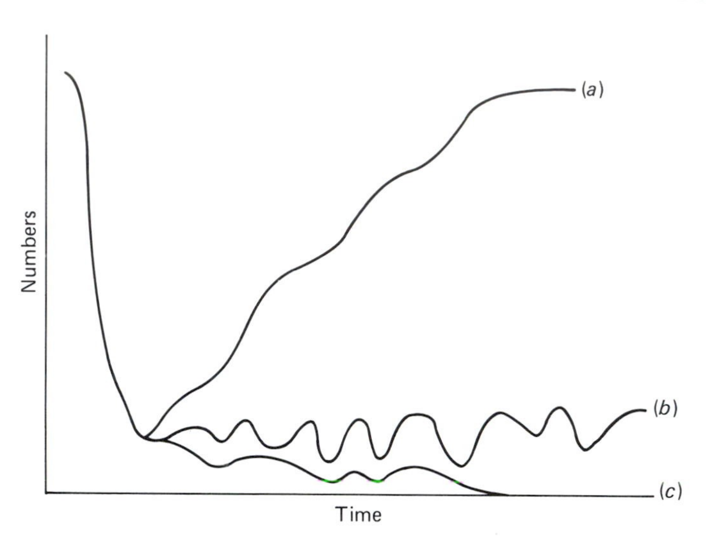

Fig. 9.4. The effect of a severe reduction in numbers of individuals in a population – the so-called bottleneck effect. (*a*) Recovery in numbers: the new population may not differ phenotypically from the original one but some genetic variability may be lost. (*b*) No recovery in numbers: if from the available seedlings, as an effect of chance only, a small random selection of adults survives at each generation, then the population may become less variable not only in qualitative traits, such as flower colour, but also in the breadth of variation in quantitative characters such as height, etc. While no population biologist disputes the importance of chance effects, it is impossible to isolate such effects from those of natural selection. (*c*) Extinction. The long-term effects of low population size in endangered species is discussed in Chapter 15.

population may be founded by one or a very few individuals, and it is likely to have only a random subset of the variation of the population from which it is drawn. As we have seen in Chapter 7, a single individual of a self-incompatible species, introduced to a new area, will be unable to reproduce sexually. In contrast, a new sexually reproducing population may result from the progeny of a single immigrant of a self-compatible species (Baker, 1955, 1967). Founder effects are particularly likely in the establishment of new populations following natural long-range dispersal to oceanic islands, and other habitats effectively isolated by ecological and geographical factors, such as lakes and isolated mountain peaks. However, in considering cases of supposed founder effect, it is important to consider the possibility that genetic drift and/or selective forces may have influenced the pattern of variation.

In a number of studies, founder effects have been proposed to account for the low level of variation in certain populations. For instance, Schwaegerle & Schaal (1979) studied variation in isozymes in 11 populations of the Pitcher Plant, *Sarracenia purpurea*, from eastern North America. A population from Ohio, growing on the 17 acre island in an artificial lake, proved particularly interesting. A population in excess of 100 000 plants was present at the time of sampling. Even though the population was very large, it was genetically depauperate relative to some of the others. This is most likely to be the result of a founder effect, as the population of Pitcher Plants was begun by planting a single individual in 1912, the population having developed through perhaps 8–15 generations.

There is also evidence for chance effects in populations of *Sarracenia* introduced from Canada to Ireland in 1906. The initial introduction of seeds and rootstock was made at Termonbarry and from these plants several new populations were founded in different parts of Central Ireland. By examining 25 enzyme systems, Taggart, McNally & Sharp (1990) studied the variation in the six extant populations, and discovered that the number of polymorphic loci was reduced in the derivative populations, which had passed through a severe founder event, only a small number of plants having been transferred to start each new population. However, if the initial populations had been small for a number of years, it is also possible that genetic drift had occurred.

Founder effects have also been invoked, for example, to explain patterns of variation in *Polygala vulgaris* (Lack & Kay, 1988) and *Nothofagus menziesii* (Haase, 1993).

Founder effects in introduced species

Weedy species are often unwittingly introduced to other parts of the world, and in some cases the populations of newly arrived species lack genetic variation and this has been attributed to founder effects, e.g. Australian populations of *Avena barbata* and *Bromus hordeaceus* (*B. mollis*) (Brown & Marshall, 1981), *Chondrilla juncea* (Burdon, Marshall & Groves,1980), *Echinochloa microstachya* (Barrett & Richardson, 1986) and *Emex spinosa* (Marshall & Weiss, 1982); Jamaican populations of *Eichhornia paniculata* (Barrett & Shore, 1990); and populations of *Striga lutea* (*S. asiatica*) in the USA (Werth, Riopel & Gillespie, 1984). However, in other cases, the newly introduced populations do not exhibit lower genetic variation relative to the proposed source or ancestral populations, e.g. Australian populations of *Echium plantagineum* (Brown & Burdon, 1983) and *Trifolium subterraneum* (Brown & Marshall, 1981); Canadian plants of *Apera spica-venti*

(Warwick, Thompson & Black, 1987); and populations of *Trifolium hirtum* in California (Jain & Martins, 1979).

Barrett & Shore (1990) provide a valuable review of isozyme variation in colonising plants. They point out that assumptions often have to be made about the source of the newly colonising species, especially if the taxon is a cosmopolitan weed. Moreover, isozyme markers only represent part of the genome. Despite having low levels of allozyme variation, five weed species introduced to Canada all exhibited substantial between and within population variation in morphology, and flowering behaviour in a garden experiment. In the case of *Sorghum halepense* and *Panicum miliaceum*, genetic exchanges between the weeds and cultivated taxa may have contributed to the variation (Warwick, 1990*a*).

Selection in populations

Processes which Darwin grouped together under this blanket term are now classified into three main types (Fig. 9.5).

The first type, so-called stabilising or normalising selection, is a process tending to produce conformity and stability. At first sight it is surprising to find selection acting in this way, as many people equate natural selection with change. Imagine, however, a genetically variable population which is well adapted to its environment. The environment may, of course, change climatically and biotically with the seasons, but we assume that it is not changing directionally and fundamentally. By sexual reproduction an array of genotypes is produced, a sample of which may exhibit a typical normal distribution for some character; most of the individuals in the array depart little from the mean, but some segregants in each tail of the distribution are markedly distinct from the mean. Stabilising selection has the effect of eliminating individuals which depart significantly from the mean, giving a bimodal distribution of eliminants. In a study of flowering in *Tephroseris integrifolia* (*Senecio integrifolius*), Widén (1991) found evidence for stabilising selection in some, but not all, years.

The second type might occur where the environment is changing in a particular direction, as it might, for instance, where a site is becoming contaminated with a heavy metal such as zinc. Selection may then occur producing a directional change in the mean values for significant genetically determined characteristics. Again, as a consequence of selection, a portion of the young is eliminated.

The third and final type is called disruptive selection. In this case, individuals of different genotype may be at a selective advantage in different

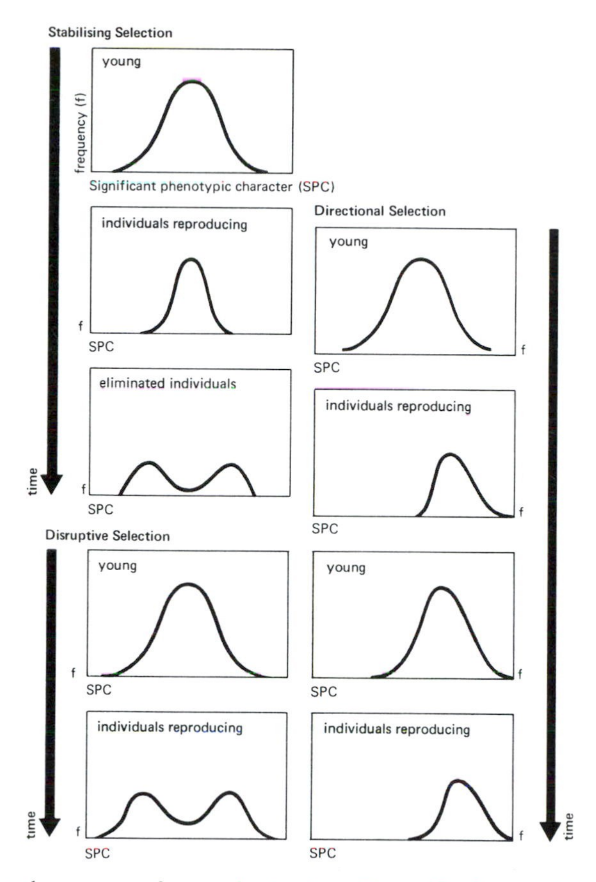

Fig. 9.5. The three types of natural selection. (From Hardin, 1966.)

places within the total area occupied by a population, giving a pattern of genetic polymorphism in a mosaic or patchy environment.

These models of natural selection provide guidelines for the design of experiments aiming at detecting selection in action in populations. Many experimental studies of natural selection make comparisons of adult material from contrasting sites. But, as we have seen from the models described above, selection involves the maturation and reproduction of only a portion of the population. The adults will be less variable (in significant phenotypic traits) than the young from which they develop. Thus, if the variation in young and adult is compared, it should be possible to detect selection at work. This idea is of course not a new one. As we saw in Chapter 5, as long ago as the 1890s, the zoologist Weldon studied variation in young and adults in Crabs.

Fitness

It is important at this point to define more carefully what is meant by 'selective advantage' by introducing the concept of 'fitness'. Fitness is relative and may be defined by the *number* of descendants left by an individual compared with those of other individuals. What is clearly important in microevolutionary change is the *relative* contribution of offspring made to the next and subsequent generations. With regard to a particular environment, the relative fitness of an alien may be estimated as:

$$\text{relative fitness of alien} = \frac{\text{performance of alien genotypes}}{\text{performance of native genotypes}}$$

In some cases where extreme habitats have been studied the calculations of fitness are simplicity itself. For instance, only heavy metal tolerant variants survive on highly contaminated soil, and seedlings unable to tolerate heavy metals in the rooting medium may almost all die on germination, making no contribution to the next generation. In most habitats, however, conditions are less harsh, and individuals of different tolerance survive and reproduce with varying degrees of success. Estimates of fitness are usually based on the comparison of the survival, growth and reproduction of plants in various experiments. As we shall see later, in situations where several estimates of 'relative fitness' have been made from different measurements of vegetative and reproductive characters of the same material, very different results have been obtained (Warwick & Briggs, 1980*a*, *b*, *c*). Wherever possible, measures of reproductive success should be obtained. A review of measurements of relative fitness is provided by Levin (1984).

Finally, we stress an important general conclusion which has emerged from genecological studies. In famous mathematical studies of selection, Fisher (1929) and Haldane (1932) showed that selection would be effective even if there was only a very small selective advantage between individuals in populations. It is now clear from the study of extreme habitats, however, that very large selective advantages must sometimes be operating in nature, and we might therefore conclude that sometimes selection may act very swiftly indeed.

Studies of single factors

The results of many genecological studies have been published in the last 40 years. One must be wary of suggesting common influences for such a varied group of experiments, but it is possible to make the following generalisations. As we have noted above, genecologists have recognised the importance of comparing plants from *extreme* habitats. If natural selection is to be studied, it will be most readily investigated by studying intraspecific variation in grossly dissimilar sites differing in toxicity, exposure, temperature, light, moisture, salinity, etc. Sites are sometimes distant from each other, sometimes side by side. Plants are collected and cultivated, and subjected to a tolerance test. In such tests a major habitat factor, in which experimental sites are presumed to differ, is investigated under garden or laboratory conditions.

Experiments of this type reveal many patterns explicable in terms of selection (see Table 9.4 for a representative sample). In judging tolerance testing it is important to assess the completeness or otherwise of the evidence. The assumption that the habitats differ in critical factors should be investigated fully by quantitative methods. For instance, if it is supposed that two sites differ in copper contamination, measurement of copper in the soil should be made and the availability of the copper *to the plant* should be examined. Bearing in mind the difficulties of extrapolation from laboratory to the field situation, tolerance tests should be as 'natural' as possible. In a large number of cases the investigation has reached the point where, for example, plants from sites A and B have been shown to be subject to different levels of factor X in the field. In a tolerance test, manipulation of factor X reveals a difference in sensitivity to X, which is consistent with the level of exposure to X each receives in the wild. In relation to a specific factor each group of plants seems to be best able to survive, grow and reproduce in its native conditions. In such circumstances the genecologist often suggests that the pattern is the result of natural selection. Commonly the evidence for this view is circumstantial, especially if the genetic basis of the difference has not been examined.

Studies of several interacting factors: *Lotus* and *Trifolium*

In cases where a genetic polymorphism has been investigated in detail, it has become apparent that the effect of many factors must be considered. This important point is beautifully illustrated by studies of cyanogenesis in plants, in which we now know that many factors interact. It is worthwhile to describe this case in some detail.

In the Sudan campaign of 1896–1900, a number of British transport animals were poisoned by eating a local species of *Lotus* (Dunstan & Henry, 1901). Chemists, interested in the losses, discovered that certain species of the genus contain cyanogenic glucosides. If leaves are bruised, the glucoside

Table 9.4. *Examples of studies of ecotypic differentiation selected to indicate the variety of tolerance tests devised by genecologists and ecologists. (See also reviews by Clements, Martin & Long, 1950; Heslop-Harrison, 1964; Antonovics, Bradshaw & Turner, 1971; and Crawford, 1989)*

1.	*Soil*
	(*i*) *Use of natural and other soils*
	Limestone soil; *Teucrium scorodonia* (Hutchinson, 1967)
	Colliery waste; *Agrostis capillaris (A. tenuis)* (Chadwick & Salt, 1969)
	(*ii*) *Natural soil plus additions*
	Mine soil plus garden soil; test used widely by Liverpool group studying heavy-metal tolerance (Bradshaw & McNeilly, 1981)
	(*iii*) *Water culture experiments*
	Effects of chromium, nickel, magnesium; *Agrostis* spp. from serpentine (Proctor, 1971*a*, *b*)
	Various heavy metals; e.g. arsenic (Pollard, 1980), cadmium (Coughtrey & Martin, 1978), zinc (Al-Hiyaly, McNeilly & Bradshaw, 1988), and see review of heavy-metal tolerance by Macnair (1993)
	Effect of aluminium (Chadwick & Salt, 1969; Davies & Snaydon, 1973*b*)
	Effects of sodium chloride (e.g. Ab-Shukor *et al.*, 1988) in *Trifolium repens*
	(iv) Soil water stress
	Pots allowed to dry out; *Pseudotsuga taxifolia* (Pharis & Ferrell, 1966)
	(*v*) *Flooding*
	Pots with different water regimes; *Veronica peregrina* (Linhart & Baker, 1973)
2.	*Light*
	Variation in light regimes using shade tubes; *Teucrium scorodonia* (Hutchinson, 1967)
3.	*Exposure*
	Plants growns in pots in exposed sites; *Agrostis stolonifera* (Aston & Bradshaw, 1966)
4.	*Effects of loss of foliage*
	Effects of grazing animals; various taxa (see Watson, 1969; Jones, Keymer & Ellis, 1978)
	Effects of mowing; *Plantago major* (Warwick & Briggs, 1980*b*)
5.	*Air pollution*
	Gas chambers, etc.; (Taylor & Murdy, 1975; Horsman, Roberts & Bradshaw, 1979; Taylor, Pitelka & Clegg, 1991)
6.	*Herbicides*
	Various tests on many taxa (see Lebaron & Gressel, 1982; Warwick, 1991)

is broken down and hydrogen cyanide is liberated. Further studies revealed that, while some plants of *Lotus corniculatus* are cyanogenic, others are acyanogenic (Armstrong, Armstrong & Horton, 1912). Dawson (1941) studied this polymorphism in the south of England. Using sodium picrate papers, which redden in the presence of hydrogen cyanide, he tested

Table 9.5. *Cyanogenesis in* Lotus corniculatus *(Dawson, 1941)*

	Numbers of plants	
Localities in England	Positive test for HCN	Negative test for HCN
Studland Heath, Dorset	77	56
Ballard Down, Dorset	95	56
Ranmore, Surrey	145	8
Crumbles, Sussex	150	5

samples from different populations; a selection of his results is given in Table 9.5. By crossing cyanogenic and acyanogenic plants, Dawson was able to show that it was likely that the presence of glucoside is dominant to its absence. The genetics of the situation is complicated, however, as *L. corniculatus* is a tetraploid.

In *Trifolium repens*, another species polymorphic for cyanide production, the genetical position is simpler. Corkhill (1942) and Atwood & Sullivan (1943) demonstrated that, in this species, glucoside presence (allele *Ac*) is dominant to glucoside absence (allele *ac*). Production of the enzyme that hydrolyses the glycoside to produce the cyanide is determined at an independently segregating locus. Presence of this enzyme (*Li*) is dominant to its absence (*li*).

The distribution of the variants in *T. repens* was examined by Daday (1954*a*, *b*), who showed that cyanogenic plants were present with high frequency in Southwest Europe. In contrast, acyanogenic plants predominate in Northeast Europe (Fig. 9.6). In intermediate sample stations different proportions of the two variants were found: there is in fact a 'ratio-cline' across Europe. Interesting observations were made also upon the frequency of cyanogenic plants at different altitudes in the Alps, and again a ratio-cline was discovered, high frequencies of cyanogenic plants being reported from low altitudes. This frequency declined with increasing elevation, until at high altitudes all the plants in the sample proved to be acyanogenic (Fig. 9.7). In Canada, similar clinal patterns have been found in populations of *T. repens*, even though the species was introduced less than a century ago (Ganders, 1990).

In interpreting his findings on the distribution of the different variants in Europe, Daday showed that there was a correlation between cyanogenesis and January mean temperatures, a decrease in temperature being associated with an increase in frequency of the acyanogenic variant. It appeared likely from this work that winter temperatures played some direct role,

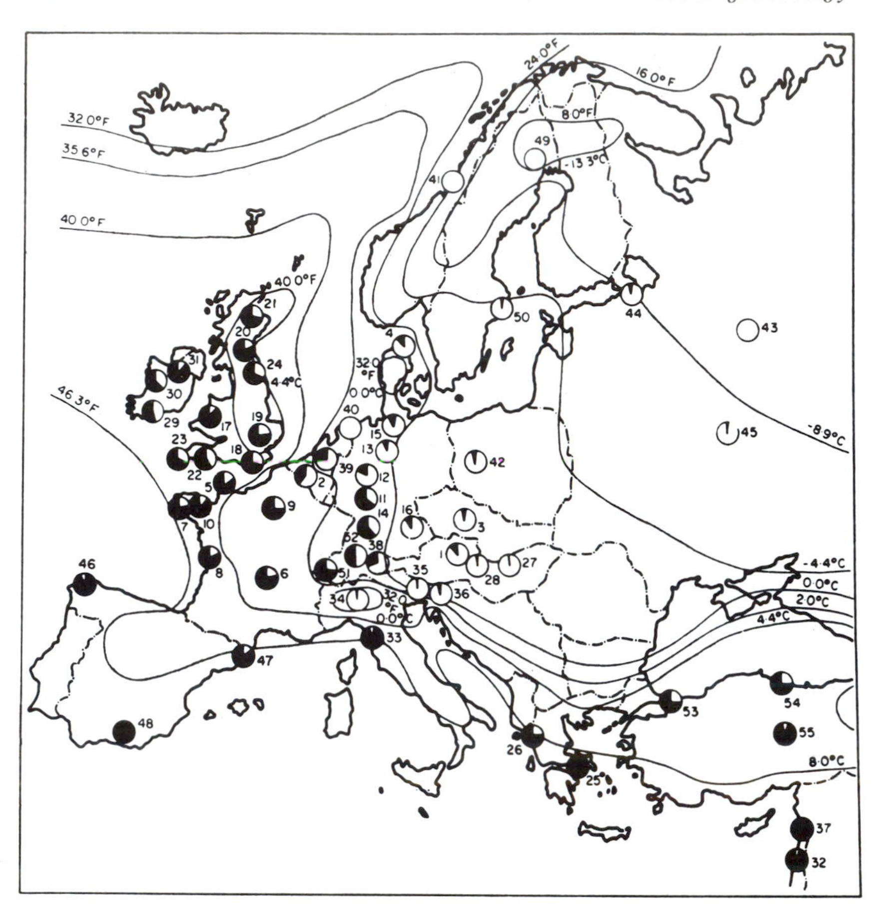

Fig. 9.6. Distribution and frequency of the cyanogenic variant in European and near eastern wild populations of *Trifolium repens*. Black section: frequency of the cyanogenic variant. White section: frequency of the acyanogenic variant. —: January isotherms. (From Jones, 1973, after Daday, 1954*a*.)

through natural selection, upon the frequency of cyanogenic plants of *T. repens*, or that the locus concerned with glucoside production is genetically linked to genes involved with fitness responses at different temperatures (Daday, 1965).

Studies by Jones (1962, 1966) shed new light on the frequency of cyanogenic plants of *Lotus corniculatus* in different English localities. He observed that while such plants were relatively free from damage by small invertebrates, many acyanogenic plants showed signs of having been grazed by slugs and snails. Following these observations, he carried out some simple experiments in which various species of slugs and snails were

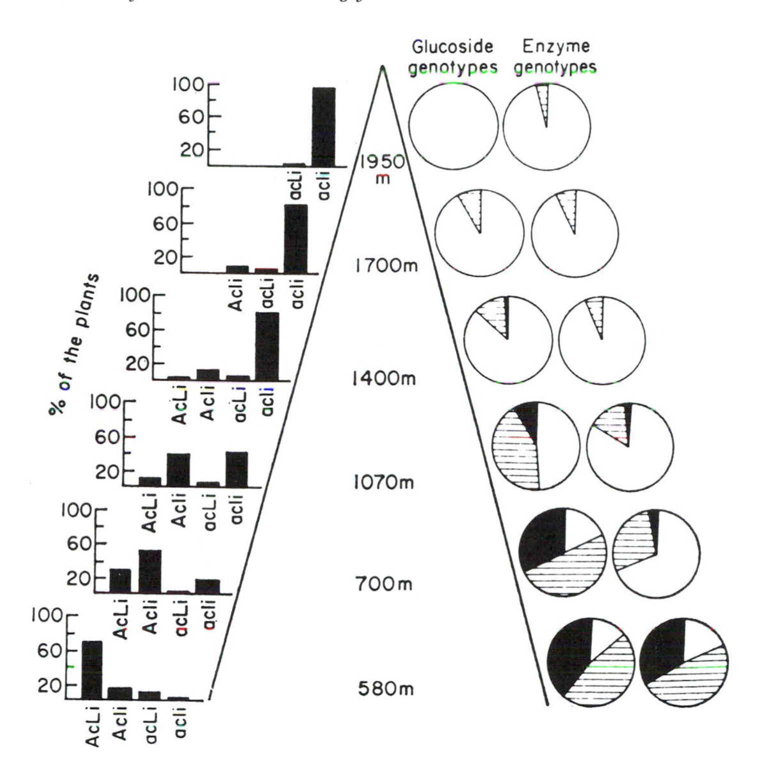

Fig. 9.7. Phenotypic and genotypic frequencies in wild populations of *Trifolium repens* from different altitudes. (From Jones, 1973, after Daday, 1954*b*).

Phenotypes (left):
AcLi – glucosides and enzyme
Acli – glucosides only
acLi – enzyme only
acli – neither glucosides nor enzyme

Estimated genotypes (right):
Black section = dominant homozygotes
Lined section = heterozygotes
White section = recessive homozygotes

confined with cyanogenic and acyanogenic plants of *Lotus*. The experiments were repeated many times and Jones obtained good evidence that two snails, *Arianta arbustorum* and *Helix aspersa*, and two slugs, *Arion ater* and *Agriolimax reticulatus*, showed selective eating of the acyanogenic plants when offered both variants. This experiment proved of great interest to genecologists and many have repeated the tests on both *Lotus corniculatus* and *Trifolium repens*. Not all their results agree with those of Jones (see, for example, Bishop & Korn, 1969). It has been argued that the particular conditions used in the test influence the results. Important factors to take into account are the food materials and how they are made available to the animals, whether they are hungry or not, and the variability in animals. To quote the opinion of Jones (1972): 'in the same way that some men like beer and others do not, individual molluscs have different palates'. The balance of evidence seems to favour the view that cyanogenic glucosides provide a defence mechanism against certain small invertebrates. In *Lotus*, however, this defence is by no means absolute (Table 9.6: Crawford-Sidebotham, 1971). For example, Lane (1962) has shown that the larvae of the Common Blue Butterfly, *Polyommatus icarus*, show no preference for acyanogenic plants of *Lotus*; in fact, they produce an enzyme, rhodanese, which converts cyanide into harmless thiocyanate. Recently, Hughes (1991) has reviewed the evidence for the defensive role of cyanogenesis against herbivores in *Trifolium repens*. While cyanide production protects against grazing by some molluscs, leather jackets, weevil larvae and some aphids, other herbivores – field crickets, grasshoppers, two species of aphid and some other mollusc species – were unable to discriminate between cyanogenic and acyanogenic plants.

How can one explain the patterns of distribution of acyanogenic and cyanogenic plants discovered by Daday in the light of Jones' findings? It seems most likely that the distributions of animals which selectively eat acyanogenic plants are correlated with climatic factors. In Atlantic regions of Europe with mild winter temperatures and at low altitude, a great number of small invertebrates likely to eat plants may be found. It is clear that one can postulate a selective advantage of the cyanogenic plants in western Europe. What is the selective advantage of the acyanogenic condition? To explain the pattern of high frequency of the acyanogenic

Table 9.6. *Summary of differential eating experiments with 13 species of slug and snail (From Crawford-Sidebotham, 1971, in Jones, 1972)*

	Lotus corniculatus	Trifolium repens
No. of species[a] showing preference for the acyanogenic form	7[b]	7[b]
No. of species[a] showing no selection or no eating of the plant	6	6

[a]These numbers do not contain exactly the same species.
[b]Slugs and snails feeding on both plants: *Agriolimax reticulatus* Müll., *Arion ater* (L.), *A. hortensis* Fér., *Cepaea hortensis* (Müll.), *C. nemoralis* (L.), *Helix aspersa* Müll., and *Monacha cartusiana* (Müll.); feeding only on *T. repens*: *Arianta arbustorum* (L.) and *Helicella virgata* (da Costa); feeding only on *L. corniculatus*: *Arion subfuscus* (Drap.); no grazing on either plant: *Agriolimax caruanae* Poll; *Milax budapestensis* (Hazay) and *Theba pisana* (Müll.).

variant at high altitude and in continental conditions, it has been suggested that frost may be important. Conditions of extreme cold will freeze the cells of plants, releasing the enzymes which break down any glucosides present. The production of cyanide through its inhibitory effect upon plant metabolism could then place the plant at a strong selective disadvantage relative to the acyanogenic variant. However, Hughes (1991) points out that there is no convicing evidence for the differential effect of frost damage.

So far our account concentrates on the interaction of climatic and biotic factors on the large scale. While the study of local variation suggests that grazing by small invertebrates is important, other factors may also be involved in *Lotus*. It is not possible to summarise all the results discovered in experiments, but they suggest that a number of interacting factors may influence patterns of cyanogenesis, for there is evidence that water stress, mammal grazing, trampling, insect damage and salt spray may all be important in influencing small-scale pattern (Keymer & Ellis, 1978; Ellis, Keymer & Jones, 1977*a*, *b*).

The classic experiments of Harper and his associates in a small 'superficially dull' pasture at Henfaes in North Wales have added greatly to our understanding of cyanogenesis and other genetic variation in *Trifolium repens*. Burdon (1980) collected many samples of *T. repens* from the field and, as a result of cultivation tests, he concluded that the population was very variable indeed. Some of this variability has been shown to be under genetic control, e.g. *T. repens* is genetically polymorphic for cyanogenesis and leaf markings. Genetic polymorphism is also suspected for other characters: for instance, some plants were more resistant than others to the fungi *Pseudopeziza* and *Cymadotheca*. Variability in the population was confirmed when Trathan, using isozyme markers, discovered that there were *c.* 50 genotypes per square metre (Harper 1983). Many previous studies of cyanogenesis have been carried out under laboratory conditions, and, realising this, Dirzo & Harper (1982*a*) investigated the distribution of different variants of the cyanogenesis polymorphism in relation to the distribution of active molluscs in the field (Fig. 9.8). Cyanogenic plants predominated in areas of high (H) and very high (VH) mollusc density, while the frequency of acyanogenic plants was greatest in areas with very low mollusc grazing. This association is statistically significant and consistent with the hypothesis that cyanogenetic plants are at a selective advantage in certain areas of the field, but not in others. While cyanogenesis confers advantage in certain parts of the field, transplant studies at Henfaes by Dirzo & Harper (1982*b*) revealed that the cyanogenic morph was more susceptible than the acyanogenic to the pathogenic rust, *Uromyces trifolii*.

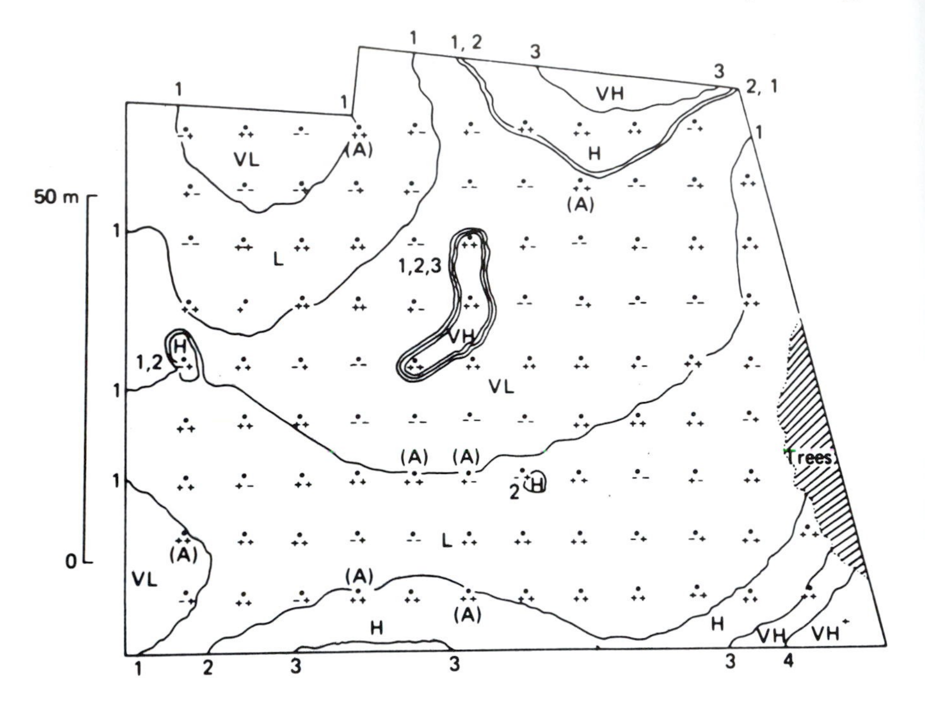

Fig. 9.8. Contour map made by a trend-surface analysis of the density of active molluscs at Henfaes field, College Farm, Aber. The values shown by the contour lines are the mean number of molluscs. The mnemonics for mollusc density are: VL, very low; L, low; H, high; and VH and VH +, very high. Sampling sites are shown by filled circles. The central zone of very high slug density and the two isolated zones of high density were mapped directly. They represent patches of *Urtica dioica* which serve as a refuge for slugs. The symbols below each site indicate the locations of the four clover morphs: + +, cyanogenic *Ac Li*; + −, glucosidic *Ac li*; − +, enzymatic *ac Li*; and − −, double recessive *ac li*. Samples close to the boundaries of zones of different mollusc density are shown by (A). (From Dirzo & Harper, 1982*a*.)

The grazing preferences of sheep were also examined, particularly in relation to the genetic polymorphism associated with leaf markings (Cahn & Harper, 1976*a*, *b*). Evidence suggested that sheep were selective. They ate more of the unmarked variants and more of the commoner morphs than the rarest. There was no evidence of a distasteful chemical or that cyanogenesis is associated with lack of markings. This could be an example of apostatic selection, in which rare morphs are at a selective advantage relative to the common variants. Harper (1983), reflecting upon the patterns detected in the Henfaes field, points out that the weather varies from year to year, leading to differences in the severity of attack by fungal pathogens, differences in sward height and grazing intensity. Thus, some of the evident pattern in this clonal species will reflect recent, as well as past, selective forces. Clearly, interactions between many factors must be considered in order to obtain a balanced picture of microevolution.

For an insight into the way the study of cyanogenesis is leading to a more complex picture, the reviews of Jones, Keymer & Ellis (1978), Poulton (1990) and Hughes (1991) should be consulted.

Reciprocal transplant experiments

By employing reciprocal transplant methods additional evidence about patterns of variation may be obtained. Suppose that two extremely different habitats (a and b) occur in the same region or side by side. Preliminary evidence of tolerance tests suggests that for the species under study each of two morphs (A and B) is best suited to its own native habitat (A in a and B in b). Clearly such a hypothesis might be put to the test by planting an appropriate number of replicates of A and B in each habitat, using either cloned or genetically uniform material for the purpose. Usually, before planting out, A and B are first grown for a period in a uniform garden, to minimise carry-over effects.

As we have already seen in earlier chapters, this type of experiment has had a long history. The early experiments by Bonnier and Clements were technical failures, but recently there has been a revival of interest in reciprocal transplant studies. These investigations are characterised by the great care taken in their execution. Davies & Snaydon (1976) marked their experimental plants with wire, and in the studies by Warwick & Briggs (1980*a*, *b*) experimental plants were grown in random array at spaced intervals in a carefully mapped area. In many cases 'native' plants perform best in their own habitats, and alien stocks do less well (Levin, 1984).

The experiments of van Tienderen & van der Toorn (1991*a*, *b*) provide an excellent example of the use of reciprocal transplants. They studied local adaptation in *Plantago lanceolata* in the contrasted habitats of pasture and hay meadow. Some background information is necessary to appreciate the context of their experiments. Evidence suggests that selection has favoured different variants in areas of different land use, in particular hay meadows and pasture (van Groenendael, 1986, and references cited therein). With their erect, tall growth habit and early fruiting to produce heavy seeds, plants from hay meadows are adapted to succeed in establishing and growing in tall swards, and reproducing before the date of hay cropping. In contrast, plants from pasture are smaller with short-leaved decumbent rosettes and bear many small inflorescences producing light seed. This represents a different syndrome of adaptive traits, related to survival and reproduction under grazing by domestic animals. From these experiments it was deduced that each variant had a different set of co-adapted traits, life history strategies and reproductive 'tactics' related to the habitat of origin. This hypothesis was put to the test in a reciprocal transplant experiment (van Tienderen & van der Toorn, 1991*a*, *b*), which provided very clear evidence of the selective disadvantage suffered by the alien transplants (Table 9.7).

This experiment is noteworthy for making several assessments of relative fitness. It is clear that, while vegetative survival of alien plants was relatively high, such plants were presented with a series of challenges as the experiment progressed and relative fitness declined during the flowering and fruiting phases.

Reciprocal transplant experiments have very clear attractions for the genecologist: they permit examination of responses to the *totality* of the environment, including competition with the native flora at each site. Such

Table 9.7. *Relative performance, under field conditions, of clonal transplants of* Plantago lanceolata *from an early mown hay meadow on clay soil and a grazed pasture on sandy soil, transplanted into hay and pasture. (From van Tienderen & van der Toorn, 1991a.) For comparative purposes the performance of plants transplanted into their 'native' site is set at 100%. 'Alien' transplants perform less well than plants in their native site. The comparative reproductive failure of the pasture variant transplanted into the hayfield is particularly significant. It is late flowering and fruiting, and is damaged by early season haymaking before it reaches reproductive maturity*

	HABITATS			
	Hay cut as usual		Pasture grazed by cattle	
Source of cloned material	pasture	hay	pasture	hay
Vegetative survival	70	100	100	90
Success in flowering	52	100	100	60
Success in producing at least one ripe fruiting spike	50	100	100	30
Total seed yield	3	100	100	40

investigations subject the plants to more natural conditions than those employed in many tolerance tests.

Experimental evidence for disruptive selection

Tolerance tests and reciprocal transplant experiments with a number of species suggest that genetic polymorphism occurs, and that different morphs may be at a selective advantage in different parts of a mosaic or patchy environment. Developing the ideas on disruptive selection outlined earlier, we may envisage a model system as follows. Across the line of contact of two dissimilar habitats, gene flow may occur between the polymorphic variants of an outcrossing species, 'well-adapted' adults on each side of the divide being cross-pollinated by pollen from the other morph. As a result of gene flow at the seed dispersal stage, seed of different genotypes will be scattered around the site. The variation in young plants, as they germinate and develop, will be subjected to disruptive selection and only the appropriate morph will survive in each area of the patchy environment. The 'potential' population represented by the young plants will be put through the sieve of selection and only well-adapted plants will survive to reproductive maturity. As this area is one of the most interesting in ecological genetics, we will consider a study which tests this model.

The effect of artificial fertilisers on crop plants was an important issue in the mid-nineteenth century, and in 1856 Lawes & Gilbert laid out the famous Park Grass Experiment at Rothamsted, which was designed to compare the yield of hay in plots treated either with artificial or natural fertilisers. For our purposes it is not necessary to give details of treatment and control plots (see Brenchley & Warington, 1969; Thurston, Dyke & Williams, 1976; and Snaydon, 1970, for details). It quickly became apparent that grassland productivity could be increased by artificial fertiliser treatments, but fortunately the experiment has continued with an early summer hay crop to the present day. One effect of fertiliser treatments, especially in plots treated with ammonium sulphate, is the lowering of the pH. Gradually plots became acid, and in 1903 (after some tentative applications of lime in the 1880s and 1890s) regular liming of the southern half of each plot was undertaken. In 1965, further subdivision of some plots was made to give four subplots, each with a different lime treatment. Looking at the experiment today one is struck by the crispness of the boundaries between plots. Clearly there is little or no sideways movement of nutrients on this more or less level site, and plots differ markedly in species composition, vegetation height and productivity. The plots were laid out on a pre-existing grassland said to have been grassland for several centuries (although there are faint traces of plough marks in part of the area). From what was likely to have been a uniform grassland, the experimental regime has produced such marked differences in adjacent plots that the patterns can be seen from the air. Snaydon and Davies realised that, because several species occurred in many of the plots, the Park Grass Experiment offered a unique opportunity to test Darwinian views on selection. Were differences due to fertiliser and lime treatment detectable within one of these common species? How quickly could changes take place?

Studying the grass *Anthoxanthum odoratum*, Snaydon and Davies discovered that, when material was collected from various plots and cultivated in a garden trial, the plants from tall vegetation plots were generally taller and heavier than plants from short vegetation plots (Fig. 9.9). Furthermore, plants showed ecotypic differentiation in relation to edaphic factors. For example, plants from acid unlimed plots, where the soil was deficient in calcium, needed much less calcium than plants from limed plots. Studies of gene flow were made and reciprocal transplant experiments carried out from acid to limed plots and vice versa, revealing that plants grew best on their native plots. Relative fitness of the alien plants was only 0.59 on the limed subplots and 0.75 on the unlimed subplots. The evidence from all these experiments suggests that population differentiation has taken place in response to fertiliser and lime treatments, as a consequence of natural selection acting over the last 120 years. Indications that important changes might be possible in an even shorter time are provided by studies of vegetation following the alteration of the lime addition regime in 1965. In experiments in 1972 it was discovered that, after only 6 years, changes could be detected in the *Anthoxanthum* populations. The highly imaginative use of this classical experiment has provided one of the most impressive studies of pattern and process in a patchy environment (Davies, 1993).

In an attempt to study selection, a number of workers have compared young and adult generations, and the essential details of one example are shown in Fig. 9.10. In most cases, the 'young' generation was represented by collecting mature seed developed in different parts of the patchy environment. Recognising that this type of study neglects the seed dispersal aspect of gene flow, in experiments on *Poa annua* Warwick & Briggs (1978*a*, *b*) investigated the seed bank flora in both lawn and adjacent flower beds. The results of these comparative experiments show that the spectrum of variation in the 'young' is different from that of the adults in the population. This evidence is consistent with the view that disruptive selection is a potent force in certain situations.

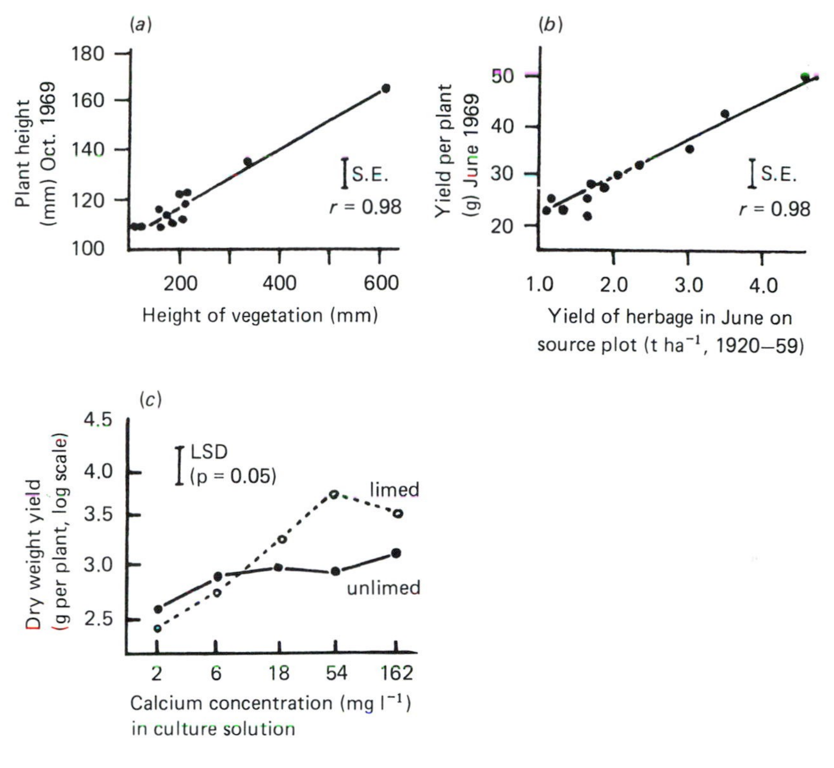

Fig. 9.9. Graphs summarising some results of studies of *Anthoxanthum odoratum* collected from the Rothamsted Park Grass Experiment and grown in a special trial. (From Snaydon, 1976.) (*a*) Relation between plant height and height of vegetation in source plots. (*b*) Relation between plant weight and yield of herbage in source plots. (*c*) Effect of calcium on dry weight of plants from limed and unlimed source plots. (Snaydon, 1970, 1976; Snaydon & Davies, 1972; Davies & Snaydon, 1973*a*. See also Davies, 1975 – response to potassium and magnesium; Davies & Snaydon, 1974 – response to phosphate.)

These experiments should not, however, be accepted as providing a complete picture of natural situations, but rather should be seen as first attempts to come to terms with the complex patterns and processes found in natural populations. The following complications should be considered in the interpretation of past and future experiments.

The model system predicts that in a particular generation the spectrum of variation in the 'young' will exceed that of the adults. Most of the comparisons so far attempted have compared adults with 'young' of the *next* generation. This comparison raises a number of problems. First, it is not clear whether the habitats under study are at equilibrium and whether the forces of selection are operating at the same intensity year by year. Furthermore, it is difficult to assess the importance of various demographic factors. Many plants are potentially long-lived and in some of the study areas it is not known how frequently individuals succeed in establishing themselves. Secondly, the model system envisages two entirely different habitats separated by an abrupt boundary. Realistically, models should take account of the evident heterogeneity within habitats in nature and the existence of transitional conditions – the so-called ecotone – between

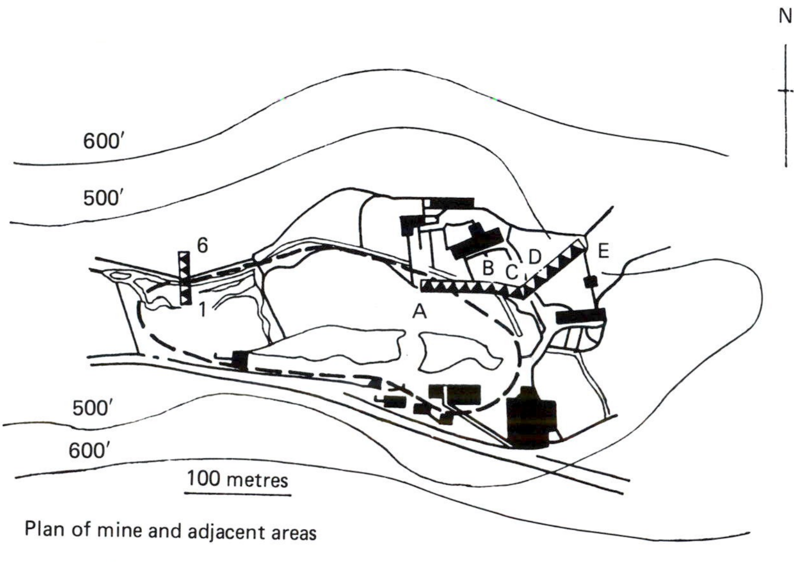

Fig. 9.10. Map of old copper mine workings at Drws y Coed, Caernarvon, Wales, showing positions of transects sampled by McNeilly (1968). In his studies of *Agrostis capillaris* (*A. tenuis*), using a water culture technique, an index of copper tolerance for adult plants and seed produced by different adults was determined for material from two transects: (i) sites 1–6; and (ii) sites A–E. Adults from the mines proved to be more copper tolerant than plants from the non-contaminated pasture adjacent to the mine. Studies of the seedlings, produced from wild collected seeds, revealed a wider spectrum of variation than in the adult plants. This pattern was particularly clear in transect A–E, where evidence for considerable gene flow of copper-tolerance genes downwind of the mine was discovered in progeny of copper-sensitive plants. This experiment is consistent with the view that strong natural selection occurs on the variable products of sexual reproduction. The only seedlings to survive to adulthood are likely to be copper tolerant on the contaminated areas and non-tolerant variants (which have been shown to be better competitors than copper-tolerant plants) on pasture areas. (From McNeilly, 1968.)

(i)

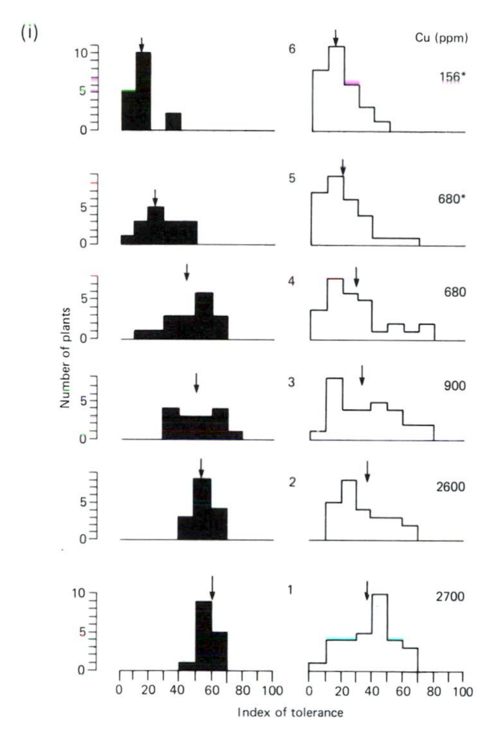

Index of tolerance for adults (black histogram) and seed (open histogram) for transect sites 1–6. (Mean values noted by arrows), with copper contents of soils. * indicates that vegetation cover suggests soils non-toxic.

(ii)

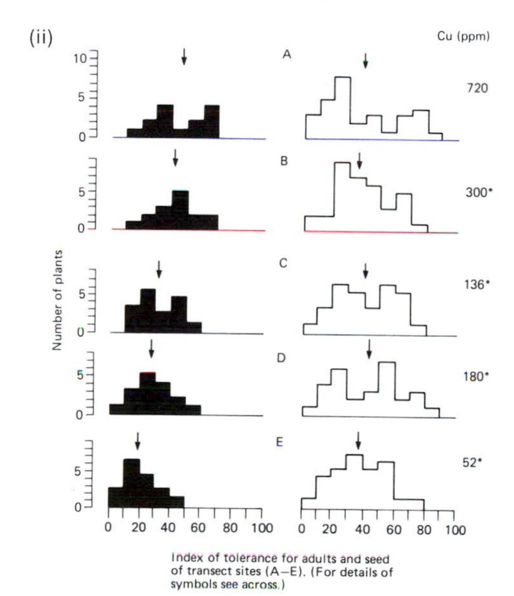

Index of tolerance for adults and seed of transect sites (A–E). (For details of symbols see across.)

habitats. In a remarkable study of a mine/pasture transition, Antonovics & Bradshaw (1970) discovered very interesting clinal patterns in *Anthoxanthum odoratum*. Thirdly, the complications raised by the existence of a seed bank in the soil have yet to be properly assimilated in experiments on disruptive selection. A huge source of potential variation may be constantly eroded away by loss of seed viability and by predation, whilst at the same time 'variation' is being added by new seed production and dispersal (Harper, 1977).

Co-selection in swards

We have already discussed some of the results of the detailed study of *Trifolium repens* in the field at Henfaes, North Wales (Harper, 1983). Another aspect of the work, involving reciprocal transplants, has yielded very significant results. Different parts of the field were dominated by one of four different species of grass and, using reciprocal transplant and other experiments, Turkington & Harper (1979*a*, *b*) and Turkington (1989) found support for the extremely interesting hypothesis that Clovers are not distributed at random over the field, specific clones being associated with each grass species. In a further series of experiments of plants from a different site, several *Lolium* genotypes were collected, together with their associated Clover plants. *T. repens* has nitrogen-fixing root nodules and the causative organism, *Rhizobium*, was extracted from each Clover plant. When the *L. perenne*, *T. repens* and *Rhizobium* were grown together in different combinations, Chanway, Holl & Turkington (1989) discovered that the best performance was achieved by the *Rhizobium*, *Trifolium* and *Lolium* combinations that had existed together in the 'wild'. These findings support the view that there is a complex pattern of co-selection at work and, clearly, it will be of very great interest to see if the phenomenon occurs in interactions of other plant species.

The speed of microevolutionary change: agricultural experiments

Some of the most convincing examples of the power and speed of selection come from studies of crop plants. Artificial directional selection has yielded spectacular results in agricultural crops (e.g. Fig. 9.11). Other experiments have studied natural selection, principally to discover the speed of change and which varieties yield best at a given site. For instance, in an investigation by Harlan & Martini (1938), 11 varieties of Barley (*Hordeum*) were mixed together in such proportions that an equal number of plants of

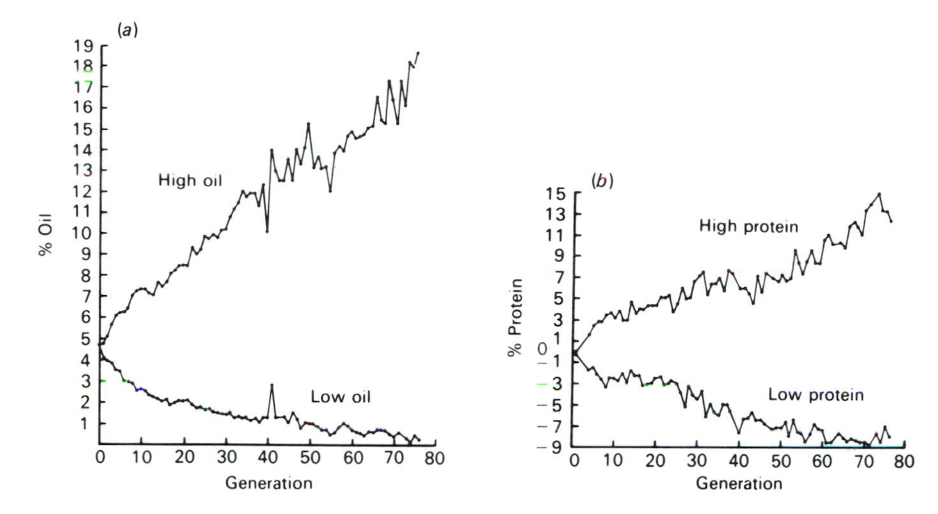

Fig. 9.11. Directional selection for high and low oil content in Maize (*Zea*) after 76 generations. (From Solbrig & Solbrig, 1979, after Dudley.)

each variety might be expected to grow. Seed samples were sent out to 10 experimental stations in different parts of the USA; all stations received the same spectrum of 'potential variation'. At each site the Barley was harvested and seed saved to sow a plot in the following season. A sample of seed was sent each year (1925–36) to Washington and the proportional representations of the 11 varieties were determined in a field trial. The results of the annual census for 1930 are shown in Fig. 9.12. Different varieties predominated in different areas, and there was rapid reduction (or even elimination) of the less well-adapted varieties. In all areas the variety which would eventually dominate the plot in a particular site was quickly evident.

Experiments of this type, where a 'standard' seed mixture is planted out in different sites, have been attempted with a number of crop plants and have yielded broadly similar results (Snaydon, 1978). Such experiments provide clear evidence that selection acts quickly and may be of great intensity. An investigation by Brougham & Harris (1967) is particularly revealing. Two varieties of *Lolium perenne* were sown together with *Trifolium repens* in an experimental plot. After establishment the sward was subjected to lax grazing for 6 months. The plot was then divided into a number of subplots, which were given either lax, moderate or continuous grazing regimes. Here we are not concerned with the detailed results of this

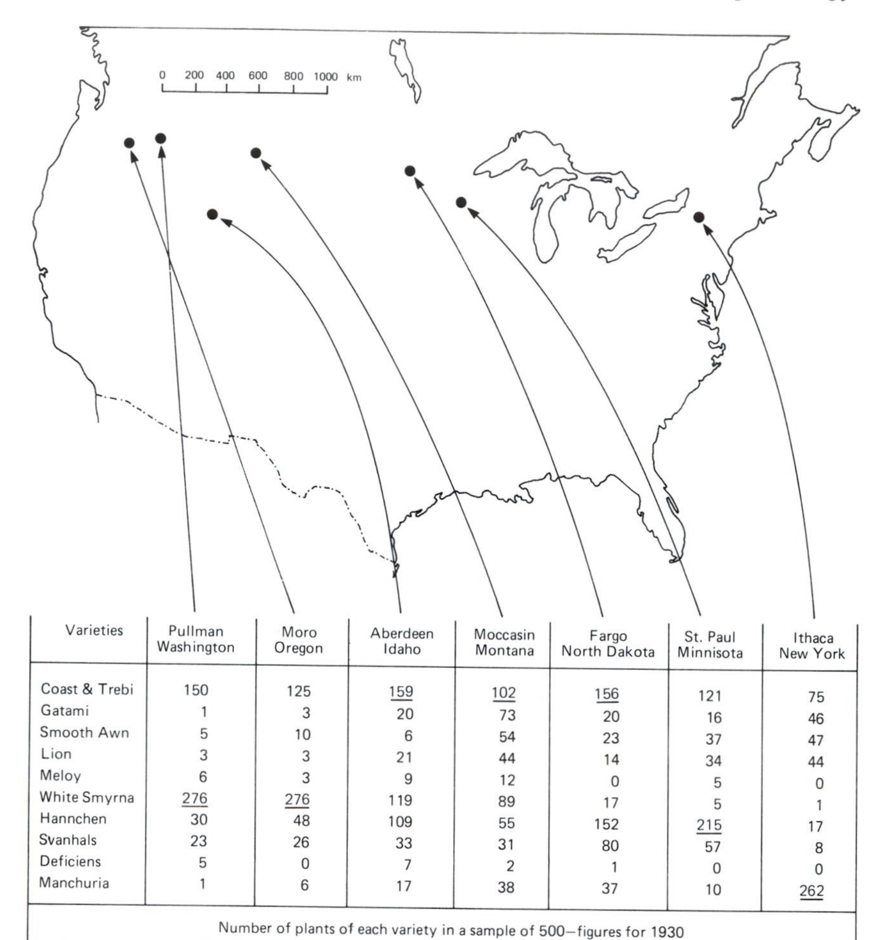

Varieties	Pullman Washington	Moro Oregon	Aberdeen Idaho	Moccasin Montana	Fargo North Dakota	St. Paul Minnisota	Ithaca New York
Coast & Trebi	150	125	159	102	156	121	75
Gatami	1	3	20	73	20	16	46
Smooth Awn	5	10	6	54	23	37	47
Lion	3	3	21	44	14	34	44
Meloy	6	3	9	12	0	5	0
White Smyrna	276	276	119	89	17	5	1
Hannchen	30	48	109	55	152	215	17
Svanhals	23	26	33	31	80	57	8
Deficiens	5	0	7	2	1	0	0
Manchuria	1	6	17	38	37	10	262
Number of plants of each variety in a sample of 500–figures for 1930							

Fig. 9.12. Diagrammatic representation of the results of an experiment with Barley (*Hordeum*) in the USA, showing rapid selection of the variety suitable to the particular site. (From Harlan & Martini, 1938.)

experiment; it is sufficient for our purposes to note that major changes in population composition were detected within 4 months of the application of the grazing regimes. The implications of this and many other experiments on agricultural stocks is that strong selective forces may act very quickly to change the composition of populations.

Rapid change in polluted sites

Heavy metal contaminated areas are found in many parts of the world, very often as a consequence of man's activities. Some plants have evolved variants able to grow on such sites. A full account of heavy metal tolerance is beyond the scope of this book, but Macnair (1993), on which we base our comments, may be consulted for a review of current knowledge. Heavy metal tolerance has been described in a number of species. However, formal genetic analysis of its control has been attempted in only a few cases. Tolerance has generally been found to be a dominant trait. For instance, single major genes – subject to the influence of modifying genes – appear to control copper tolerance in *Mimulus guttatus* (Macnair, 1983) and arsenate tolerance in *Holcus lanatus* (Macnair, Cumbes & Meharg, 1992). Contaminated sites differ in the suite of heavy metals contaminating the soil. Some studies have suggested that tolerance to one heavy metal confers tolerance to other metals. In contrast, other investigations point to the independent evolution of tolerances at individual sites. Further studies of co-tolerance to different heavy metals are required. While some progress has been made on studying the physiology of metal tolerance, much has still to be learnt of the mechanisms involved.

Studies of heavy metal tolerance in the field have provided fascinating insights into the speed of microevolutionary change. About 1900, a factory refining copper was opened at Prescot, Southwest Lancashire. Dust rich in copper was produced by the processing of the metal and the area around the factory became contaminated by aerial pollution. At the time of the study, by Wu, Bradshaw & Thurman (1975), total copper levels in lawn soils were as high as 10 800 parts per million. Investigations revealed that *Agrostis stolonifera* plants in lawns were copper tolerant to varying degrees and that, while the vegetation cover was complete on the 15-year-old lawns laid with normal turf (mean copper tolerance 42%), it was patchy on a new (7-year-old) lawn, which had a mean tolerance of 32%. In the establishment of this new lawn, repeated sowing with commercial seed had failed to achieve complete cover. When commercial seed stocks were tested for metal tolerance on contaminated soil, most seedlings died, but a few revealed their metal tolerance by growing normally (Bradshaw & McNeilly, 1981). From this study it was deduced that selection was taking place each time seed was sown on the contaminated new lawn area. Only copper-tolerant varieties survived, and such was the toxicity of the ground that only a few survivors were likely from any seed batch. The possibility that all the survivors were of a single genotype was examined by studying plants in cultivation and by isozyme analysis, and it was discovered that several tolerant genotypes were present in the contaminated lawns. As well as the lawns, the boundary area developed over 4 years from rough grassland was also examined. It had a mean copper tolerance of 21%. This splendid example of the detailed study of a man-disturbed site suggests that the different grasslands can be seen as a time series, the copper tolerance increasing, as a multi-stage process, with the age of the lawn. It has not yet reached the mean value of *c.* 70% copper tolerance typical of plants growing on the mine spoil. Another point of great significance is that there are many species to be found in the areas surrounding the factory, but only *Agrostis stolonifera* and *A. capillaris* have the appropriate variation in copper tolerance to respond to the selection pressure exerted by the highest levels of copper contamination from aerial fall-out.

The notion that selection might be constrained because of lack of the appropriate variation has received support from other experiments on metal tolerance. Electricity pylons are coated with zinc compounds to protect them against corrosion. Over a period of time zinc is leached from the metal structures (Fig. 9.13) and, in areas of acid soils, it is a serious source of local contamination (mean total zinc 1250–6500 $\mu g\,g^{-1}$ under pylons versus 170–320 from control sites 10–50 m from the pylon). Pylons represent a series of replicated situations in space. Is the genetic outcome the same at each site? Tests reveal that zinc-tolerant *Agrostis capillaris* plants occurred at a number of sites but not all (Al-Hiyaly *et al.*, 1993).

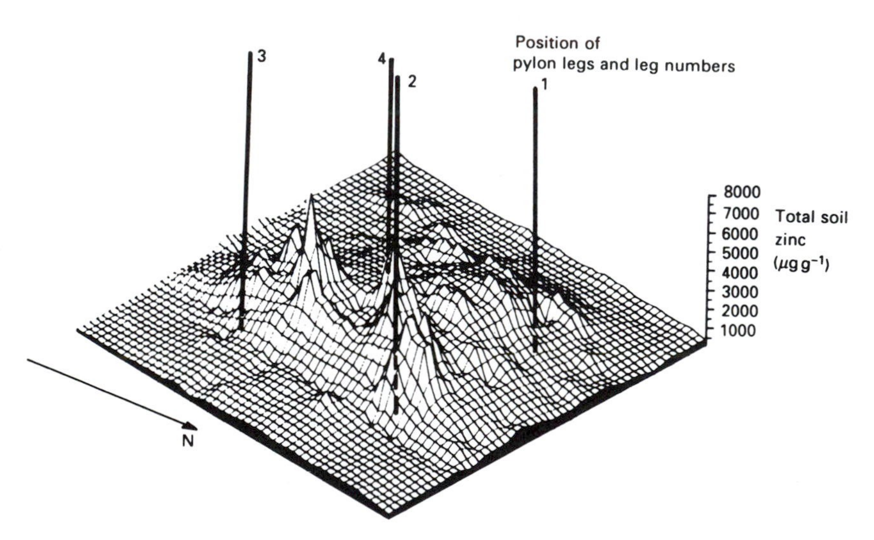

Fig. 9.13. Total soil zinc concentration in 399 soil samples from an electricity pylon in North Wales. Data smoothed by extending sampling scale × 2.5; intermediate values obtained by averaging. (From Al-Hiyaly, McNeilly & Bradshaw, 1988.)

Samples of seed were collected from several pylon sites. In the cases where tolerance had been detected in adults collected under the pylon, the local seed population responded positively to selection for zinc tolerance. However, at a site where no zinc-tolerant plants occurred under the pylon, no response to selection was detected in seedlings raised from local seed. This confirms that the absence of zinc-tolerant plants under the pylon was due to the lack of the appropriate genetic variation in the vicinity. This finding supports the view that lack of appropriate variation may limit evolution and indicates 'the stochastic nature of the evolutionary process because of the randomness in the occurrence of the necessary variability' (Al-Hiyaly *et al.*, 1993).

A study by Meharg, Cumbes & Macnair (1993) of populations of the grass *Holcus lanatus* provides a different perspective on the availability of 'tolerance genes' in uncontaminated areas. On arsenate contaminated mines around Dartmoor, Southwest Britain, plants were all arsenate tolerant. However, at 39 uncontaminated sites distant from mines in southern Britain, there was a high frequency of arsenate tolerant plants. The significance of this finding is, as yet, unclear.

Other studies of sites of known history provide circumstantial evidence for rapid change in populations under natural selection. Thus, a factory producing smokeless fuel from coal was opened in a country district in West Yorkshire in 1926, and a good deal of sulphur dioxide pollution was produced from the works. Ayazloo & Bell (1981) discovered that plants of the grasses *Dactylis glomerata*, *Festuca rubra*, *Holcus lanatus* and *Lolium perenne* growing in the vicinity of the works were significantly more tolerant of sulphur dioxide than samples of the species further from the factory. The rapidity with which selection pressures change is revealed in another experiment on air pollution. In 1975, the Sports Turf Research Institute of Britain set out plots of *L. perenne* in a polluted area of Manchester. Wilson & Bell (1986) raised plants from the original seed mixture and collected samples each year from the plots. Then, the sulphur dioxide tolerance of plants raised from the original seed – an unselected population – was compared separately with the tolerance of those collected each year (1976–82) in the polluted site. They discovered that the samples taken from the urban plot in 1979 and 1980 – which suffered significant pollution with sulphur dioxide – were statistically significantly more tolerant of sulphur dioxide than the 'original' material. However, the samples for 1981 and 1982 were not significantly different in their sulphur dioxide tolerance from the controls. As sulphur dioxide concentration in the areas declined in the early 1980s, due to control measures and industrial recession, it seems clear that the selection pressure favouring sulphur dioxide tolerance was relaxed during this period.

Studies of the effects of another atmospheric pollutant have yielded interesting results. Ozone is an important component of air pollution, especially in country districts down wind of conurbations. It is formed when oxides of nitrogen, resulting from combustion, including the burning of fossil fuels and vehicle exhaust emissions, react with oxygen in sunlight. *Plantago major* is a very common weed and it is exposed to different levels of ozone in different parts of Britain. In a standard test, populations from southern Britain exposed to high levels of ozone have proved to be much more resistant than those from more northerly areas (Reiling & Davison, 1992). However, the pattern of resistance is changing, as two of these populations have increased ozone resistance over a 5-year period that included two summers with high ozone levels (Davison & Reiling, 1995).

Microevolution in arable areas

So far, in this chapter, we have been examining grasslands of various sorts – hay, pasture and lawns. Investigations of arable weeds have also produced some very interesting insights into the action of natural selection (Baker, 1965, 1974, 1991; McNeill, 1976; Warwick, 1990*a*, *b*).

Barrett (1983) has reviewed the phenomenon of crop mimicry by weeds. 'The selective forces imposed by agricultural practices have resulted in the evolution of agricultural races of weeds... Such associations can involve a system of mimicry, whereby the weed resembles the crop at specific stages of its life history and, as a result of mistaken identity, evades eradication. Mimetic forms of weeds are most likely to be selected by hand-weeding of seedlings or by harvesting and seed cleaning procedures'. In the past, before thrashing and seed cleaning were improved, many weed species fruited at the same time as the crop and contaminated seed crops and, thus, such species as *Agrostemma githago* were a regular component of the weed flora before the widespread use of herbicides. Barrett draws attention to a number of other important examples. For instance, the grass *Echinochloa crus-galli* is a weed complex with many different variants, including mimics of rice. In a comparison of rice mimics and other weedy variants the precision of the mimicry became evident: e.g. anthesis in the mimetic variant usually coincides with the flowering of rice, presumably because genotypes flowering earlier than the crop would be visible and, therefore, weeded out. Rice has now become an important crop in California, where modern methods of agriculture are being used, rather than the labour-intensive methods of the Far East; it will be interesting to study the effect of mechanised farming on the crop mimic. Barrett draws attention to other strategies found in arable weeds: e.g. *Aethusa cynapium* has tall and dwarf genetic variants (Weimarck, 1945). The dwarf variant of the corn field, which survives below the level of cutting at harvest, flowers and fruits after harvesting, a good deal later than the tall variant which is found in field margins and on waste ground.

The development of herbicide-resistant variants of common weeds provides another example of the evolution of weeds in relation to agricultural practices. Resistance has been confirmed in more than 100 weed species (Warwick, 1991; Cousens & Mortimer, 1995). In most cases tolerance to a single herbicide has been established, but as many herbicides are often used it is not surprising that multiple resistance has been detected, e.g. a triple-resistant variant of *Chenopodium album* from Hungary (Solymosi & Lehoczki, 1989). Simple Mendelian control of resistance which is nuclear encoded has been discovered in some situations, for example resistance to paraquat of *Conyza bonariensis* involves a single dominant allele. However, in some cases genetic control of herbicide resistance by nuclear genes is more complex. In other cases, herbicide resistance is determined by genes of the chloroplast genome, e.g. many studies have been made on the triazine group of herbicides. Our account is based on Warwick (1991). Triazines powerfully inhibit photosynthesis and resistance is usually, but not always, maternally inherited. Molecular studies reveal that resistance in *Poa annua* is due to the loss of the herbicide binding site, as a result of a point mutation (Barros & Dyer, 1988). The pattern of distribution of resistance – western USA, eastern Canada, various parts of Europe, etc. – suggests that such variants arise independently (that is to say, polytopically) in different places. Resistance arises after a few years' use of the triazine, at doses of this long-acting group of herbicides that achieve weed control for the whole growing season (without using other herbicides), and especially in areas where the same crop is grown year after year. Studies of variation in resistant populations, using isozymes, have revealed that initially, at least, they are not as variable as normal populations. It seems possible that the initial selection pressure exerted by the herbicide is so great that the population goes through an extreme bottleneck, maybe of a single resistant individual. The spread of the resistant plants may also be subject to founder effects. Resistant plants generally, but not always, pay a cost of resistance, being less fit than sensitive variants in triazine-free soils.

The regular application of herbicide does not necessarily lead to populations of herbicide-resistant plants, as was discovered in research by Holliday & Putwain (1977, 1980). When seed stocks of *Senecio vulgaris* were screened for resistance to simazine – one of the triazine group – by sowing seed on to soil containing the herbicide, almost all the seedlings died, a surprising result considering that some of the sites had been treated with simazine for a number of years. An analysis of the ecological situation revealed an explanation. First, simazine applications were made once in the year and, after a time, the herbicide degrades in the soil. Moreover, some simazine-resistant *S. vulgaris* plants arising in the populations may have been killed, because other herbicides were also used on the plots to keep down the weeds. Furthermore, there appeared to be a bank of simazine-sensitive seed in the soil, which germinated when brought to the soil surface, and could grow and reproduce in the autumn when herbicide activity was low or absent. Thus, the population dynamics of sites treated with herbicides may be interestingly complex, and intermittent or seasonal factors will not necessarily lead to directional or disruptive selection.

Weed control is a characteristic of man-made habitats, and there is increasing evidence that weeding may act as a selection pressure favouring variants of common weeds that are capable of precocious development. Thus, *Senecio vulgaris* from Botanic Gardens and other well-weeded sites have a quicker rate of development than samples taken from poorly weeded or non-weeded sites (Kadereit & Briggs, 1985; Theaker & Briggs, 1993). There has recently been much interest in the evolution of life histories. It is perhaps significant that these studies have been particularly in vogue in the 1980s and employ 'monetarist' language to model and measure investments/costs/benefits/trade-offs, etc. Stearns (1992) defines the key concept: 'A trade-off exists where a benefit realised through a change in one trait is linked to a cost paid out through a change in another'. In the case of *Senecio vulgaris* from Botanic Gardens, the 'benefit' of precocious reproduction, in the face of frequent regular weeding of the flower beds, incurs costs because, in comparison with material from poorly weeded or non-weeded habitats, such early reproduction is associated with shorter stature, fewer leaves and earlier onset of senescence.

Adaptive and non-adaptive characters

In this chapter, we have examined the adaptive significance of various morphological traits and physiological characteristics. In many cases we do not know whether particular characters are adaptive or not.

In a study of *Spergula arvensis*, New (1958, 1959) showed that a ratio-cline existed in the British Isles, populations in the north and northwest having significantly higher proportions of the variant with smooth seed-coat than those in the south and east, where papillate seed-coats contributed quite high frequencies (Fig. 9.14). Examination of herbarium material for Europe as a whole confirmed the general tendency for the smooth variant to be characteristic of higher latitudes. Some 20 years later, New (1978) was able to show that the ratio-cline for seed-coat pattern had not significantly changed in Britain (though there *was* evidence for a significant change in proportion of strongly glandular-hairy plants in the populations). Cultivation experiments showed conclusively that the variant with smooth seeds was less tolerant of high temperatures and low humidity than the variant with papillate seeds. In her early work, New concluded that the apparently non-adaptive seed-coat character, which she had shown to be determined by a single gene without dominance, must be correlated with genotypic differences in physiology which were themselves clearly of selective importance in different climates. However, the most

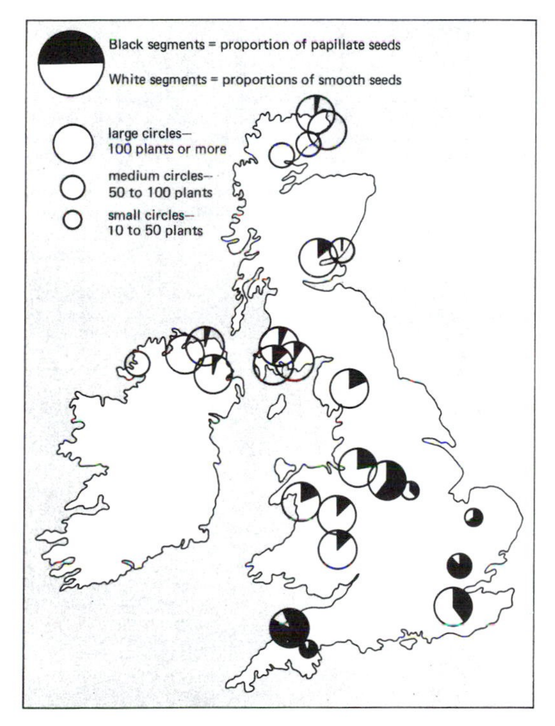

Fig. 9.14. Distribution in the British Isles of variants of *Spergula arvensis* with smooth and papillate seed-coats. (From New, 1958.)

recent paper (New & Herriott, 1981) shows that papillate seeds germinate more easily than smooth ones under dry conditions, so that an apparently non-adaptive character has, after all, been shown to have direct adaptive significance. There seem to be two lessons to be drawn from this unusually well-investigated case: first, that assessing characters as non-adaptive may merely reflect ignorance; and secondly, that actual polymorphisms are unlikely to be explained in terms of the operation of any single factor.

Turning to a second example, Lords and Ladies (*Arum maculatum*) shows remarkable polymorphism for the spotted versus unspotted leaves, and for purple versus yellow spadix (Fig. 9.15). Attempts to demonstrate that the spadix colour affects the quantity or quality of small insects trapped by the remarkable pollination mechanism have been unsuccessful; nor has any clear advantage been demonstrated in the presence of anthocyanin spots in the leaves (Prime, 1960).

Plants vary in leaf shape, size and indumentum. The adaptive signifi-

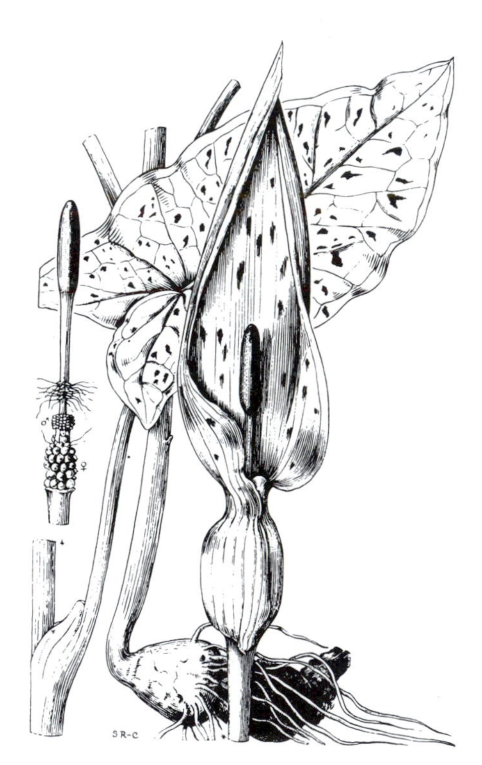

Fig. 9.15. *Arum maculatum*, spotted variant, showing characteristic leaves and inflorescence (× 0.5). (From Ross-Craig, 1973.)

cance of such variation may be sought in terms of differences in water relations, gas exchange and photosynthesis rather than leaf-shape *per se* (Johnson, 1975; Givnish & Vermeij, 1976; Givnish, 1979). However, the evolutionary responses of plants to herbivory should also be considered. For instance, there is evidence that hairs may protect the plant against attack by certain herbivores (Levin, 1973; Harborne, 1993). Also, it has been suggested that herbivory might have been an important factor in the evolution of leaf-shape, size and coloration (see review by Brown & Lawton, 1991). On the basis of studies by Niemela & Tuomi (1987) and others, Brown & Lawton have speculated as follows:

> Evidence that some insects recognize and respond to leaf shape ... suggest(s) that narrow, irregular hollows on the leaf blades of certain Moraceae mimic caterpillar feeding damage. These irregular 'incisions' occur on some of a plant's leaves, but not others, enhancing resemblance to feeding damage ... [f]alse damage could protect the plant in several ways. Ovipositing females may avoid it because real damage signals the presence of competing larvae, or induced biochemical defences in leaves; false damage may also be avoided because real damage serves to attract parasitoids and predators ... A wide range of plants outside the Moraceae also have leaves with herbivore-like indentations and holes ... The hypothesis that 'pseudo-damage' is mimetic and protective against herbivores (including vertebrates) deserves more attention, along with the related suggestion that variegated colour patterns on certain leaves may serve a similar function, for example by mimicking leaf-mines.

Brown & Lawton consider a number of other ways in which leaf size and shape could have evolved in relation to herbivory. These include: 'mimicry, not only of leaves of other plant species but also grazed leaves and inanimate objects; crypsis; physical barriers to being eaten [hairs, teeth, spines etc.]; interspecific differences in leaf morphology to reduce recognition by herbivores; very small or highly divided and dissected leaves that reduce feeding efficiency; and different adult and juvenile foliages'. These hypotheses need to be rigorously tested in well-designed experiments. Likewise, further work is essential to understand the adaptive significance of leaf variation in relation the physiological and mechanical demands imposed by different habitats. We return to this topic in Chapter 14.

Patterns of variation in response to seasonal or irregular extreme habitat factors

In many of the examples we have been studying, habitats remain extreme throughout the year; for instance, in a soil contaminated by copper, metal ions may leach out into the soil solution all the year round. It is clear, however, that some habitats are subject to severe conditions only at certain critical times. What patterns of variation are found at such sites?

One of the most familiar concerns a variable water table, the habitat being dry in some parts of the growing season and wet or flooded at others. Cook & Johnson (1968) studied intraspecific variation in the heterophyllous species *Ranunculus flammula*, which has lanceolate aerial leaves and linear leaves under water. They collected plants from a number of habitats in Oregon, and grew them under aquatic and terrestrial conditions. They discovered that certain populations, which were likely to experience the most unpredictable regimes in the wild, could produce the extreme heterophyllous leaf types in the appropriate conditions, whereas, in contrast, plants from habitats which were less frequently flooded showed very little heterophylly in the experiments. This investigation raises a number of questions about the modification of the phenotype.

In Chapter 6 we discussed the phenomenon of phenotypic plasticity, and showed that the phenotype is the product of the interaction of genotype and environment, and that any given genotype will produce different phenotypes in different environments. Each genotype is likely to have a characteristic breadth of phenotypic plasticity, itself under genetic control. In extreme sites, plants with a narrow range of responses may be selected, whereas in less extreme or variable sites plants with a wider spectrum of responses might be at a selective advantage. Let us examine why this might be so. Imagine plants growing on an exposed sea cliff. Plant A is genetically programmed to produce a tall phenotype, while B is of the genotype appropriate to a dwarf plant. In exposed conditions B is dwarf, but so too is A, and A can be said to be a 'phenocopy' of B. In the unusual event of a period with little wind, plant A may grow tall, only to be severely damaged in the next storms. Plant B, however, is constrained genetically within a narrow range of phenotypic possibilities in height, and under unusual conditions it does not grow 'too tall' for the habitat. It is easy to see how selection might operate to favour plants of genotype B in such circumstances. However, given unpredictable or highly heterogeneous habitats, it is apparent that a plant with a wide phenotypic plasticity could be at a selective advantage. Genecologists have not, in general, given enough attention to phenotypic plasticity.

Physiological plasticity should also be examined. An intriguing example of this phenomenon has recently been published. There is variability in the expression in cyanogenic plants of *Trifolium repens* (Till, 1987; Hughes, 1991). Moreover, some cyanogenic plants of *Lotus corniculatus* are of stable phenotype, whilst others are cyanogenic only at certain times of the year and under some conditions. Concerning *Lotus*, Ellis, Keymer & Jones (1977*b*) are inclined to see this physiological flexibility as adaptively significant, the cyanogenic mechanisms effective against herbivory being 'switched off' at just those times when grazing pressure is likely to be low and the risk of damage to the plant by other factors is at its highest.

Concluding remarks

Research carried out in the last four decades has transformed the subject and in concluding our brief survey we wish to make three important points. First, many of the elementary texts on ecological and population genetics give the impression that the subject has an elegant structure very nearly complete in every detail. By a judicious mixture of algebra and experiment drawn from animals and plants in the laboratory and field, the reader may be seduced into thinking that the various aspects of the subject are beginning to come together in a splendid synthesis. However, in our view, the student of evolution must be prepared to face with some humility the complications of both natural and artificial ecosystems, and be prepared to test and improve the present conceptual framework.

Secondly, it seems likely that genecology will be absorbed wholly, or in part, in the developing subject of population biology (Antonovics, 1976). We applaud this development. No investigation can proceed without a taxonomic framework, however, and it is hoped that many will still approach the subject from a thorough knowledge of the taxonomic background, for the history of the subject reveals how much the experimenter is indebted to the taxonomist who has an 'eye' for significant patterns of variation.

Finally, we might suggest what could be most important growth areas for the subject in the future. Harper (1977, 1983) points out that Wallace and Darwin had different views about the forces at work in natural selection, Wallace being concerned to emphasise the effects of inanimate nature – climate, soil, etc. – whilst Darwin stressed the competitive interaction of organisms in his account. Risking a generalisation, it seems that genecologists have, like Wallace, been mainly interested in tolerance of soil and climatic factors. Such a view may give too static a picture of microevolution. Plants are endlessly facing new situations as ecosystems develop. As we saw in our account of the experiments at Henfaes, biotic factors – herbivores, pests and diseases – play a crucial role in determining pattern and influencing process (Burdon, 1987; Cockburn, 1991). With some honourable exceptions, genecologists have not yet come to terms with the complex picture of the plant in its ecosystem. However, some notable advances have recently been made, for instance, in our understanding of host–parasite relations. Thus, Alexander *et al.* (1996) have explored the ecological genetics of the dioecious host *Silene latifolia* (*S. alba*) and the fungus *Microbotryum violaceum* (*Ustilago violacea*). The fungus produces spores in the flowers, and normal growth of ovules and pollen is prevented. Insects visiting diseased flowers carry the spores to healthy plants, and systematic spread of the fungus typically leads to host sterilisation. In studying the spread of the disease on local and regional scales, investigators have discovered that host plants differ greatly in resistance. Efforts are now being made to understand the detailed ecological genetics of both the host and parasite. An exciting period of research lies ahead as we explore – with a widening repertoire of techniques, most notably the new molecular tools – the genetic aspects of the complex interactions between plants in communities, and the intricacies of the relationships between plants, animals and disease organisms.

Progress may also be made in our understanding of the historical basis of regional differences in populations. In essence, present-day floras in northern Europe and North America are the descendants of plants that migrated north, perhaps by different pathways from a refugium or refugia in the south. Clearly, founder effects, historic population bottlenecks and the complex selection processes acting on migrating populations are likely to have influenced patterns of variation in present-day populations. The development of molecular methods is allowing greater insight into this fascinating area. For instance, regional differences in populations of *Abies alba* in the eastern Alps of Austria may owe their origin to different migration paths in the Alps, one from the east and the other from the west (Briettenbach-Dorfer *et al.*, 1992). A study of variation in Scandinavian populations in *Hippocrepis emerus* (*Coronilla emerus*), using allozyme and DNA fingerprint techniques, suggests the possibility of an early westerly immigration route into Norway and a later eastern route into Sweden (Lönn, Prentice & Tegelström, 1995). Also, differences in eastern and western populations of *Gentiana pneumonanthe* in Norway support a hypothesis of two different immigrations (Oostermeijer *et al.*, 1996).

10

Species and speciation

The species concept

The word 'species' has different meanings for different biologists. Consider first the species described by taxonomists. Naming, description and classification are based largely upon morphological details of herbarium specimens and to a lesser extent on living material that has been cultivated or collected from the wild. This is supplemented by geographical and sometimes ecological information. The aim of the taxonomist is to provide a convenient general-purpose classification of the material, a classification which will serve the needs of biologists in diverse fields. It is quite obvious that in order to communicate experimental findings to others, by word of mouth, in the literature and through databases, the experimentalist, like any other botanist, must be able to name plants unambiguously. To this end, an International Code of Botanical Nomenclature was agreed earlier this century. The development of this Code has a fascinating history (Smith, 1957). By 1900, four rival codes of practice were employed in different herbaria. Discussions of the problem occupied taxonomic sessions at International Botanical Congresses in Vienna (1905), Cambridge (1930), and Amsterdam (1935), and the successive Congresses, now at approximately 5-yearly intervals, are the occasion for continued revision of the Code. The agreements leading to a unified Code must be recognised as a major achievement.

One meaning of the word 'species' is now clarified. We may say that species are convenient classificatory units defined by trained biologists using all the information available. Clearly there is a subjective element in their work, and we must therefore face the fact that there will sometimes be disagreements between taxonomists about the delimitation of particular species. But there is a very large measure of agreement, for all except 'critical groups', in regions where the flora has been studied for many years.

Other species definitions

Since the time of John Ray, whose own attempt at a definition of species we discussed in Chapter 2, there has been no universally agreed definition; different definitions have been devised by biologists working in different specialist fields. (A thorough review is provided by King, 1993.) For instance, in order to deal with both modern and fossil lineages, Simpson (1961) proposed an evolutionary species concept: 'An evolutionary species is a lineage (an ancestor-descendant sequence of populations), evolving separately from others and with its own unitary evolutionary role and tendencies'. In an attempt to include ecological elements in a species definition, van Valen (1976) considered species to be: 'A lineage (or closely related set of lineages) which occupies an adaptive zone minimally different from that of any other lineage in its range and which evolves separately from all lineages outside its range'. Many biologists are fascinated by the possibility of discovering the phylogenetic history of groups of organisms, and Cracraft (1983) considers that: 'A species is the smallest diagnosable cluster of individual organisms within which there is a parental pattern of ancestry and descent'. As we shall see in Chapter 14, the study of plant phylogeny has greatly increased with the development of cladistic methods, which allow the computer generation of phylogenetic trees. This has generated further interest in species concepts (see Davis, 1995; Luckow, 1995).

The definitions we have examined are important in the context in which they developed, but they have not proved particularly useful in the practical definition of species in contemporary organisms. During the 1920s and 1930s, Turesson, Clausen and others, including the zoologist Dobzhansky (1935, 1937), suggested a number of new terms for units at or about the level of the species based upon breeding behaviour, an idea with a long history as we have seen. It was, however, the eminent zoologist Mayr who produced the most often quoted definition of what he called the 'biological species concept', a definition which is now found in botanical as well as zoological works. Biological species are 'groups of actually or potentially interbreeding natural populations which are reproductively isolated from other such groups' (Mayr, 1942). As we shall see in Chapter 13, this concept has been much criticised. A number of biologists pointed out that the concept of 'potential' interbreeding was very difficult and, later, Mayr (1969) removed the words 'actually or potentially' from his definition. Then, in response to further criticisms, Mayr (1982) presented yet another version: 'A species is a reproductive community of populations (reproductively isolated from others) that occupies a specific niche in nature'. This definition too has caused argument, principally concerning the difficulties of defining the concepts of 'niche' and 'reproductive community'.

Mayr's biological species concept, in its different forms, makes it clear that a key step in speciation is the acquisition of reproductive isolation. Thus, groups of related plants, which are distinct at the level of biological species, do not interbreed when growing in the same area in nature. They are said to pass the test of sympatry, that is of growing together without losing their identity through hybridisation. The mechanisms which keep biological species separate have been closely studied for many years, and will be examined in detail in Chapters 11 and 12, but, in general, as Table 10.1 shows, isolating factors fall into three groups.

In some cases pollination may be prevented; for instance, biological species found in the same area may grow in slightly different habitats, or flower at different times of day or in different seasons, and/or, for reasons of flower structure or pollinator behaviour, cross-pollination may not be successfully achieved. Even if cross-pollination occurs, pollen may fail to grow down the style. It is also possible, if plants are regularly and automatically self-pollinating, that cross-pollination may be prevented or its frequency may be greatly reduced.

Experimentalists also recognise a group of so-called 'post-zygotic' mechanisms. The seed from a cross between two biological species may fail to develop properly as a consequence of incompatibility between embryo, endosperm and maternal tissues (Valentine, 1956). In other cases, where hybrids are produced, they may show various signs of defective development or reduced fertility. More pronounced difficulties have been detected in other situations, where hybrids may be viable but sterile, or they may be weak as well as sterile. Finally, defective progeny, in a cross between two putative biological species, may be discovered only in the F_2 or later generations.

In an important review of isolating mechanisms, Levin (1978*b*) suggests that entities distinct as biological species are not generally separated by only one isolating mechanism and, furthermore, that barriers to crossing are not encountered simultaneously, but may be seen as a series of resistances which have to be overcome, if crossing is to be effected. Thus, potentially, a number of barriers – both pre- and post-zygotic – may be discovered between biological species.

The notion of a biological species began to catch the imagination of experimentalists, particularly following the publication of a number of important books on microevolution. The most important of these are:

Table 10.1. *A classification of isolating mechanisms in plants (Levin, 1978*b)

Premating	
Spatial	
1. Ecological	
Reproductive	
2. Temporal divergence	
(a) Seasonal	
(b) Diurnal	
3. Floral divergence	
(a) Ethological	
(b) Mechanical	
Postmating	
4. Reproductive mode	
5. Cross-incompatibility	
(a) Pollen–pistil	*Pre-zygotic*
(b) Seed	*Post-zygotic*
6. Hybrid inviability or weakness	
7. Hybrid floral isolation	
8. Hybrid sterility	
9. Hybrid breakdown	

Dobzhansky (1941), Huxley (1940, 1942), Mayr (1942), Simpson (1944), Stebbins (1950), Clausen (1951) and Heslop-Harrison (1953). From this ferment of discussion the following ideas emerged:

1. As the definition of biological species (and more or less equivalent groupings described in the 1920s and 1930s) involved a test of breeding behaviour, experimentalists considered that entities defined thereby were more objective than the 'species' of the herbarium taxonomist (cf. Gregor, 1931; Müntzing, Tedin & Turesson, 1931; Clausen, Keck & Hiesey, 1939).
2. The species and classifications produced by taxonomists – the so-called 'alpha taxonomy' – should be modified in the light of experiments to give a more perfect system eventually leading to an 'omega taxonomy' in which all the knowledge of biologists reached proper synthesis (Turrill, 1938, 1940).
3. Crossing experiments and other information, such as chromosome numbers, might reveal something of the phylogeny of groups, and the existing classificatory systems, which are a mixture of convenient

arrangement and phylogenetic speculation, could be modified to allow classification to reveal the evolutionary pathways leading to present-day patterns (Darlington, 1956, 1963).

These ideas were attractive to some botanists and provided a stimulus for research of various kinds. To give the correct historical perspective, however, it is important to state that they were not accepted by all, and a great deal of argument ensued. While some biologists were concerned to try to apply the biological species concept in taxonomy, others saw the concept as a model, useful for thinking about the processes involved in reproductive isolation. We will examine the present views of the relation of experimental and taxonomic categories in Chapter 13, after presenting some of the ideas and results produced by the work of experimentalists who came to be known as 'experimental taxonomists' or 'biosystematists' (Camp & Gilly, 1943).

We may now turn our attention to the origins of species. The plural 'origins' is important, as biologists now accept that there are a number of modes of speciation, which may conveniently be grouped under two headings, namely 'gradual speciation' and 'abrupt speciation'.

Gradual speciation

Population geneticists point to four groups of processes of importance in the separate evolution of different geographically isolated 'sub-populations' derived from a single original local interbreeding population.

1. *Mutations.* As genetic mutations are the result of random events, it is likely that patterns of genetic change will be different in the derived isolated populations.
2. *Other effects of chance.* As we have seen in Chapter 9, chance may influence the variation in populations. Some of the derived populations, dispersing to new territories, may be subject to founder effects and, if in any population the numbers of individuals becomes small, then genetic drift may occur (Wright, 1931) and alleles might be completely lost by accidents of sampling, or in other cases rare alleles might by chance become more frequent, without regard to their adaptive value.
3. *Selection.* As derived populations will be subject to different climatic, edaphic and biotic factors, the action of selection is likely to lead to genetic differentiation between the derived populations.

4. *Migration.* The extent and direction of change could be greatly influenced by the degree to which derived populations are effectively isolated from gene flow.

By virtue of the independent small cumulative changes possible in different isolated sub-populations of an original interbreeding population, it seems probable that, given a long enough period of geographical and genetic isolation, genetic differentiation in a group of derived populations of common origin might proceed first through an ecotypic phase and then, later, with the gradual evolution of genetic isolating mechanisms, derivatives at the rank of biological species might be formed. As we shall see in Chapter 11, such gradual speciation is thought to be highly important in the evolution of biological species.

Abrupt speciation

As we saw in Chapter 2, one of the controversies stimulated by Darwin's work concerned the possibility of 'saltations', or abrupt changes in the course of evolution. Darwin himself saw evolution as a continuous, gradual process, with no place for sudden events, but those who argued for saltations have been vindicated. There is now overwhelming evidence for abrupt speciation. Thus, chromosome repatterning and changes in chromosome number may occur. Genetic events may change the breeding system. These abruptly arising changes in some individuals of a population may lead, as we shall see in Chapter 12, to the sympatric origin of new species. In many cases, polyploidy is involved in abrupt speciation, but other chromosome changes are possible and likely to be very important. As this chapter is designed to provide an introduction to a more thorough treatment in later chapters, at this point we shall confine our attention to polyploidy and will consider other models of abrupt speciation later.

According to Rieger, Michaelis & Green (1976), the term 'polyploid', for a plant containing more than the normal number (two) of sets of chromosomes, seems to have been first defined by Strasburger (1910). Winkler (1916), in an important paper, describes what is probably the first clear case of experimental production of polyploids in the Tomato, *Lycopersicum esculentum* and the related Nightshade, *Solanum nigrum* (Fig. 10.1). Before this, a great deal of interest had centred on the work of De Vries with Evening Primrose (*Oenothera*) (to which we have already alluded in

Chapter 6) and, in particular, upon a 'mutation' which was called '*gigas*' because it was generally larger than the parent *O. glazioviana* (*O. lamarckiana/O. erythrosepala*). This '*gigas*' mutant had been shown to possess twice the normal somatic chromosome number of 14, and there was argument between De Vries and Gates as to the significance of this difference, Gates (1909) holding the view that the chromosome doubling was itself a cause of the differences in morphology between '*gigas*' and the normal plant. Winkler's demonstration that his experimentally produced polyploid Tomatoes, with double the normal chromosome complement, differed also from the 'parent' diploid in similar ways to the '*gigas*' mutant strongly supported Gates' interpretation, and subsequent work showed that artificial polyploids generally differed in the larger size of all their parts, ranging from mean cell size to the size of the whole plant.

Soon after Winkler's paper, Winge (1917) made an important contribution in distinguishing between this kind of polyploidy where, at least in theory, the simple doubling of the chromosome number in a single individual was all that was involved (autopolyploidy), and a more complicated situation where polyploidy succeeded hybridisation (allopolyploidy). (The terms auto- and allopolyploidy were coined by Kihara & Ono in 1926.) A number of tests of Winge's hypothesis were made. For instance, Clausen & Goodspeed (1925) crossed a diploid species (*Nicotiana glutinosa*; $2n = 24$) with a tetraploid (*N. tabacum*; $2n = 48$) and produced a fertile hexaploid ($2n = 72$). Following the production of hybrids between Radish (*Raphanus sativus*; $2n = 18$) and Cabbage (*Brassica oleracea*; $2n = 18$), spontaneous chromosome doubling occurred to give a fertile tetraploid *Raphanobrassica* ($2n = 36$) (Karpechenko, 1927, 1928).

Autopolyploidy can be explained as follows. A diploid plant receives a haploid set of chromosomes (a genome) from each parent. Its constitution can be represented as AA. If the plant is subject, for example, to temperature shocks, the regular process of mitosis may be disturbed and, instead of two cells each with the diploid number of chromosomes, a single diploid cell with four times the haploid number may be formed (AAAA):

$$\mathrm{AA} \xrightarrow{\text{doubling}} \mathrm{AAAA}$$

In this way polyploid cells arise and may give rise to polyploid branches on diploid plants. Experimentally polyploid cells can be produced with the drug colchicine, which acts as a spindle inhibitor preventing regular disjunction of chromosomes. Thus, chromosome replication in colchicine-treated material is not combined with the proper division of the products into two daughter nuclei, and the formation of a nuclear membrane around all the replicated chromosomes yields a polyploid cell, which may be involved in the production of autopolyploid seeds (AAAA). An autopolyploid plant may also be produced by different means, namely by the fusion of two unreduced gametes (AA + AA).

Let us now consider the origin of allopolyploids. Two related diploid species, which have diverged by gradual speciation from a common ancestor, may be different both chromosomally and genetically, and may be represented as AA and BB. A hybrid between the two species, AB, may very well be highly infertile, as there is insufficient homology between the A and B genomes for proper pairing at meiosis. The hybrid plant may produce a polyploid branch, of genetic constitution AABB, by the process outlined above. However, in the AB hybrid a very small but significant percentage of unreduced (AB) eggs and pollen may be produced, together with unbalanced haploid meiotic products. On fusion of unreduced gametes, a plant

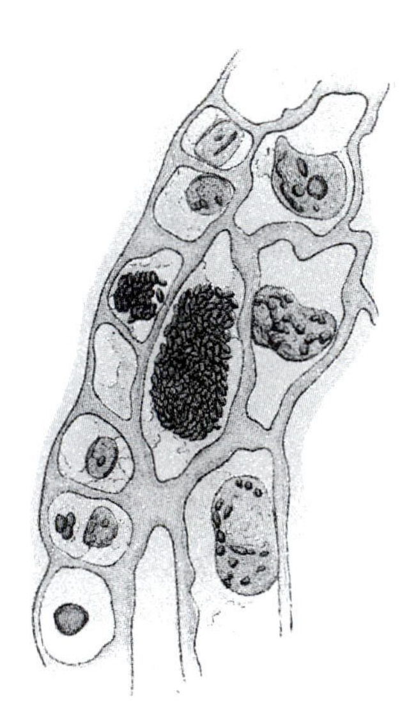

Fig. 10.1. One of the earliest studies of polyploidy was made by Winkler (1916), who investigated this complex phenomenon in experimentally produced chimaeras between different species of *Solanum*. The figure shows one of his drawings of high-polyploid cells side by side with diploid cells in the tissue of the anther wall in one of his experimental chimaeras (× 1000).

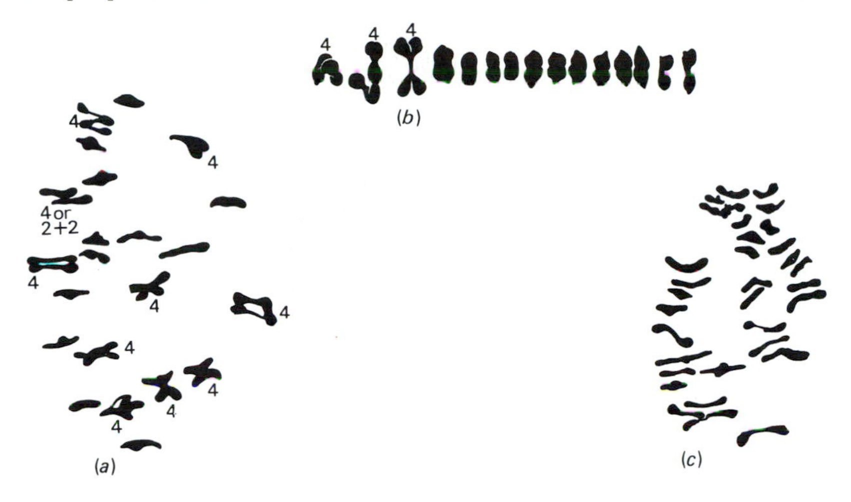

Fig. 10.2. (*a*) Meiosis (metaphase I) in autotetraploid Watercress ($2n = 4x = 64$) prepared by colchicine from *Rorippa nasturtium-aquaticum* (*Nasturtium officinale*) ($2n = 2x = 32$) (× 1600). (b) Meiosis (metaphase I) in *Primula kewensis* ($2n = 4x = 36$) (× 2500). Note the three quadrivalents. (*c*) Meiosis (metaphase I) in wild tetraploid Watercress *Rorippa microphylla* (*N. microphyllum*) ($2n = 4x = 64$) (× 1600). ((*a*) and (*c*) from Manton, 1950; (*b*) from Upcott, 1940.) In tetraploids, a range of different cytological behaviour is found. In autotetraploidy (of type AAAA), quadrivalents are frequently found, as in (*a*). In allotetraploids (of type AABB), where each chromosome has a pairing partner, normal bivalent pairing is found as in (*c*). Sometimes a mixture of quadrivalents and bivalents is discovered as in (*b*). These complex situations are discussed in Chapter 12.

with the constitution AABB is produced, in which the chromosome number has been effectively doubled:

$$\text{AA} \times \text{BB} \longrightarrow \text{AB} \longrightarrow \text{AABB}$$

In this simple case we are dealing with a tetraploid with twice the normal diploid number. Such plants are sometimes referred to as 'amphidiploids' or 'amphiploids'. Other kinds of polyploids with extra genome sets are described with the appropriate term – 'triploid', 'hexaploid', etc. The level of 'ploidy' can be represented as the multiple of the 'basic number' x, which is the haploid number of the presumed original diploid or diploids; thus a triploid can be represented by $3x$, a tetraploid by $4x$, etc. (If this notation is used, it is then possible to retain n and $2n$ to indicate the functional haploid and diploid numbers, as distinct from the presumed polyploid relationships within a whole genus or group of species.)

Meiosis in the new allopolyploid is more normal than in the diploid hybrid (AB), as genomic pairing – A with A and B with B – can occur. If, however, there is still a high degree of homology between A and B genomes (they were derived from a common ancestor by gradual speciation), then more complex pairing of the chromosomes may occur and groups of three and four chromosomes may be found (Fig. 10.2).

Allopolyploid derivatives are reproductively isolated from their parents, as can be seen by examining what happens when an allopolyploid AABB (with gametes AB) is crossed with one of its 'parental' species, AA (with gametes A). Triploid individuals of constitution AAB are produced. Even though A genomes may pair at meiosis, there is no pairing partner for the B genome, and a highly irregular meiosis occurs, which leads to infertility in the hybrid. An isolating mechanism now exists between diploids and their derived allopolyploid (Fig. 10.3). It is by the abrupt origin of an isolating mechanism in this way that new biological species arise by the process of

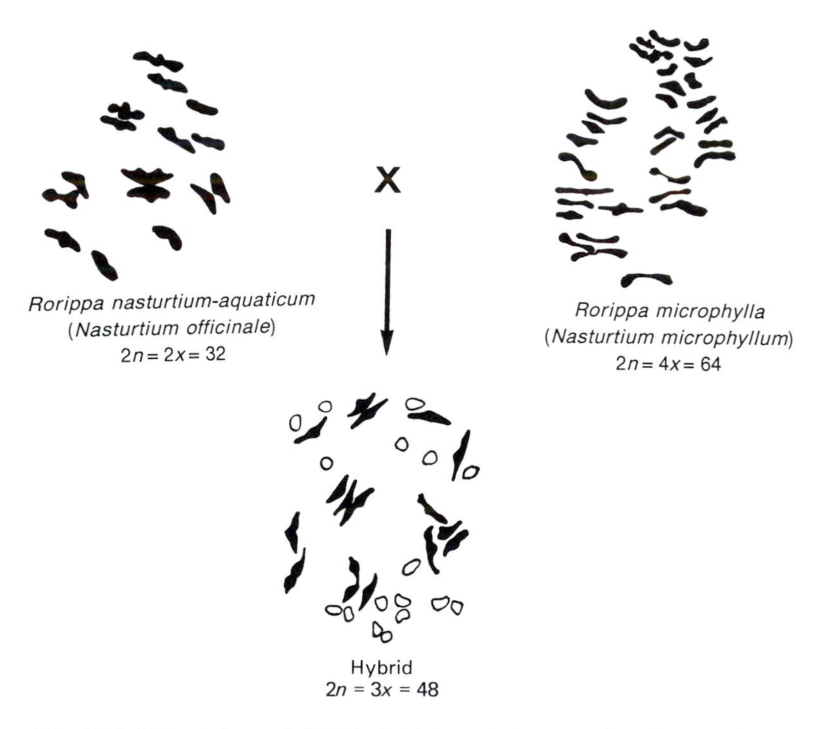

Fig. 10.3. Meiosis in a triploid hybrid (× 1600). Note the mixture of bivalents (black) and univalents (white) at metaphase I of meiosis in the triploid. (From Manton, 1950.)

polyploidy. New species may originate where the two parental species occur together, that is, sympatrically. Moreover, as new polyploids are produced by single, abrupt events, we have here a mechanism whereby, within a man's lifetime, new, fertile species may arise.

In the chapters that follow, we shall examine in more detail these two contrasting modes of speciation and try to assess their relative importance in evolution as a whole.

11

Gradual speciation and hybridisation

In Chapter 10 we presented a simple model of gradual speciation. Two populations derived from a common ancestor and occupying different geographical areas (i.e. allopatric) pass through a period of independent change yielding derivatives of different morphology, which are reproductively isolated from each other. In such cases the existence of isolating mechanisms is revealed if the taxa come to occupy the same area (i.e. become sympatric).

Allopatry may arise in many different ways. We note a few of the many possibilities. Studies of climate change in the post-glacial period, both in Europe and North America, point to the enormous importance of species migration in establishing 'new' isolated populations. Thus, geographical isolation of daughter populations may result from the destruction of land bridges (e.g. the opening of the Irish and North Seas following post-glacial sea-level changes). New isolated populations may also result from long-range dispersal of propagules to 'islands' of different sorts, whether they be oceanic islands, isolated mountain peaks, landlocked lakes or specialised rock outcrops (e.g. serpentine). In the longer perspective, species distributions may be fragmented by continental drift and associated mountain-building processes.

To take account of the complexities of different situations likely to be important in nature, a group of models have been devised incorporating a range of different assumptions (Grant, 1971). These models differ in the relative importance they attach to the effects of selective forces and to chance events. For instance, population establishment or subsequent development may involve severe reductions in numbers, allowing the possibility of genetic drift at some period in the independent evolution of daughter populations. In constructing model systems, different assumptions might be made about differentiation. Thus, morphological differentiation in daughter populations might proceed in tandem with the sort of genetic changes which yield eventual isolating mechanisms or, alternatively, morphological change and 'reproductive isolation' may evolve at different rates.

Evidence for gradual speciation

In considering the evidence for various models of gradual speciation, the extended time-scale of hundreds of generations presents an immediate difficulty. It seems that the details of the processes of gradual change from a single ancestral population to two biological species must remain unknown. However, it has been suggested (for example, by Clausen, 1951) that, if groups of different taxa are examined, they may be at different stages in speciation; thus, population differentiation might be at the ecotypic stage, as, for example, in *Potentilla glandulosa*, whilst other populations, perhaps given subspecific or specific rank by taxonomists, may be at a later stage in speciation. By examining a range of different types of situation, from ecotypes to island endemics, from local races to vicariads (two similar taxa occupying different geographical areas), a composite picture of gradual speciation might be built up.

To test the ideas of Clausen, it is necessary to discover the degree of reproductive isolation between collections of different taxa – taxa which, on account of their distribution, morphology, etc., are likely to share, either closely or remotely, a common ancestor. Plants from different areas, whose cytotaxonomic background is well understood, are brought together and, by appropriate crossing experiments, the viability and fertility of any resulting hybrids are determined. In this way, the genetic basis of ecotypic differentiation in *Potentilla glandulosa* was investigated in crosses between alpine and coastal plants, and between plants from subalpine habitats and those from the foothills in California. Clausen & Hiesey (1958) detected considerable genetic differences between the ecotypes.

Also, working with several different genera, botanists have made very extensive and elaborate crossing programmes between species and subspecies, the results of which are set out in crossing polygons, in which the fertility of F_1 (and sometimes F_2) hybrids is revealed in diagrammatic fashion as a geometric figure (Clausen, 1951, and Radford *et al.*, 1974, provide a range of examples).

Crossing experiments with species of *Layia*

We may take as an example of such investigations the experiments of Clausen (1951) and associates, who studied the results of crossing different subspecies of *Layia glandulosa*. Evidence of partial genetic barriers is provided by some of these crosses (Fig. 11.1). They also examined the cytogenetic and crossing relations of different species of the genus *Layia* (Fig. 11.2). The species fell into two major groups – those with a chromosome number $n = 7$ and those with $n = 8$. The degree to which crossing was possible between the different species was examined in experiments (Clausen, Keck & Hiesey, 1941). As a result of these experiments Clausen (1951) concluded that '*Layia* is an excellent example of a genus in various stages in evolutionary differentiation'. Between the races or ecotypes there are 'moderate barriers to interbreeding... At an even more advanced stage, the species have become so distinct that the hybrid is

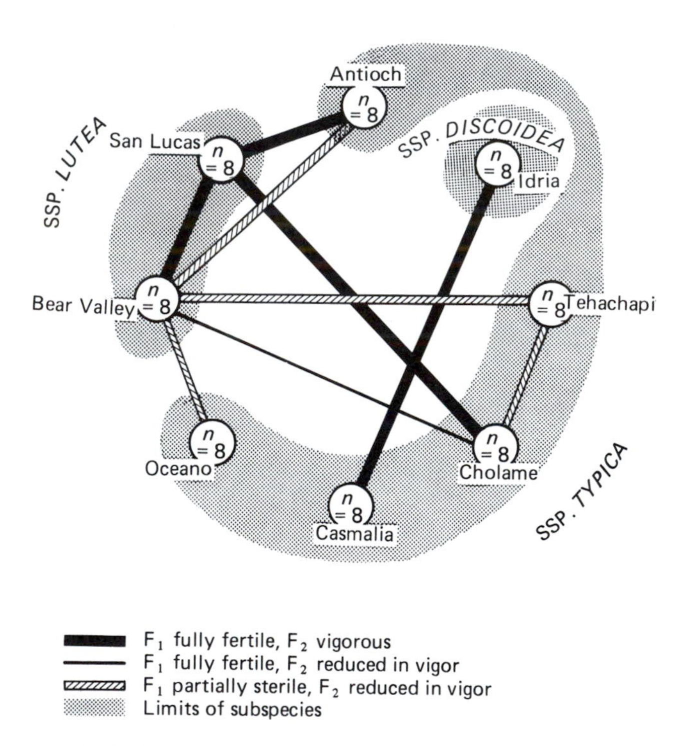

Fig. 11.1. Diagram of intraspecific crossings within *Layia glandulosa*. Used by permission of the publisher, Cornell University Press, from Clausen (1951), *Stages in the evolution of plant species*. © 1951 Cornell University Press.

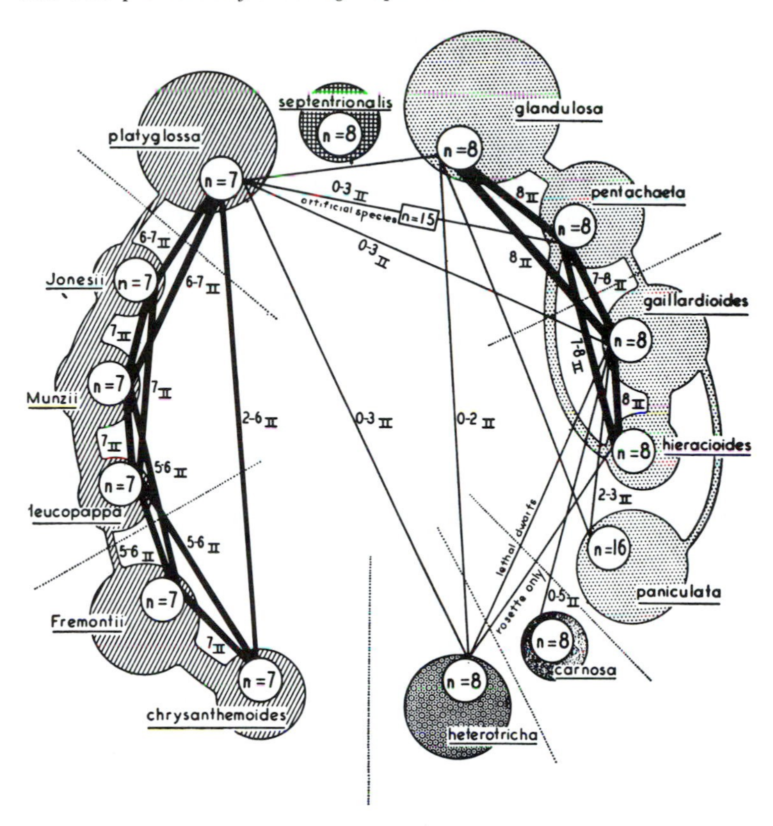

Fig. 11.2. Crossing polygon of *Layia* showing cytogenetic relationships. The black lines connecting species indicate successful hybridisations, and the width of the lines, the approximate degree of pairing between the parental chromosomes in the hybrid. The widths of shaded connections between species indicate degrees of gene interchange possible. Dotted radials indicate major morphological discontinuities within the genus. (From Clausen, Keck & Hiesey, 1941.)

much weaker than its parent species, or the species are unable to produce a viable hybrid'.

The interpretation of crossing experiments

While crossing experiments provide valuable information, they require cautious interpretation. First and foremost, only small numbers of plants are used in many crossing programmes and, in the extreme case, a single plant is taken as a representative of a population! Crossing experiments are often halted at the F_1 stage. A large amount of garden space may be needed to raise F_2 families. There are other grounds for caution. Glasshouses are frequently used in breeding experiments. Insect pests and fungal diseases may suddenly reach epidemic proportions during a series of crossing experiments, and success or failure of crossing may be influenced by such pests and diseases. The effects of bagging flowers must also be considered. Geiger (1965) reviews the relevant literature, which indicates that the temperature inside pollen/insect-proof bags might be up to 15 °C higher than ambient temperatures in the daytime and 1–2 °C lower at night. As pollen sterility and other effects may be induced at high temperatures, failure in crossing may be due to these external factors, rather than to intrinsic differences. Thus, pollen viability may decline quite rapidly, both inside and outside bags, and therefore it may be vital to use fresh pollen in crossing experiments (see Stone, Thomson & Dent-Acosta, 1995). Furthermore, the 'fertility' of pollen in hybrids is commonly estimated, not by direct study of germinability or in a crossing test, but by staining with acetocarmine or other stain. Fully formed grains with nuclear staining are assessed as 'good' pollen, and misshapen, undersized, inadequately stained grains are judged to be 'bad'. The reliability of staining as an indicator of fertility is rarely, if ever, put to the test. In assessing crosses, the possible complicating factor of genetic incompatibility must be considered. Also, artificial crossing experiments do not often assess possible pre-zygotic isolating factors, but concentrate on the results of experiments which often involve crude surgery. Will such experiments reveal what would happen if isolated populations became sympatric in nature and only the normal agencies of pollen transfer were to operate?

Studies of *Layia* using molecular methods

Given some of the potential problems of crossing experiments, molecular tools now provide an important opportunity for an independent assessment of the models of gradual speciation. Many of the historic results of microevolutionary research, first examined using cytotaxonomic methods, are now being re-examined. For example, further evidence concerning the evolution of the genus *Layia* has come from studies of a number of populations of each of the six species that have the chromosome numbers $n = 7$ (Warwick & Gottlieb, 1985). Using electrophoresis of leaf extracts, 13 enzyme systems were analysed.

Before looking at these results we must consider the general question of how to estimate the similarity or difference between and within species. From the results of studying electrophoretic variability in enzymes, a statistic devised by Nei (1972) – the genetic identity – may be employed (Avise, 1994, may be consulted for details of the formula). Genetic distance values range from 1.0 to 0.0. If two populations are identical, then the genetic identity value will be 1.0. If there are no alleles in common, the value will be 0.0. In comparisons of populations of the same species, mean genetic identities are usually about 0.90, while comparing congeneric species gives values of about 0.67 (Crawford, 1989). These values indicate that species of a genus are genetically more different from each other than populations of a single species.

In the study of *Layia*, Warwick & Gottlieb discovered that the genetic identity for populations within species was 0.96 or higher, except in the case of *L. platyglossa*, where the populations were more variable, which had a genetic identity value of 0.88. (This finding of greater variability in *L. platyglossa* confirms the findings of Clausen, 1951.) The mean genetic identity values produced by comparing the six species are given in Fig. 11.3. The genetic distance of *L. platyglossa* from the other species is similar to that reported in other comparisons of congeneric species. Concerning the high genetic identity between the three endemic species *L. jonesii*, *L. munzii* and *L. leucopappa*, Warwick & Gottlieb note: 'Their high genetic identity suggests that they are more closely related to each other than to other *Layia*

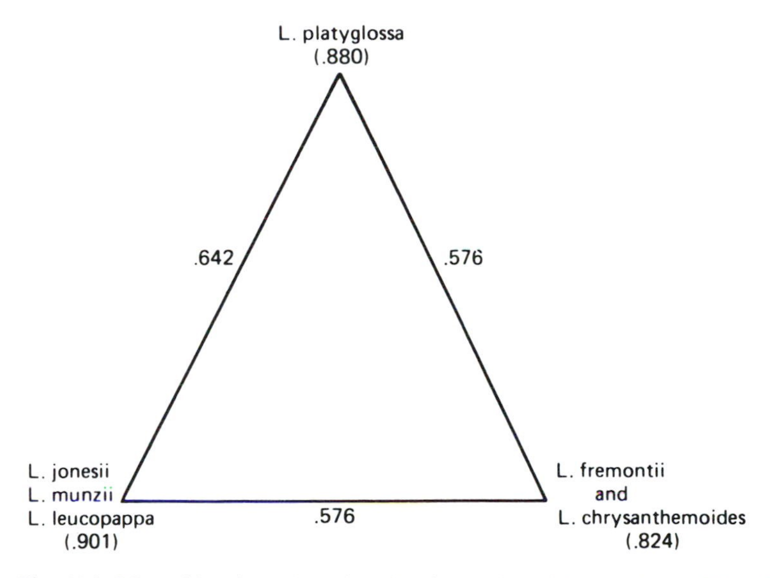

Fig. 11.3. Mean identity values for the six species of *Layia*. (From Warwick & Gottlieb, 1985.)

[species] and that their divergence was relatively recent, presumably a consequence of recent (Pleistocene) isolation on different newly exposed soils'. Considering all their findings they conclude: 'the genetic identity values were consistent with those expected on a model that the species diverged gradually as they adapted to geographically separate habitats'.

Studies of a number of other genera reveal patterns of genetic divergence consistent with a model of allopatric gradual speciation, e.g. *Limnanthes* (McNeill & Jain, 1983), *Lisianthius* (Sytsma & Schaal, 1985), *Hosta* (Chung *et al.*, 1991) and *Streptanthus* (Mayer, Soltis & Soltis, 1994).

Uncertainties about the concept of gradual speciation

A fundamental issue must be faced in considering models of gradual speciation. There will always be uncertainty about the magnitude and sequence of past events. Founder effects, and genetic and chromosomal mutations are likely in the history of populations. In reality, gradual speciation is dependent upon these abrupt events. If these events have small effects, and many such changes are 'needed' for speciation, then models of gradual speciation may have some validity. It is clear, however, that the division of speciation into gradual and abrupt modes may be difficult, especially if the material is not studied cytologically. Such studies might provide clues as to the possibility of significant chromosome differences between isolated populations, differences likely to have had an abrupt origin. We discuss abrupt speciation further in Chapter 12.

Evidence supports the view that allopatric speciation is one means by which speciation occurs in plants and the results are of especial interest to taxonomists, who have named various subspecies or species according to the degree of morphological distinctness exhibited by plants. It is interesting, however, that crossing polygons do not reveal any consistent linkage between the degree of morphological differentiation and crossing behaviour. In some cases, there is a suggestion that different entities recognised as taxonomic species are reproductively isolated from other such groups, but it is abundantly clear that there is no necessary correlation between the presence of sterility barriers and morphological differences. Thus, in *Elymus glaucus*, which taxonomists treat as a single species (Snyder, 1950, 1951), there would appear to be sterility barriers between different collections (Fig. 11.4). On the other hand, certain taxonomic species of *Mimulus* (Vickery, 1964) appear to have fully fertile F_1 hybrids (Fig. 11.5). Clearly, the factors controlling morphological differentiation and post-zygotic isolating mechanisms may not necessarily evolve in step.

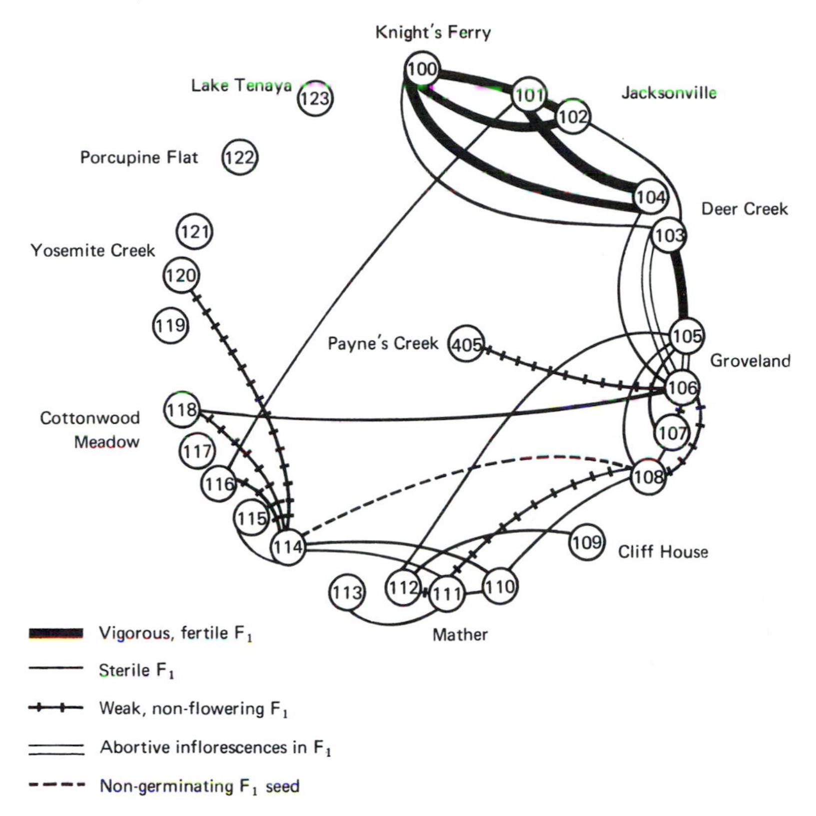

Fig. 11.4. A summary of hybridisation between populations of *Elymus glaucus* collected along a 75-mile transect in the Sierra Nevada. The diagram shows developmental behaviour and pollen fertility of the hybrids. Snyder concludes from the cytological behaviour of the hybrids at meiosis that much of the sterility is caused by small structural differences in the chromosomes and by specific genes. He also suggests that hybridisation in nature between *Elymus glaucus* and species of the related genera *Agropyron*, *Hordeum* and *Sitanion* might be responsible for much of the variability. (From Snyder, 1950, 1951.)

Natural hybridisation

So far we have been examining artificial hybridisations. Experimental investigation of natural hybridisation between different taxonomic species has also contributed to our understanding of speciation. The genus *Geum*, widespread in the temperate regions of the world, will serve to illustrate a number of important points.

The two most widespread European species are *G. rivale* and *G. urbanum*.

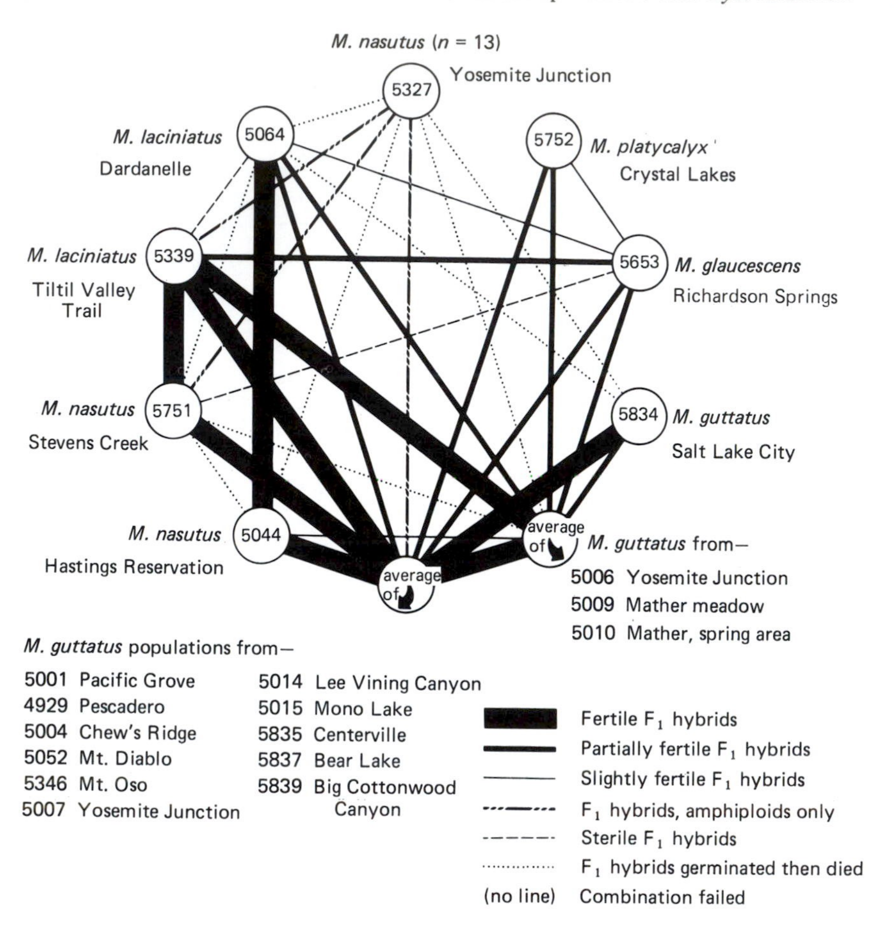

Fig. 11.5. Crossing polygon indicating fertility of F_1 hybrids between different populations of the *Mimulus guttatus* complex in North America. (From Vickery, 1964.)

The latter, which has a somewhat 'weedy' tendency to which we shall refer later, has also become widely naturalised in North America. Figure 11.6 shows that the general appearance of the flowers of these two species is very different. *G. rivale* is a typical 'bee' flower and species of Bumble Bee (*Bombus*) are recorded as the commonest visitors. The purplish colour and the somewhat concealed entrance to the hanging flower are features shown by many 'bee' flowers. Contrast with this the smaller, open, erect, yellow flower of *G. urbanum*, which shows no specialisation for the visits of particular insects and seems to be frequently self-pollinated.

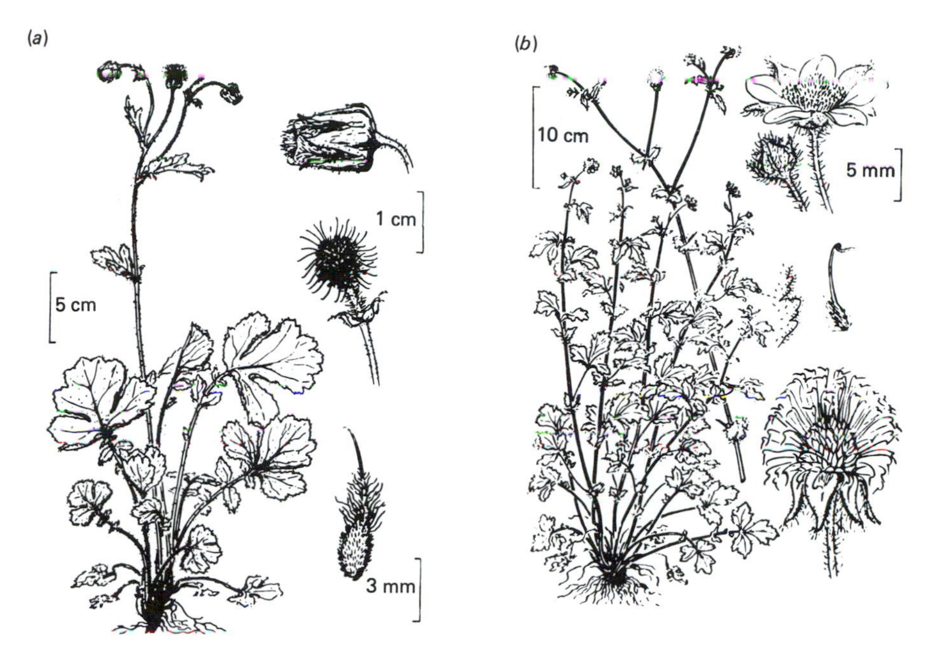

Fig. 11.6. (*a*) *Geum rivale*. (*b*) *G. urbanum*. (From Roles, 1960.)

Over much of Europe these two species are sympatric (Fig. 11.7). They are, however, usually effectively separated by ecological differences. *G. rivale* often grows in damp, shady places in South and Central Europe and is more or less confined to the upland regions. It is absent from much of the Mediterranean region, but it occurs in Iceland, from which *G. urbanum* is absent. Over most of lowland Europe, however, *G. urbanum* is the common plant, growing particularly in hedgerow and woodland communities affected by man. It has long been known that plants with somewhat intermediate characters sometimes occur in abundance in woods and scrub where the two species meet; such obviously hybrid plants were called *G. intermedium* by Ehrhart as early as 1791. These hybrids have attracted much attention, mainly because the two parent species look so different, and in some places hybrid swarms are found in which there is a remarkable range of variation. Such a hybrid population formed the basis of a detailed genetic study by Marsden-Jones (1930), who was able to work out to some extent the inheritance along Mendelian lines of several of the characters determining the differences between the two species. This work was followed, and greatly enlarged, by the Polish botanist Gajewski, whose study of the genus *Geum*,

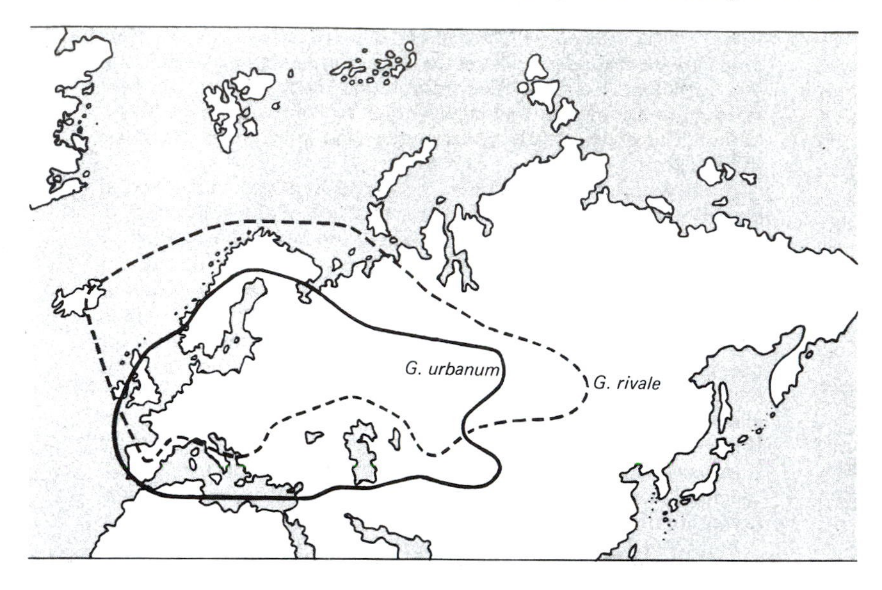

Fig. 11.7. Map showing the distribution in Eurasia of *Geum rivale* and *G. urbanum*. (From Gajewski, 1957.)

published in 1957, is the fruit of many years of experimental study and is of outstanding value.

The situation described in detail by Marsden-Jones for a wet wood of Alder (*Alnus glutinosa*) at Bradfield, Berkshire, could be paralleled in a good many places in lowland England. On the other hand, Gajewski, who studied *Geum* mainly in Poland, emphasises that large hybrid populations are rare, and that even where the two species are growing close together, there are often very few intermediate plants. What is the cause of this apparent difference in efficiency of isolation between England and Poland?

The first relevant point is that artificial F_1 hybrids between the two species can be made, though not with ease, and that such plants are highly fertile, the F_2 showing a range of segregates as might be expected if (as Marsden-Jones' detailed genetic experiments bore out) many genes are involved in the specific differences. Judged purely in terms of the theoretically possible gene flow, therefore, *G. rivale* and *G. urbanum* would fall within a single biological species. As a matter of fact, Gajewski's work demonstrated that all 25 species in the subgenus *Geum* (to which our two species belong) will hybridise with each other, and that most of these hybrids are at least partially fertile. In practice, however, a good many species-pairs or species-groups are allopatric and hybridisation does not take place in nature (Fig. 11.8).

We have already seen that ecological preference will normally separate, at least partially, mixed populations of the two species. Moreover, the difference in type and colour of the flowers would ensure a segregation of insect pollinators. To this difference should be added a rather obvious difference in the times of the beginning of flowering, which in Britain differ by 3 or 4 weeks, *G. rivale* being the earlier. This would mean that, even in mixed populations, seed set on the early flowers would be necessarily 'pure' in the case of *G. rivale*. Such considerations point the way to at least a tentative answer to our question. Gajewski records that he grew seeds taken from plants of each species growing in a mixed stand

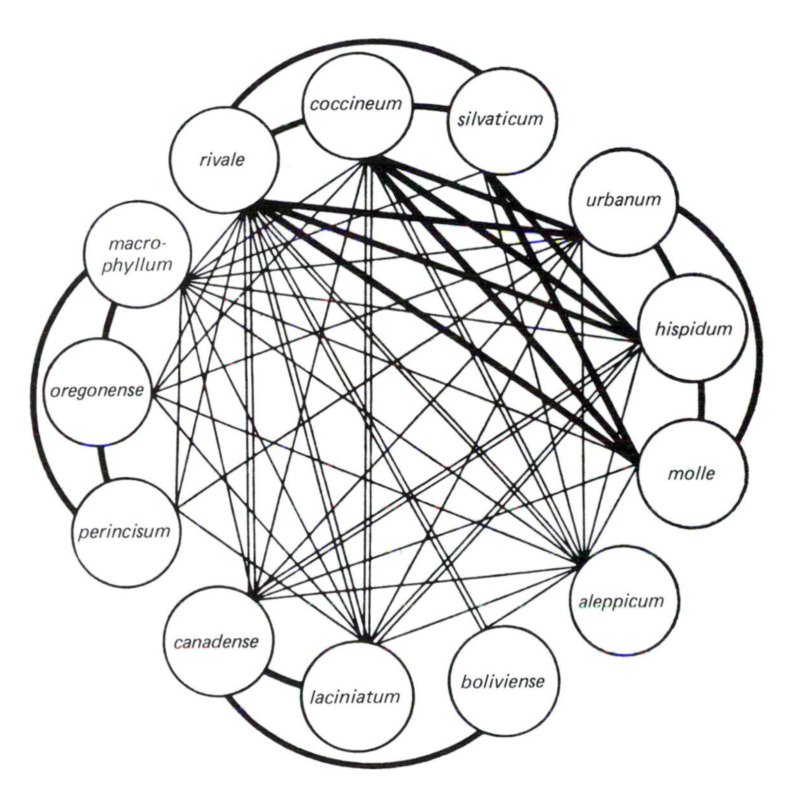

Fig. 11.8. Fertility in F_1 hybrids among different hexaploid *Geum* species. Thick lines indicate fertile hybrids; double lines partially fertile hybrids; and thin lines sterile hybrids. (From Gajewski, 1957.)

in a Polish locality where hybrid plants were rare, and found that the progeny were 'pure' with no detectable sign of hybridisation. Clearly factors such as the preferences of insect visitors, coupled with differences in flowering time, effectively prevent more than a minimum of gene flow between the species in Poland. What is different about the English conditions?

The most important difference lies probably in the complex history of man's interference with the vegetation. Gajewski's observations were made, partly at any rate, in the great forest nature reserve of Białowieza, in eastern Poland, where the forest is as little affected by human activity as anywhere in lowland Europe. Here such hybrids as he recorded were single individuals on roadsides and in forest rides, where *G. urbanum* probably owes its existence to its having accompanied man into these new habitats. *G. rivale* behaves as the original, native species. In England, on the other hand, most woodland is present as small 'islands' in a 'sea' of agricultural land, and the disturbed marginal habitats suitable for *G. urbanum* have clearly been enormously extended over the centuries by the human activities of drainage, forest management and clearance, hedgerow planting, etc.

Although this difference in vegetation history may be the most important factor in determining the local frequency of hybrid *Geum* populations, we should bear in mind the possibility that the other isolating factors may also be less effective in some circumstances than others. Is it possible, for example, that the separation in flowering time between the two species is less effective in the relatively mild climate of England than in the more continental one of Poland? We do not have any detailed information on this point, but field investigations of this and many other relevant questions about pollination would clearly be very interesting in the areas where the hybrids grow.

If our general thesis is correct, we are dealing here with a partial breakdown, brought about by man's activities, of naturally effective ecological isolation. This kind of explanation has been extended by Gajewski, admittedly more speculatively, to cover the recent evolutionary history of both species. He pictures *G. urbanum* as originally evolved in geographical isolation from *G. rivale*, perhaps in Southeast Europe. Certain adaptations, among them the unspecialised type of pollination (and often self-pollination) and the efficient, small, animal-dispersed fruits, made *G. urbanum* an effective 'weed' of marginal woodland habitats created by man. In this way, the species became sympatric with *G. rivale* over much of Europe. In this new situation, the advantage lies with the 'weedy' species, for most vegetational change brought about by man will favour it rather than its relative.

We have dealt at some length with the *Geum* example, because it provides an excellent paradigm of what we take to be a general phenomenon. Man's activities in the exploitation and management of various areas have led to a variable degree of breakdown of ecological isolation. In areas long settled by man there is hardly a square inch which has not in some way or other been influenced by human activity. Burning, forest clearance, drainage, grazing and other agricultural practices, mining, and all aspects of urban industrialisation from road building to atmospheric pollution – all these activities have contributed to changes in vegetation. Man has 'hybridised the habitats', in the famous phrase of Anderson (1949). Thus, the acreage of semi-natural forests, wetlands, mountain communities and coastal vegetation types has gradually diminished. 'Islands' of apparently unchanged vegetation sometimes survive on land less fitted for some form of agriculture or other development, but on inspection these too have been exploited by man for food, fuel (peat, wood), building material (wood, thatching material), sport or in other ways.

Man's activities result not only in complex patterns of breakdown of ecological isolation, but also in changes of geographical isolation. Evidence suggests that by man's deliberate or accidental transport of plants, the 'natural' distribution of some plants has been greatly changed. In particular, weed species have been carried across the world, and plants of horticultural interest and agricultural importance have been widely disseminated.

Thus, in considering the course of gradual speciation, a further complication is beginning to emerge. In some areas man's activities are likely to have influenced, perhaps decisively, not only the past and present distributions of plants, but also the population variation at different sites. Therefore, our models of gradual speciation must take account of the fact that in relatively recent times (geologically speaking) man has become the dominant factor in landscape management, greatly influencing patterns of distribution in many areas of the world. The breakdown of geographical and ecological isolation may lead, as in the *Geum* example, to hybridisation, and the consequences of such hybridisation are likely to differ in different circumstances. What consequences might we expect in various circumstances?

The consequences of hybridisation: some theoretical considerations

Consider the case of two populations, A and B, which have become sympatric after a period of allopatry. The populations may have changed genetically in isolation to such an extent that, in sympatry, they are unable to cross freely, hybridisation being a rare event leading to infertile products (both pre- and post-zygotic factors may be important). The two populations are in effect behaving as two biological species. It may be, on the other hand, that the two populations have come to differ in morphology and ecological requirements, but crossing experiments reveal only partial, or no, major barriers to interbreeding. As daughter populations of A and B differ in ecological preferences, the fate of any hybrids produced in nature is likely to be influenced decisively by ecological factors.

If no intermediate habitats are found between those preferred by taxa A and B, then hybrids between A and B are likely to be at a selective disadvantage. In comparison with crosses A × A and B × B, the hybrids derived from the crosses A × B and B × A may yield fewer (or no) viable offspring. Selection, in favouring the progeny of A × A and B × B, will act against the products of hybridisation. Any partial isolating mechanism minimising A × B and B × A will be subject to selection. The perfecting of partial isolating mechanisms under the impact of selection may then take place, such a change being referred to as the 'Wallace effect' (Grant, 1966) after Alfred Russel Wallace, the co-founder with Darwin of the theory of evolution by natural selection, who speculated on this subject.

If habitats intermediate between those preferred by A and B occur, then in such habitats hybrids between A and B may be 'fitter' (i.e. contribute a greater number of offspring to future generations) than the progeny of A × A and B × B. Any partial isolating factors developed in the once allopatric populations are likely to be overcome and the distinctness of the two populations may be lost, locally or regionally, in a mass of hybrids and backcrosses.

These two models, which are in reality extremes of a spectrum of possibilities, will now be examined in further detail. It is clearly very difficult to study natural populations to discover the validity, or otherwise, of the Wallace effect, but there is a certain amount of circumstantial evidence which supports the model.

One line of evidence comes from studies of 'experimental' populations. In the Fruit Fly, *Drosophila*, studies have shown that intensification of reproductive isolation may result from artificial selection (e.g. Koopman, 1950; Knight, Robertson & Waddington, 1956). Rice & Hostert (1993) have reviewed 40 years of laboratory experiments on speciation in animals.

A most elegant botanical example was studied by Paterniani (1969), who grew two varieties of Maize (*Zea mays*) together in an experimental garden. To explain Paterniani's experiment, some details of the stocks used are necessary. One variety had white, flint cob characters (genetically determined with alleles *yy SuSu*) whilst the other had yellow, sweet cob characters (*YY susu*). Thus, each stock had one dominant and one recessive marker gene. Pollen of the white, flint stock had the genotype *y Su*, while pollen of yellow, sweet had the genotype *Y su*. Any cross between the two stocks would yield kernels with yellow colour in the white, flint parent and flint (often called starchy) kernels in the yellow, sweet parent. Because these markers give pronounced effects, visible in the developing cob, it is possible to determine the degree of hybridisation between varieties without raising progeny, enabling the experiment to be carried out in a relatively short time. Using the experimental stocks, planted in such a way as to maximise the opportunity for hybridisation between varieties, Paterniani investigated what happened when selection for reproductive isolation was carried out. This was achieved by selecting at each of a number of generations the cobs which showed the *least* hybridisation. From such cobs, grains of the phenotypes white, flint and yellow, sweet were used to provide seedlings for the next generation. At the start of the experiment the degrees of outcrossing were 35.8 and 46.7% for white, flint and yellow, sweet stocks, respectively. After six generations of selection, the levels of intercrossing were 4.9% for white, flint and 3.4% for yellow, sweet. The factors involved in this isolation were examined, and it was discovered that the number of days from sowing to flowering was probably the most important factor. The original stocks flowered on average at the same time, but by the end of the experiment the mean flowering time for white, flint was 5 days earlier, while the mean for yellow, sweet was 2 days later. By selecting against hybridisation the mean flowering times of the two stocks had been separated by about 7 days (Fig. 11.9). This experiment suggests that similar effects on flowering time might be expected in wild populations where hybrids are at a selective disadvantage.

The evolution of flowering time differences between sympatric populations, where hybrids are inviable, receives support from computer simulation models. For instance, van Dijk & Bijlsma (1994) have modelled a somewhat different situation in *Plantago media* where there are diploid and tetraploid plants which, when crossed, produce 3*x* inviable hybrids. They discovered that flowering time difference could arise but stress that coexistence has to last long enough for it to develop. The results of a number

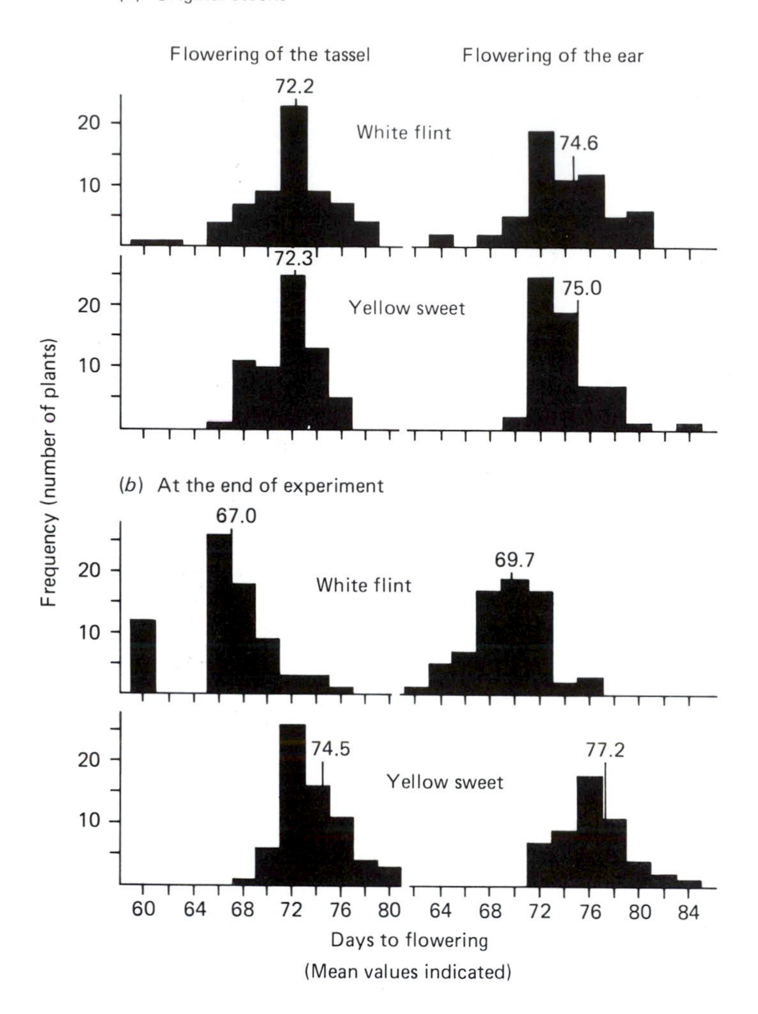

Fig. 11.9. Histograms showing the number of days to flowering for the original stocks of two varieties of Maize (*Zea*), and for the stocks after the selection experiment had run for six generations. Maize has separate male and female flowers on the same plant. Hence the separate recording of tassel (male) and ear (female). Details in text. (From Paterniani, 1969.)

of experimental and computer simulation experiments carried out by other biologists are reviewed in this paper.

Results of disruptive selection in polymorphic populations

The situation where genetically polymorphic species exist in a patchy or mosaic habitat would seem to offer the possibility of speciation (Thoday, 1972). Several studies of plants have been made on mine debris containing heavy metal residues, sites which often exist as 'islands' in a 'sea' of pasture. Some species occur both on the mine and in the adjacent pasture. As we have seen in Chapter 9, there is evidence for gene flow between pasture and mine plants, followed by disruptive selection. Furthermore, it is clear that pasture plants cannot grow on the mine debris and mine plants grow less well in the pasture (Bradshaw, 1976; Bradshaw & McNeilly, 1981). It seems that crosses within habitat type (mine × mine and pasture × pasture) will produce fitter progeny than the crosses mine × pasture and pasture × mine. Therefore, it has been argued that selection would favour any variants arising in the population which would restrict 'gene flow' by promoting within-habitat crosses. Experiments by Antonovics (1968) and McNeilly & Antonovics (1968) working with *Agrostis capillaris* (*A. tenuis*) and *Anthoxanthum odoratum* showed that, although there was some overlap in flowering times between pasture and 'mine' plants, those at the mine edge flowered about 1 week earlier, a difference maintained in cultivation. Further, plants of both species growing on mine debris, though normally self-incompatible, were found to be capable of a degree of self-fertility.

Both these traits may be seen as barriers to free gene flow and could be interpreted as the first steps in the speciation process. It is unclear, however, how much selfing actually takes place in wild populations. The studies of Lefèbvre (1973) suggest that, in *Armeria maritima* growing on mine debris, the potential for self-fertilisation exists, but in a study of a particular mine population, it was discovered that outbreeding was the rule.

These situations take us beyond the simple model of gradual allopatric speciation, with which we opened this chapter, for, as far as we know, the different polymorphic variants of each species have not developed in geographic isolation, but in closely adjacent areas. More studies are required to determine whether reproductive isolation – as a result of flowering time differences or changes in breeding behaviour – is the product of many genetic changes over a long period or one or a few changes occurring abruptly. A model involving the abrupt origin of reproductive isolation through a simple genetic mechanism has received support from the studies of crosses between copper tolerant plants of *Mimulus guttatus* from Copperopolis, California and non-tolerant material (Macnair & Christie, 1983). Crosses revealed the existence of a gene associated with copper tolerance, either closely linked or pleiotropic, which interacted with genes in the non-tolerant material, resulting in the death of some F_1 offspring.

Other crossing experiments have yielded interesting evidence. If isolating mechanisms are initiated in allopatric populations but perfected in sympatric situations, then it follows that crossing within groups of sympatric and allopatric taxa might be revealing. Such studies have been carried out by Grant (1966), who studied nine species of *Gilia* (Fig. 11.10). Five of the species were sympatric, having overlapping distributions in the foothills and valleys of West California. The other four species were allopatric in maritime habitats in North and South America. As Table 11.1 shows, crossing between sympatric taxa yields fewer seeds per flower than that between allopatric taxa. A similar study of other *Gilia* species gave the same sort of result. However, crosses between the *G. splendens* and *G. australis* groups revealed the opposite pattern, with sympatric populations giving greater seed yield when crossed *inter se*, than allopatric populations similarly crossed. Thus, two of the three situations examined experimentally provided some support for a Wallace effect.

Population studies of wild plants provide a another area of evidence concerning the Wallace effect. A modified population containing *Phlox glaberrima* (mean pollen size 55 μm) and *P. pilosa* (mean pollen size 30 μm) was studied by Levin & Kerster (1967). In order to investigate the effect of flower colour on interspecific pollinations made by Lepidoptera, transplants of red-flowered variants of *P. pilosa* were added to the naturally occurring white-flowered variants of this species at a site in Cook County, Illinois. Thus, two colour variants of *P. pilosa* were available for interspecific pollinations with the red-flowered *P. glaberrima*. Evidence of pollen grains on a sample of *P. pilosa* stigmas suggested that, while 30% of the red flowers of *P. pilosa* received alien pollen, only 12% of the white ones bore such pollen. It would appear that the flower colour differences in the unmodified population of *P. glaberrima* (red) and *P. pilosa* (white) might act as aids to pollinator discrimination and thus reduce interspecific pollinations. It is possible to devise models of selection for reproductive isolation between two daughter populations that have come to differ in flower colour, one population (A) being polymorphic for flower colour, while population B has only one colour phase, the flower colour of B being one of the colours possible in A. If pollinator activities result in cross-pollination,

Table 11.1. *Comparative crossability of species with different geographical relations in the Leafy-Stemmed Gilias (Grant, 1966)*

Geographical relation of parental species	No. of combinations of parental species	Mean number of seeds per flower		Mean of means
Foothill species inter se (*sympatric*)	9	0.0	0.1	0.2
		0.0	0.1	
		0.0	0.4	
		0.0	1.2	
		0.0		
Maritime species inter se (*allopatric*)	5	7.7		18.1
		16.7		
		19.6		
		21.9		
		24.8		

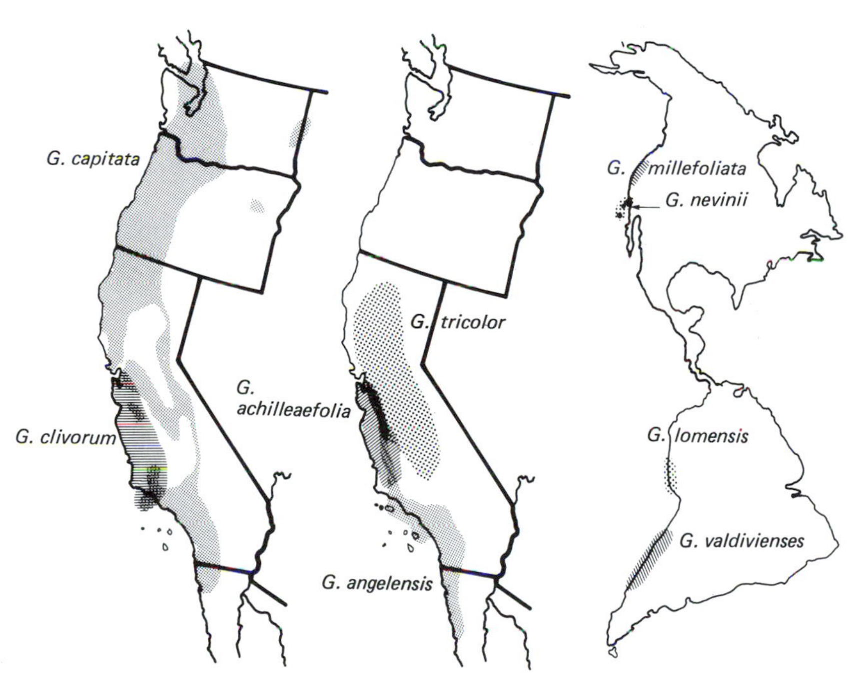

Fig. 11.10. Geographical distribution of the nine species of *Gilia* studied by Grant in America. (From Grant, 1966.)

A × B, between similar colour variants and this leads to less fit progeny, then the balance of colour variation in population A could change. Ultimately A and B in sympatric situations could come to be characterised by different flower colour. These cases provide some support for the Wallace effect. Levin (1985) has carried out further experiments on character displacement in *Phlox*, and his paper makes reference to other examples where 'the presence of a related species may have been the stimulus for divergence in flower colour'.

Introgression and other patterns of hybridisation

We can now turn our attention to what happens where hybrids might be at a selective advantage. In 1949, the American geneticist Anderson published a book entitled *Introgressive hybridisation* in which he postulated that in some cases when species hybridise and the environment has been disturbed naturally or by man's activities, incorporation of the genes of one species into another might occur, as a consequence of hybridisation and repeated backcrossing. This process, for which the shorter term 'introgression' is now used, Anderson claimed was much more widespread and important in evolution than had been previously thought (Fig. 11.11). Anderson (1949) considered that 'the raw material for evolution brought about by introgres-

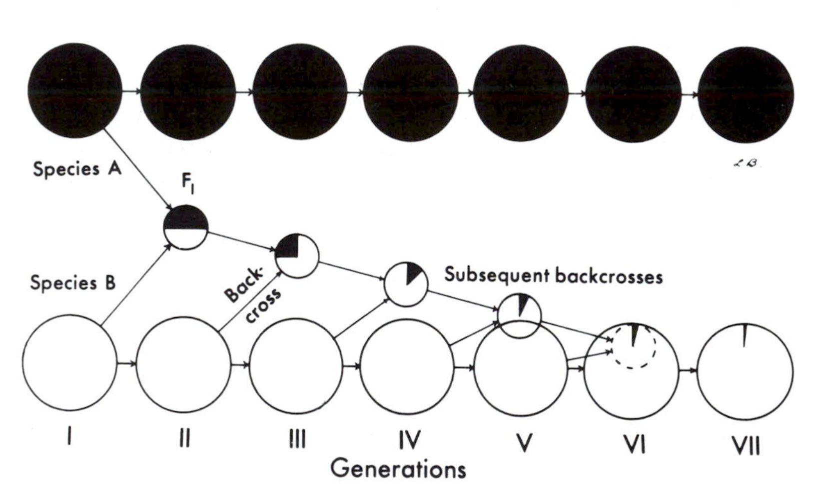

Fig. 11.11. Diagram illustrating introgression between two species. Backcrossing of the F_1 hybrid to species B ultimately results in the absorption of some genes from species A into at least some individuals of species B. Reprinted by permission of John Wiley & Sons, Inc. from Benson (1962), *Plant taxonomy*. © 1962.

sion must greatly exceed the new genes produced directly by mutation'. However, a significant number of botanists have not accepted this view. In their recent review, Rieseberg & Wendel (1993) quote sceptics who are of the opinion that the 'ultimate contributions made by hybrids must be very small or negligible' and that introgression is a kind of 'evolutionary noise' being a primarily local phenomenon with only transient effects.

Anderson invented several simple methods of displaying variation in hybridising populations; his hybrid index method is illustrated in Fig. 11.12. By devising a suitable scale of numerical values an investigator can make a rapid survey of field collections. In the interpretation of complex populations, interspecific hybridisation is often assumed to be the cause of the pattern. But it would seem more intellectually honest to see the hybrid index method as a means of describing the degree of separation of plants of different morphology. As we shall see below, other evidence is necessary to assess whether interspecific hybridisation is actually involved. Furthermore, it is essential to have some clear idea of the variation to be expected in the supposed parental species. To this end samples should be collected from sites where the two 'pure' parents are not in contact with each other. (We may note in parenthesis that the collection of 'pure' parents is often a highly subjective and difficult task, especially in cases where the 'parental' taxa are broadly sympatric.) The hybrid index method yields a scale of variation, with pure parental colonies gaining the highest and lowest scores, and plants of intermediate morphology having intermediate scores.

The method has the disadvantage that the variation of individual plants is effectively 'lost' and, in the display of the results, plants with the same hybrid index score may differ phenotypically. Anderson's pictorialised scatter diagram technique, illustrated in Fig. 11.13, provides an attractive method of display which overcomes this problem to some extent. Material from complex and putative pure parental populations is scored for a number of features, and two quantitative characters are used to generate an ordinary scatter diagram. The figures for each plant serve to determine the location of a spot on the diagram. Spots may be of different shape or colour, and are decorated with appropriate arms to show the qualitative and quantitative characteristics of each specimen.

In the excellent example of studies of *Primula vulgaris* and *P. veris* populations, Woodell (1965) explored the variation in 'parental' populations at Marley Wood and Dickleburgh (Fig. 11.13*a*) and in a complex population at Boarstall Wood (Fig. 11.13*b*). The results of Woodell's study are also displayed as a hybrid index (Fig. 11.13*c* and Table 11.2). They raise a point of especial interest. In scoring corolla diameter 0 to 9, and calyx

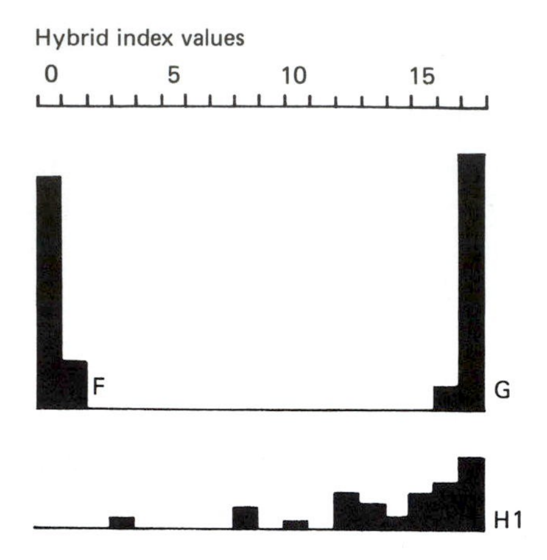

Fig. 11.12. Hybridisation between *Iris fulva* and *I. hexagona* var. *giganticaerulea* in Louisiana, USA, as an example of the use of the hybrid index. First, a list is compiled of the differences between the two taxa in, for example, flower colour. Next, one species is arbitrarily chosen to be at the low end of the index scale, and the other at the upper end. Specimens from natural populations are then scored character by character giving the appropriate score as outlined on the following scale:

	Tube colour	Sepal blade colour	Sepal length	Petal shape	Exertion of stamens	Stylar appendage	Crest
Like *I. fulva* score	0	0	0	0	0	0	0
Intermediates score	1	1, 2 or 3	1, 2	1	1	1	1
Like var. *giganticaerulea* score	2	4	3	2	2	2	2

Plants exactly like *I. fulva* score 0 for each character, giving a grand total of 0. Total score for plants like var. *giganticaerulea* is 17. Intermediate plants score between 1 and 16. Riley's results shown here are for three populations (sample size 23). Colonies F and G are more or less pure parental species. Colony H1, on the other hand, contains many hybrid plants. In the main these hybrids resembled var. *giganticaerulea* rather than *I. fulva*. (From Riley, 1938.) A number of botanists (e.g. Stebbins, 1950; Grant, 1971) have considered this example to be a clear case of introgressive hybridisation. However, Randolph, Nelson & Plaisted (1967) re-examined the problem of field hybridisation of *Iris* species in Louisiana. Although they established that hybridisation occurred not infrequently and in one case were obliged to interpret a particular population as being a 'new' taxon of hybrid origin, they came to the conclusion that their results provided no evidence that introgression had altered significantly the status of the three cross-compatible species of Louisiana Irises as stable taxonomic units.

tooth length 0 to 7, while giving other characters scores 0 to 4, the characters of corolla diameter and calyx tooth length have been given greater weight in the calculation of the index. The justification of weighting and the ways this might be achieved have received particular attention from various experimentalists: for example, Gay (1960) and Hathaway (1962).

Another method of displaying the variation of individual plants is provided by the polygonal graph method of Hutchinson (1936), later elaborated by Davidson (1947). An excellent example is provided in the study by Moore (1959) of *Viola* hybrids (Fig. 11.14, p. 298).

Various multivariate methods of studying field collections have also been devised, which are most conveniently carried out using computer packages. Figure 11.15, p. 299, shows the pattern of variation revealed by principal component analysis of populations of *Quercus robur*, *Q. petraea* and a complex population thought to contain hybrids (Rushton, 1978, 1979). It is not our intention to discuss how such analyses are carried out; the point we wish to make is that students of evolution are continuing to make use of techniques for displaying interesting phenotypic variation in populations, the latest methods yielding patterns of dots arranged around axes of a highly derived kind. While many botanists are fascinated by multivariate grouping and clustering techniques, others, not necessarily the mathematically incompetent, find this trend in data processing unsatisfactory. It yields diagrams of over-rich complexity from which it is impossible to judge the phenotypic characteristics of any particular specimen. Perhaps the valuable insights offered by complex multivariate techniques need to be combined with displays which reveal the characteristics of the plants under study.

A character which is often scored in field samples, and sometimes in herbarium studies, is pollen stainability, as judged by staining with acetocarmine or other stains. As we have mentioned earlier, pollen stainability sometimes masquerades as pollen fertility in the report of investigations. Since the relation of pollen stainability to fertility is almost never investigated, it would seem necessary to exercise caution in interpretation. The rationale behind the study of pollen stainability, in the context of natural hybridisation, is as follows. Two populations, as a consequence of genetic differences developed in their period of allopatric isolation, may produce primary hybrids of low fertility, and, in association with morphological intermediacy, low pollen stainability may therefore be taken as evidence of hybridisation.

Table 11.2. *(a) Characters used in construction of the hybrid index* (Fig. 11.13)

	veris									*vulgaris*
Character score	0	1	2	3	4	5	6	7	8	9
1. Corolla diameter (mm)	12–14	15–17	18–20	21–23	26–26	27–29	30–32	33–35	36–38	39–41
2. Calyx circumference (mm)	24–22	21–19	18–16	15–13	12–10					
3. Calyx tooth length (mm)	2–3	3–4	4–5	5–6	6–7	7–8	8–9	9–10		
4. Throat pattern	C	C(P)	CP	(C)P	P					
5. Calyx hair	C	C(P)	CP	(C)P	P					
6. Pedicel hair	C	C(P)	CP	(C)P	P					
7. Leaf hair	C	C(P)	CP	(C)P	P					

Note: In characters 4–7, C = *P. veris*: C(P) = putative backcross to *P. veris;* CP = putative F_1; (C)P = putative backcross to *P. vulgaris:* P = *P. vulgaris*. The minimum score, representing 'pure' *P. veris*, would be 0, that representing 'pure' *P. vulgaris* would be 36.

(*b*) *Pollen fertility of all plants sampled from Marley Wood, Dickleburgh and Boarstall Wood, as judged by staining with acetocarmine*

Species	Locality	No. of plants	Mean of fertility	Individual values
P. vulgaris	Marley Wood	60	99.97	99, 99, 58 at 100
	Boarstall Wood	60	99.98	99, 59 at 100
P. veris	Dickleburgh	42	99.93	98, 99, 40 at 100
	Boarstall Wood	39	99.95	99, 99, 37 at 100
P. veris × *vulgaris*	Boarstall Wood	21	42.95	9, 17, 20, 24, 31, 33, 33, 35, 38, 40, 45, 46, 50, 50, 51, 60, 60, 62, 63, 65, 70
Putative backcross to *P. veris*	Boarstall Wood	6	88.17	63, 70, 98, 99, 99, 100
Putative backcross to *P. vulgaris*	Boarstall Wood	12	77.00	2, 62, 63, 72, 75, 80, 85, 95, 95, 96, 99, 100

(*c*) *Variation in hybrid* Primula *populations*

Species	Corolla diameter	Calyx circumference	Calyx tooth length	Leaf length breadth ratio
P. veris thrum	—	—	*	—
P. veris pin	—	—	—	—
P. vulgaris thrum	*	—	*	—
P. vulgaris pin	—	—	—	—

As an increase in variability of the parental taxa is likely to follow introgressive hybridisation, Woodell examined by means of a statistical test whether the two parental taxa were more variable in the hybrid population than the pure populations. In only three cases (marked *) was there statistically significantly greater variation in the hybrid population and Woodell concluded that there was little evidence of introgression.

Genetic investigations of hybridisation

Interpretation of the variation displayed in Andersonian pictorialised scatter diagrams and hybrid index histograms in terms of F_1, F_2 and backcross derivatives is often attempted by botanists. However, Baker (1947, 1951) insisted that intermediacy does not necessarily indicate hybridity and that the hypothesis of introgression should be put to the test, employing the techniques then available for genetic analysis. Artificial F_1, F_2 and backcrosses should be produced from pure parental stocks; comparison of synthesised plants with wild ones may then be made, as in the case of *Geum* outlined above. Furthermore, the range of F_1 variation should be examined and the genetics of the different traits studied.

Thus, in the case of the *Primula* investigation already discussed, the existence of earlier experimental hybridisation studies makes the interpretation more secure. Artificial F_1 hybrids have been made between *P. veris* and *P. vulgaris*, but only with *P. veris* as the female parent, the reciprocal cross giving empty or imperfect seeds. Artificial F_1 hybrids are vigorous with pollen stainability that is *c.* 30% of the parental types (some cytological irregularities have been discovered in meiotic studies of hybrids) and backcrosses have been made (Valentine, 1975*b*, and references to earlier studies cited).

Sometimes, as part of genetic studies, progeny trials are carried out using seed collected in experiments and in the wild (see Heiser, 1949*b*, for examples). As the pollen parent is almost always unknown in seed stocks collected from nature, care must be exercised in interpretation (Baker, 1947).

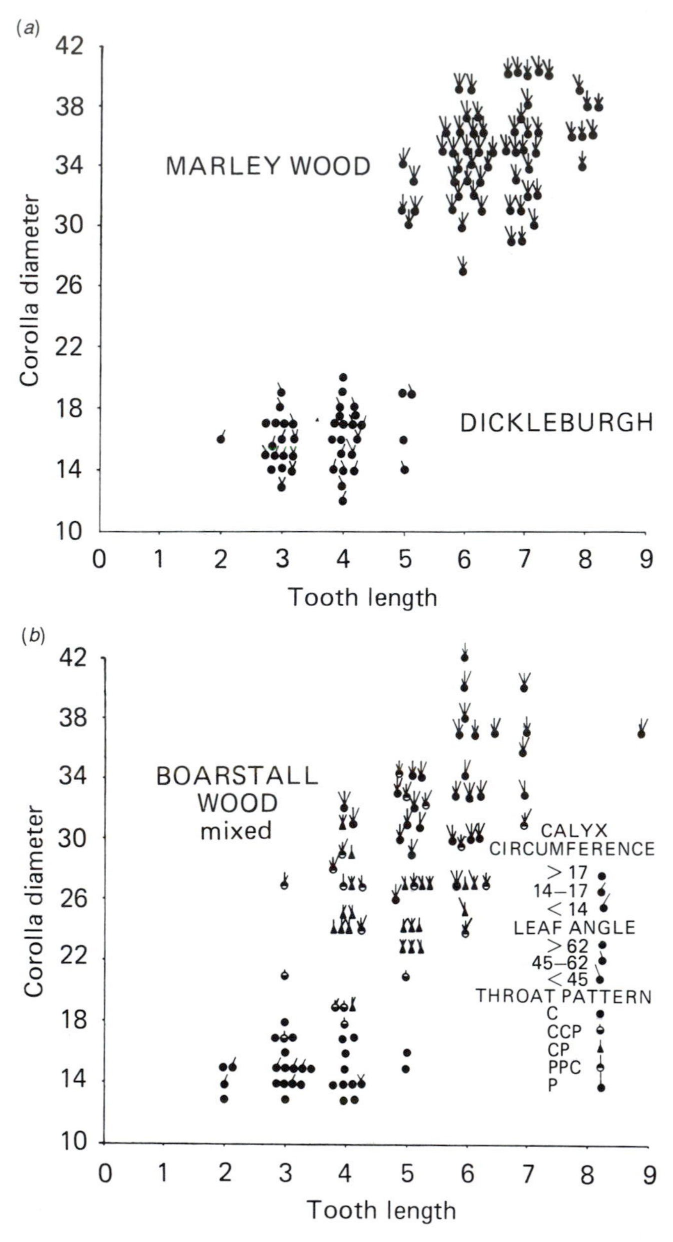

Fig. 11.13. Scatter diagrams and hybrid index histograms representing populations of *Primula vulgaris*, *P. veris* and hybrids in English woodland and meadow sites. (From Woodell, 1965.) (*a*) Scatter diagram of pure populations of *Primula vulgaris* from Marley Wood, Berkshire, and *P. veris* from Dickleburgh, Norfolk. (See (*b*) for key to symbols.) (*b*) Scatter diagram of hybrid population from Boarstall Wood, Buckinghamshire. (*c*) Frequency distributions of scores of plants using hybrid index. (For characters employed in calculating indices see Table 11.2.)

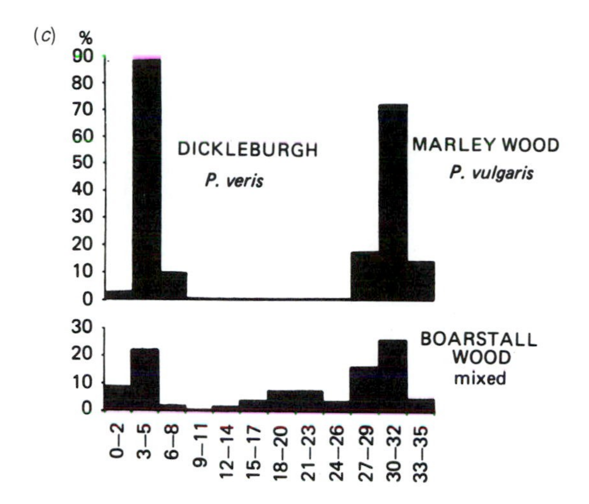

If field collections are the sole evidence available, as they must sometimes be, the proper course is to make a provisional hypothesis, making it completely clear how far the situation has been explored, and in particular whether artificial hybrids have or have not been made experimentally.

Chemical studies of hybridisation

In the 1960s, with the development of chromatography and other biochemical techniques, new sources of evidence became available for interpreting cases of interspecific hybridisation. Chemical investigations of plant variation have provided important insights at many levels. The reviews by Alston & Turner (1963*a*), Smith (1976) and Ferguson (1980) may be consulted for the history of the development of this important subject. For reviews of recent studies, see Stace (1989) and Stuessy (1990).

The examination of chemical characteristics has proved particularly helpful in interpreting population variation where hybridisation is suspected. Often species A and species B differ in chemical characters, and, as a general rule, hybrids A × B and B × A generally have an 'additive' pattern for a particular class of secondary chemical compounds (i.e. the sum of chemical constituents A and B, and sometimes some 'hybrid' compounds as well). In such chemical taxonomic investigations many classes of secondary plant products have been examined, e.g. terpenes, alkaloids, phenolics, etc.

There are many examples of the use of chemical methods in studying complex variation. For example, Alston & Turner (1963*b*) studied the flavonoids present in population samples of four *Baptisia* species, in an

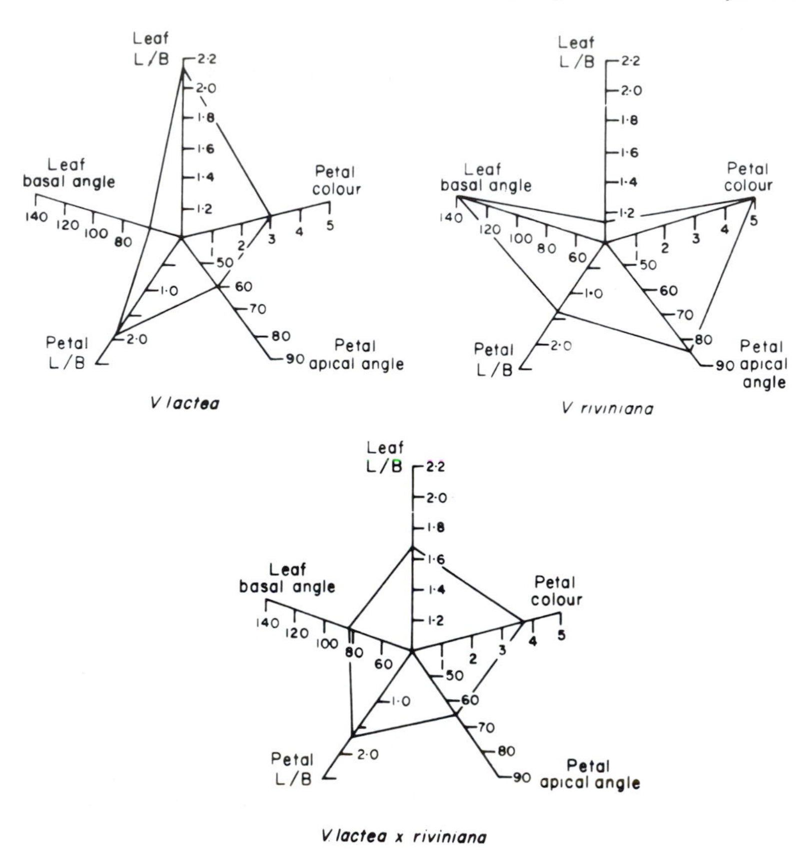

Fig. 11.14. Polygonal graphs of five quantitative characters of *Viola lactea*, *V. riviniana* and their hybrid. (From Moore, 1959.)

attempt to resolve patterns of natural hybridisation. Another elegant example is provided by the investigations of Fröst & Ising (1968), who used chemical markers in their study of hybridisation between the widespread northern species of *Vaccinium myrtillus* and *V. vitis-idaea*. The sterile F_1 hybrid between these two species has long been known as *V.* × *intermedium.* A study of the phenolic compounds of leaf extracts by two-dimensional chromatography was undertaken by Fröst & Ising using Scandinavian material (Fig. 11.16). They discovered differences between *V. vitis-idaea* and *V. myrtillus* at two localities. While the variation within *V. vitis-idaea* was small, *V. myrtillus* showed considerable differences between sites. Generally speaking *V.* × *intermedium* (which may or may not have been produced from the particular individuals of the parents studied

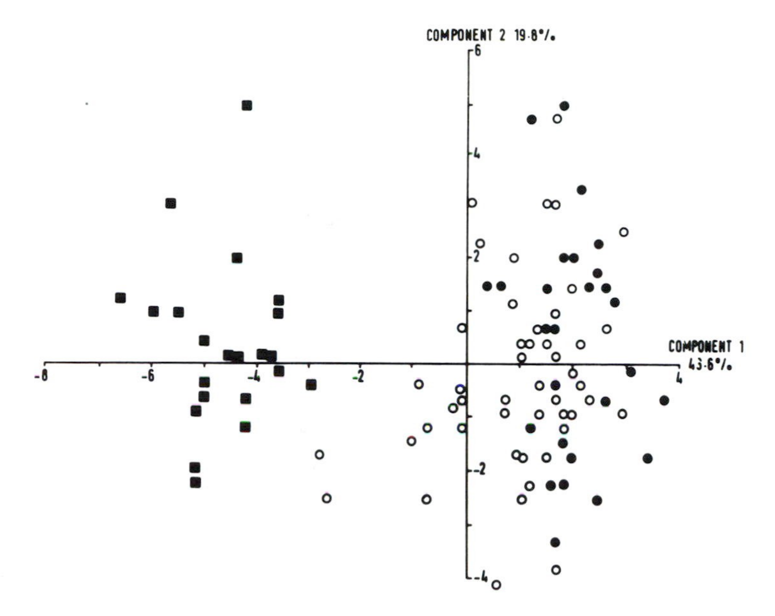

Fig. 11.15. Hybridisation between the two species of Oak, *Quercus robur* and *Q. petraea*, in Britain (see Morris & Perring, 1974). This scatter diagram illustrates the use of principal component analysis in the separation of *Q. robur* (dots) from *Q. petraea* (squares), and the intermediate nature of a putatively hybrid population (open circles). (From Rushton, 1978, where further details can be found.)

chromatographically) had the 'spots' of both parental taxa, but the patterns were not identical at the two sites. As the genotypes of parental stocks may differ, it is obvious that F_1 variation is to be expected, although Ritchie (1955*a*, *b*) found F_1 plants in British populations to be homogeneous morphologically.

Critical tests of the hypothesis of introgression

In a critical review of the situation in the early 1970s, Heiser (1973) concluded that 'if the hypothesis of introgression is to be accepted, there must be evidence of the transfer of genetic material from one species into another'. For example, species A, through introgression of alleles or genes from B, may come to show a different pattern of variation from that in sites where A exists alone. Flower colour, leaf-shape and other 'markers' may signal the presence of genes from species B in species A. Alternatively, species A may become more variable in quantitative characteristics, a

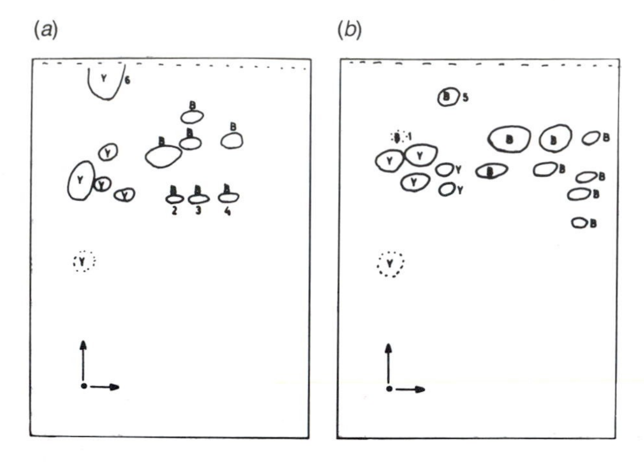

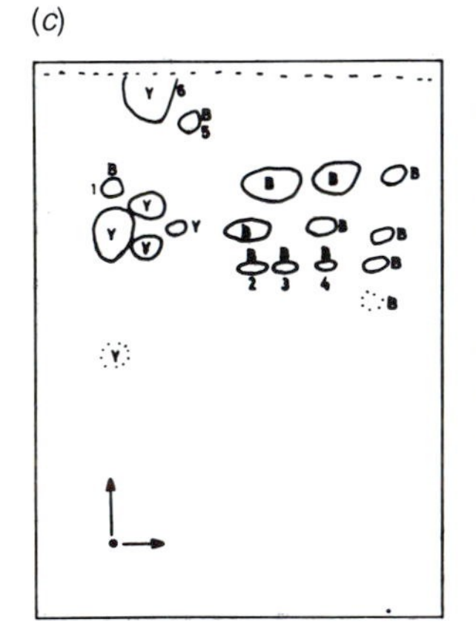

Fig. 11.16. Chromatograms of *Vaccinium (a) V. vitis-idaea. (b) V. myrtillus. (c)* The almost completely sterile hybrid, *V.* × *intermedium.* Note the largely additive effect in the hybrid pattern. (From Fröst & Ising, 1968.)

possibility examined, for instance, by Woodell in *Primula* (Table 11.2).

Heiser (1973) points to a major difficulty in the interpretation of patterns of variation. Usually, botanists do not consider any hypotheses other than introgression, even where other explanations are equally plausible. Introgression is essentially a down-grade process, in which populations developed in isolation come together with local or regional blurring of pattern. Various other explanations for the variability of taxon A in the direction of taxon B might be devised, which do not necessarily involve present or recent introgression. For instance, as taxa A and B are likely to have developed from a common stock, mutations in A, which are independent of any involvement with B, might appear to be introgressants. Other patterns of variation may be primary (up-grade) situations of great complexity.

Dobzhansky (1941) suggested that 'remnants of the ancestral population from which two species differentiated might have the appearance of hybrids'. Other possibilities include segregation in polyploid species (Gottlieb, 1972).

An example of different interpretations is provided by studies in the genus *Juniperus*, where introgressive effects have been claimed over a wide geographical range. For example, Hall (1952) described a case in *J. virginiana*. This North American species has a number of distinct 'races', where it meets *J. horizontalis* to the north, *J. scopulorum* in the west, *J. ashei* in the southwest and *J. barbadense* in the south (Fig. 11.17). Hall considered that these variants resulted from allopatric introgression involving the other species, but this picture has been criticised by Barber & Jackson (1957), who think that other explanations (e.g. ecotypic

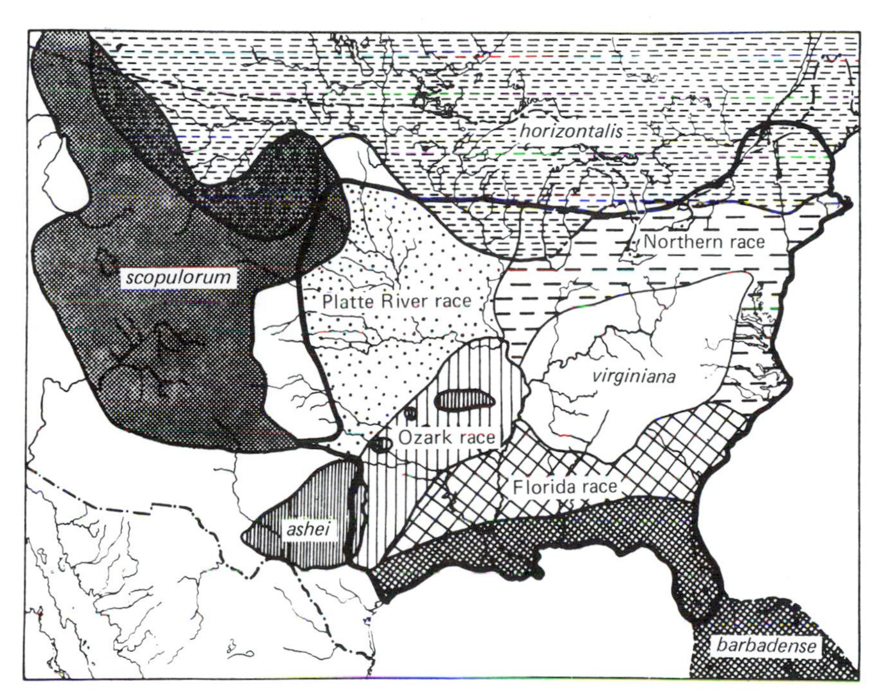

Fig. 11.17. Distribution in the southeastern USA of *Juniperus virginiana* and other partially sympatric species of *Juniperus*. Hall claimed that 'races' of *J. virginiana* were the result of introgression from the other species, but this interpretation has been challenged. (From Anderson, 1953, modified from Hall, 1952.)

differentiation) should be considered. This case has continued to intrigue botanists, and a large number of chemical studies have been undertaken. While there is some evidence for introgression between *J. virginiana* and *J. scopulorum*, studies of other taxa have not supported this interpretation for the other variants (see Flake, Urbatsch & Turner, 1978, and references cited therein).

Heiser (1973) concluded that while introgression does occur, in most cases it is highly localised and rarely dispersed or widespread. Critical tests are difficult in the absence of genetic information.

Recent studies of introgression using molecular tools

Having outlined the difficulty of deciding between conflicting hypotheses, Heiser (1973) concludes his review with the prediction that 'with the new tools from biochemistry and genetics now available ... we may expect some contributions towards a solution of the problem in the future'. This statement can now be seen as prophetic, as Rieseberg & Wendel (1993), in their recent review of introgression, reveal how the use of molecular markers is transforming our understanding of introgression.

They consider some of the difficulties in basing an analysis of introgression on morphological characters alone. 'Morphological characters typically have an unknown, but presumably complicated genetic basis' and 'a nonheritable component that is difficult to estimate'. Furthermore, ' there are often few morphological characters differentiating hybridising taxa, and these characters are often functionally or developmentally correlated'. For progress to be made in the critical testing of hypotheses concerning introgression, it is essential to have genetic markers. With the development of molecular tools, for the first time we have 'large numbers of independent molecular markers that allow the detection and quantification of even rare introgression' (Rieseberg & Wendel, 1993). Progress has been made in analysing not only local and dispersed introgression but also where introgression between two species has produced a stabilised introgressant (Fig. 11.18).

A total of 165 proposed cases of introgression are listed by Rieseberg & Wendel. Their list is not meant to be exhaustive, but concentrates on well-documented cases. (Undoubtedly a much longer list could have been prepared, if it had included all cases of presumed introgression.) Some 85% of the examples listed are from the dicotyledons, and nearly all the different growth form types are represented. More than 90% of the examples come from temperate zones, with 25% of the total being from California. In 65 of

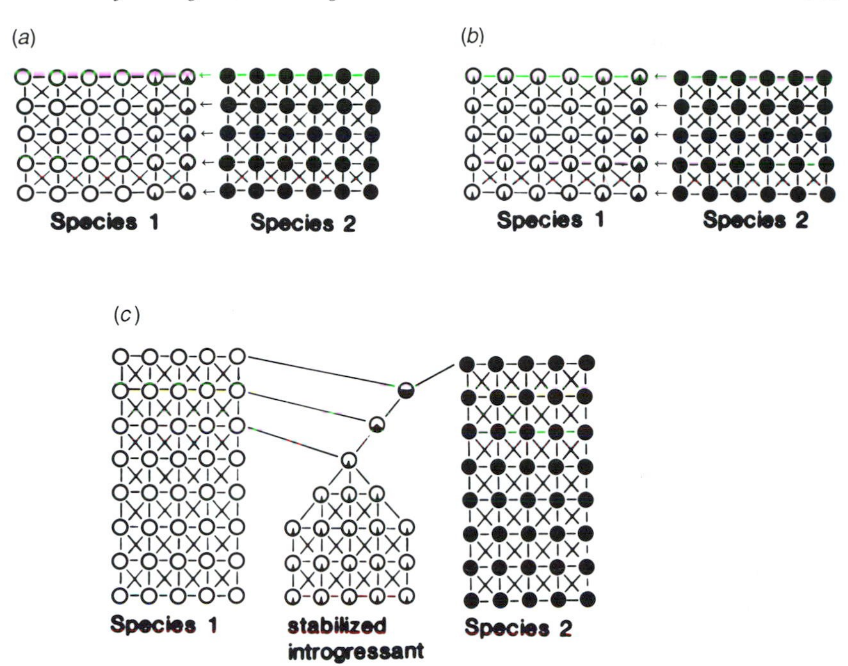

Fig. 11.18. Localised introgression, dispersed introgression and the origin of a stabilised introgressant. Open circles = populations of species 1; closed circles = populations of species 2; black lines = crosses between populations; and arrows = direction of introgression. (*a*) Unidirectional localised introgression from species 2 into species 1. (*b*) Unidirectional dispersed introgression. (*c*) Origin of a stabilised introgressant. (From Rieseberg & Wendel, 1993.)

the 165 cases, the authors consider that the evidence for introgression is strong.

There are 37 examples of the study of introgression using chloroplast DNA, and of these, in the opinion of Rieseberg & Wendel, 29 cases provide 'robust' demonstrations. In some cases, complex results are obtained, introgression being indicated by these cytoplasmic markers, but not confirmed by nuclear markers (isozymes, ribosomal DNA) studied in the same investigation. In an increasing number of cases, critical tests of hypotheses involving introgression have been made using molecular tools, e.g. *Brassica* (Palmer, 1988; Song, Osborn & Williams, 1988), *Carduus* (Warwick *et al.*, 1989), and *Gossypium* and *Helianthus* (see Rieseberg & Wendel, 1993, for details). To illustrate the power of the new techniques, we examine in some detail introgression in *Iris*.

Introgression in Louisiana Irises

Hybrids have long been known in *Iris* in the state of Louisiana, North America. Indeed, as we have seen above (Fig. 11.12), Anderson (1949) took, as one of his major examples of introgression, the patterns of hybridisation found along the banks of the Mississippi river in populations of two perennial clonally spreading *Iris* species: *I. fulva* ($2n = 42$; found in shaded understorey areas along the banks of the water channels, the so-called bayou) and *I. hexagona* ($2n = 44$; growing in open freshwater marshes and swamps). The two species have overlapping flowering times and common pollinators (Arnold & Bennett, 1993). Notwithstanding the lowered fertility of interspecific crosses and reduced seed germination in their progenies, there are many plants of intermediate phenotype, especially in areas where man has hybridised the habitat. Whether this pattern of variation is caused by introgressive hybridisation has been disputed (for example, see Randolph, Nelson & Plaisted, 1967).

Using a range of molecular tools, Arnold and his associates have discovered species-specific genetic markers for the presumed parental taxa – *I. fulva* and *I. hexagona* – and have examined the distribution of these markers in plants of different phenotype (Arnold & Bennett, 1993). Fig. 11.19 shows a model for the distribution of the two species and their hybrids (Viosca, 1935), and Fig. 11.20 indicates the proportion of *I. fulva* and *I. hexagona* markers discovered in a sample of 42 plants. Introgression is elegantly confirmed. In the swamp cypress area, between the bayou and the road, there was a mixture of variants and a few *I. hexagona* plants. On the other side of the road, in the marsh area, a range of variants was found.

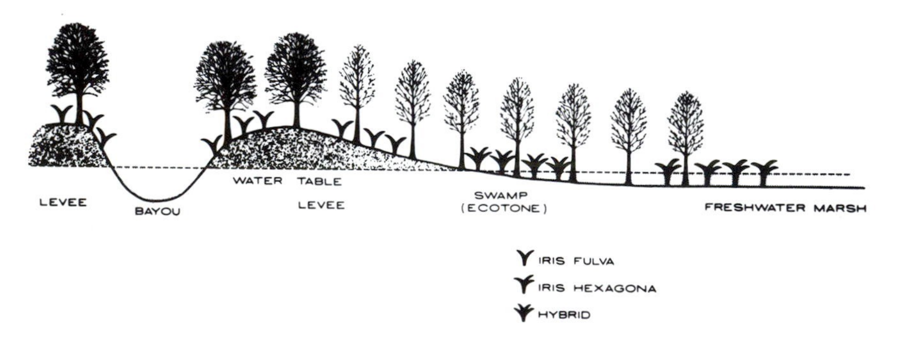

Fig. 11.19. Habitat associations for *Iris fulva*, *I. hexagona* and natural hybrids. This illustration advances the hypothesis that natural hybrids between *I. fulva* and *I. hexagona* occupy an ecotone between the parental habitats. (After Viosca, 1935, from Arnold & Bennett, 1993.)

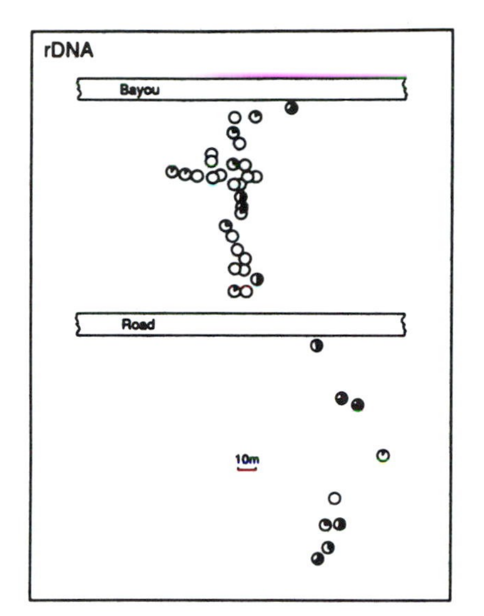

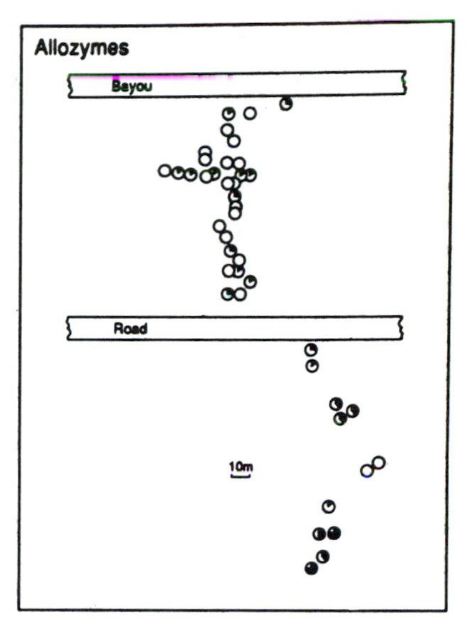

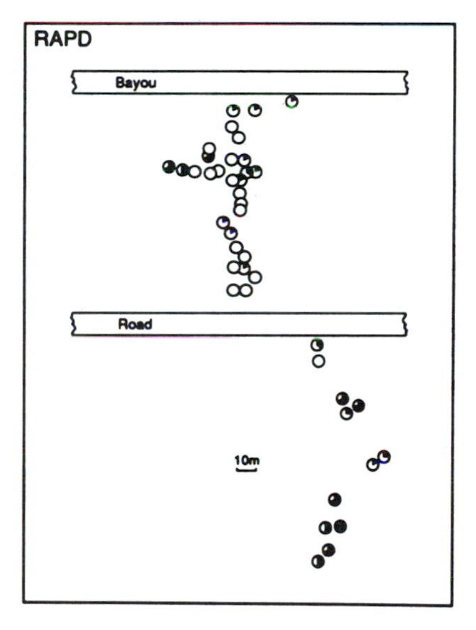

Fig. 11.20. Genetic variation in a sample of 42 individuals from the Bayou L'ourse population. Each circle represents an individual plant. The filled and open portions of each circle represent the proportion of *Iris fulva* and *I. hexagona* markers, respectively. The top left panel, top right panel and bottom panel illustrate the rDNA, allozyme and RAPD variations, respectively. There are three missing data points (plants) for the rDNA analysis. (From Arnold & Bennett, 1993.)

While there were no pure *I. fulva* plants, the frequency of '*fulva*' markers increased in samples taken further and further out into the marsh. In a further study of many of the same samples, it was discovered that *I. fulva* and *I. hexagona* have different chloroplast DNA profiles (Fig.11.21). As chloroplast DNA is maternally inherited, it is apparent that many of the hybrids found between the bayou and the road have *I. hexagona* (and its hybrid derivatives) as the female parent, and have been cross-pollinated by *I. fulva* pollen carried by Bumble Bees (Arnold, 1992).

Besides facilitating the analysis of localised introgression in *Iris*, analysis of molecular markers has also made it possible to confirm that dispersed introgression also occurs. Thus, marker genes for *I. fulva* have been found in *I. hexagona*, 10 km from the nearest *I. fulva* population. Likewise, *I. hexagona* markers have been detected in *I. fulva* plants 25 km from the nearest colony of *I. hexagona* (Arnold & Bennett, 1993).

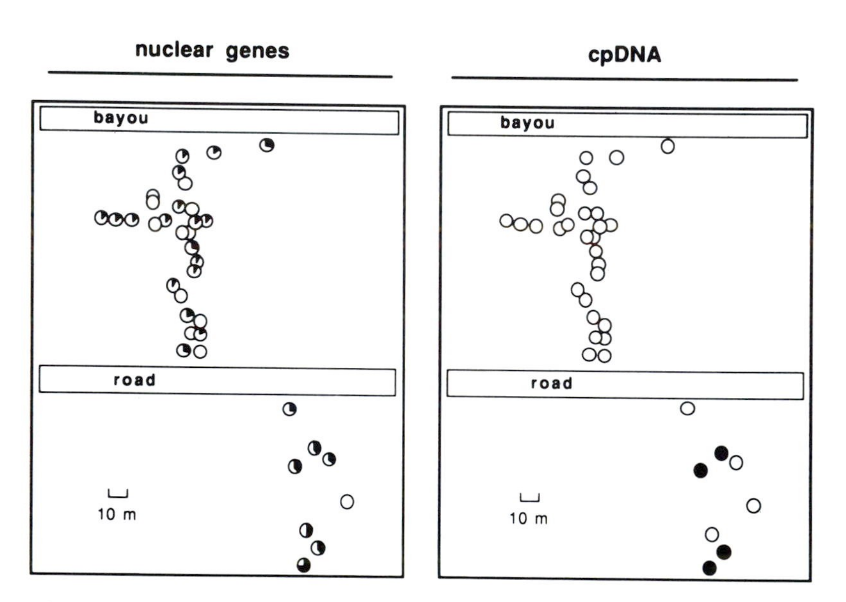

Fig. 11.21. Asymmetrical introgression between *Iris fulva* and *I. hexagona* likely to result from pollen flow. (After Arnold, 1992, from Avise, 1994.) Each circle represents a single plant. Left: relative proportion of *I. fulva* (shaded) and *I. hexagona* (unshaded) nuclear markers. Right: similar representation for maternally transmitted chloroplast DNA markers. Note in particular the population between the road and bayou, where multilocus nuclear genotypes suggest the presence of advanced generation hybrids or backcrosses, despite the apparent absence of seed dispersal that would be registered by chloroplast DNA from *I. fulva* in this area. (After Arnold, 1992, from Avise, 1994.)

Studies of hybridisation in *Iris* have shed light on yet another interesting question: can a stable hybrid species be produced by introgressive hybridisation? Arnold and associates, using the evidence of species specific molecular markers, have discovered that *I. nelsonii* is a fixed derivative of the hybridisation of *I. fulva*, *I. hexagona* and *I. brevicaulis* (Arnold, 1993). *I. nelsonii* grows in deep water which is heavily shaded and is, therefore, ecologically separated from its parental species. New stabilised introgressants have also been detected in *Helianthus*, etc. (see Rieseberg & Wendel, 1993). In the next chapter, we will discuss the possibility that they may have had an abrupt origin.

Concluding remarks

Using the new molecular approaches, it will be very interesting to re-examine yet more cases of 'introgression' published since Anderson's classic formulation of the concept and especially in those cases where the necessary historical details are available, whether hybrid populations are stable or dynamic. An excellent example is provided by the investigation of hybridisation and introgression between the two introduced species *Carduus nutans* ($2n = 16$) and *C. acanthoides* ($2n = 22$) in Grey County, Ontario. The first studies of the population were made in the 1950s, and involved cytological and morphological studies of adult plants and progenies of field-collected seed (Moore & Mulligan, 1956, 1964). Warwick *et al.* (1989) revisited the original and other sites, and studied population variation using molecular and chemical markers. The results suggest that introgression may well be bidirectional and that a stable zone of hybridisation had changed little in the past 30 years.

Major advances in the study of hybridisation and introgression are likely in the near future, especially if the use of newer techniques is combined with morphological, genetical and ecological studies (Rieseberg, 1995). There are many interesting questions still to be examined. For example, how far are hybrids intermediate in character expression? A review of past studies reveals that hybrids are often a 'mosaic of both parental and intermediate morphological characters rather than just intermediate ones, and that a large proportion of first (64%) and later generation hybrids (89%) exhibit extreme or novel characters' (Rieseberg & Ellstrand, 1993). To increase our understanding of this important area, further studies of hybrids are required, using traditional and molecular techniques.

More studies of hybridising populations are needed to test the hypothesis that introgression might result in an increase in genetic diversity. Few detailed studies have yet been made, but in hybridising populations of *Pinus contorta* and *P. banksiana* there is evidence for a small increase in allozyme diversity (Wheeler & Guries, 1987). Other studies are reviewed by Rieseberg & Wendel (1993).

Another area of great interest is whether adaptively significant genes are being transferred through introgression. There is evidence that molecular marker genes are transferred, but, as they may be selectively neutral, the possible transfer of genes of selective importance remains an open question. For instance, can evidence be found for the introgressive transfer of genes for disease resistance? This important question is discussed by Rieseberg & Wendel, and it is clear, too, that genetic linkage mapping could provide important insights to the problem. Another significant area for future research concerns the measurement of relative fitness of parental plants and individuals of different genotypes arising from their introgressive hybridisation. Arnold & Hodges (1995), reviewing recent progress, note that 'hybrids are not uniformly unfit' but may have 'lower, equivalent or higher levels of fitness relative to their parental taxa'. Looking to the future, we predict that the reciprocal transplant techniques developed by genecologists will be very important in estimating the relative fitness of the various genotypes in hybrid swarms. Another key question is how to unravel the long-standing patterns of hybridisation in habitats disturbed by man. Can ancient introgression be detected or long-standing patterns of hybridisation unravelled in such genera as *Salix*, *Quercus*, *Helianthus*, etc.? Furthermore, it has often been contended that very subtle patterns of introgression may occur, and that such cryptic effects may be very important in evolution (Rieseberg & Wendel, 1993). This suggestion needs to be thoroughly investigated. Further advances might also be expected in the analysis of introgression between crops and their weedy relatives (Heiser, 1973).

12

Abrupt speciation

In contrast to the gradual processes whereby two species may diverge under geographical and ecological separation from a single ancestral species, new species may suddenly arise by abrupt speciation, most commonly by polyploidy. First, we shall review this topic, and then, later in this chapter, refer to other possible modes of abrupt speciation.

How common is polyploidy?

Botanists have been studying the chromosome numbers of plants for more than 100 years and the results of their work have been published in a multitude of books and scientific papers. In an attempt to make this body of information accessible to biologists, 'chromosome atlases' have been produced (e.g. Tischler, 1950; Darlington & Wylie, 1955; Löve & Löve, 1961). The most up-to-date reference work available for the chromosome numbers of flowering plants world-wide, Bolkhovskikh *et al.*, (1969) includes counts published up to the end of 1967. Many new chromosome counts are still being published. The *Index to plant chromosome numbers 1986–87* was published by Goldblatt & Johnson (1990), which also gives details of lists for earlier years. As part of the *Flora Europaea* project, a verified list of chromosome numbers of European plants has been published (Moore, 1982).

An examination of the available information makes possible some important generalisations. First, while polyploidy is apparently rare in animals, only being found in a small number of groups (White, 1978), it is a common phenomenon in plants. Calculations of the numbers of polyploid species in flowering plants and the proportion of polyploids in different regions are rendered difficult by differences of opinion between taxonomists as to what constitutes a species. Further, since only a small fraction of the world's flora, mostly in temperate regions, has been studied cytologically, extrapolation to tropical floras is difficult. Also some, perhaps many, chromosome numbers published in the literature are incorrect (Fig. 12.1).

When the information for different higher plant groups is examined, a number of patterns emerge. The chromosome numbers of species in many genera show well-developed series, having simple multiples of a minimum or 'basic number'. For instance, in *Rumex* subgenus *Rumex* there is a series with the basic number $x = 10$, which runs from $2n = 2x = 20$ for *R. sanguineus*, through $2n = 4x = 40$, in, for example, *R. obtusifolius*, up to $2n = 20x = 200$ in *R. hydrolapathum*. Other excellent examples of polyploid series based on a single basic number are provided by the species of the genus *Chrysanthemum sens. lat.* ($x = 9$) with $2n = 18$, 36, 54, 72, 90 and 198 (Tahara, 1915), and *Solanum* ($x = 12$) with $2n = 24$, 36, 48, 60, 72, 96, 120 and 144. However, in some cases the basic number may not be so obvious (Gibby, 1981). For instance, the Swede (*Brassica napus*; $2n = 38$) is not based on $x = 19$, but is, in fact, a so-called 'dibasic' tetraploid (Stebbins, 1971), originating from the cross between Turnip (*B. rapa* (*B. campestris*); $2n = 2x = 20$) and Cabbage (*B. oleracea*; $2n = 2x = 18$).

In some genera every step in a polyploid series is 'occupied'. In others, we find that all the extant species have high chromosome numbers, and it is generally supposed that these plants are of ancient polyploid origin, the lower multiples of the base number having been lost. Khandewal (1990) discovered a variant of the fern *Ophioglossum reticulatum* with $2n = 1440$ (believed to be 96-ploid). The highest chromosome number in the dicotyledons is that of *Sedum suaveolens* ($2n = c.$ 640, *c.* 80-ploid: Uhl, 1978), while in the monocotyledons Johnson *et al.* (1989) discovered that *Voaniola gerardii* has $2n = c.$ 596, which is *c.* 50-ploid.

Stebbins (1971) has calculated that some 30–35% of flowering plants are 'straightforward' polyploids fitting into polyploid series. However, as we shall see later in this chapter, some polyploids do not fit into simple series of multiples of a basic number. Working on the premise that plants with chromosome numbers in excess of $x = 13$ are polyploids, Grant (1971) came to the conclusion that 47% of the angiosperms are polyploids, the figure for dicotyledons being 43% and that for monocotyledons 58%. Reviews of the subject have suggested an even higher percentage. Goldblatt (1980) considers that many species with $n = 11$, $n = 10$ and even $n = 9$ have polyploidy in their ancestry, giving a figure of at least 70% for the incidence of polyploidy in the monocotyledons, and Lewis (1980*a*) concludes similarly that perhaps 70–80% of dicotyledonous species may be of polyploid origin.

	Chromosome numbers in TISCHLER (1950)	Correct	Correct chromosome numbers in 1969
M. sparsiflora MIKAN	$2n = 18$	!	$2n = 18$
M. sylvatica EHRH. ex HOFFM	$2n = 14$		$2n = 18$
	$2n = 18$	!	
	$2n = 24$ (= *M. alpestris*)		
	$2n = 32$ (= *M. decumbens*)		
M. alpestris F. W. SCHMIDT	$2n = 14$		$2n = 24, 48, 72$
	$2n = 24$	!	
M. lithospermifolia HORNEM.	$2n = 48$ (= *M. alpestris*)		$2n = 24$
M. suaveolens W. & K.	$2n = 72$ (= *M. alpestris*)		$2n = 24, 48$
M. decumbens HOST			
ssp. *kerneri* (D.T. & SARNTH.) GRAU	$2n = 16$		$2n = 32$
M. rehsteineri WARTM.	$2n = 22$	!	$2n = 22$
M. palustris (L.) NATHH.	$2n = 64$		$2n = 66$
	$2n = 18$		
	$2n = 42$		$2n = 44$
M. laxa LEHM. ssp. *caespitosa*			
(C.F.) SCHULTZ	$2n = c.$ 80		$2n = 88$
M. stricta LINK	$2n = 36–40$		$2n = 48$
M. ramosissima ROCHEL	$2n = 48$	!	$2n = 48$
M. arvensis (L.) HILL	$2n = 24$		$2n = 52$
	$2n = 48$ (= *M. ramosissima*)		
	$2n = 54$		
M. discolor PERSOON	$2n = c.$ 60		$2n = 72$

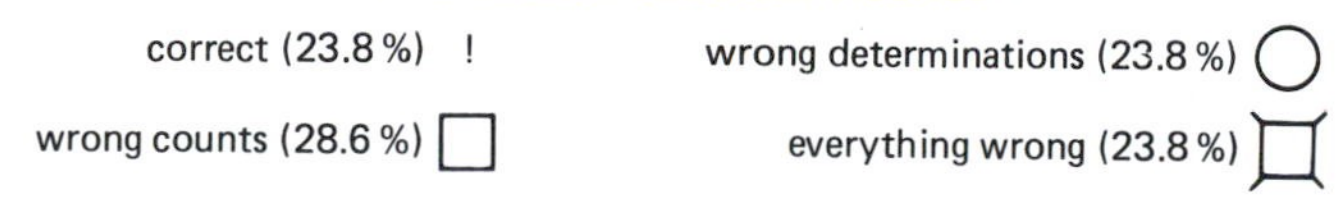

Fig. 12.1. Chromosome numbers in *Myosotis* checked by Merxmüller and Grau. (Reproduced by kind permission of the International Association for Plant Taxonomy from Merxmüller, 1970.) Cytologists rarely admit in print that the information in chromosome lists is often wrong. Chromosome numbers, especially in the early literature, may be wrongly counted and are in any case often based on a single cell preparation. Furthermore, no voucher specimen is available in many cases so that verification of the identification is impossible. Such checking is essential especially in groups where the taxonomy is not straightforward. This survey by Merxmüller and Grau provides a clear warning example against uncritical acceptance of published information.

To provide a more objective basis for considering the frequency of polyploidy, isozymes have been examined to determine which chromosome numbers indicate polyploidy. Species in the Compositae (Asteraceae) with chromosome numbers $n = 4$, 5 and 9 were selected for investigation. It is possible that $x = 9$ is the original base number of the group and that 4 and 5 are the result of a reduction in chromosome number, involving a process which we will consider later in this chapter. Alternatively, plants with $n = 9$ could be allopolyploid derivatives of crosses between species with $n = 4$ and $n = 5$. Polyploids have multiple representation of genomes and, therefore, multiple representation of isozyme markers should be found if $n = 9$ is an allopolyploid. Gottlieb (1981*b*) studied the electrophoretic patterns of isozyme markers in a selection of the Composites with $n = 4$, 5 and 9, and found no multiplicity of isozymes in $2n = 9$, concluding that they were not allopolyploids. Further studies are needed in other groups. The importance of polyploidy in angiosperms is clearly beyond question, but there are difficulties in estimating its frequency.

Studies of lower plants permit the following cautious generalisations (see Lewis, 1980*a*, *b*, *c*, for review articles on the incidence of polyploidy in all groups). Polyploidy is apparently rare in fungi and gymnosperms, but is recorded in algae and many bryophytes, and is particularly common in ferns and their allies. Grant (1971) calculates that, on the basis of chromosome numbers in excess of $x = 13$, 95% of fern species are polyploids.

Experimental studies of polyploids

Since the early years of this century, attempts have been made to determine the ancestry of particular polyploids. As we shall see, early investigations were cytogenetic. Then, in the 1960s, chemical approaches provided valuable insights and, more recently, molecular investigations have greatly increased our understanding of the origin, variability and evolutionary potential of polyploids.

Early cytogenetic studies

As we saw in Chapter 10, Winge's (1917) hypothesis that fertile derivatives could be derived from sterile hybrids by polyploidy was tested in a number of experimental studies.

A most famous case of allopolyploidy was provided by *Primula kewensis*, which was discovered amongst seedlings of *P. floribunda* at Kew in 1899.

The proposition that *P. kewensis* was a hybrid between *P. floribunda* and *P. verticillata* was put to the test by making the hybrid experimentally, using *P. floribunda* as the female parent. *P. kewensis* was morphologically intermediate between the parental stocks and had the same chromosome number $2n = 18$ (Digby, 1912; Newton & Pellew, 1929). Although meiosis was regular in the hybrid, the plants were sterile, presumably because of genetic imbalance. This sterile hybrid was vegetatively propagated and widely distributed as an ornamental garden plant. On three occasions, however, hybrid plants were observed to set good seed (in 1905 at the nurseries of Messrs Veitch, in 1923 at Kew Gardens, and in 1926 at the John Innes Horticultural Institution). In each case the progeny proved to be tetraploid and fertile. Moreover, in one original hybrid plant the investigators discovered that vegetative cells from the fertile stem were tetraploid, the parent plant itself being largely diploid with sterile inflorescences, showing that a sterile hybrid had become fertile by somatic doubling. The fertile *P. kewensis* behaves as a new species, morphologically similar to the sterile hybrid stocks from which it was derived, but distinct from both parents. Recently, in the account of *Primula* for *The new Royal Horticultural Society dictionary of gardening* (Huxley, Griffiths & Levy, 1992), some doubt has been expressed about the ancestry of *P. kewensis* proposed by Newton & Pellew and the case should be re-examined, using molecular markers.

It can readily be seen that, if the primary hybrid and allopolyploid derivatives arise in cultivation, they may be detected and their ancestry investigated. Many other examples of allopolyploids arising in experiments and in cultivation from sterile hybrids are known (e.g. Darlington, 1937; Grant, 1971, 1981; Lewis, 1980*a*) and essentially similar means have been used to deduce ancestry.

Resynthesis of wild polyploids

In the reconstruction of the origin of allopolyploids arising in experiments or in cultivation, the parental stocks may be obvious or the number of candidates limited. As there are often many diploid taxa in a genus, unravelling the ancestry of wild polyploids is altogether a more formidable undertaking. It is very instructive to examine the famous experiments of Müntzing (1930*a*, *b*), who studied the origin of the weedy species *Galeopsis tetrahit* ($2n = 4x = 32$). First he made a careful study of six diploid species ($2n = 2x = 16$); from these, *G. pubescens* and *G. speciosa* were selected as closest in morphology to *G. tetrahit*. F_1 hybrids ($2n = 2x = 16$) were produced between *G. pubescens* (female) and *G. speciosa* (male), which proved to be highly, but not absolutely, sterile. After self-fertilisation these F_1 plants yielded strongly variable F_2 progeny, amongst which was a triploid plant ($2n = 3x = 24$), presumably arising from the union of one reduced and one unreduced gamete. This highly sterile triploid was backcrossed to *G. pubescens* and yielded one seed which germinated and grew to give a plant with $2n = 4x = 32$ chromosomes (presumably derived from a cross between an unreduced gamete from the triploid plant and *G. pubescens* pollen). Morphologically this tetraploid derivative was very like *G. tetrahit*. In a study of the 'status' of the experimentally produced tetraploid, Müntzing (1930*a*, *b*) discovered that there was no difficulty in crossing it with wild stocks of *G. tetrahit*. Moreover, fertile offspring were produced. Thus, it is highly likely that the ancestors of present-day *G. tetrahit* arose from stocks ancestral to present-day *G. speciosa* and *G. pubescens*.

While the approach employed by Müntzing to the problem of the resynthesis of a wild polyploid is conceptually sound, the method suffers from the weakness that the investigation is dependent upon a chance event which produces the polyploid derivative from diploid stocks. Also, in the *Galeopsis* experiments, the one triploid plant produced in the F_2 progeny might have been overlooked, and the *single* seed produced on crossing this plant with *G. pubescens* might have failed to grow. What was needed was a method of generating polyploids at will from diploid stocks.

As long ago as 1904 the Czech botanist Nemeč reported that chromosome doubling in root cells could be induced by treatment with chloral hydrate or other narcotics. Blakeslee & Avery (1937), studying the effects of various substances and treatments on stem tissue, reported that the alkaloid colchicine had the property of inducing chromosome doubling. Different methods of application were devised. Seeds could be soaked in dilute colchicine solutions, or the alkaloid could be applied to plants as a lanolin mixture, in agar blocks, by application of drops of solution to bud tissue, or by atomised sprays. The cytological effect of colchicine is illustrated and discussed in Fig. 12.2. The discovery that colchicine treatment could induce polyploidy was a major breakthrough in experimental taxonomic studies, and we will give an example of its use in our discussion of the origin of Bread Wheat (*Triticum aestivum*) later in this chapter.

Auto- and allopolyploidy

In Chapter 10 we discussed two types of polyploidy, auto- and allopolyploidy. It is instructive at this point to consider early views on their

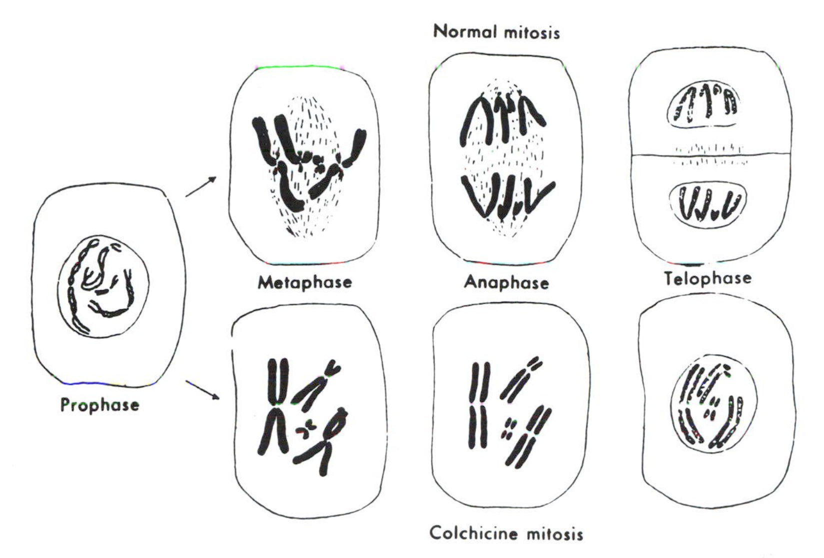

Fig. 12.2. Diagram of the difference between normal mitosis (above) and mitosis that has been changed by the action of colchicine (below). (From Müntzing, 1961.) Note the characteristically widely spaced chromatids in the metaphase after treatment with colchicine. The normal mitosis gives rise to two cells, each with four chromosomes. By the action of colchicine, in which the mechanism of movement of the chromosomes is anaethetized, one cell with eight chromosomes is formed. If a cell with eight chromosomes is removed from colchicine, a spindle may form at the next nuclear division producing daughter cells which are 'polyploid'. If the cells remain in colchicine, additional C-mitoses may take place with further increments of polyploidy. For example, Onion (*Allium*) roots left in colchicine for four days have cells with more than 1000 chromosomes (Levan, 1938)! On removal from colchicine, competition between cells of different number occurs and a new thick root may grow out of diploid cells. Experimental treatments with colchicine are likely to yield mosaics of cells, some polyploid, but others, not actively dividing at the time of treatment, may remain diploid. For further information on colchicine the monograph by Eigsti & Dustin (1955) may be consulted. Of particular interest is the detailed history of studies of the effects of colchicine, especially the question of priority in understanding the significance of chromosome doubling. Biologists intending to use colchicine for experiments should be aware of its carcinogenic properties.

cytological behaviour and the light this sheds on problems of ancestry.

Autopolyploidy involves the multiplication of the same chromosome set. Thus a diploid, which has two like chromosome sets (genomes), could give rise to an autotetraploid with four such sets by chromosome doubling. Such a change could be represented symbolically as follows:

AA ——————→ AAAA

Normal sexual reproduction in diploids involves the production of gametes by meiosis, a process in which the homologous pairs of chromosomes become associated together and eventually separate after an exchange of a portion of genetic material in crossing-over. This regular pairing at meiosis is dependent upon there being two, and two only, of each homologous chromosome, forming a bivalent. In the autotetraploid, four members of each homologue are present. Evidence suggests that chromosome pairing is only possible between two homologues at any particular point on the chromosomes, but the proximity of four homologues, and the fact that pairing may begin at several different points during the pairing process, results in the association of, and chiasma formation between, three or four chromosomes. Unpaired single chromosomes (univalents) may also remain. Groups of three or four chromosomes lead to chromosome structures known as multivalents, such associations being easily recognised in favourable cytological material.

Univalents may be segregated in different (unbalanced) numbers in the two products of the first division of meiosis, an event contributing to chromosome imbalance and infertility. Multivalent production sometimes results in a failure of normal separation of chromosomes: for example, bridges of chromosome material may be stretched across the division figures at anaphase I of meiosis, as multivalents 'attempt' disjunction. It is easy to see how such meiotic irregularity may lead to sterility in gametes. This may be detected on the male side in the production of a high proportion of irregular-sized, misshapen pollen grains.

In contrast, the allopolyploid is the product of the addition of unlike chromosome sets, usually following hybridisation between two species. Thus, two diploid taxa AA and BB may yield an infertile hybrid AB, and the production of unreduced gametes from such a plant will give an allotetraploid of formula AABB. In contrast to the autopolyploid situation discussed above, the typical allopolyploid is usually fertile. It is not difficult to see why this should be so. Multivalents are less likely to be formed, since each chromosome can pair with its exact partner and no other. This lack of correspondence, which ensures proper pairing in the allopolyploid preventing the association of A with B genomes, is the likely cause of sterility in the primary hybrid AB.

Thus, as studies of polyploids progressed in the 1920s and 1930s, it seemed possible, for a time, that the study of meiosis in polyploids would provide an easy way of detecting ancestry. Multivalent associations would indicate autopolyploidy and bivalent pairing would indicate an allopolyploid origin. Such a simple classification was quickly abandoned as intermediate meiotic situations were found (Darlington, 1937). The distinction made in Chapter 10 between autopolyploidy and allopolyploidy, though useful and clear enough in the extreme cases, now seems to be misleading when applied to the evolution of groups of polyploid taxa. The difficulty can be appreciated if we consider what we mean by a hybrid individual with A and B genome sets. We have seen in the previous chapters that ordinary diploid sexual species with some degree of outbreeding are genetically very variable. The genomes of any two individuals of such a species are most unlikely to be identical. It is, therefore, a conventional oversimplification to represent such an individual as having identical genomes contributed by each parent. In order to represent the origin of a polyploid derivative, it would be better to write the following in such cases:

$$\mathrm{AA'} \xrightarrow{\text{doubling}} \mathrm{AAA'A'}$$

As soon as we do this, we see the nature of the difficulty. Is this situation to be described as autopolyploidy or as allopolyploidy? Clearly the answer hinges on our definition of these terms. If we restrict allopolyploidy to those cases where a *sterile species-hybrid* (represented by AB) gives rise to a fertile polyploid derivative (AABB), then all the other cases where the parents of the diploid belong to the same species would be described as autopolyploid. This is a very unsatisfactory definition, for it obscures the essential similarity in the two situations. A better solution would be to use, as Stebbins (1947) suggested, a third term, 'segmental allopolyploidy', for all the intermediate cases where the parent diploid possesses some measure of chromosomal and genetic difference between its genome sets, but where its parents are sufficiently similar to be assigned to the same species.

Many polyploids are of the segmental allopolyploid type and because of their genomic constitution multivalents and univalents are formed at meiosis. It is important to consider in more detail the implications of multivalent production. In a very useful review of polyploidy Gibby (1981) writes:

> Multivalents in themselves do not lead to infertility, for so long as segregation of the chromosomes during anaphase is balanced, then multivalent-forming polyploids are potentially fertile. The presence of trivalents and to a lesser extent chain quadrivalents may result in infertility following mis-orientation at metaphase I, but ring quadrivalents can give numerically equal segregation. It is the presence of univalents that leads to chromosome imbalance and infertility. In autotetraploid rye, which shows the presence of trivalents and univalents as well as quadrivalents

and bivalents, selection for improved fertility results in a decrease in the number of univalents and an increase in quadrivalent frequency (Hazarika & Rees, 1967).

Later in the chapter we will consider the important question of the genetic control of chromosome pairing and the evolutionary potential of autopolyploids.

With regard to the behaviour of odd-numbered polyploids, Gibby writes that their fertility 'is usually reduced as a result of the presence of univalents or odd-numbered multivalents'. However, 'some triploids are fertile, but the gametes they produce are diploid, triploid etc. or aneuploid, and these give rise to progeny with a variety of chromosome numbers'.

Genome analysis

Based on the models of chromosome pairing we have just outlined, the cytological study of polyploids and their hybrids has frequently yielded valuable evidence on ancestry. The basic idea of genome analysis may be appreciated by considering an example. If an allopolyploid ($2n = 4x = 28$) of genomic constitution AABB is crossed with a plant thought to be an ancestral diploid ($2n = 2x = 14$), a triploid ($2n = 3x = 21$) is formed. Suppose at metaphase I of meiosis seven bivalents and seven univalents are seen in division figures. We could deduce that the diploid and tetraploid shared a genome in common, say genome A, by employing the following argument. In the triploid, the A genome from the tetraploid would form bivalents with the A genome originating from the diploid; the B genome would have no pairing partner and remain as univalents. A search could be made for the donor of the B genome by making triploids from crosses between the tetraploid and other diploids. In searching for such a plant, crosses might be made which would yield a triploid in which there was no pairing at meiosis, and the parental diploid could be supposed to have a genome different from either A or B, say C. This example is set out diagrammatically in Fig. 12.3.

A very convenient and familiar example of genome analysis is provided by the common Polypody Fern (*Polypodium*), widespread in many parts of Europe and with related species in North America. First, we will consider the situation in Europe (Manton, 1950; Shivas, 1961*a*, *b*). In Central Sweden, where Linnaeus knew the plant, it is not a variable species: to Linnaeus this was *Polypodium vulgare*. This rather narrow-leaved, mainly northern European plant is then *P. vulgare* L. *sensu stricto*. If we now look at *Polypodium* in southern Europe, we find a rather

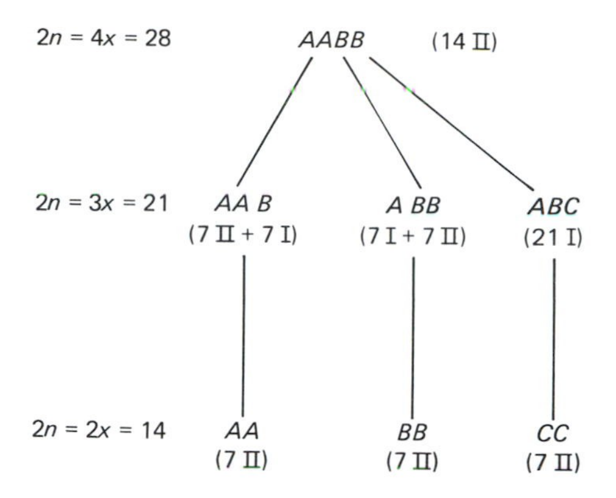

Fig. 12.3. Genome analysis of a presumed allopolyploid. Meiotic pairing noted in brackets (I = a univalent; II = a bivalent).

obviously different looking plant, with roughly triangular leaves, which is clearly adapted to a Mediterranean climate of mild, damp winters and hot, dry summers, producing new fronds in the autumn and withering in the summer. This species was called, appropriately, *P. australe* by Fée in 1850. (It has recently been shown that *P. cambricum* is the earliest name for this taxon. We have retained Fée's name here, as most botanists still use it.) Not only are the two *Polypodium* species separable on general appearance, ecological requirements and geographical distribution, but there are also quite precise characters of the reproductive structures which serve to distinguish them. Further, the cytological situation is clear: *P. australe* is a diploid with $2n = 74$, while *P. vulgare* is a tetraploid with $2n = 148$. In many parts of northwestern and western Europe, however, *Polypodium* plants do not divide readily into these two taxa, and a third taxon, somewhat intermediate between the other two, is common. This is the allohexaploid, *P. interjectum* which has $2n = 222$, and overlaps in morphology and distribution with both its parent species. It was the presence of the allopolyploid which confused the traditional taxonomy; no clear recognition of the three taxa had been made before Manton and Shivas carried out their experimental and cytological investigations. Typical fronds are illustrated in Fig. 12.4 and the results of the cytological investigations are set out in Fig. 12.5. The evidence suggests that *P. interjectum* is the allopolyploid derivative (AABBCC) of the cross between *P. australe* and *P. vulgare*. Furthermore, there appears to be no common genome between *P. australe* (CC) and *P. vulgare* (AABB). Genome analyses showed, however, that the European tetraploid has a genome in

Fig. 12.4. Variants of *Polypodium vulgare*. (*a*) Diploid ($2n = 2x = 74$) from Cheddar, England. (*b*) The triploid hybrid ($2n = 3x = 111$) between diploid and tetraploid from Roches, Switzerland. (*c*) Tetraploid ($2n = 4x = 148$) from North Wales. (*d*) The pentaploid ($2n = 5x = 185$) hybrid from Bolton Abbey, Yorkshire, England. (*e*) Hexaploid ($2n = 6x = 222$) from Ireland. (*f*) The tetraploid ($2n = 4x = 148$) hybrid between diploid and hexaploid from Istanbul, Turkey. (Fronds × 0.2.) ((*a*)–(*e*) from Manton, 1950; (*f*) from Shivas, 1961*a*.)

common with the North American *P. virginianum*, but the source of the other genome is unknown (see Lovis, 1977).

Recently, the origin of the North American *P. virginianum* has been investigated using genome analysis (Haufler & Zhongren, 1991). Evidence

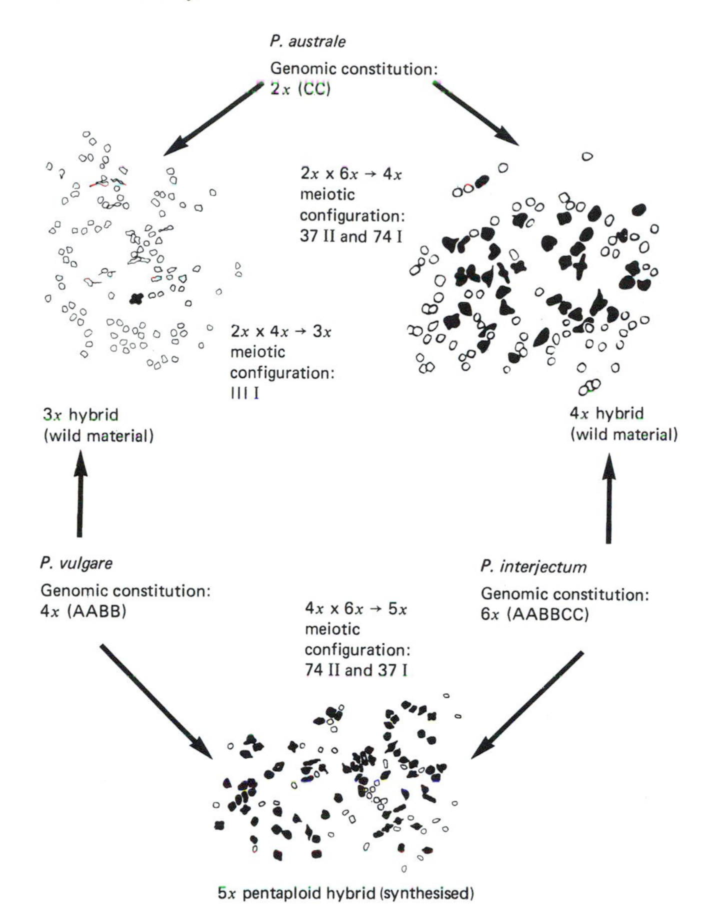

Fig. 12.5. Diagrammatic summary of the results of cytotaxonomic study of *Polypodium* based on the work of Manton (1950) and Shivas (1961*a*, *b*). *P. australe* Fee is now called *P. cambricum* L.

suggests that it is the allopolyploid ($2n = 4x = 148$) derivative of hybridisation between two diploids, a southern species P. *appalachianum* ($2n = 2x = 74$) and a species with a northern distribution, *P. sibiricum* ($2n = 2x = 74$). Naturally occurring sterile triploid hybrids were also detected, where the allopolyploid was sympatric with its diploid parental species. The genomic relationships between the North American and European species have yet to be properly examined.

Genome analysis: uncertainties about ancestry

Genome analysis has been used to investigate relationships in a number of other genera, e.g. *Viola* (Moore, 1976: Fig. 12.6). It has also proved to be of enormous value in studying the origins of important crop plants such as the Turnip and its relatives (*Brassica*), Cotton (*Gossypium*), Banana (*Musa*), Tobacco (*Nicotiana tabacum*), Potato (*Solanum tuberosum*) and Grape (*Vitis*) (Simmonds, 1976; Zohary & Hopf, 1993; Smartt & Simmonds, 1995). However, experiments with Wheat have revealed complications in the analysis of pairing behaviour in polyploids. In order to understand these, it is necessary, first, to outline what has been discovered about the ancestry of Bread Wheat.

After decades of experiments, observations and speculation, there seemed to be general agreement on the origin of Bread Wheat (Riley, 1965: Fig. 12.7). McFadden & Sears (1946) were successful in crossing tetraploid Wheat ($2n = 4x = 28$) with *Aegilops squarrosa* ($2n = 14$) and showed that, when the triploid product was treated with colchicine, the synthetic hexaploid so formed resembled certain hexaploid Wheats. Sarkar & Stebbins (1956), after studying the patterns of variation in *Triticum* and *Aegilops*, considered that *A. speltoides* was the most likely donor of the B genome. The logic of the argument used in their studies is of general interest. Given a hybrid, knowing the morphology of one of its parents, and considering that most hybrids are intermediate between their parents in quantitative characteristics, then it should be possible to pick out the 'missing' parent from an array of taxa. This technique was called the method of 'extrapolated correlates' by its inventor Anderson (1949), and it

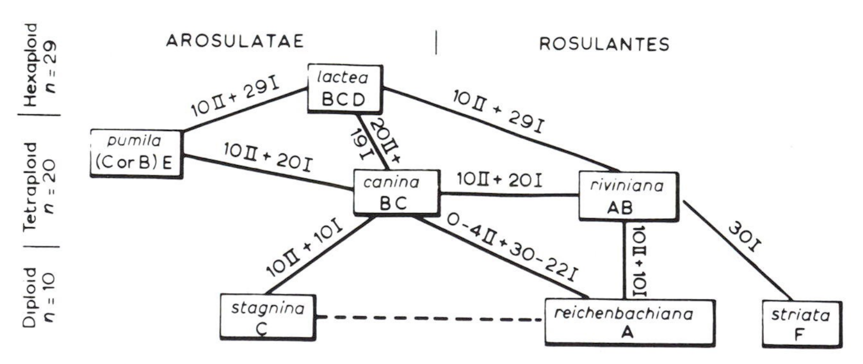

Fig. 12.6. Genomic constitutions of species of *Viola* subsection *Rostratae*, and chromosome pairing in hybrids. Unsuccessful crosses are shown by broken lines. *Viola stagnina* is now called *V. persicifolia*. (Based partly on Moore & Harvey, 1961, from Moore, 1976.)

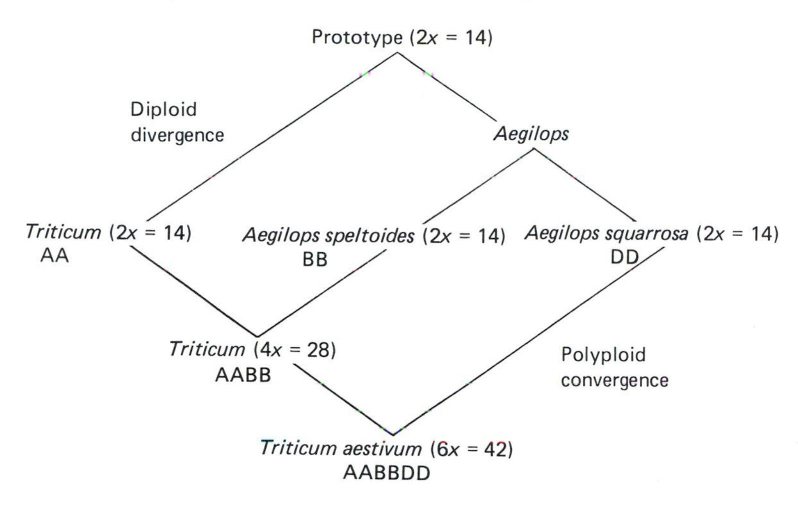

Fig. 12.7. Ancestry of Bread Wheats (*Triticum aestivum*). (From Riley, 1965.) (Note that in some of the literature on Wheat, *Aegilops squarrosa* is called *Triticum tauschii*.)

has been used in many studies of introgression and other situations involving hybridisation.

In the case of Wheat, the variation in diploid and tetraploid *Triticum* was reasonably well known. That an *Aegilops* species was implicated in the ancestry seemed to be likely on cytological grounds. Early studies of karyotypes supported this idea (Riley, Unrau & Chapman, 1958). The Wheat 'story' displayed in Fig. 12.7 has been widely quoted in the literature, often without giving the evidence. However, studies of meiosis (Kimber & Athwal, 1972), seed proteins (Johnson, 1972) and chromosome staining patterns (Gill & Kimber, 1974: Fig. 12.8) have now cast doubt on *A. speltoides* as the source of the B set.

In a recent paper, Breiman & Graur (1995) review all the available evidence. Molecular studies reveal that the tetraploid *Triticum dicoccum* (called *T. turgidum* in some recent literature) shares the same chloroplast and mitochondrial DNA as *T. aestivum*, and therefore, given the maternal inheritance of these genomes, it may be concluded that *T. dicoccum* was the female parent in the cross that gave rise to Bread Wheat (Fig. 12.9). However, studies of restriction fragment profiles have failed to identify a diploid with the same mitochondrial DNA as *T. dicoccum*. It seems very likely from these results that the donor of the B nuclear genome also contributed the cytoplasmic genomes. Reviewing the question of the

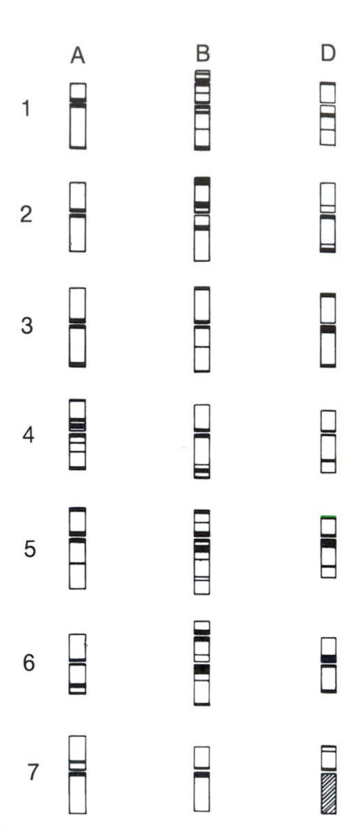

Fig. 12.8. Chromosome staining patterns in the three genomes (A, B, D) which make up the chromosome complement of Wheat (*Triticum*). (From Gill & Kimber, 1974.) *A. speltoides* does not have the same pattern of Giemsa staining as the B genome in Wheat.

identity of the BB ancestor, Miller's (1987) conclusions still stand: 'several possibilities exist for the origin of the B genome: the original donor may now be extinct, the donor may be a yet-undiscovered diploid species, the genome may be derived from more than one source, or a rearrangement of the DNA may have occurred since its incorporation into the tetraploid'. Furthermore, it is possible that very considerable hybridisation may have taken place at the tetraploid level, and that this may have modified the genomes to the point where the tracing of ancestry may be very difficult (Zohary & Feldman, 1962; Pazy & Zohary, 1965). Thus, the early ancestry of Wheat remains an open question. The whole affair is more a cautionary tale than a 'simple story', reminding us of the complexities likely to be involved in the unravelling of ancestry in polyploids.

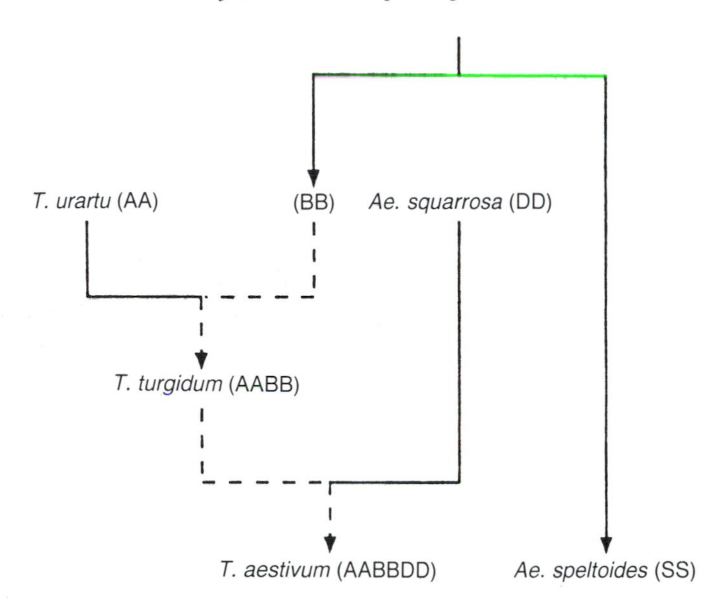

Fig. 12.9. Schematic representation of recent studies on the origin of Bread Wheat (*Triticum aestivum*). The cytoplasmic pathway is indicated by dashed lines. The diploid BB ancestor has not yet been identified. (From Breiman & Graur, 1995.)

Genetic control of chromosome pairing: the implications for genome analysis

Studies of meiosis in hexaploid Wheat have provided important information which has implications for genome analysis. Pairing patterns in hybrids between diploid taxa suggested that some homologies between Wheat genomes might exist. Indeed, the scheme for the evolution of Wheat proposed that the taxa responsible for the formation of Wheat had a common ancestor, and it might be supposed that, while some chromosome differentiation has taken place, some degree of residual homology remains between the genomes, i.e. the genomes could be said to be homoeologous. On the basis of these considerations *T. aestivum* would be classified as a segmental allohexaploid. When meiosis in hexaploid Wheat is examined, however, pairing is seen to be strictly bivalent in character; pairing behaviour does not appear to reflect the presumed origin!

Experiments by Riley & Chapman were designed to investigate this phenomenon. In hexaploid Wheat (and many other polyploids) plants may survive and reproduce in the absence of a full chromosome complement, and plants with $2n = 40$, with a pair of chromosomes missing, can be produced experimentally. With $2n = 6x = 42$, 21 such types – known as nullisomics – are viable and were produced. Riley & Chapman (1958) discovered that, if a certain pair of chromosomes was absent (chromosome 5 of genome set B), then multivalent pairing occurred at meiosis. Studies of the other 20 nullisomics revealed strict bivalent pairing. They deduced that chromosome 5B carried a gene (*Ph*1) which enforced bivalent pairing. In the presence of 5B, homologies were overruled and bivalents were produced; when 5B was absent, multivalents were produced reflecting the segmental allopolyploid nature of *T. aestivum*. It has since been discovered that pairing behaviour is affected by many genes in Wheat. Gale & Miller (1987) may be consulted for a review of later investigations.

It seems possible that 'diploidising' mechanisms may also occur in Oats (*Avena* ($2n = 6x = 42$); Rajhathy & Thomas, 1972), the grass *Festuca arundinacea* ($2n = 6x = 42$; Jauhar, 1975), and perhaps in Cotton (*Gossypium*) and Tobacco (*Nicotiana*) (Kimber, 1961; Riley & Law, 1965). For a review of recent work, particularly the control of pairing behaviour in the *Lolium*/*Festuca* complex, see Evans (1988).

These findings, of special concern to plant breeders, are also of general interest. In the past it had been supposed that pairing behaviour was simply a rather mechanical affair, dictated by chromosome homologies alone. Such a view forms the rationale behind genome analysis. Now that the situation in Wheat and other plants is beginning to be revealed in its complexity, the basis of genome analysis no longer seems so secure. How can we ever rule out the possibility that some degree of genetic control is being exercised in the pairing behaviour of polyploids? The experiments necessary to demonstrate the presence of a diploidising genetic system are time consuming and costly. They may not be technically feasible in plants with small rather undifferentiated chromosomes, since identification of different nullisomics is a necessary part of the analysis.

Another complexity must be raised at this point. It has been known for some time that certain polyploids producing some multivalent associations at meiosis are nevertheless fertile, e.g. *Agrostis canina*, *Arrhenatherum elatius*, *Dactylis glomerata* and *Tradescantia virginiana*. After studying *Anthoxanthum odoratum* ($2n = 4x = 20$), Jones (1964) concluded that multivalent associations need not necessarily indicate a chaotic meiosis. It is possible to devise models of multivalent formation and separation which will yield balanced chromosome products, such systems being under genetic control. It is envisaged, therefore, that in polyploids genetic control of meiotic behaviour may arise, taking the form either of a diploidising mechanism or of a system of co-ordinated multivalent formation and separation, both mechanisms yielding fertile offspring.

Given the possibility of genetic control of meiotic behaviour, the uncritical use of genome analysis must be avoided (see de Wet & Harlan, 1972). Perhaps the method is best used in association with other experimental approaches, such as karyotype analysis and crossing experiments.

Studies of karyotypes

In cases where chromosomes are large, an examination of karyotypic differences is often of great value in understanding the origin of particular polyploids. An elegant example, elucidating the evolutionary relationships of common species, is provided by the work of Jones (1958) with the widespread European grasses *Holcus lanatus* and *H. mollis*. *H. lanatus* is uniformly diploid and fertile, with $2n = 2x = 14$. *H. mollis*, on the other hand, has plants with four different chromosome numbers $2n = 28$, 35, 42 and 49. In many areas in Britain, *H. mollis* is represented by the sterile pentaploid with $2n = 5x = 35$, which reproduces entirely vegetatively. Jones studied the chromosomes of *H. mollis* and *H. lanatus*. In the latter, he found a particular chromosome with the basic set of seven, which was conveniently recognisable by carrying a 'satellite'. Tetraploid *H. mollis* ($2n = 4x = 28$) also had a pair of satellited chromosomes, but these were much shorter and easily distinguished from those of *H. lanatus*. In pentaploid *H. mollis*, a pair of short satellited chromosomes and a single long satellited chromosome were recognised. In a triploid hybrid between the two species found in the wild, both long and short satellited chromosomes were found. Jones concluded that the only simple sequence of events which would yield a pentaploid of the appropriate karyotype, would be the prior origin of the triploid hybrid and then a backcross of an unreduced gamete with a normal gamete of the tetraploid parent *H. mollis*. The evidence is presented diagrammatically in Fig. 12.10. These investigations have revealed the remarkable fact that populations of *H. mollis* in Britain are largely composed of a complex pentaploid hybrid in the parentage of which *H. lanatus* is involved. Moreover, it is not possible to distinguish morphologically between the chromosome races of *H. mollis* (Jones & Carroll, 1962). However, recent studies have revealed a different picture in France. Here pentaploid *Holcus* appears to have originated from the cross *H. lanatus* ($2n = 2x = 14$; reduced gamete) with *H. mollis* ($2n = 4x = 14$; unreduced gamete) without the triploid intermediary. Furthermore, the pentaploid is, perhaps, not entirely sterile (Richard *et al.*, 1995).

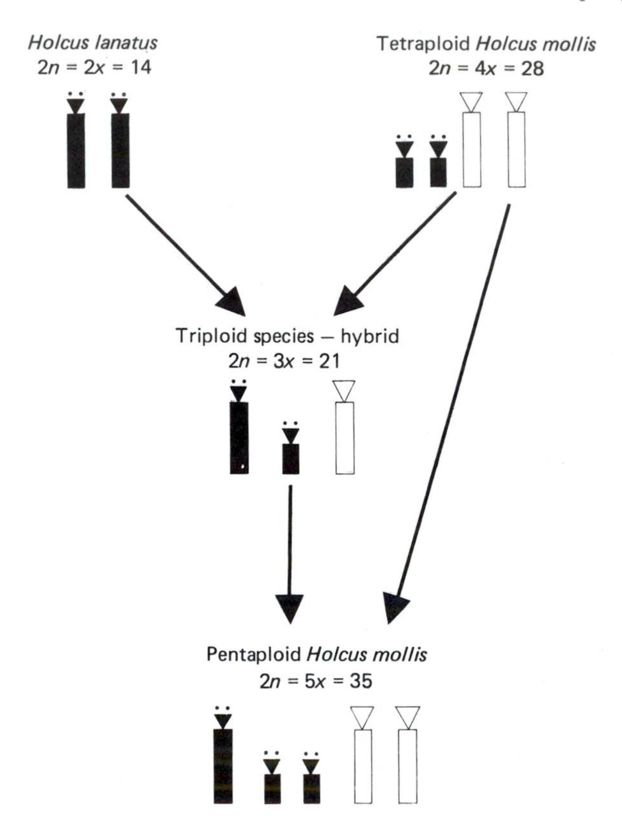

Fig. 12.10. Karyotype analysis of *Holcus*. (From Jones, 1958.)

Another interesting example of the use of karyotypic information concerns the origin of the widespread polyploid weedy grass *Poa annua* ($2n = 4x = 28$). Nannfeldt (1937) suggested that this plant was the allopolyploid derivative of the cross between the diploids *P. supina* and *P. infirma*. In contrast, de Litardière (1939) suggested that *P. annua* could be an autopolyploid derived from *P. infirma*. Crosses between *P. infirma* and *P. supina* were made by Tutin (1957) and yielded *P. annua*-like plants. Tutin suggested that during the Pleistocene period *P. supina*, a perennial mountain species, was probably driven down to lower altitudes and came into contact with the ephemeral grass *P. infirma* in the northern Mediterranean region. Crossing between the two taxa could have occurred, especially where mountains are close to the coast, giving rise to the tetraploid plant, which is now found world-wide.

Koshy (1968) made a detailed study of the karyotype of *P. annua* and showed that there are three particularly distinctive chromosomes, each present as a pair (Fig. 12.11). Nannfeldt (1937) showed that the karyotypes of *P. supina* and *P. infirma* were rather similar. Koshy draws the following conclusions: *Poa annua* does not have four identical sets of chromosomes, as would be required in an autotetraploid, nor does it have the sum of the karyotypes of *P. infirma* and *P. supina*. Either *P. annua* has undergone structural changes since its formation from *P. infirma* and *P. supina*, or it may be derived from either *P. infirma* or *P. supina* forming an allopolyploid derivative with another, as yet unknown, species of *Poa*. Clearly, more studies will be needed to sort out the ancestry of *P. annua*.

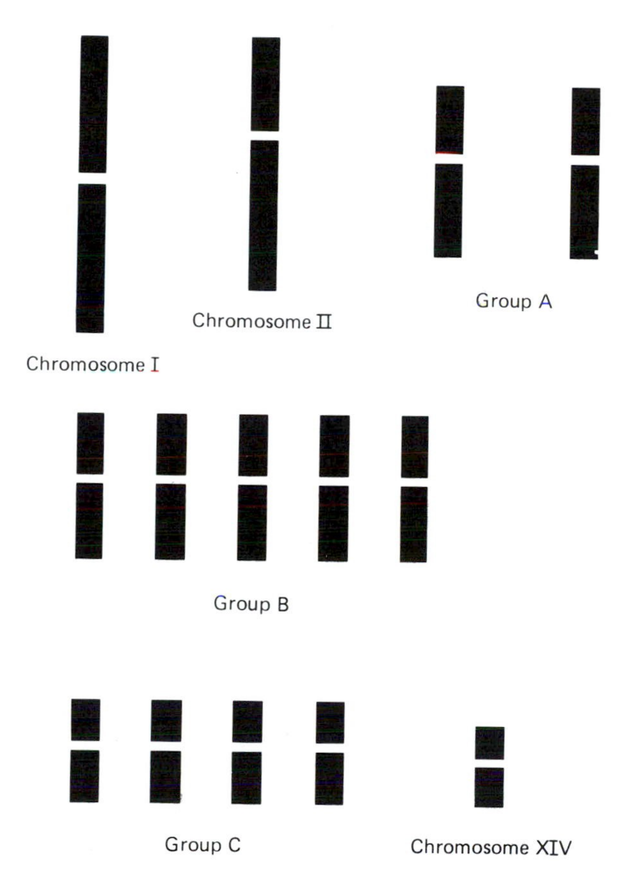

Fig. 12.11. Idiograms of the karyotype of *Poa annua*. (From Koshy, 1968.) Note that *P. infirma* is sometimes called *P. exilis*.

Chemical studies

Often species differ in chemical characters, and hybrids between them, and the polyploid derivatives from such hybrids, generally have an 'additive' pattern of secondary chemical compounds. Thus, the secondary chemical constituents found in species AA and species BB are likely to be present in the allopolyploid derivative AABB, providing a strong line of evidence with which to consider the ancestry of particular polyploids. A particularly elegant example is provided by studies of North American *Asplenium* species (Fig. 12.12).

In many experiments in the early 1960s, no attempt was made to identify the secondary chemical compounds separated by various techniques, and deductions were made on the basis of patterns. For example, Stebbins *et al.*, (1963) found chromatographic evidence in support of the view that the tetraploid, *Viola quercetorum* ($2n = 4x = 24$), is a polyploid hybrid deriva-

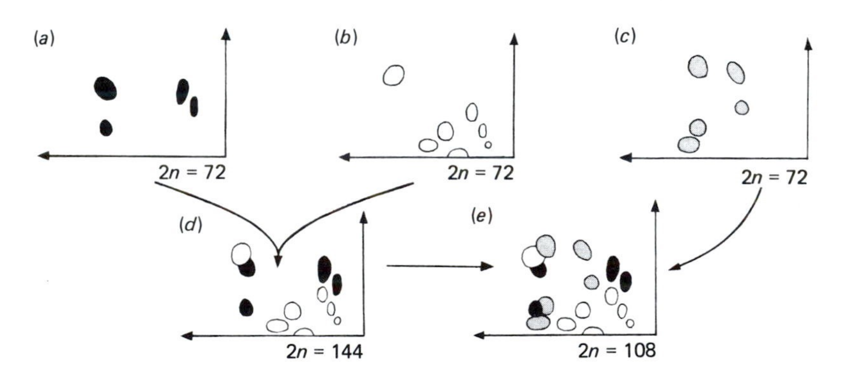

Fig. 12.12. Diagrammatic representation of two-dimensional chromatograms of three species of *Asplenium* and their hybrids. The flavonoid pattern for *A.* × *kentuckiense* appears to combine the profiles of three diploid species. This evidence adds weight to the hypothesis that *A.* × *kentuckiense* is the trigenomic allopolyploid derivative from the hybridisations outlined in the diagram. (*a*) *A. rhizophyllum*. (*b*) *A. montanum*. (*c*) *A. platyneuron*. (*d*) Bigenomic allopolyploid *A. rhizophyllum* × *A. montanum*. (*e*) Trigenomic allopolyploid *A.* × *kentuckiense* (*A. rhizophyllum* × *A. montanum* × *A. platyneuron*). (From Smith & Levin, 1963, in Heywood, 1976.) A much more detailed study of the flavonoids in various *Asplenium* taxa has been undertaken which confirms the earlier findings (Harborne, Williams & Smith, 1973). Evidence from a study of isozymes corroborates the findings of the chemotaxonomists (Werth, Guttman & Eshbaugh, 1985*a*). Furthermore, detailed investigations of the allotetraploid. *A. pinnatifidum* (*A. montanum* × *A. rhizophyllum*) indicates that this widespread species has been formed more than once from its diploid progenitors (Werth, Guttman & Eshbaugh, 1985*b*).

tive of the cross between the diploid ($2n = 2x = 12$) taxa, *V. purpurea* subsp. *purpurea* and *V. aurea* subsp. *mohavensis*. In many of these classic studies a single individual served to represent a population or a taxon. Later investigations, however, suggested that it was important to consider variation within as well as between taxa.

While many chemotaxonomists still study 'patterns' of variation in unknown substances in extracts of plants, other studies of variation, many of which concern ancestry, have involved an attempt to identify the chemical compounds. For instance, the Red Horse-Chestnut (*Aesculus* x *carnea*; $2n = 4x = 80$), which originated sometime before 1818 in Europe (Li, 1956), has long been considered to be the allopolyploid derivative from the European *A. hippocastanum* and the introduced North American species *A. pavia* (both taxa are diploid; $2n = 2x = 40$). Chromatographic studies of phenolic compounds (Hsiao & Li, 1973) offer support for this hypothesis.

Autopolyploids: reassessment of their evolutionary potential

Recently, great advances in our understanding of polyploidy have come from the investigation of isozymes, DNA profiles, etc., particularly in those cases where classic examples of polyploidy have been re-examined using modern methods.

In their review of polyploidy Soltis & Soltis (1993) state: 'In only the past five years, molecular data have provided a dramatic reshaping of long-standing views regarding autopolyploid evolution. Traditionally, autopolyploid evolution was viewed as maladaptive... this view was based on the assumption that autopolyploidy is accompanied by pairing difficulties, resulting in the frequent formation of multivalents and a concomitant reduction in pollen and seed fertility'.

Lumaret (1988) has considered, in detail, the situation in tetraploid *Dactylis glomerata*, which is generally considered to be cytologically autopolyploid. She writes:

> Evidence of regular meiosis in natural tetraploid *Dactylis* is provided by the occurrence of exclusively bivalent and quadrivalent pairing, substantial frequency of ring bivalents, and of zigzag as well as chain quadrivalents. These configurations are known to reduce perturbations in anaphasis [anaphase] by restoring the balance between chromosome sets. Meiosis in natural hexaploids was similarly shown to be rather regular and involve essentially bivalent and quadrivalent pairing. This is not the case in raw colchicine-induced autopolyploids of several diploid species and in their artificial allopolyploid hybrids. Such tetraploids show a

substantial frequency of cells with laggards, univalents and trivalents. Such a difference in pairing behaviour between raw autopolyploids and natural ones is one of the more striking results in *Dactylis* research. It suggests that selection for sexual fertility might act to stabilize meiosis by eliminating unbalanced configurations and increasing the frequency of disjunctional quadrivalents. The mechanism responsible for this results remains unknown but seems ... to be distinct from any diploidisation process.

Studies using isozymes have also provided new insights into autopolyploidy. Autopolyploids are characterised by having several representatives of the same genome, and the tetrasomic inheritance shown by tetraploids (and so-called polysomic inheritance in polyploids with higher chromosome numbers) can be used, more decisively than in cytological investigations, as a means of identifying them. Recent studies have detected 19 cases of autopolyploidy. In six detailed studies, comparing autopolyploids with their diploid progenitors, they exhibit increased heterozygosity and allelic diversity. 'Thus, rather than the traditional view of autopolyploidy as maladaptive, molecular data provide strong genetic arguments for the potential success of autopolyploids in nature' (Soltis & Soltis, 1993).

Polytopic multiple origin of polyploids

Many botanists have perhaps subconsciously assumed that a particular polyploid is formed only once. But, if related diploid taxa regularly come into contact, could allopolyploid derivatives be produced repeatedly and polytopically (i.e. in different sites)?

Very clear evidence supporting the hypothesis of polytopic origin of allopolyploid species in nature comes from the study of introduced species of *Tragopogon* and their hybrids. Three European species of this Composite genus occur in North America as weeds of roadsides and disturbed ground: *T. dubius*, *T. pratensis* and *T. porrifolius*. All these are diploid species with $2n = 12$ (Ownbey, 1950), and highly sterile F_1 hybrids between all the pairs are known in Europe. In the western USA where Ownbey studied them, he found it very easy to detect these hybrids by their failure to set good heads of seed. In four separate localities, however, he found small groups of fertile plants with the intermediate characters of the hybrids. These proved to be tetraploid with $2n = 24$. Morphologically, one of the polyploids appeared to have arisen from the cross *T. dubius* × *porrifolius* and the other from T. *dubius* × *pratensis*. Since these fertile allopolyploids are both morphologically distinct and genetically isolated from their parent species, Ownbey described them as new species: *T. mirus* and *T. miscellus* (Fig. 12.13).

Successive studies involving different techniques support Ownbey's ideas about the origin of *T. mirus* and *T. miscellus*. These investigations include the study of karyotypes (Brehm & Ownbey, 1965; Ownbey & McCollum, 1953, 1954), biochemical characteristics (Belzer & Ownbey, 1971), isozymes (Roose & Gottlieb, 1976) and DNA markers (literature reviewed by Soltis & Soltis, 1993).

The investigations of Ownbey and his associates suggested another interesting possibility, namely that *T. mirus* had arisen independently in three separate areas and *T. miscellus* had been produced twice in separate localities. Recently, using molecular approaches, it has been possible to make critical tests of these suggestions. From the study of allozymes, Roose & Gottlieb (1976) obtained evidence that *T. mirus* had at least three independent origins, but their results were inconclusive concerning the possibility of multiple origins of *T. miscellus*. Studies of chloroplast DNA, which is maternally inherited, have proved very informative, as it has proved possible to determine which of the diploid species was the maternal parent in the production of a particular sample of the polyploid (Soltis & Soltis, 1989). From the DNA profiles produced by cleavage with restriction enzymes (restriction fragment length polymorphisms; RFLPs), it was discovered that populations of *T. miscellus* in the Pullman area in Washington State had *T. dubius* as the maternal parent. In contrast, all the other populations of *T. miscellus* had *T. pratensis* as the maternal parent. Evidence from RFLP analysis clearly indicated that all the populations of *T. mirus* had *T. porrifolius* as the female parent. However, an analysis of ribosomal DNA demonstrated that *T. mirus* had two independent origins (Soltis & Soltis, 1991).

More recently, a thorough search has revealed further populations of both *T. mirus* and *T. miscellus* and, combining the information from DNA and enzyme electrophoresis, indicates a minimum of five and a maximum of nine independent origins of *T. mirus*, and a minimum of two and a maximum of 21 for *T. miscellus*. Thus, the evidence for the multiple origin of *Tragopogon* allopolyploids on a local scale is very strong, and indeed recurrent formation of *T. mirus* would appear to have occurred in the Palouse region in Washington State and adjacent Idaho (Soltis *et al.*, 1995). As both polyploid species have been discovered in Arizona, and *T. miscellus* in Wyoming and Montana, it will be interesting to determine whether these populations of the allopolyploids were also produced independently.

Novak, Soltis & Soltis (1991) have recently published an account of the history of *Tragopogon* in the western USA. Evidence suggests that *T. porrifolius* and *T. pratensis* were introduced about 1916, with *T. dubius* arriving a little later (*c.* 1928). A field survey of populations revealed dramatic increases in the distribution of the new allopolyploids, since Ownbey's pioneering investigations. Moreover, it might be assumed that the appropriate pair of diploid species found in association with a 'new' polyploid were likely to be descended from its parents. However, in mixed populations containing *T. mirus* and both parental diploids, the diploids present were not always of the correct genotype to have produced the allopolyploid. Clearly, active dispersal of the plants was occurring. The survey also suggested that, as there were many primary diploid hybrids, there was the potential to produce further new polyploid derivatives.

The survey also provided information concerning population size of older populations. For instance, in 1949 two small populations of *T. miscellus* in Moscow, Idaho, each had 30–35 individuals. In 1950, one of these had been reduced to seven plants, whereas the other had increased to 75. A third population found at Moscow in 1950 was destroyed by the construction of a house and lawn. Clearly, each episode of polyploidy is likely to produce a single individual and not all small populations are likely to survive. It is to be hoped that further population studies will be carried out, for there is no doubt much to learn about the factors which lead to the production of new polyploid individuals the demographic and ecological properties of parental populations and their interaction with their allopolyploid derivatives. Finally, the survey produced the revelation that several of the 25 non-native *Tragopogon* species obtained by Ownbey for his chromatographic work have formed hybrids in the garden where they were grown. Furthermore, a number of these species have escaped into the wild to join the three that have provided the raw material for one of the most fascinating situations of recent microevolution. We can be sure that Linnaeus, whose work on hybrid *Tragopogon* (described in Chapter 2) gained him the Imperial Academy of Sciences prize in St Petersburg in 1760,

Fig. 12.13. *Tragopogon* species in the USA. (*a*) Bottom row: flowering heads of the diploid species of *Tragopogon*, *T. pratensis* (left), *T. dubius* (centre) and *T. porrifolius* (right). Top row: flowering heads of the polyploid hybrid species, *T. miscellus* (left) and *T. mirus* (right). (From Ownbey, 1950.) (*b*) Occurrence, generalised, of the European diploid species *T. dubius* as a naturalised weed in the western USA, and localities where its hybrid polyploids with *T. pratensis* (*T. miscellus*) and with *T. porrifolius* (*T. mirus*) have been found. The diploid species *T. pratensis* is extensively naturalised throughout the area of *T. dubius*, while *T. porrifolius* occurs chiefly in the western part of the area. (From Stebbins, 1971, based on unpublished data of Ownbey.) Both *T. mirus* and *T. miscellus* have also been discovered in Arizona (Brown & Schaack, 1972).

(*a*)

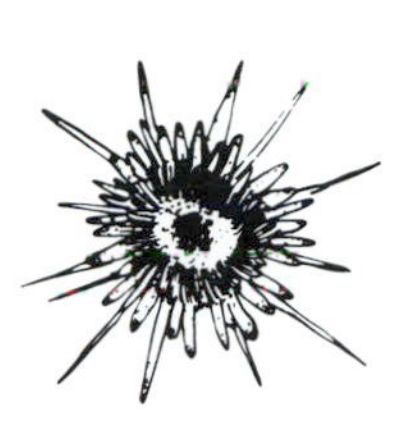

miscellus $2n = 24$

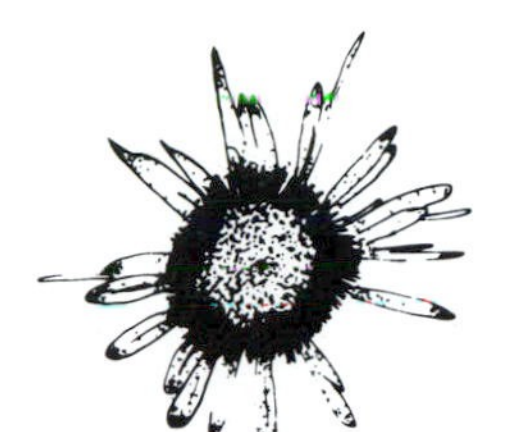

mirus $2n = 24$

pratensis $2n = 12$

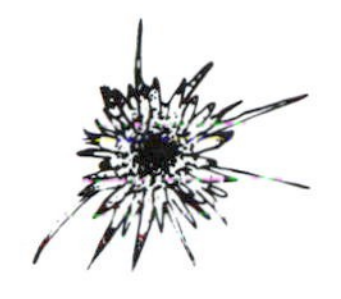

F_1 $2n = 12$

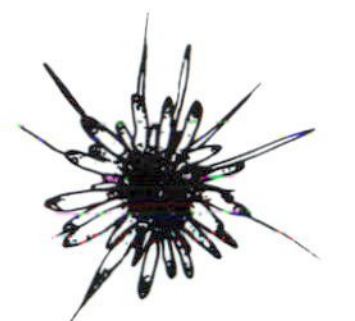

dubius $2n = 12$

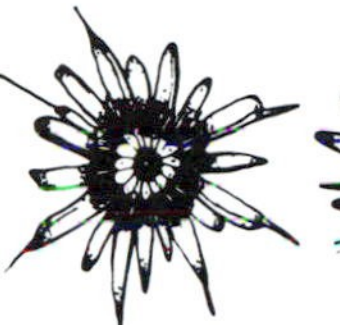

F_1 $2n = 12$

porrifolius $2n = 12$

(*b*)

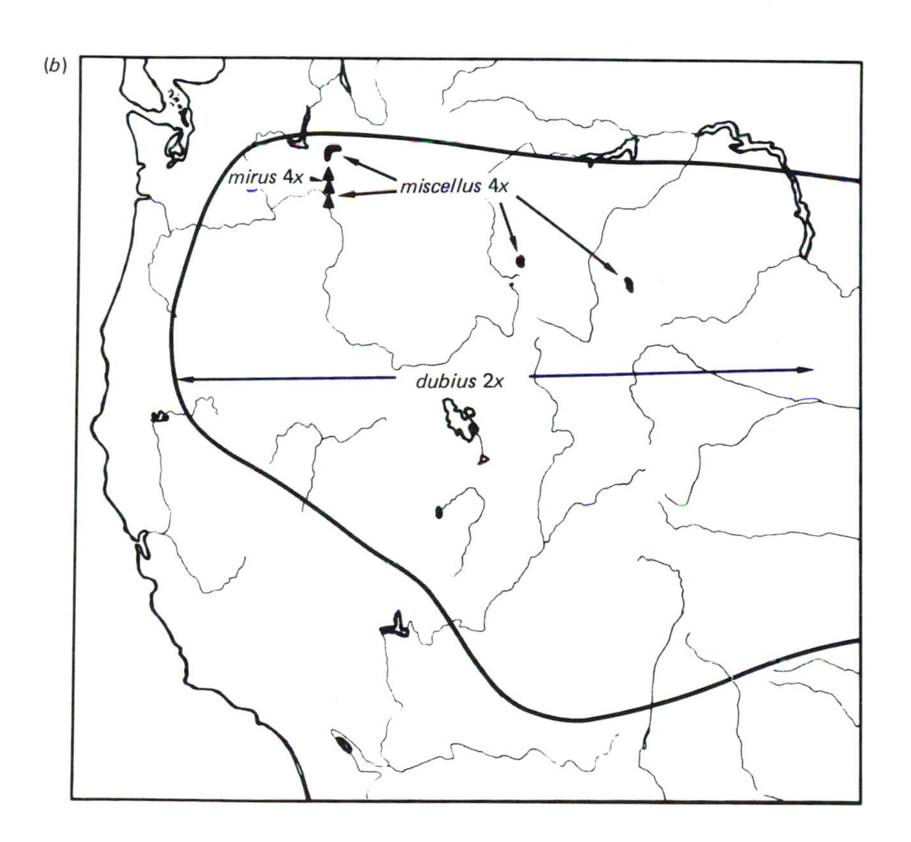

would have been particularly interested in the results of these investigations!

The question now arises: is the recurrent formation of allopolyploid derivatives in *Tragopogon* exceptional, or is the phenomenon found in other polyploid groups? Soltis & Soltis (1993) have recently reviewed the evidence, and report over 40 cases of multiple origin of polyploids, in both auto- and allopolyploids. These include a number of ferns, as well as more than 25 cases in flowering plants. Soltis & Soltis (1993) conclude that 'perhaps the most important contribution of molecular data to the study of polyploid evolution in plants is the documentation that a single polyploid species may be formed recurrently'.

The recurrent formation of polyploids is such an important finding that we will consider briefly two European examples.

The allopolyploid *Senecio cambrensis* ($2n = 60$) was first recorded on roadsides in North Wales in 1948. It is similar to the sterile hybrid (*S.* × *baxteri*; $2n = 30$) resulting from a cross between a species introduced into Britain (*S. squalidus*; $2n = 20$) and the native species, *S. vulgaris* ($2n = 40$). An artificial allopolyploid which is closely similar to the wild plant was synthesised by treating the synthetic hybrid *S. vulgaris* × *squalidus* with colchicine (see Rosser, 1955, for an account of experiments by Harland). Resynthesis has also been successfully accomplished by Weir & Ingram (1980). In 1982, *S. cambrensis* was discovered in another area – on demolition and redevelopment sites in Leith, Edinburgh (Abbott, Ingram & Noltie (1983*a*, *b*). Natural dispersal from North Wales is very unlikely. Examination of the chloroplast DNA of the *S. cambrensis* from the two sites revealed significantly different profiles, indicating the polytopic origin of this new allopolyploid species (Harris & Ingram, 1991). Support for this conclusion was provided from studies of the variation in allozymes of *S. cambrensis* and its parental species (Ashton & Abbott, 1992).

Concerning the ancestry of *S. cambrensis*, another hypothesis has recently been proposed (Lowe & Abbott, 1996). Taking into account chloroplast DNA profiles, they suggested that the Welsh *S. cambrensis* may have arisen not by allopolyploidy but directly from introduced material of the Canary Island endemic *S. teneriffae*. The close similarity of the two taxa is indicated by rDNA studies, crossing experiments and morphological comparisons, and both species have the hexaploid chromosome number $2n = 60$. However, isozyme studies reveal that *S. teneriffae* has isozyme markers not present in *S. cambrensis* and reviewing all the available evidence, Lowe & Abbott consider that both Welsh and Scottish material of *S. cambrensis* have arisen as allopolyploid derivatives from hybridisation between *S. vulgaris* and *S. squalidus*.

There has also been some debate about the origin and taxonomic affinities of *S. squalidus*. Material was brought from Sicily in the seventeenth century and grown in the Oxford University Botanic Garden, from which it escaped. The plant in now widespread in England and Wales, and is spreading in Scotland (Stace, 1991). The results of a recent molecular taxonomic survey of British plants of *S. squalidus*, and its close diploid European relatives, are in agreement with the hypothesis that the material introduced to the Oxford Garden from Mount Etna might have been hybrid in origin, collected, perhaps, in the zone of hybridisation (*c*. 1200 m) between *S. chrysanthemifolius* (found up to *c*. 1000 m) and *S. aethnensis* (1400–2600 m) (Abbott, Curnow & Irwin, 1995).

The second example, so far unresolved with regard to 'multiple' origins, is provided by studies of the famous polyploid '*Spartina townsendii*'. Huskins (1930) provides the classic account of this plant, but the paper contains many ambiguities, and the situation has been re-examined by Marchant (1963, 1967, 1968) who discovered the inaccuracy of Huskins' chromosome counts. Taking all the evidence at present available, it is likely that the fertile allopolyploid was produced on tidal mud at Southampton Water in southern Britain, when the North American species *S. alterniflora* with $2n = 62$ (introduced in the nineteenth century, most probably in ship's ballast) hybridised with the native *S. maritima* with $2n = 60$. However, populations of so-called *S. townsendii* consisted of a sterile hybrid, the fertile allopolyploid or mixtures of the two. *S.* × *townsendii* is the correct name for the sterile hybrid ($2n = 62$) the first confirmed record for which is 1870. The fertile allopolyploid ($2n = 120$, 122, 124), the earliest certain record for which is 1892, is now called *S. anglica*. The story of *Spartina* is curiously incomplete. Repeated attempts to produce the artificial hybrid between the supposed parents have been unsuccessful, and the attempted experimental production of polyploid derivatives from *S. alterniflora*, *S. maritima* and *S.* × *townsendii* by treatment with colchicine has also failed (Marchant, 1968).

In an attempt to discover whether *S. anglica* might have been formed more than once, isozymes and seed proteins have been examined by electrophoresis (Raybould *et al.*, 1991*a*, *b*). The isozyme phenotypes discovered in these investigations were those expected if the species was an allopolyploid derivative of *S. alterniflora* and *S. maritima*. Reviewing all the information about the variation and origin of *Spartina*, Gray, Marshall & Raybould (1991) concluded: '*S. anglica* is almost totally lacking in genetic variation. This may result from a narrow genetic base following a single origin or from a multiple origin from uniform parents'.

Another interesting problem remains to be examined. In an area of

Southwest France, near Bayonne, where *S. maritima* is found and *S. alterniflora* had been accidentally introduced, a variant of *Spartina* (*S.* × *neyrautii*) was discovered in 1892. It has the same chromosome number and meiotic behaviour as *S. townsendii*, but there are morphological differences between the two plants. Raybould *et al.* (1990) confirmed that the two plants had the same isozyme phenotypes. It seems possible that the two sterile hybrids discovered last century in Southwest France and South Britain represent reciprocal hybridisations. Investigation of maternally inherited marker genes (e.g. chloroplast or mitochondrial DNA profiles) may throw light on this intriguing problem.

The examples we have considered – *Tragopogon*, *Senecio* and *Spartina* – all involve at least one introduced parental species. It is important to stress that multiple origins have also been established in wild native plants. For instance, in studying the complex genus *Draba* in Scandinavia, Brochmann, Soltis & Soltis (1992*a*, *b*; Soltis & Soltis, 1993) discovered evidence for recurrent origins, in the narrow endemic octoploid *D. cacuminum* (3) and in *D. norvegica*, (6), *D. lactea* (6) and *D. corymbosa* (16) (number of times given in brackets).

Returning briefly to the *Polypodium vulgare* complex mentioned earlier in the chapter, molecular studies have provided important new evidence on the evolution of the group (Haufler, Windham & Rabe, 1995). It has been established that there are seven diploid species in the group and seven polyploids, whose ancestry has been confirmed by isozyme analysis. Also, studies of chloroplast DNA have revealed which diploids contributed cytoplasmic genomes to particular polyploids. In three of the five allopolyploids there was evidence of 'reciprocal parentages and, therefore, evidence of recurring origins' (Haufler, Soltis & Soltis, 1995).

The origin of new polyploids: the role of somatic events and unreduced gametes

In an important review, Harlan & de Wet (1975) have pointed out that, in many accounts of polyploidy, the mechanism generating new polyploids is left as a shadowy area or it is said that polyploids arise by 'hybridisation followed by chromosome doubling'. This rather ambiguous assertion could imply either that somatic doubling of chromosomes occurs in the primary diploid hybrid, giving rise to the polyploid derivative, or that unreduced gametes are involved. Harlan & de Wet (1975) and de Wet (1980) review the copious literature on the subject, and conclude that in very few cases does somatic doubling seem to be implicated (as in *Primula kewensis*) and that unreduced gametes are of supreme importance. This is also the view of Thompson & Lumaret (1992) in their recent review of polyploidy.

In high polyploids, with their multiple representation of genomes, variation in chromosome number is sometimes found (e.g. *Spartina anglica*, which has three chromosome numbers, $2n = 120$, 122 and 124). It is much more likely that these polyploid derivatives had their origin in meiotic 'events' rather than by somatic doubling.

Harlan & de Wet (1975) provide a useful review of the early literature on the frequency with which polyploid individuals arise. For instance, some Maize stocks produce more than 3% unreduced eggs, but tetraploid Maize does not compete well with diploid Maize under field conditions.

Bretagnolle & Thompson (1995) have recently provided a very thorough review of the mechanisms of the formation of unreduced gametes and their role in the evolution of polyploids. There has been considerable progress in the study of unreduced gametes. Thus, the development of a new technique – flow cytometry – has made it possible to estimate, by the examination of progenies, the frequency of unreduced pollen and eggs. Quantification of stained nuclear DNA is carried out by passing cell suspensions through a laser beam. If appropriate standards are employed – samples of leaf cells of individuals of known chromosome number – then it is possible to estimate the 'ploidy' level of individual plants of unknown chromosome number. Maceira *et al.* (1992) devised a procedure to determine the occurrence and frequency of $2n$ pollen in diploid *Dactylis glomerata*. Progenies of controlled crosses ($4x \times 2x$) were examined for $4x$ progeny (arising from $2n$ egg from the tetraploid + $2n$ unreduced pollen from the diploid). *Dactylis* is mostly but not exclusively outcrossing, and allozyme markers were therefore used to distinguish between selfed and crossed $4x$ progeny. It was estimated that the average production of $2n$ pollen was 0.98%. Six genotypes produced exceptionally high frequencies of $2n$ pollen, ranging from 8–14%. Using the same method, De Haan *et al.* (1992) examined the occurrence and frequency of $2n$ eggs in the same species. On average the $2n$ egg frequency in fertile controlled crosses ($2x$ seed parent $\times$ $4x$ pollen parent) was 0.5%. The frequencies in individual plants were all less than 3.5%, but one plant proved exceptional with 26%.

Given this level of production of $2n$ gametes it might be expected that tetraploids would be produced either by the fusion of $2 \times 2n$ gametes from diploid plants or by a two-step process involving fusion of $2 \times 2n$ gametes, one from a triploid , the other from a diploid.

Two questions must now be faced. First, are the figures for the production of $2n$ gametes in *Dactylis* typical of diploid species in general or are they unusually high? Further surveys are necessary to answer this question. Secondly, what is the fate of newly formed polyploid plants?

The persistence of polyploids

In their recent review of polyploidy, Thompson & Lumaret (1992) make the important point that 'the rate of successful establishment of polyploids is an entirely different matter from their rate of spontaneous origin', for it seems highly likely that single polyploid individuals may frequently be produced in populations. First, a 'new polyploid derivative' is likely to encounter competition from its diploid progenitor(s). However, many polyploids are hybrid in origin. Thus, they may have a different ecological niche from their parental species and may be able to establish themselves in the wild (Fowler & Levin, 1984). A new polyploid will encounter problems of reproduction. Gibby (1981) notes that polyploidy is commoner in perennial herbs that are capable of vegetative reproduction. Thus, a new perennial polyploid individual might survive for a time. However, a new polyploid is likely to suffer 'minority-type disadvantage' in sexual reproduction (Levin, 1975). For instance, a new allopolyploid may receive pollen from its parental diploids, leading to the production of sterile triploid progeny. Given this disadvantage, how is the very high incidence of polyploid species in the flowering plants to be explained? Evidence suggests that polyploidy is often associated with a change in the breeding system.

Polyploids may have a higher level of self-compatibility than their related diploids. For instance, as we saw in Chapter 7, Shore & Barrett (1985) have shown that in *Turnera* diploid and tetraploid plants are distylous and self-incompatible, while hexaploids are homostylous and self-compatible. Self-fertility is also higher in polyploid species of *Paspalum* (Quarin & Hanna, 1980) and *Solanum* (Marks, 1966). In *Primula* (Section *Aleuritia*), diploid species ($2n = 18$) are distylous and self-incompatible, while polyploids in the group ($4x$, $6x$, $8x$ and $14x$) are all homostylous and self-compatible (Richards, 1993). Given the strength of minority-disadvantage faced by a lone polyploid amongst numerous diploid relatives, it is likely that self-fertile polyploid derivatives would be at a selective advantage (Thompson & Lumaret, 1992).

Given the commonly observed correlation between polyploidy and apomictic reproduction, it is clear that newly formed polyploids may be released from minority-disadvantage if they are capable of vegetative or seed apomixis. We now examine the implications of such changes in breeding behaviour.

Self-compatibility.

We have seen in Chapter 7 how repeated self-fertilisation may lead to homozygous derivatives and inbreeding depression. In comparison with the diploid, in a polyploid the march to homozygosity with inbreeding is not quite so rapid and this is clearly of great significance in the evolution of polyploid species. Thus, in a diploid plant of genotype *Aa*, selfing produces progeny distributed in the familiar Mendelian ratio 1*AA*:2*Aa*:1*aa* and 50% of the progeny are homozygous. In a tetraploid plant of genotype *AAaa*, in which the alleles are located near the centromere on different chromosomes and where the four homologous chromosomes separate at random in pairs, the ratio on selfing is:

1/36*AAAA*:8/36*AAAa*:18/36*AAaa*:8/36*Aaaa*:1/36*aaaa*

Segregation follows Mendelian principles but with a different ratio, yielding only 1/18 homozygous derivatives; 94.4% are heterozygotes of various genotypes.

Models may be constructed in which selfing proceeds for several generations and where the genotype frequencies are not influenced by selection. Thus, to reduce the percentage of heterozygotes to less than 1% from a population initially wholly heterozygous (*Aa* in the diploid and *AAaa* in the autotetraploid) will take seven generations for the diploid but it will take 27 generations for the tetraploid. About 46 generations would be needed to achieve less than 1% heterozygosity in the autohexaploid (Parsons, 1959). Clearly inbreeding effects are less severe in polyploids.

Apomixis.

The 'Achilles' heel' of many 'new' polyploids is defective meiosis and consequent sterility. Thus, hybridisation, whether between diploid or polyploid species, frequently produces more or less sterile plants. It is easy to see which classes are likely to show defective meiosis: those which have odd numbers of genomes (e.g. $3x$, $5x$); those which lack genomic homology and show multivalent formation with univalents; and those with abnormal, defective segregants because of genomic incompatibility. In sexual reproduction such hybrids are clearly at a selective disadvantage. However, the fitness of hybrids would be increased if they were able to reproduce apomictically, either by vegetative apomixis or agamospermy. Clearly, the ease with which variants derived from sexually reproducing species can become agamospermic will depend upon the genetics of apomixis (examined in sexual × apomictic crosses between related species).

The genetics of apomixis is a highly technical subject and here we attempt only a very brief introduction. In a very influential paper, Powers (1945) produced a purely hypothetical model that involved three rare simultaneous changes – the formation of the unreduced embryo sacs, the failure of fertilization and their parthenogenetic development. These changes, which can occur separately in plants, were considered to be controlled by three recessive genes or groups of genes. Asker & Jerling (1992) discuss all the recent research on the genetics of apomixis, and point out that homozygosity for recessive apomixis genes is not compatible with the fact that apomicts are polyploid and mostly highly heterozygous. They consider that models involving the monogenic control of apomixis should be tested. Nogler (1984), in an important review, concluded that monogenic control of apospory might occur in *Poa*, *Hieracium*, *Potentilla*, etc. In some cases, apomixis would indeed appear to be controlled by a single dominant gene (e.g. apospory in *Panicum maximum*: Savidan, 1978). However, the results from studies of other apomicts (e.g. members of the Rosaceae) show 'no obvious indication of a monogenic inheritance of apomixis and the dominance relationships are not clear' (Asker & Jerling, 1992). Obviously, the genetic control of apomixis may be different in different groups of plants, and further research is necessary to determine the genetics in different cases.

Many polyploids are known to reproduce apomictically, whilst the related diploid taxa are sexual. Thus, within the variable species *Ranunculus ficaria*, sterile variants with bulbils (vegetative apomixis) are triploid ($2n = 3x = 24$) or tetraploid ($2n = 4x = 32$), whilst diploid plants ($2n = 2x = 16$) set seed by normal sexual means (Taylor & Markham, 1978). In the case of *Sorbus* (Fig. 12.14), three widespread and variable species in Europe are diploid and sexual, whilst other more restricted taxa, some of which have leaf-shape and other characters intermediate between two of the diploid species, are triploid or tetraploid and reproduce apomictically. These apomictic taxa are likely to have been derived by hybridisation and polyploidy from the diploids (Liljefors, 1953, 1955). Support for this view comes from studies of secondary chemicals (Challice & Kovanda, 1978) and isozymes (Proctor, Proctor & Groenhof, 1989). In *Alchemilla*, very few of the hundreds of microspecies distinguished in Europe show any trace of sexuality and most have high polyploid chromosome numbers; they look like ancient polyploids derived from sexual ancestors which are now extinct (Walters, 1972, 1986*a*).

Apomixis has been seen as 'an escape from sterility' (Darlington, 1939). However, the positive adaptive significance of apomixis must be stressed. We might therefore look upon the polyploid and apomictic lines more in

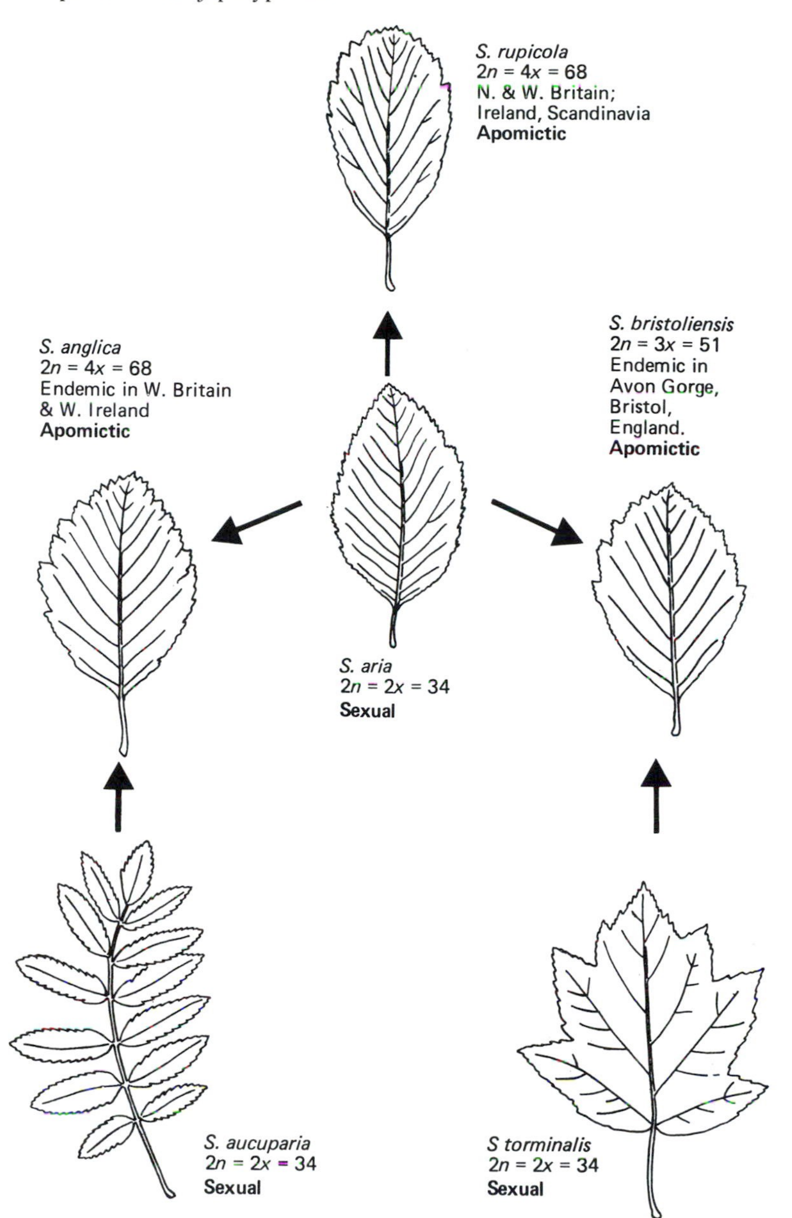

Fig. 12.14. Sexual and apomictic species in *Sorbus* (× 0.33). A typical leaf of each is shown. The arrows indicate possible origins of the apomicts from the sexual species.

terms of positive adaptation to particular kinds of habitat, than in terms of a negative escape from sterility. This is particularly true now that many polyploids are considered to reproduce by facultative rather than obligate apomixis (Asker & Jerling, 1992).

It is possible to point to certain evolutionary situations and types of habitat in which the advantages of a quick, safe method of reproduction and dispersal outweigh the advantages of outbreeding in maintaining genetic variability. In such situations, autogamy or apomixis could well be favoured. The most obvious kind of habitat in which quick, safe reproduction is essential is the cultivated or ruderal habitat created by man. The loss of outbreeding mechanisms in many modern polyploid weeds is therefore understandable in terms of the great selection pressures in favour of quick possession of artificially bared ground, whether arable field or 'waste land'.

If we look at the natural open habitats in which bare ground is continually made available by agencies of erosion and catastrophic destruction, we can find here in many cases the presumed native habitats of familiar weeds. In temperate regions these natural open habitats are mostly on coasts or mountains, though river banks, rocky gorges and other local features may also provide them. An excellent example of a weed species which has related native variants in coastal and mountain habitats is the Bladder Campion, *Silene vulgaris*, and many other examples may be found in the review papers on the evolution of weeds (Baker, 1974; Warwick, 1990*b*).

Gene flow between diploids and polyploids

The simple picture of an allopolyploid species arising suddenly and achieving at one bound fertility and genetic isolation is unlikely to be the whole story in the complex polyploid evolution of plants. In particular, the genetic isolation of allopolyploids must be questioned. Jones & Borrill (1961) report very interesting work on the important pasture grass *Dactylis*, which illustrates the problem. The common *Dactylis glomerata* is a variable tetraploid with $2n = 4x = 28$. Cytologically the plant is autopolyploid (Lumaret, 1988). Crossing *D. glomerata* and a diploid, *D. glomerata* subsp. *aschersoniana*, resulted in a triploid hybrid which is male sterile, but partially fertile as a female parent. Backcrosses of the triploid with subsp. *aschersoniana* were less successful than with the tetraploid *D. glomerata*. Female gametes produced by the triploid ranged in chromosome number from $n = 7$ to $n = 23$; in general those with $n = 14$ could function well in the backcross to the tetraploid. Thus, hybrid tetraploids (or near-tetraploids) could arise relatively easily and moreover showed a high fertility equal to that of wild *D. glomerata*.

Jones & Borrill then ask the important question: 'How likely is gene flow in natural populations?'. There is only fragmentary evidence here, but they quote the statistics of Zohary & Nur (1959) who reported that a deliberate search of an area in Israel where both diploid and tetraploid populations occur resulted in seven triploid plants in a total of 4000 examined. The intercrossibility of these triploids with diploids and tetraploids was studied and Zohary & Nur discovered that, considering the fertility of the product, gene flow was likely only from diploid to tetraploid level.

Levin (1978*b*) discusses a number of examples of gene flow in polyploids. Although this is usually a unilateral occurrence from diploids via triploid to tetraploids, he presents the evidence from natural and experimental situations, e.g. in *Solanum*, *Viola* and *Papaver*, in which gene flow occurred in the reverse direction from tetraploid via triploid to diploid.

Recently, Stace (1993) has emphasised the important role of 'rare events' in the evolution of polyploid groups. Thus, in his studies of the *Festuca* and *Vulpia* groups, there is evidence of rapid introgression from tetraploid to hexaploid levels via highly sterile pentaploid intergeneric hybrids, through the rare production of euploid (in this case triploid) gametes. Stace cites a number of other published cases (e.g. *Ranunculus*, *Centaurium*, *Euphrasia* and *Pilosella*), which suggest that introgression can occur between species of different ploidy level, through the production of euploid gametes by highly sterile hybrids.

In considering the possibilities of gene flow between diploid and polyploid plants, models can be constructed which do not assume the involvement of plants with odd numbers of genomes – $3x$, $5x$, etc. For instance, in studying experimentally the variation of *Betula*, Johnson (1945) suggested that gene flow might occur directly between diploids and tetraploids via unreduced gametes of *B. pendula* ($2n = 28$) which could fuse with the normal ($n = 28$) gametes of *B. pubescens* ($2n = 56$) to give hybrid plants with a tetraploid chromosome number. Elkington (1968) has also considered this possibility for species of *Betula*.

Not only is there evidence in particular cases of gene flow between plants with a different ploidy level, but there is also the important possibility of new hybridisation at the polyploid level. A case of particular interest was described by Fagerlind as early as 1937 in the common European genus *Galium*. The white-flowered *G. mollugo* and the yellow-flowered *G. verum* are both represented in Southeast Europe by diploid plants ($2n = 2x = 22$). These diploids are completely intersterile. In central and northern Europe, however, the common representatives of both species are tetraploid ($2n = 4x = 44$), and the hybrid between them shows almost normal meiosis and some degree of fertility. We are in this case forced to conclude that an effective sterility, evolved at the earlier diploid level, has been broken down in the tetraploid. Such cases open up possibilities of 'reticulate' evolution in polyploids which could be extremely difficult to elucidate.

Polyploids: their potential for evolutionary change

Polytopic origins of both auto- and allopolyploids and gene flow between and within different polyploid levels may further increase the variability of polyploids. Thus, the ecological amplitude and geographical distribution of polyploids may exceed that of parental diploid taxa. For example, diploids in *Achillea millefolium* have a rather restricted distribution in southern Europe at the present time, in contrast to the more widespread tetraploid and hexaploid plants (Ehrendorfer, 1959).

It is also possible to view polyploidy as a more efficient means of conserving variation. In diploid organisms of the heterozygous genotype *Aa* (where *A* is dominant to *a*), the allele *a* is sheltered from the immediate effects of selection and may survive in the population even if *aa* is deleterious. However, at meiosis gametes containing either *A* or *a* are produced and selection may act on *a* at this stage. The diploid state offers some 'shelter' for recessive alleles; polyploids, with their several genomes giving multiple representation at the *A* locus, have a greater capacity to store variation. Moreover, as a hybrid is likely to be heterozygous for many loci, chromosome duplication in the allopolyploid will produce fixed heterozygosity at these loci.

Natural polyploids, once formed, are likely to change with time, through the action of mutation and selection, and as a consequence of chance events. In experimentally produced polyploids, there is a good deal of evidence that fertility may increase as 'raw' polyploids are subjected to artificial selection (Stebbins, 1950; de Wet, 1980).

Using molecular techniques, there have recently been a number of interesting advances in our understanding of the evolution of polyploid genomes. Polyploids have several copies of genomes and evidence from the study of isozymes suggests that gene silencing occurs. Soltis & Soltis (1993) write: 'Given sufficient time, the number of genes expressed in a polyploid could be sufficiently reduced to return the polyploid to a level of expression similar or identical to that of the diploid ancestor i.e. the polyploid would become genomically diploidized'. They then develop an idea first stressed by Ohno (1970). 'The polyploid process provides an abundance of redundant genes that can subsequently undergo mutation and potentially evolve new functions'.

Gene silencing.

Evidence for gene silencing comes particularly from studies of ferns. Put simply, in homosporous ferns the mean basic chromosome number is $n = 55$, suggesting the importance of polyploidy in their evolution and the possibility of multiple representation of genomes in many ferns. If this is the case, then multiple representation of genes should result in multiple banding patterns of isozymes in electrophoresis. However, many studies have revealed diploid expression in the banding patterns of isozymes. This result seems so surprising that it has even been questioned whether ferns are indeed high polyploids, as such high numbers could, perhaps, have been achieved by aneuploidy. However, in their review Soltis & Soltis note that recently 'multiple copies of defective genes encoding the CAB [chlorophyll-binding] proteins have been detected in the genome of the isozymically diploid fern *Polystichum munitum* ($n = 41$) by Pichersky, Soltis & Soltis (1990)'. Thus, at the present time the balance of evidence favours the hypothesis that the homosporous ferns are indeed polyploid and that gene silencing has occurred. But further studies are needed in this very interesting field. Caution in interpretation is needed as duplicate genes have been found in some plants following chromosome mutation (Gottlieb, 1982).

To provide a balanced picture of this complex area we note that evidence for gene silencing has not always been detected in polyploids. Many angiosperm families – e.g. Salicaceae – are regarded as ancient polyploids, yet they show polyploid expression of genes in electrophoresis (see Soltis & Soltis, 1993, for further details).

Chromosome repatterning.

There is evidence from a number of sources that chromosome repatterning may occur in polyploids (Soltis & Soltis, 1993). Informative results have been obtained using a new technique called genomic *in situ* hybridisation (GISH), which exploits the fact that homologous portions of DNA will 'hybridise'. A study by Bennett, Kenton & Bennett (1992) examines hypotheses concerning ancestry and questions relating to chromosome patterning. On the basis of the karyotypes, it has been proposed that *Milium vernale* ($2n = 8$; with eight large L chromosomes) is one of the parents of the allopolyploid grass *M. montianum* ($2n = 22$; which has eight L chromosomes matching those of *M. vernale* + 14 small chromosomes). To provide a molecular test of this hypothesis, a total DNA extract of *M. vernale* was made into a probe conjugated to fluoresceinated avidin, which was hybridised *in situ* to the root-tip chromosomes of *M. montianum*. The probe hybridised to the eight L chromosomes (indicated by yellow fluorescence), supporting the theory that *M. vernale* was the donor of the chromosomes (see cover illustration). Clearly, GISH provides a powerful tool for the analysis of polyploid ancestry. The results of the experiment also showed the potential of the technique for studying chromosome repatterning. Soltis & Soltis (1993) comment: 'significantly, however, a few regions along the L chromosomes clearly lacked the hybridization signal of the *M. vernale* genomic probe, suggesting that either (1) these sequences arose after the formation of the alloploid, or (2) a limited intergenomic transfer of DNA from chromosomes representing the second genome present in the alloploid (unidentified) to the 'L' chromosomes donated by *M. vernale*'.

Evidence for chromosome repatterning has also been discovered in studies of allotetraploid Tobacco and hybrids (Parokonny *et al.*, 1994), the GISH technique revealing numerous intergenomic translocations. Using the same technique in a study of hexaploid Oats ($2n = 6x = 42$; AACCDD), Chen & Armstrong (1994) discovered evidence for intergenomic interchanges between the A and C genomes. Rapid genetic changes have been detected in newly formed artificial polyploids in *Brassica* (Song *et al.*, 1995). While more studies are needed, including wild plants, it seems very likely that traditional views of polyploidy are seriously incorrect. Formerly, as for example Wagner (1970), polyploids were considered as invariant products produced by single events leading to evolutionary dead ends. Given the evidence of recurrent polyploid formation, subsequent gene silencing and potential for genome evolution, molecular studies are 'prompting a dramatic rethinking of polyploid evolution' (Soltis & Soltis, 1995).

Distribution of polyploids

One of the phenomena which has excited interest since the early days of cytogenetic studies on plants is the increase in frequency of polyploidy with increasing latitude in the Northern Hemisphere (Fig. 12.15). The earliest reference to this correlation seems to have been made by the Swedish cytologist Täckholm (1922), in his impressive study of the genus *Rosa*, and Hagerup (1931), Manton (1950), Morton (1966), Favarger (1967) and Stebbins (1984) have provided important reviews.

Three hypotheses, not mutually exclusive, have been advanced to explain this phenomenon:

1. It has been proposed that polyploids have greater resistance to the more severe conditions, particularly cold, at high latitudes. Only few experi-

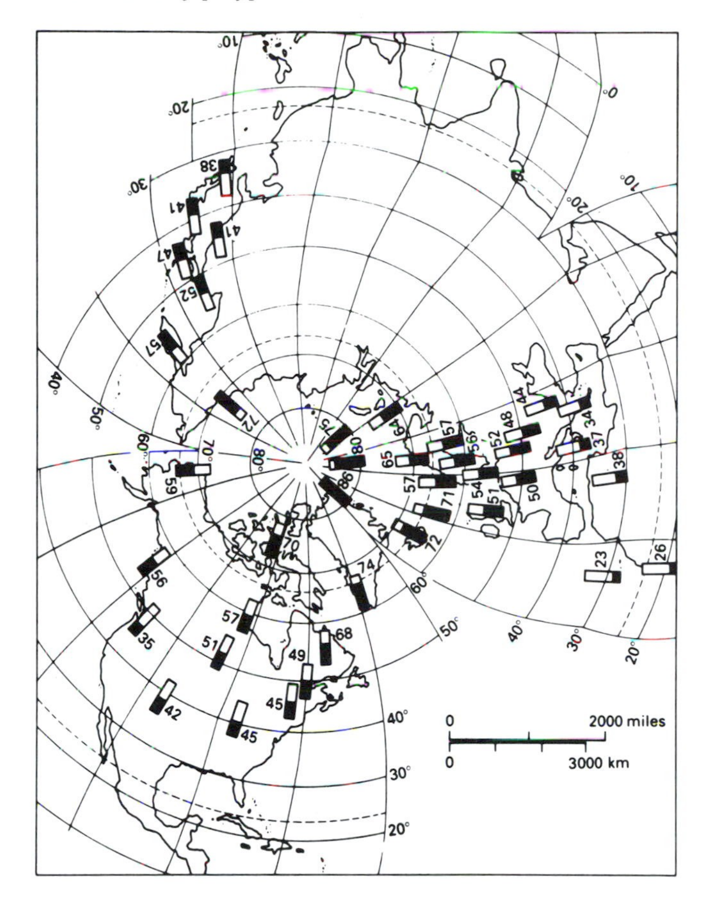

Fig. 12.15. Frequency of polyploids (given as percentage figures and indicated as black portion of column) in the floras of various territories in the Northern Hemisphere. (From Löve & Löve, 1974.)

mental tests of this proposition have been made. It is interesting, however, that in a study of diploid and tetraploid Rye (*Secale cereale*) the polyploid was less able to withstand warmer temperatures (Hall, 1972). Further investigations of the tolerances of diploids and related polyploids are needed.

2. It has also been stressed that the frequency of polyploids may be related to the breeding behaviour (Crawford, 1985). Polyploidy is often associated either with a self-compatible breeding system or seed/vegetative apomixis. At high latitude the growing season is very short and plants with these breeding systems may, because of the higher level of reproductive assurance they provide, be at a selective advantage relative to plants with other breeding systems.

3. To account for the observed differences in the frequency of polyploids at different latitudes, Stebbins (1984) proposed a secondary contact hypothesis. This hypothesis relates to present-day distribution of polyploids with the events of the recent geological past (see Pennington, 1974; Ritchie, 1987; Roberts, 1989). Stebbins writes: 'During the latter half of the Pliocene and beginning of the Pleistocene epochs, a period of 5–6 million years, many plant genera responded to the increasingly harsh conditions along the crests of newly rising mountain ranges as well as at high latitude by evolving races and species, chiefly diploid, having increased tolerance of cold, and in the north, of long arctic nights. With the onset of the Pleistocene glaciations, alpine populations colonized lower altitudes and northern populations moved southwards. Many secondary contacts between different variants were repeatedly established and broken. Hybridization between previously separated populations, accompanied or followed by either polyploidy or introgression at diploid levels, generated new races and species, some of which became adapted to the new conditions prevailing in regions vacated by the ice. These new races and species, that now form the bulk of the arctic-alpine flora, originated during the entire period of a million years or more during which glaciers advanced and retreated, but some of them probably date from the beginning of the final recession, about 10 000 to 14 000 years ago'.

Perhaps Stebbins' secondary contact model might also explain more ancient polyploidy? For instance, the whole of the subfamily Maloideae (Pomoideae) of the Rosaceae is characterised by the basic number $x = 17$, and diploid sexual species of *Sorbus*, for example, have $2n = 2x = 34$. The other subfamilies of Rosaceae contain the basic numbers $x = 8$ and $x = 9$, and it seems very reasonable to speculate on an ancient allopolyploid origin for the Maloideae from $x = 8 + 9$. Similar cases can be found in other flowering plant families, and Manton (1950) has shown that the tropical fern flora contains many cases of 'ancient' polyploidy.

Reverting to the general correlation between polyploidy and latitude, three important points need to be stressed. First, many more chromosome counts are needed world-wide to test Stebbins' ideas, especially from tropical and subtropical regions differing in the extent to which 'intermingling of floras' has been induced by volcanic activity, etc. (Morton, 1966). Secondly, it is by no means an invariable rule that within groups of related taxa in the Northern Hemisphere the diploids are the most southerly in distribution. Indeed, Lewis (1980*a*) gives a number of examples where the reverse is the case. Finally, our interpretations involve the assumption that, within any group, polyploidy is essentially a one-way process, in which diploid or low-polyploid taxa are the parents of the higher allopolyploids. Whilst there is good reason to think of many of the patterns we see in this light, are we right in assuming that such an interpretation is valid for every case? Until recently, writers on evolution of higher plants have assumed that it was; ancient polyploid species, such as those that make up the genus *Equisetum*, certainly look like evolutionary relics that are eventually doomed, and there is no obvious widespread mechanism for descending a polyploid series as there is for ascending it.

How important is polyhaploidy?

Raven & Thompson (1964) drew attention, however, to the possible significance of the occasional phenomenon of 'polyhaploidy' – the production of functional plants of reduced chromosome number (e.g. $2x$ plants from $4x$ individuals, etc.) from polyploids by parthenogenetic development. In a classic example of the significance of polyhaploids, de Wet (1968) and de Wet & Harlan (1969) have described what they call diploid-tetraploid-dihaploid cycles in the *Bothriochloa-Dichanthium* complex. In the words of Jerling & Asker (1992): 'In natural populations, tetraploid facultative apomict plants predominate, but a small part of these populations consists of sexual diploids. By fusion of unreduced gametes the diploids give rise to sexual autotetraploids, allowing gene exchange with apomictic tetraploids. By haploid parthenogenesis, the tetraploids give rise to new diploids (dihaploids), which are often fertile and vigorous'. Polyhaploids may be important in other apomictic groups, for Nogler (1984) detected a number of vigorous and fertile dihaploids in the progeny of hybrids of aposporous *Ranunculus auricomus* plants.

Ornduff (1970) and de Wet (1971) consider the possible evolutionary implications of polyhaploidy. While polyhaploids have been detected in experiments and sometimes in wild populations, it is very difficult to evaluate their evolutionary significance.

The delimitation of taxa within polyploid groups

Polyploids are formed by the addition of like or unlike genomes, and the degree of morphological distinction between presumed ancestral diploids and derived polyploid may be considerable or very slight. All will depend upon the degree of morphological difference between the diploid taxa contributing genomes to the polyploid in question. In experimental polyploids, an increase in chromosome number yields larger nuclei and cells with larger diameters and volumes. The members of a polyploid series may differ in mean pollen and stomatal cell size, and measurement of samples of such cells may be a reliable way to separate plants with different chromosome numbers. In using this method, e.g. on herbarium specimens, it is important to investigate cell sizes in plants whose chromosome number has been determined. While the study of cell sizes has been helpful in distinguishing chromosome races in some polyploid groups, other groups do not show so-called '*gigas*' effects in the plants with higher chromosome numbers (Stebbins, 1950; Davis & Heywood, 1963; Lewis, 1980*a*).

Before we leave the subject of polyploidy, we might consider some of the now classical experimental investigations of species in the light of our present knowledge. The case of *Leucanthemum vulgare* (*Chrysanthemum leucanthemum*), for example, which in Chapter 3 we used to illustrate the biometricians' interest in 'local races', is now known to be complicated by the widespread occurrence of plants with different chromosome numbers which are to some extent morphologically separable on a number of quantitative characters (Favarger & Villard, 1965; Marchi *et al.*, 1983). In a similar way we now know that the kind of difference which Burkill found between Cambridge and Yorkshire populations of *Ranunculus ficaria* (Chapter 3, Table 3.8) is to be found between diploid and tetraploid plants of this common and variable species. Turning to the work of Turesson, we find again that part at least of the variability which he was able to detect in common European species (such as *Achillea millefolium*, *Caltha palustris* and *Dactylis glomerata sensu lato*) is certainly attributable to the occurrence within these Linnaean species of more than one chromosome number. This does not, of course, in any way cast doubt upon his demonstration of ecotypic differentiation; it merely emphasises that species recognised on grounds of morphology are often highly complex entities when studied experimentally. Finally, there is the case of *Erophila verna*, the common variable annual weed which, as we saw in Chapter 2, Jordan studied in such detail. A number of cytologists have studied this group, discovering the following chromosome numbers: $2n$ = 14, 24, 28, 30, 32, 34, 36, 38, 39, 40, 52, 54, 58, 64 and 94 (see Winge, 1940).

Abrupt speciation

So far in this chapter we have discussed abrupt speciation by means of polyploidy. There is now evidence that new species may arise sympatrically by other processes within populations of their parental species. Influenced by the results of experiments with animals, several models of abrupt speciation in plants have been proposed, e.g. involving changes in chromosome number, chromosome repatterning, changes in breeding behaviour and the stabilisation of hybrid derivatives. King (1993) provides a very full account of modes of speciation in animal populations, with some discussion too about plant species and their origin. Levin (1993) has also written an important review of aspects of abrupt sympatric speciation in plants. We now consider a number of examples where processes other than polyploidy have resulted in sympatric speciation.

Changes in chromosome number

We have, so far, considered the idea of a basic number of chromosomes and multiplication of this number in the formation of polyploids. However, sometimes, a genus has more than one basic number. In accounting for different basic numbers in the next few pages, we will show, in outline, how one number may arise from another and how this process represents a mode of speciation quite different from polyploidy. First, a model of how chromosome changes may occur is presented, and then a number of experimental studies are considered. For more complete accounts, see White (1978) and King (1993).

The model is based on the behaviour of centromeres. We have already discussed the role of these structures in nuclear division. With the exception of certain diffuse centromere types (see below), the chromosomes of plants have a defined centromere which, together with the spindle fibres, ensures proper chromosome disjunction at meiosis and mitosis. It is important to stress that centromeres cannot arise *de novo*, but are formed from pre-existing centromeres. Another element of the model is the fact that plants with fewer or more chromosomes than the normal complement may be viable. Many polyploids are tolerant of aneuploidy, a tolerance which may owe its origin to homologies between the 'different' genomes. Some duplication of genetic material is likely, which may cushion the plant against chromosome loss. Loss or gain of chromosomes may be much more damaging in diploids. Loss of a chromosome implies that the plant may have a portion of the normal DNA missing from its genotype and such plants are likely to be inviable. Given the sensitivity of diploids to chromosome loss (and perhaps gain also) changes in the base number requiring, as they do, losses or gains, seem difficult to explain.

However, a simple model suggests a possible mechanism. As we saw in

Chapter 6, chromosomes are composed of the genetically active euchromatin, which appears to be essential for the viability of the organism, and heterochromatin (the role of which is less clear) which is probably less essential. The model supposes that chromosomes composed of heterochromatin may be lost from or added to the genome, with perhaps some effect on viability and fertility, but without the major effect of the loss or gain of a chromosome composed largely of euchromatin. Thus, given a plant with, say, $n = 4$ chromosomes, reciprocal translocations between two of the chromosomes may produce a situation where one of the chromosomes becomes entirely heterochromatic. In gamete formation, the small, derived heterochromatic chromosome may be lost from the chromosome complement or, by misdivision of the chromosomes, may be present more than once. A possible mechanism for the generation of plants with new base numbers is now at hand. Clearly there are enormous hurdles to surmount before gametes with reduced or increased chromosome numbers can give rise to a population with a different basic number. Many aneuploid individuals must fail to survive, but evidence suggests that changes in the basic number have occurred in the evolution of plants and constitute an important class of abrupt changes (Fig. 12.16). Species with changed chromosome numbers may then give rise to polyploid derivatives, based on the 'new' reduced or elevated number.

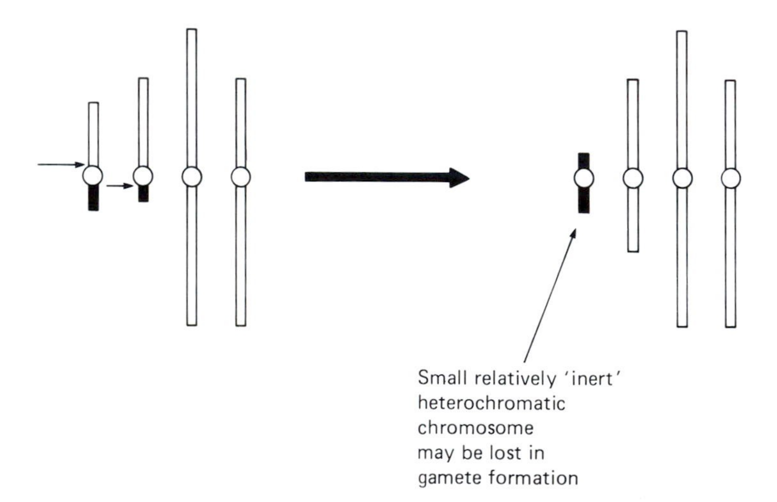

Fig. 12.16. Diagram of translocation, illustrating the possible mode of origin of a new basic chromosome number ($x = 3$) from $x = 4$ by loss of a small heterochromatic new chromosome. (From Moore, 1976.)

An excellent example of an aneuploid change is provided by the studies of Kyhos (1965) on three species of the Composite genus *Chaenactis*. He studied the yellow-flowered *C. glabriuscula* ($2n = 12$), a western North American plant of mesic habitats, and the white-flowered Californian desert species, *C. stevioides* and *C. fremontii* (both $2n = 10$). By examining meiosis in the hybrids, both artificial and natural, between these three species, Kyhos was able to make a careful study of chromosome associations. He deduced that *C. stevioides* and *C. fremontii* had been independently produced from *C. glabriuscula* by processes involving chromosome translocations and loss (Fig. 12.17).

A number of other examples of changes in chromosome number have been reported in the literature, e.g. *Crepis fuliginosa* ($n = 3$) derived from *C. neglecta* ($n = 4$) or its near ancestor (Tobgy, 1943); *Crepis kotschyana* ($n = 4$) derived from an ancestor like *C. foetida* ($n = 5$) (Sherman, 1946); and *Haplopappus gracilis* ($n = 2$) from plants with $n = 4$ (Jackson, 1962, 1965). Further examples of both descending and ascending basic aneuploidy are given in Grant (1981).

Aneuploid reduction may also occur in polyploids; a number of case histories are considered by Grant (1981). In some species both polyploids and aneuploids derived from them are found: for example, in the extreme case of *Claytonia virginica*, there are diploids with $2n = 12$, 14 and 16 and

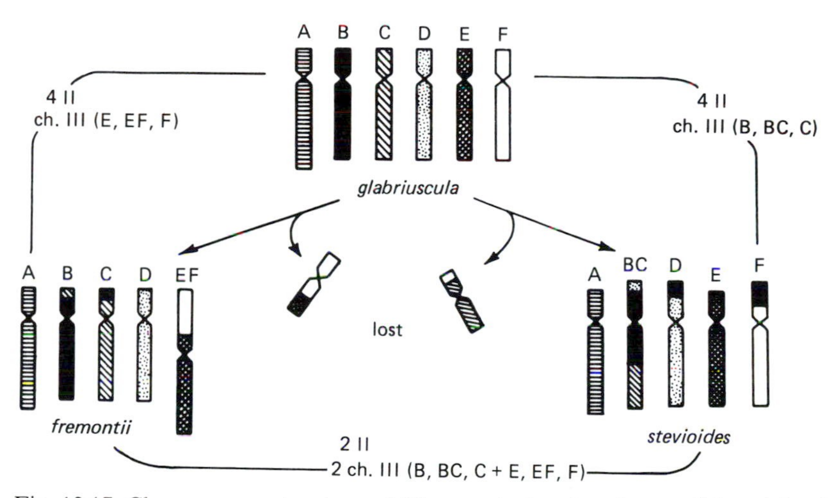

Fig. 12.17. Chromosome structure of *Chaenactis* showing the possible origin of related species by chromosome translocation and loss. (After Kyhos, 1965, from Moore, 1976.)

polyploids with the numbers $2n = 17$ to 37 inclusive, 40, 42, 44, 46, 48, 50, 72, 81, 85, 86, 87, 91, 93, 94, 96, 98, 102, 103, 104, 105, 110, 121, 173, 177 and 191 (Lewis, 1976). Some of the variation in chromosome number may be the result of aneuploidy, but it is possible that allopolyploidy has occurred repeatedly in this group, in which there is said to be relatively little taxonomic variation. Another taxonomic species with a spectacular array of chromosome numbers is *Cardamine pratensis*, with $2n = 16$, 24, 28, 30, 32, 33–37, 38, 40–46, 48, 52–55, 56, 57, 58, 59–63, 64, 67–71, 72 and 73–96 (see Lövkvist, 1956).

In most plant groups a definite centromere is present in each chromosome. As we have seen, centromeres are involved with the spindle fibres in chromosome disjunction, and proper separation of chromosome arms lacking centromeres or chromosome fragments is not possible. In certain groups, however, notably the Juncaceae (e.g. *Luzula* and *Juncus*) and Cyperaceae (e.g. *Carex* and *Scirpus*), centromeric activity is not localised; the plants are said to have diffuse centromeres (Godward, 1985). In plants with this type of centromere, fragmentation and fusion of the chromosomes may occur; such plants may be viable and a whole series of chromosome numbers may be generated thereby (a condition known as agmatoploidy). Even chromosomes of *Luzula* fragmented by radiation can take part in mitosis (La Cour, 1952).

In the large genus *Carex* there is an exceptionally long series of chromosome numbers from $2n = 12$ to $2n = 114$ (Luceño & Castroviejo, 1991), with every number represented between 12 and 43. Both aneuploidy and agmatoploidy are probably involved (Davies, 1955), but, according to Luceño & Castroviejo (1991), no cases of allopolyploidy have yet been detected. Luceño & Castroviejo (1991), studying cytological variation in *C. laevigata* in the Iberian peninsula, have found a wide range of chromosome numbers (from $2n = 69$ to $2n = 80$), which they consider reflect not only chromosome fragmentation but also chromosome fusions.

Similarly, in the genus *Luzula* there is not only evidence of orthodox polyploid series but also Nordenskiøld (1949, 1951, 1956, 1961) has described in the *L. campestris* and *L. spicata* groups how diploid, tetraploid and octoploid races have chromosomes in descending order of size, a situation which is interpreted as being due to chromosome fragmentation, rather than multiplication of chromosome sets. While investigations confirm the notion of the importance of fragmentation, this cannot be the whole story, for plants with the same chromosome numbers have different DNA contents (Barlow & Nevin, 1976).

Chromosome repatterning

Besides changes in chromosome numbers, there is evidence that chromosome rearrangements are of great importance in abrupt speciation in plants. Zoologists have studied the phenomenon much more intensively than botanists, and several authors consider that chromosome changes are paramount in the evolution of animal species (see, for example, White, 1978; Jones, 1978; King, 1993, for a thorough review of the subject). The idea of chromosome repatterning has an interesting history. For instance, Goldschmidt (1940, 1955) favoured models of speciation which involved wholesale repatterning of the karyotype, the expression of genes in their new positions being modified by neighbouring loci. The end point of these changes has sometimes been referred to as 'a hopeful monster'. However, research in various organisms does not support the idea of the tens or hundreds of changes, and the ideas of Goldschmidt fell into disrepute. Recent studies suggest that perhaps one or a few chromosome translocations or inversions may be sufficient to produce a post-zygotic isolating factor between parental and derived taxa. With regard to the frequency of change, the rates of chromosome mutation have been examined in a number of plants (Parker, Wilby & Taylor, 1988). For instance, in *Scilla autumnalis* 5 to 10% of mature plants have unique rearrangements (Ainsworth, Parker & Horton, 1983), and frequent chromosome rearrangements have been detected in *Rumex acetosa* (Parker, Wilby & Taylor, 1988).

For many years Lewis and associates have studied the evolutionary relationships between a number of diploids in the genus *Clarkia* (for a review of these studies, see Lewis, 1973). The results of their work, which are based on the analysis of meiosis in hybrids, are set out in Fig. 12.18. In some cases, the evidence suggests an aneuploid origin of derived taxa with a reduction in chromosome number. However, as in the case of the origin of *C. lingulata* ($n = 9$) from *C. biloba*, ($n = 8$), evidence suggests that chromosome rearrangement may be important, as well as an increase in chromosome number, for hybrids between these two species are sterile. In our discussion of gradual speciation (Chapter 11), we saw that allopatric differentiation often yielded derivatives showing significant genetic divergence, the mean genetic identity between such species being $I = 0.67$. An electrophoretic study of *C. lingulata* and C. *biloba* revealed that progenitor-derivative species had a very high genetic identity ($I = 0.88$), indicating that a newly arisen species can be nearly identical genetically to its progenitor. (See Gottlieb, 1986, in which there is a full review of speciation in *Clarkia*.)

Speciation does not always involve a change in chromosome number. Evidence in support of this mode of speciation has been obtained by

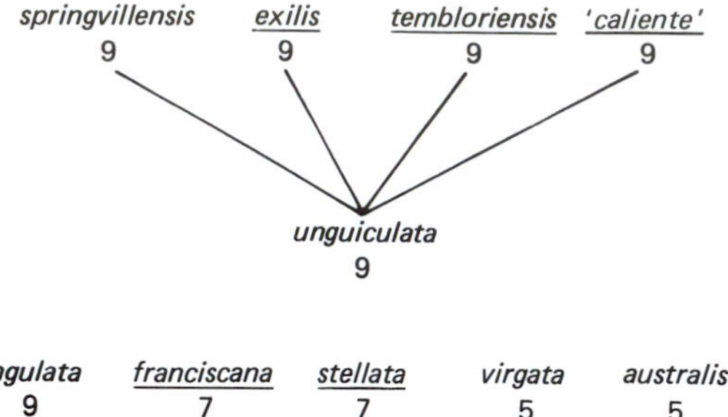

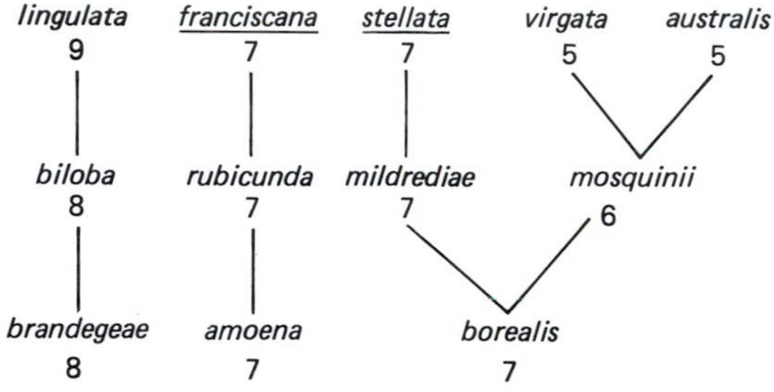

Fig. 12.18. Diagram showing the relationship of species of *Clarkia*. Haploid chromosome numbers are indicated. Predominantly self-pollinating species are underlined; all others are normally outcrossed. (From Lewis, 1973.)

studying a number of pairs of species, one the supposed progenitor and the other the derived species. As judged by isozyme analysis, each of these pairs also had high genetic similarity. For instance, the self-compatible *Stephanomeria malheurensis* arose from the self-incompatible *S. exigua* subsp. *coronaria* (Gottlieb, 1973, 1979), and hybrids between the two species are sterile as a consequence of chromosome rearrangements. Other examples of progenitor-derivative species pairs sharing the same chromosome number are provided by *Gaura longiflora* and *G. demareei* (Gottlieb & Pilz, 1976), and *Lasthenia minor* and *L. maritima* (Crawford, Ornduff & Vasey, 1985).

Speciation following hybridisation

It has been postulated that in some circumstances hybrids between species could be stabilised and become a new species (Grant, 1981). Species isolated by chromosomal differences might, through hybridisation, produce a derivative itself reproductively isolated from both parents. It has been claimed that such derivatives could be the outcome of introgressive hybridisation. Rieseberg & Wendel (1993) have provided a thorough review of suggested models and the available evidence in a number of case histories. For example, the hypothesis that *Stephanomeria diegensis* is derived from *S. exigua* and *S. virgata* has received support from studies of isozymes; markers from both parents are present in the derivative (Gallez & Gottlieb, 1982). It has been proposed that certain species of *Crepis*, *Elymus*, *Gilia*, *Helianthus*, *Iris* and *Salvia*, etc. have evolved by diploid hybridisation (for details, see Rieseberg & Wendel, 1993).

Critical testing of such models is now possible with the development of molecular tools. For instance, Heiser (1949*a*) postulated that a distinct variant of *Helianthus bolanderi* had originated through the introgression of genes from *H. annuus*. However, molecular studies have failed to confirm this suggestion (Rieseberg, Soltis & Palmer, 1988; Rieseberg, Soltis & Soltis, 1988).

In contrast, as we saw in Chapter 11, molecular evidence supports the suggestion, based on morphological characters, that *Iris nelsonii* is the stabilised hybrid derivative between *I. fulva*, *I. hexagona* and *I. brevicaulis*. With regard to its origin and characteristics Arnold & Bennett (1993) write: 'clonal reproduction by rhizomatous habit ... may have facilitated the establishment of *I. nelsonii*', which occupies a different niche from its parents, namely heavily shaded areas with deep water.

Minority disadvantage

The importance of aneuploidy and repatterning as modes of abrupt speciation has been clearly established. However, unless the new derivative occupies an unoccupied specialised niche (as has been found in studies of *Iris nelsonii*), minority disadvantage must also apply, as it did in situations where new polyploids arise. Isolated individuals are at a reproductive disadvantage amongst the more numerous parental plants. How do these variants, produced in the first place as deviant individuals, come to form populations of a new species? Lewis (1973) has speculated on the likely steps involved in the evolution of species of *Clarkia*, calling the rapid process 'saltational speciation' due to 'catastrophic selection'. The outline quoted below is from White (1978), who has condensed and slightly modified the detailed account of Lewis (1973).

> An exceptional drought reduces a normally outcrossing population to very few plants or eliminates it. The few survivors of founders that re-establish the population undergo self-pollination. Extensive chromosomal rearrangements occur; structural heterozygotes are partly sterile. Chance formation of a chromosomally monomorphic population [occurs] from homozygous combinations of rearranged chromosomes. The 'neospecies' is genetically isolated from the parent species by hybrid sterility and because it is self-fertilizing.

Whether the process of saltational speciation, as envisaged in *Clarkia* by Lewis, is a general phenomenon has yet to be determined.

The last decade has seen major advances in our understanding of abrupt speciation. Further progress is certain, particularly if molecular tools are used in conjunction with more traditional approaches to make critical tests of the various models of speciation. It seems highly likely that GISH techniques, which we discussed earlier in the chapter, may play a significant role in such studies. Bennett (1995) has provided an important review of recent progress and future prospects. Not only has GISH provided the means of testing hypotheses concerning the ancestry of polyploids, it has been used in studies of species hybridisation and introgression (see studies of *Festuca/Vulpia* intergeneric hybrids by Bailey *et al.* 1993) and the genomic evolution of sibling species (e.g. *Gibasis* (Commelinaceae): Parokonny *et al.* 1992). Bennett notes that resolving genomes would appear to depend upon differences in non-coding, widely dispersed repeat DNA sequences. Angiosperm groups differ hugely in their genome size, the proportion of repeat sequences being highest in species with large genomes. However, the technique has been successfully applied to species from different taxonomic groups, including those with low genome size. Hence, it seems very likely that the GISH method, and other *in situ* hybridisation techniques using small parts of the genome, have the potential to test a wider range of models of genome and karyotype evolution.

13

The species concept

The 'species concept' in biology, as we saw in Chapter 2, has engaged the attention of most of the famous naturalists in earlier centuries, from Ray through Linnaeus to Darwin himself, and the argument continues today. Oversimplifying greatly, we could say that both Ray and Linnaeus thought that the genus and the species were 'God-given', a part of the natural order which could be named and described; Darwin, on the other hand, impressed by change and variation, saw both as mere convenient abstractions; and the majority of biologists today see the genus as an abstraction but the species as 'real'.

The comment of many logicians and philosophers on this kind of controversy tends to be impatient; they dismiss arguments about the reality of taxa as being conducted naïvely and in the wrong framework of conceptual thought, and point to the universality of the problem. Perhaps we can re-frame the question along the following lines. When we look at nature, are the 'units' we recognise and name already there to be recognised, or have we 'made' them in the process of looking? The 'naïve realist' view is that they *are* all out there; but it does not need much training in the philosophy of science to see how unsatisfactory this view must be. Recognising even an individual tree or dog is a complex mental process, which cannot be independent of our previous experience. How much more so must the recognition of genera and species be 'subjective' in this sense. We *learn* to recognise both genera and species (and, for that matter, families as well); even if we wanted to, we could not begin without the accumulated wisdom and experience of our ancestors.

In recent years the growth of interest in different 'folk taxonomies' – the classifications made, apparently quite independently, by primitive peoples isolated from the main cultural influences – has stimulated this question about the reality of taxa. If it can be shown that the biological units which are named and classified by a particular tribe in an obscure language are roughly equivalent to those recognised in modern biological taxonomy, the case for the 'reality' of taxa is strengthened. Gould (1979) surveys the results of a number of studies of this kind and is particularly impressed by the work of Berlin, Breedlove & Raven (1974, and earlier references cited), who investigated and analysed the plant classifications of the Tzeltal Indians of Mexico. These authors initially interpreted their data as showing relatively poor correspondence between folk names and Linnaean species but, after a more extensive study which recognised unsatisfactory procedures in their earlier work, Berlin reversed his earlier view and agreed that 'there is a growing body of evidence that suggests that the fundamental taxa recognised in folk systematics correspond fairly closely with scientifically known species'. Gould sides with Mayr in quoting the latter: 'species are the product of evolution and not of the human mind'.

We have clearly not heard the last of this kind of argument. It may be that the correspondence between folk names and Linnaean species is especially close for higher animals, less close for angiosperms and relatively poor for 'lower' organisms, whether animals or plants. Common sense dictates that 'any fool can recognise a tiger', but it does not follow that we should recognise, say, the several species of *Hypericum* from each other without explicit training. A difficulty hinted at, but dismissed perhaps too easily, by Gould concerns the nature of any inherent limitations and prejudices, which are present in the human mind and which therefore we all have in common. Do we construct similar hierarchical classifications apparently independently because, to use the modern computer jargon, our brains are all programmed in the same manner? Retreating from this complex, speculative field we now consider the impact of the 'biological species concept' on plant taxonomy.

The biological species concept

As we saw in Chapter 10, Mayr (1942) produced the most often quoted definition of the biological species. Biological species are 'groups of actually or potentially interbreeding natural populations which are reproductively isolated from other such groups'.

Many zoologists have followed Mayr in equating the taxonomic with the biological species in a variety of groups of animals. As we have seen, the biological species concept has also been applied to plants. How far has it been successful? Clearly, the biological species concept has been a most important model, drawing attention to the crucial role of isolating mechanisms in speciation. But some botanists have wished to replace the taxonomic species with biological species. How far has this endeavour been successful? Reviews of the concept are provided by Sokal & Crovello (1970), Raven (1976), Holsinger (1984), Jonsell (1984), Ruse (1987) and King (1993). Mayr (1982) has provided a further detailed account of the concept, but in our view he has not succeeded in silencing his critics, who, as we shall see, point to major problems in applying the concept in botanical taxonomy. Jonsell (1984) regards the concept 'as a hypothesis open to testing'. He asks the very important question: 'has this hypothesis been carefully and unconditionally scrutinised? Or has this model been taken for granted to such an extent that there has been more of a trend to seek its confirmation, to make it fit the scheme, and to regard nonconforming observations as exceptions or anomalies?'.

Detailed analysis of the biological species concept in terms of its practical value in taxonomy has been provided by Sokal & Crovello (1970), who force attention on the impracticability of actually applying the concept in any concrete case. They consider each word or phrase in Mayr's definition:

Groups.

At the outset, setting the limits of the groups to be investigated presents difficulties. How many taxa do we involve in the experimental study? In the case of *Elymus* investigated by Snyder (1951), as we saw in Chapter 11, hybrids between *E. glaucus* and species belonging to other genera such as *Agropyron*, *Hordeum* and *Sitanion* may well have influenced the variation pattern. We have no way of telling, by looking at the taxonomic information before an experimental study is begun, where we should place the limits.

Actual interbreeding.

As it is prudent to check any cases of presumed hybridisation seen in the field, 'interbreeding' is often tested by setting up artificial crossing experiments. Many experiments have been carried out, the results of which have in some cases been set out as crossing polygons. It is clear from our previous comments that such crossing experiments provide information on interbreeding under some particular experimental conditions. How far can generalisations be made about groups from the reproductive behaviour of small samples? How far can they reveal the likely behaviour in the field? Moreover, they do not provide a proper test of whether pre-zygotic mechanisms might operate in nature.

Potentially interbreeding natural populations.

This phrase presents obvious difficulties. How does one proceed if two populations are allopatric? Holsinger (1984), summarising Mayr's views, states that: 'To decide whether two allopatric populations that are somewhat different morphologically are part of the same species ... we assess the degree of difference between closely related non-interbreeding entities that occur sympatrically and determine whether the differences are less or greater than those between allopatric populations. If less, we regard them as the same species. If greater, as different species'. For many botanists this type of reasoning is not a proper basis for testing hypotheses, and 'potential interbreeding' is seen to be an unworkable criterion. In fact, in response to criticism, Mayr removed the words 'actually or potentially' in a later rewording of his definition (Mayr, 1969).

Reproductively isolated from other such groups.

Finally, how is 'reproductive isolation' to be assessed? Inferences have often been made. For example, as we saw in Chapter 12, breeding barriers have been detected between diploids and their related polyploids, and have been inferred in other cases where a polyploid series is found. While it is clear that there may be a minimum of two biological species – a parental diploid being in one biological species, while an allopolyploid derivative is in another – can we be confident that this is the full picture in the absence of any experiments? Clearly, it is possible that each may contain a number of biological species.

In the assessment of reproductive isolation, experiments have often been carried out to supplement the evidence from field and herbarium studies. But, in interpreting crossing experiments, what level of pollen stainability of F_1 and F_2 hybrids indicates an effective barrier to crossing 50%, 5% or some other figure? Clearly, subjective decisions enter into the assessment. It is not at all surprising that some botanists who have produced elaborate 'crossing polygons' when summarising their results have not tried to recognise 'biological species' in their material.

The views of botanical taxonomists

While many zoologists have accepted the biological species concept, botanical authors have retained a healthy scepticism. Stace (1980) is perhaps typical. He writes that 'there have been many attempts to define a species, none totally successful', and proceeds to list four criteria which are used either singly or together by 'most taxonomists'. Only one of these is concerned with interbreeding, and even this has to be qualified at the outset with the phrase 'in sexual taxa'. Like most botanists, Stace recognises it as 'a fact of life' that our species will be 'equivalent only by designation and must therefore be regarded to a considerable degree as convenient categories to which a name can be attached'. He goes further: 'it is not realistic to consider species which are well differentiated on phenetic, genetic and distributional grounds as ideal or normal, and those whose taxonomic recognition poses great difficulty as non-ideal, abnormal or atypical'. This is a clear rejection of all attempts to apply a biological species concept in the taxonomic context. For a well-documented review of the controversy, which comes down uncompromisingly against any single species definition, we recommend a paper by Levin (1979). Heywood (1980) expresses a similar view.

Not all botanists would agree with Stace, Levin and Heywood. Löve over many years argued for a biological species concept to apply to all organisms, and frequently acted upon his principles in, for example, elevating to the rank of species variants which differed cytologically even if their morphological differences were negligible.

With regard to apomictic taxa, even Löve was obliged to recognise that they cannot be dealt with in this way. He says (1962): 'to classify them on the basis of reproductive isolation would lead to a confusion even greater than that created by the morphological-chorological method of study of these groups, since every individual is reproductively isolated from all its relatives'. When Löve wrote this opinion, it was assumed that the concept of biological species did not apply to apomictic plants. However, ideas about apomixis have now changed. As we saw in Chapter 7, it is now generally believed that obligate apomixis is rare and that most agamospermic groups are facultatively apomictic, and thus able to interbreed to some extent. Logically, some apomictic plants fall within the biological species concept, and we can immediately appreciate the complexities involved in trying to define biological species in such groups.

With regard to species definitions in practical taxonomy, our own views are close to those of Stace. Neither in theory nor in practice can we adopt as our definition of species any single criterion or even group of criteria. The taxonomic process provides us with a hierarchical system of categories by means of which we can name, and therefore discuss, our material. It is not reasonable to assume that taxa must be equivalent; nothing in nature looks simple and there is no reason why we should expect it to be simple. Of course we tend to look for simplifying hypotheses, which enable us to understand what were previously independent phenomena, but we should not complain when the phenomena remain diverse.

Without doubt, theoreticians will surely continue to employ the biological species concept in their model-building. But, as the records in databases of current literature reveal, experimentalists are interested in testing hypotheses concerning the patterns and processes involved in speciation. They are not trying to 'perfect' taxonomy by defining biological species to replace the species of the taxonomist.

14

Evolution: some general considerations

In choosing to focus most of our book on the study of the nature and causes of variation within and between species, we have largely excluded two areas of enquiry which the reader might expect us to cover. One of these is the traditional field of comparative morphology, which compares and contrasts the structure of a series of representative 'types' of organisms from the algae to the flowering plants. Most textbooks of botany include a survey of this kind, and at least an outline knowledge of the most important differences between alga, fungus, moss, fern and flowering plant is required of anyone who would claim to be a botanist. We, therefore, feel that we are justified in assuming such knowledge. The second excluded area is closely linked, in an important way, to the first – the study of the evolution of the plant kingdom as a whole, and particularly the relevance of our knowledge of the structure of fossil plants to our understanding of evolution. It is quite impracticable to give any detailed account of palaeobotany here, but as certain important questions about the variation of plants can only usefully be considered in relation to the studies of both modern and fossil structures, we will indicate some of these questions in the hope that the interested reader might examine them further.

The variation of plants can be approached in two quite different ways. We can focus our attention on the static patterns of variation (this, as we have seen, was the traditional manner); or we can trace the variation in time, which is more difficult. This change of emphasis was one of the main results of Darwinism, and the growth of experimental studies of the processes of evolution is the natural development which we have outlined in the main part of this book. We have seen that it is possible to argue, cautiously and tentatively, from the existing, static patterns of plant variation to the dynamics of evolutionary processes and that, within strict limits, an experimental approach is possible to some at least of the key questions in the understanding of evolution. It is now time to summarise briefly the picture we have and to see it in terms of evolution as a whole.

One difficulty is apparent from the outset. By far the greater part of the technical literature on biological evolution is written by zoologists using animal species for their studies and, whilst it is true that the basic principles of genetics apply to plants and animals alike, we have seen that in many important respects a typical 'higher plant' differs from a 'higher animal'. The plant is stationary and more plastic in its morphology, with a localised but more random dispersal of its reproductive structures than the higher animal (Bradshaw, 1972; Walbot, 1996). Given the high incidence of hybridisation and polyploidy in higher plants, the course of botanical evolution is necessarily a special study with its own peculiar complications. In one respect, of course, the study of the evolution of plants is immeasurably simpler than that of animals. The botanist is not involved in those complex and necessarily controversial areas of evolutionary speculation concerning the nature and origin of nervous and mental activity, which culminate in the study of man himself as the product of evolution. We might, however, note recent questioning (e.g. Gould, 1994) about the idea of progressive evolution, which can be seen as a product of Victorian optimism applied to the *Scala Naturae*, which we discussed in Chapter 2. Much of this questioning centres around the growing interest in catastrophic extinction, with subsequent rapid speciation amongst the survivors, and is an increasingly live issue amongst palaeobotanists. Raup (1991) is full of new ideas on the significance of extinction in the evolutionary process. While the evidence for periods of rapid evolution is surely relevant to plants as well as animals, it remains the case that, if botanists differ amongst themselves on philosophical attitudes to the broad questions of human evolution, they do so as laymen rather than as scientists with a special knowledge of plants.

The fossil record

We can now turn to look briefly at some aspects of botanical evolution as a whole. The first of these concerns the evidence from fossil remains. Darwin rightly saw that fossil evidence provides overwhelming support for organic evolution as a historical event – indeed, it is the only reasonably direct evidence we have, or are ever likely to have. It is, nevertheless, a matter of some difficulty to reconstruct the course of evolution, even in short lengths, from the fossil record. The main reason is the difficulty of interpreting the *absence* of any particular kind of organism from the fossil record, in view of the obviously fragmentary nature of that record. Nowhere is this difficulty greater than in tracing the record of the flowering plants as a whole. The apparently sudden origin of the angiosperms is still the mystery which it was to Darwin a century ago. Most of the main kinds of floral specialisation, in terms of which our modern flowering plant families are defined, can be found represented in Cretaceous fossils. After reviewing the evidence, Stewart & Rothwell (1993) conclude 'that rapid diversification, if not the origin, of angiosperms occurred in the Lower Cretaceous'. However, Hughes (1994) states: 'The still unsolved problem of angiosperm origins ... remains ... as wide open as ever. It is currently possible to publish serious papers proposing origins scattered over 150 million years of Triassic to Cretaceous time'.

Broadly speaking, there have been, for many years, two hypotheses about the first angiosperms. Some authors consider that they had small simple unisexual flowers similar to the modern catkin-bearing plants, the Amentiferae (Wettstein, 1907). Others conclude that they had hermaphrodite petal-bearing flowers. The modern Magnoliaceae and related families are thought to be similar to the basic hermaphrodite archetype (Bessey, 1897, 1915) from which all the 'simpler' flower types, including wind-pollinated catkin-like flowers, were derived by reduction and fusion of parts. Recent studies have revealed that fossil pollen typical of the catkin-bearing plants is not present in the initial radiation of the flowering plants (Friis & Crepet, 1987). At first sight this finding might appear to favour the second hypothesis, but a variety of floral variants have been found, and Hickey & Doyle (1977) conclude that the 'woody order Magnoliales should not necessarily be considered archetypic for the angiosperms as a whole, but represents only one of several lines of specialization'. Thus, there is a 'shift away from the concept that the Magnoliales might represent *the* ancestral type from which all other angiosperms evolved' (Stewart & Rothwell, 1993).

The question then arises as to whether there was a single ancestral angiosperm. Some writers seem unwilling to accept a multiple origin for the group (see Stewart & Rothwell, 1993), but others ask the question, is the requirement of monophylesis a reasonable one? One of the most consistent of these has been the Dutch botanist Meeuse (1966, 1987), who stresses the evidence for the gradual acquisition of angiosperm characters in separate lines of gymnosperm evolution, and concludes that modern angiosperms must represent the end-points of several quite different lineages (see also Hill & Crane, 1982; Stewart & Rothwell, 1993). Most recently, Hughes (1994) states boldly:

> The angiosperms are assumed to be of polyphyletic origin until proved otherwise. The main concept of the angiosperms as a monophyletic group depends on the belief

that the curious and interesting double fertilisation and eight-cell female gametophyte was constant in detail throughout 250 000 species and that such a structure could not have evolved twice even in 100 million years. It is extraordinary that such an obstructive piece of conjecture could have been so little challenged and thus could have remained as dogma for a century. It is time that alternative views should in their turn be tested to the full, regardless of which view may prevail in the end.

Hughes (1994) takes the view that 'the most satisfactory way to demonstrate monophyly is surely to assume the opposite as a general case and then steadily prove the single origin of the selected taxon through fossil evidence. Sadly the plain assumption of a monophyletic origin ... is almost automatic among the majority of botanists and paleobotanists'.

Diversification of the angiosperms

Stebbins (1974) provides a speculative account of flowering plant evolution, considering not only flowers and fruits approached from an ecological perspective, but also vegetative habit and form. He asks the question: in what kind of habitats would adaptive radiation proceed most rapidly? His conclusion is that the early angiosperms, which showed rapid diversification in the Cretaceous fossil record, were probably relatively small shrubs inhabiting somewhat intermediate or marginal communities, not, as often pictured, large trees with single, terminal flowers similar to the modern Magnoliaceae. In a review of the radiation of the angiosperms, Crane (1987) considers that early flowering plants were probably weedy herbs or shrubs colonising predominantly open environments. Later, they came to dominate the vegetation of many areas of the world, especially when large canopy-forming trees developed. A number of hypotheses have been proposed to account for the rapid success of angiosperms. As we discussed in some detail in Chapter 7, Whitehouse (1950) proposed that the early angiosperms were self-incompatible, and the enforced outbreeding imposed by this breeding system contributed to the success and rapid diversification of the early angiosperms.

The co-evolution of animal and plant groups is increasingly seen as playing a crucial part in the success of the angiosperms (Nitecki, 1983; Friis, Chaloner & Crane, 1987; Harborne, 1993). Three main areas have been identified. For convenience, we treat each separately, but it is clear that co-evolution may have involved concurrent responses.

First, a very important co-evolutionary area concerns the responses of angiosperms to herbivores. As they came to dominate greater areas of the world in the Cretaceous, flowering plants were subject to predation by an increasing number of herbivores. From a position of early rarity and habitat restriction, evidence suggests that they became the primary producers of biomass in a widening range of habitats. Investigations of properties of many so-called 'secondary chemical compounds' in contemporary plants provide evidence that such chemicals have adaptive value as defence mechanisms against herbivores. A number of models of the co-evolutionary interactions between herbivores and plants have been developed. For instance, some have considered the interaction as an 'arms-race', as successive changes in the chemical defences of plants 'under attack' are followed by insect counteradaptation (Spencer, 1988). We now recognise that the chemical defences of plants are at least as complex and varied as the more traditional structural ones such as basal meristems – that allow recovery after grazing – as well as spines and prickles (see Harborne, 1993, and references cited therein). Also, plant mimicry is now being seen as of adaptive significance in reducing predation by animals (Wiens, 1978). For instance, members of the genus *Lithops* are so well camouflaged in their stony background in African deserts that they are referred to as 'living stones', and Australian Mistletoes of the genus *Amyema* are practically invisible on their hosts as they mimic its foliage (Barlow & Wiens, 1977). Concerning chemical protection, very complex mimetic systems involving secondary compounds have been described (Stowe, 1988). The wild Potato, *Solanum berthaulthii*, has evolved an effective deterrent to various aphids. It produces an alarm pheromone known only from a small group of aphids. At the onset of grazing, the alarm mechanism is triggered, protecting the plant not only from aphid damage, but also from virus infections carried by the aphids (Gibson & Pickett, 1983). Even the inferior ovary – traditionally interpreted as an adaptation to specialist animal pollination – may also be important in the protection of vital structures against insect attack (Grant, 1950).

Secondly, the diversification of angiosperms in the Cretaceous period coincides with the rise of modern insects, the pollinators of many angiosperm flowers. Undoubtedly very powerful selection was exerted by the behaviour of insects, birds and bats visiting flowers, leading to the complex relationships of mutual advantage arising from the foraging for nectar or pollen. Of crucial significance was the evolution of faithful visiting by a particular animal, leading to the precision pollination of plants with their co-evolved specialised flower type. It has been suggested that odour preceded colour as the pollinator attractant in the early angiosperms, and that such odours originated from chemicals protecting against herbivore attack (see van der Pijl, 1960; Pellmyr & Thien, 1986). In these circumstances, evolutionary change could be rapid for both plant and animal visitor (Takhtajan, 1969; Armstrong, Powell & Richards, 1982; Harborne, 1993; Proctor, Yeo & Lack, 1996). Such changes could give rise to adaptive radiation, as different members of a group co-evolved with different pollinators. The classic studies of Grant & Grant (1965) on the diversification in the genus *Phlox* in relation to pollen vectors are illustrated in Fig.7.14. The co-evolution of insects and plants has produced very complex systems, as, for example, in Orchids of the genus *Ophrys*, where the plants have evolved to provide false clues to attract male pollinators, not only by producing mimetic flowers resembling the female of the species, but also specific female sex pheromones to lure the males without providing any reward (see review by Dettner & Liepert, 1994).

Linked with herbivory is a third major area of animal – plant co-evolution, namely the evolution of specialised disseminules – either fleshy or adhesive – dispersed by animals (Wing & Tiffney, 1987; Harborne, 1993).

The interrelationships of animals and plants may be very complex. For instance, at different times and places, ants may act as herbivores, seed dispersers and pollinators (Huxley, 1991). Furthermore, in some cases ants inhabit specialised structures in the stem, leaves or petioles of plants in tropical areas and they may actively prune vines and protect their 'host' from attack by herbivores (Huxley & Cutler, 1991).

In order to give a balanced picture of the important factors in the rise of the angiosperms, we ought to note that some authors are critical of the view that reproductive superiority was the key to their success. Midgley & Bond (1991) consider that plant-animal interactions are over-emphasised, and that the gymnosperms were out-competed by the angiosperms, which had superior growth rates and more rapid cycles of reproduction and seedling establishment.

Microevolution and macroevolution

In our studies of microevolution, we have identified a number of important processes determining patterns of variation in contemporary plants. In a wholly orthodox neo-Darwinian interpretation it might be assumed that as random mutation, recombination, chance events and selection are sufficient to explain microevolution of plants, such processes are also likely to be sufficient to explain the distant past. However, such an assumption may not be justified.

The balance of forces evoked by an orthodox interpretation of evolution has been challenged by Kimura (1983) and others, who have proposed a neutral theory of evolution (see Futuyma, 1986). They assert that nucleotide substitution at the DNA level does not stem from Darwinian selection. Rather, changes are the result of mutation followed by chance fixation of essentially neutral mutations. These may spread through populations by random genetic drift, even though they confer no selective advantage. Such changes may have a major impact in the long-term. In contrast, strict Darwinians maintain that selection operates at all levels including DNA, and are concerned to discover how selection maintains genetic variation in populations. Other biologists are content with the notion that at least some evolutionary changes are driven by random genetic drift rather than natural selection. The debate continues. Cockburn (1991) provides an excellent account of the neutral theory and the discussion on whether genes, molecules, cells, organisms and populations are all 'units of selection', and whether one level is paramount.

Textbooks tend to dismiss or ignore Lamarkian ideas, but an increasing body of evidence points to the inheritance of acquired characters in certain circumstances (see review by Landman, 1991). For instance, it has been suggested that chloroplasts and mitochondria evolved *within* the cells of early organisms. However, various lines of evidence, including molecular investigations, offer support to the hypothesis that mitochondria and chloroplasts represent the descendants of free-living organisms that were assimilated into a primitive non-photosynthetic host and co-existed. The 'acquired characteristics' were then inherited as both organelles contained genetic information (see Lewin, 1992). Many biologists are now considering 'whether chloroplast origins were monophyletic or polyphyletic and ... which alga or algae provided the initial symbiont or symbionts' (Whatley, 1992).

Another phenomenon must be mentioned at this point. In the 1940s, McClintock investigated puzzling 'mutants' in the pigmentation of grains in Maize cobs. Molecular studies of bacteria, yeast, *Drosophila*, as well as Maize, have recently investigated the underlying mechanisms behind these changes. DNA sequences called transposable elements or transposons have been identified that can move from one site to another (Lewin, 1990, 1997). Most transposons can insert themselves anywhere on a chromosome. While it seems possible that transposons may be 'useless, selfish' DNA, sometimes they carry adjacent sequences with them and their presence at a 'new site' may control the activity of genes near or at the insertion site. Unlike the translocations of chromosomes, transposition may not only result in a copy being transferred to another site on the chromosome, but the transposon sequence may also sometimes remain at its original position. It has yet to be discovered how far transposons influence the variability of wild plants. Evidence suggests that transposons may be molecular parasites, but study of DNA sequences suggests that some may have lost their virulence and become beneficial genes. Zeyl & Bell (1996) write: 'Forty years ago, it became widely recognised that the eukaryotic cell has evolved as a community of symbiotic microbes of various kinds. Today, we may be on the brink of a similar realization, the interpretation of major components of the eukaryotic genome as a community of domesticated elements with independent ancestries as parasites'.

In considering whether the present is a sufficient guide to the past, we may mention briefly the complete revolution in our views on the position of continents and oceans, in relation to the origin of the flowering plants. In the past, it was believed that the continents had always been fixed in their present positions. The proposal that the continents were once joined into a supercontinent Pangaea and subsequently drifted to their present positions was made by Wegener in 1912. At first his ideas were ridiculed, but advances in plate tectonics offer overwhelming support for the theory of continental drift. Figure 14.1 shows the beginning of the breakup of Pangaea in the late Triassic, with the development of the Tephys sea separating Laurasia from Gondwana. There has been much speculation about the origin of the angiosperms in the Cretaceous period, by the end of which Gondwana had broken apart to make the separate land masses of South America, Africa and peninsular India. After a lifetime studying the evolution and classification of the flowering plants, Cronquist's (1988) 'gut reaction' is that they arose in Gondwana, but Doyle (1978) states that 'it would be premature to localize the origin of angiosperms on one side of the Tephys sea rather than the other'.

A new area of interest concerns patterns of extinction. Raup (1991) speculates that perhaps 5 to 50 billion species may have existed on the planet since its origin. Perhaps 50 million exist at the present day. There have been several periods of mass extinction of species and higher groups in the past. Did these extinctions occur simply by natural selection, as we envisage it working in contemporary populations? A number of factors may have contributed to mass extinction episodes, e.g. sea-level and climate changes, and major episodes of vulcanism. However, a radical suggestion was made by Alvarez *et al.* (1980), who proposed that the mass extinction at the Cretaceous–Tertiary boundary 65 million years ago was caused by the impact of an extra-terrestrial body, an asteroid of *c.* 10 km diameter, which produced so much atmospheric debris and acid rain that photosynthesis

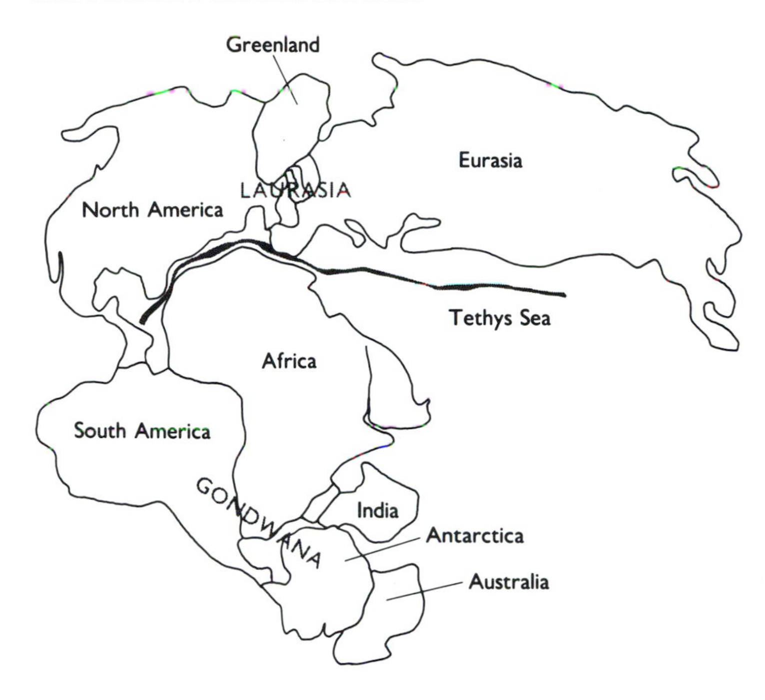

Fig. 14.1. The breakup of Pangaea during the Mesozoic Era. In the Late Triassic, rifting began to separate the northern region of Laurasia from southern Gondwana, and the Tethys Sea began to open between the two areas. During the Jurassic Period, Pangaea was torn in two by rifting, as North America separated from South America and began to break away from Africa. (From Avers, 1989.)

was severely effected both on land and at sea. At the northern edge of the Yucatán peninsula in Mexico, a crater of the appropriate size and age has been discovered. At more than 80 sites around the globe, a layer of clay rich in the element iridium has been found sealed in the rocks of the Cretaceous–Tertiary boundary. Iridium is much more common in asteroids and meteorites than on the earth's surface. Certain groups – dinosaurs, some cephalopods, planktonic foraminifera, etc. – became extinct. Mammals (rat-size or smaller) and dinosaurs had co-existed for 100 million years but, with the extinction of the dinosaurs, mammals came to dominate the vertebrate world (Gould, 1994). There is no evidence for catastrophic destruction of angiosperm groups (Hughes, 1994), but clearly the loss of thc dinosaurs and the further evolution of the mammals must have greatly influenced the course of flowering plant evolution. Considering the implications of this mass extinction, it is clear that some species were in a sense unlucky, being in the wrong place at the wrong time. It is not appropriate to examine the enormous literature that has developed on the subject . The pros and cons are still being debated (Hallam, 1994), and indeed it is being considered whether other mass extinctions may have had a similar cause (see Raup, 1991). Clearly, while there is still much uncertainty, we have to take seriously the possibility of a major force in evolution outside the realms of ordinary experience.

In our brief review we must also consider questions about the power of selection. As Bradshaw (1989) has pointed out it is 'tempting to follow Darwin's lead and to 'see no limit to its power'. However, as we have seen, it has become increasingly clear that selection cannot act unless the appropriate variation is available. Another constraint has also been proposed that may have had very considerable influence on the course of flowering plant diversification. Gould & Lewontin (1979) make the very important point that at any particular point in evolutionary history the *Bauplan* – the way that an organism is constructed – represents a constraint on evolution of that lineage. They also point out that evolutionists often 'atomise' the organism into a number of traits, each of which is explained in terms of a structure optimally designed by natural selection for its particular function; any sub-optimal functioning is then assumed to be due to the balance between competing demands. Each investigation ends with the telling of a plausible story invoking the force of selection. Plausibility is then often the only test of validity invoked by the evolutionist. Gould & Lewontin take the view 'that organisms must be analysed as integrated wholes, with Baupläne so constrained by phyletic heritage, pathways of development and general architecture that the constraints themselves become more interesting and more important in delimiting pathways of change than the selective force that may mediate change when it occurs'. Botanists might reasonably point to a number of writers on plant evolution who have entered similar *caveats* over the years: an excellent example is provided by Manton (1950) in the final chapter of her now classic work on the Pteridophyta. The difficulty with all such views, which, in the extreme case, suggest that in some way the direction of evolution is determined by factors internal to the organism, is that they do not allow any single unified explanation; but, as Gould & Lewontin plausibly argue, we should not be looking for simplicity, rather supporting 'Darwin's own pluralistic approach to identifying the agents of evolutionary change'.

We saw in Chapter 9 that, under certain conditions of strong selection, evolution could proceed so rapidly that we could detect its operation in higher plant populations. If we assume that the main course of evolution has proceeded on the basis of mutation, recombination, chance events and selection, we would then expect that in certain situations evolution, as measured by change in the form of the successive generations, would be rapid. The question arises as to whether continual change is the norm. In his classic hypothetical evolutionary tree published in the *Origin*, Darwin illustrated situations where some lineages continued to diverge over many generations, while others remained relatively unchanged or went extinct (Fig. 14.2). Broadly speaking, this picture is confirmed by the fossil record, for certain modern plant genera appear in essentially the same form as ancient fossils and must have survived as virtually unchanged lineages over many millions of years (Fig. 14.3). With these ancient 'conservative' stocks we can contrast the new species, such as the grass *Spartina* discussed in Chapter 12, which has originated, as it were, in the last second of time on a

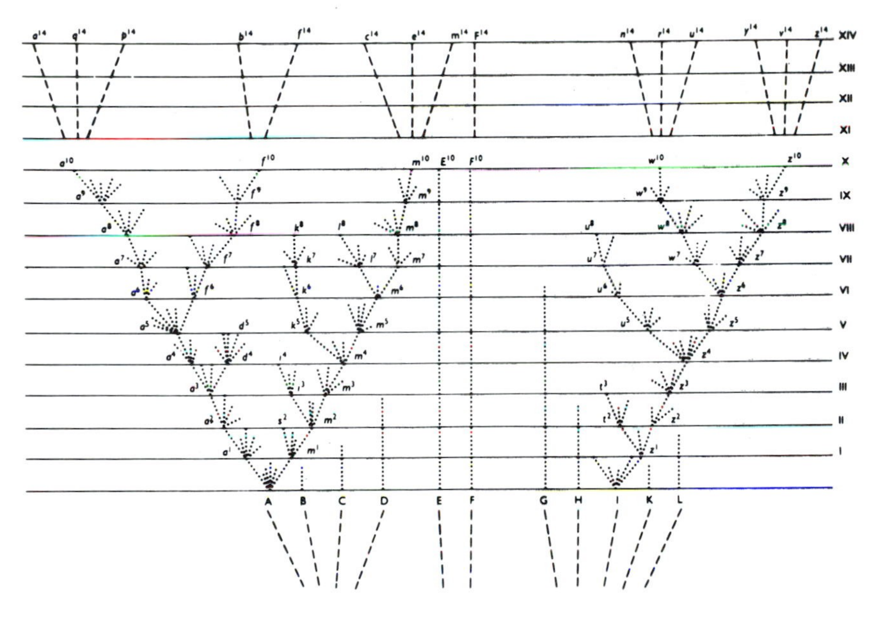

Fig. 14.2. A hypothetical evolutionary tree from Darwin's *On the origin of species*, illustrating descent with modification through natural selection. Eleven different species (A–I, K, L) of a genus (shown at the bottom of the illustration) diverge over time, shown by horizontal lines labelled with Roman numerals and each representing 1000 or more generations. Some of the original species, such as A and I, diverge more than others and produce new varieties, such as a^1, m^1 and z^1. These new varieties in turn diverge and perhaps give rise to new species, such as p^{14}, b^{14} and n^{14}, after thousands of generations. Darwin believed that some species diverge enough to produce new species. Others, such as E and F, remain relatively unchanged or, more likely, go extinct (B, C, D, G, H, K, L). (From Avers, 1989.)

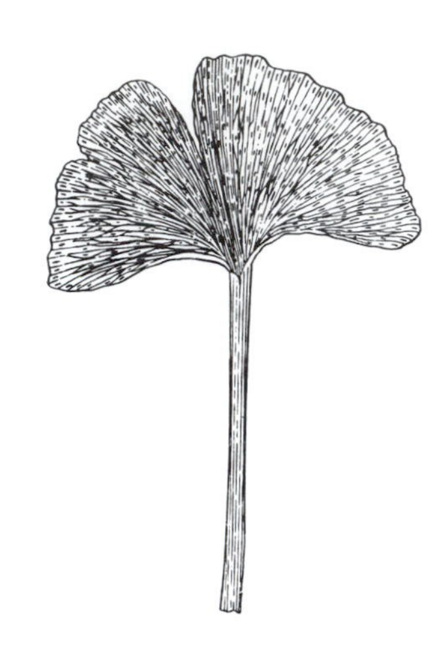

Fig. 14.3. A fossil leaf of *Ginkgo* from Jurassic rock (left) and a leaf of living *G. biloba* (right). Details of the taxonomy and former distribution of *Ginkgo* and related plants are given in Tralau (1968). (Fossil: photograph from Natural History Museum, London.)

geological time scale. Clearly the crucial and as yet unresolved question, both for the experimentalist and the palaeontologist, is the *balance* of gradual and abrupt events (Williamson, 1981). Botanists, to whom the idea of abrupt evolutionary change has never seemed particularly heretical, have always found a place for such views in their speculations on evolution, as the writings of Willis (1922, 1940, 1949), Good (1956) and Lamprecht (1966) bear ample witness. They may therefore be forgiven if they approach with a sense of *déjà vu* arguments between palaeontologists and evolutionists, provoked by Gould & Eldredge (1977, 1993), who have re-awakened the issue of the relative importance of abrupt change in evolution. They challenge the view held by some that continual change is the norm, considering that, as reflected in the fossil record, most species have changed very little or else fluctuated mildly in morphology with no apparent direction. These two polar views are illustrated in Fig. 14.4. While evidence in favour of the punctuated equilibrium model has been found for a number of animal groups (Gould & Eldredge, 1993), little is known about plants. In the light of the fact that the earth has been subject to orbital variation leading to climatic oscillations on ice-age time-scales (Bennett (1997), further studies of plants are needed to estimate the relative importance of gradual and punctuated change – and extinction – on different time-scales.

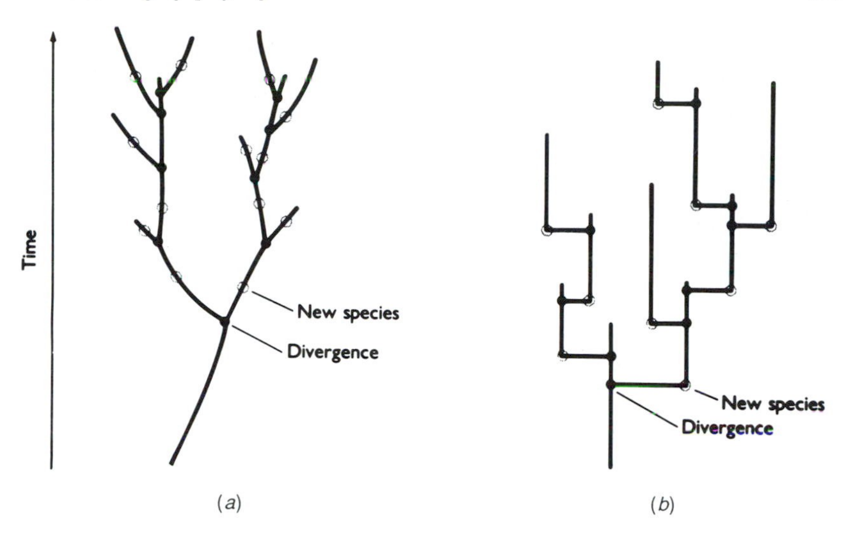

Fig. 14.4. Speciation by gradual divergence (*a*), as compared with the pattern of evolutionary stasis and bursts of speciation postulated by the punctuated equilibrium model (*b*). (From Avers, 1989.)

The devising of phylogenetic trees

At this point it is appropriate to turn to another question. Given that the fossil record of the flowering plants is inadequate, can the evolution of the angiosperms be unravelled by studying living plants? In particular will the cladistic methods, now being used to produce phylogenetic trees from molecular biological data sets, allow us to determine the course of evolution? In order to understand these comparatively new approaches, it is important to understand the context in which they developed. First, we must examine the different sorts of classifications that botanists make, and consider the extent to which they might reveal evolutionary pathways. Then, we trace the development of numerical methods for the study of variation leading to the cladistic approaches important in molecular phylogenetics.

Gilmour (1937, 1940, 1951; Gilmour & Walters, 1963) has stressed that different classifications have been developed for different purposes. Two types of classification may be devised. First, there are the special-purpose classifications, sometimes called artificial classifications, which are based on one or a few characters. Thus, plants may be divided into trees, shrubs and herbs, the characters height and woodiness having been chosen *a priori*, i.e. before the assignment to class was made. As we saw in Chapter 2,

Linnaeus produced a famous artificial classification, his so-called Sexual System being based on the number of parts in the flower (the number and form of the stamens, and the number of pistils). In this classification species of different families are sometimes placed in the same group. The only prediction we can make is that the member of each class are alike in the chosen characteristic. An artificial classification, based on, for example, height and woodiness is, of course, useful in some contexts, e.g. for gardeners, or for ecologists interested in the distribution of different life-forms.

By examining the same set of organisms, a natural classification may be devised, which groups together plants having many attributes in common, to provide a general-purpose classification that serves a much wider range of botanical purposes than the special-purpose one. Gilmour notes that the high information content of such a classification allows many predictions to be made about the properties of members of its constituent groupings. In the post-Linnaean period, natural classifications were increasingly developed using the hierarchical classification of species/genus/family, etc. A good deal of the taxonomic framework we use today is pre-Darwinian. It has been argued that, in the post-Darwinian period, the craft of taxonomy was not affected; biologists continued to fit newly discovered organisms into an increasingly natural system. As before, close study of the material allowed taxonomists to use their expertise and intuition in recognising taxa and placing them within the classificatory system. However, the classification itself was reinterpreted and revised in order to reveal the presumed phylogenetic pathways. Despite the poor fossil record of the flowering plants, for many biologists 'The idea that there is only one true phylogeny is a lure that few can resist' (Sneath, 1988) and, in an attempt to find this 'true' phylogeny, many different arrangements of groups have been proposed at every level of the taxonomic hierarchy. For many, the practice of taxonomy seemed to require its practitioners to produce phylogenetic schemes for every group. Indeed, some took the view that phylogenetic studies were a necessary route to classification. For example, Holttum (1967) wrote: 'If organisms have reached their present state by a process of evolution, it follows that they have a built-in classification, and man's problem is to find it. This is quite a different problem from that presented to the classifier of man-made objects'.

Taxonomists have always been selective in the characters they use, and such selectivity is known as weighting. Weighting has always been practised in making classifications (Davis & Heywood, 1963), but in the post-Darwinian period it often involved characters thought to be important indicators of evolutionary pathways. For example, recognising that floral parts are often less variable than vegetative structures, taxonomists often weighted these characters *a priori* in constructing classifications. Sometimes, after a thorough study of the material some characters – often those of presumed importance in evolution – were given special emphasis; this can be called '*a posteriori*' weighting. Finally, some characters were not used at all. In the post-Darwinian period many classifications purporting to reveal the course of evolution have actually involved one or few characters, and are therefore in Gilmourean terms artificial classifications. Put simply, two positions emerged in the 1940s. They each have their modern adherents. First, there are those who consider that, with increasing information, classifications may be successively modified from the earliest 'alpha-' to a more perfect 'omega-taxonomy' reflecting the way the plants evolved. Such a view was presented by Turrill (1940) in an influential volume of essays published in *The new systematics* edited by Huxley. Turrill's vision of a gradually evolving system of natural classification has proved an attractive idea to many taxonomists. Stace (1989), for example, states that an 'omega-taxonomy' 'is, almost by definition unattainable, but it is the distant goal at which taxonomists should aim'. However, there has been serious criticism of the idea. As we have seen, generations of biologists have shaped a natural classification for organisms, and this broad map of variation is essential for biological science. 'Correct' and stable names are necessary as a means of communication and information retrieval, from books and computer databases, and frequent changes seriously impair the usefulness of the system. These pragmatic considerations were stressed in particular in the work of Gilmour, who pointed out (Gilmour & Gregor, 1939; Gilmour & Heslop-Harrison, 1954; Gilmour & Walters, 1963) that biosystematists and others interested in expressing their views on microevolutionary change could devise and operate separate 'special-purpose' classifications to reveal the patterns that interested them. For this purpose Gilmour and his colleagues devised the '-deme' terminology (see Glossary for details). The very limited acceptance of this special terminology (see Briggs & Block, 1981; Walters, 1989*b*) should not obscure the very important contribution made by Gilmour and his associates. Their insistence that all classifications must be judged in terms of their usefulness, and that the same material can be the subject of different classifications for different purposes, remains of crucial importance. In such thinking there can be no place for a single, 'correct' omega-taxonomy.

The use of computers in taxonomy

In the mid 1960s, the use of computers by biologists increased and 'numerical taxonomy' developed. The following account draws on a number of major reviews of the subject (Sokal & Sneath, 1963; Sneath & Sokal, 1973; Dunn & Everitt, 1982; Stace, 1989; Stuessy, 1990; Pankhurst, 1991). Recognising that natural classifications are of fundamental importance to biologists, the practitioners of numerical taxonomy sought to produce such classifications by using explicit, repeatable and objective methods. Phylogenetic speculation was to be strictly separated from procedures devised to examine relationships, which were to be evaluated purely on the basis of resemblances between living plants, to give so-called 'phenetic' classifications. No weighting of characters was to be allowed. Phylogenetic interpretation of the resulting relationships was not ruled out, but it did not play a part in the procedures used to estimate relationships.

A typical numerical taxonomic investigation involves the following stages (Heywood, 1976). A number of decisions and problems must be faced at every step.

1. A number of specimens (OTUs; operational taxonomic units) are chosen for study. How OTUs are to be chosen has been the subject of much debate.
2. The material is scored for a much wider range of characters than is usual in taxonomic practice. A total of 60 characters would be a minimum number, 80–100 + the preferred range (Stace, 1989). 'New', as well as more conventional characters, are considered (e.g. details of plant and seed surfaces revealed by scanning electron microscopy). There has been a great deal of discussion on what constitutes a character. Numerical taxonomists attempt to use unit characters, i.e. characters that cannot logically be subdivided. Thus, the 'character' leaf shape would be subdivided into many component unit characters such as length, breadth, etc. In scoring the OTUs homologous structures should be examined, and this requirement may cause problems. For instance, petal-like structures in flowers are not all homologous. Comparative study has revealed that the 'petals' in *Anemone* are not homologous with those of *Ranunculus*. In the former they are regarded as modified sepals, but not in the latter. Stace (1989) discusses this and other examples. Stuessy (1990) may be consulted for a discussion of the problems to be faced in trying to distinguish between homologous structures, which are

modifications of the same organ, and analogous structures, which are similar looking structures evolved as a result of natural selection acting on different organs.

3. Binary coding of characters (+ / −) is the next step and this is straightforward in many cases, but problems arise in some circumstances. For instance, plants of different flower colour may be represented in the OTUs under study. It is possible to convert red, white or blue petals into three binary situations red versus not red, white versus not white, etc., but some have pointed out that having three binary choices weights flower colour relative to the other characters (Stace, 1989). For a very thorough review of coding see Sneath & Sokal (1973).
4. A table of data is then prepared (Fig. 14.5) as a prelude to assessing overall similarity, using one of a number of different numerical means of estimation (Fig. 14.6). The results are displayed as a cluster diagram or as a tree-like diagram called a dendrogram (Fig. 14.7). It is important to note that the vertical axis indicates degree of similarity, not time. Taxa possessing the greatest number of shared characters are clustered together: 'none of these [features] is individually either necessary or sufficient to define the group' (Stace, 1989). The different ways of estimating similarity lead to different dendrograms.
5. In making a classification from the resulting dendrogram, numerical taxonomists recommend that ranks appropriate to different levels of similarity are chosen.

Characters	Taxa (OTU's)			
	A	B	C	D
1	+	+	−	NC
2	+	+	+	+
3	+	+	+	−
4	−	+	NC	NC
5	+	+	+	+
6	+	+	−	+
7	+	+	−	NC
8	NC	−	+	+
9	+	+	+	+
10	+	+	+	−
11	+	NC	−	NC
12	+	+	+	−

Fig. 14.5. Coded data table ($t \times n$ table). See text for further explanation. (After Sneath, 1962, from Heywood, 1976.)

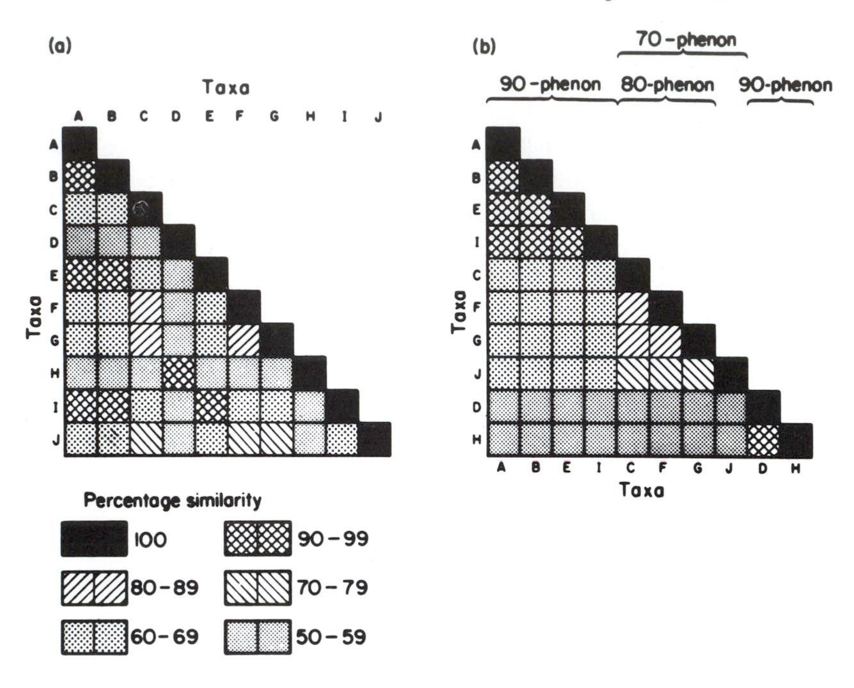

Fig. 14.6. (*a*) Schematic diagram showing a matrix of hypothetical similarity coefficient between pairs of groups (taxa); the magnitude of the coefficients is shown by the depth of shading. (*b*) The same coefficients arranged by placing similar taxa next to each other; this gives a triangle of high similarity values. Phenons are groups of desired rank. (After Sneath, 1962, from Heywood, 1976.)

The influence of numerical taxonomy

A great deal of effort has been expended in developing and refining the different numerical taxonomic methods available for estimating and displaying relationship. Although many research papers have been produced, numerical taxonomy is not a practical alternative to traditional taxonomy. In the past, taxonomists may not have had access to computers; others have proved critical of the methods. However, it seems to us that numerical methods are important in the repertoire of techniques available to taxonomists and, in the future, they may be most useful, for example, where variation within a species or species complex is being studied in detail (Jardine & Sibson, 1971).

While the practical impact of numerical taxonomy may have been slight in many taxonomic institutions, nevertheless, it caused taxonomists to reflect critically on their concepts and working methods, and the relation-

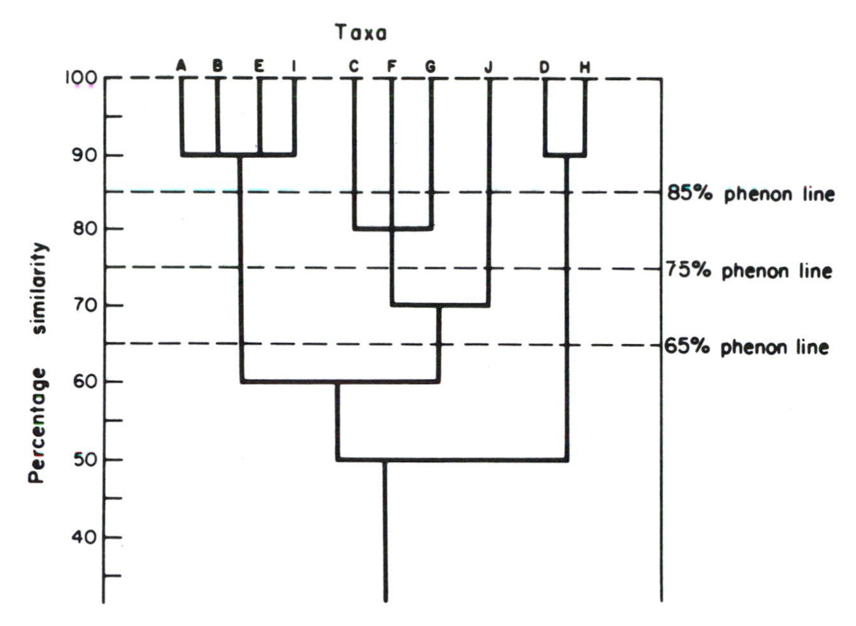

Fig. 14.7. A dendrogram representing the hypothetical hierarchy of groups (taxa), based on the data in Fig. 14.6. The ordinate indicates magnitude of similarity coefficient at which stems join to form higher ranking groups. Horizontal lines delimit groups of equal rank (per cent phenon lines). (After Sneath, 1962, from Heywood, 1976.)

ship between classification and phylogenetics. Also, it had another much more important effect. It introduced many taxonomists to computers, and some of the practitioners of numerical taxonomy were deflected from the strict confines of their original subject to begin, with others, to be interested in modelling phylogenetic pathways.

Cladistics

Cladistic investigations aim to replace the phenetic schemes of the numerical taxonomists, and also the intuitive phylogenetic trees devised by generations of biologists, with branching networks of ancestor/descendant relationships produced by precise models with clear assumptions and procedures (Wiley, 1981; Simpson, 1986; Stuessy, 1990). Some of the theoretical work necessary for the development of cladistics was carried out in different disciplines: for instance, models were devised to minimise the amount of cable needed in telephone systems. In 1950, a book by Hennig was published – *Grundzüge einer Theorie der phylogenetischen*

Systematik – that became the basis for what Cronquist (1987) has called 'a new and messianic school of taxonomic thought, later (Mayr, 1969) named cladism'. Many found the book very difficult, and it was not until an English translation was published in 1966, with the title *Phylogenetic systematics*, that Hennig's views were more widely known. Many biologists, for example the botanists Zimmermann (see Donoghue & Kadereit, 1992) and Wagner, also contributed to the development of cladistics. Stuessy (1990) may be consulted for a historical review of the development of the subject, which has proved to be highly controversial, with acrimonious debates and personal attacks. While some biologists – particularly certain molecular biologists – report a 'growing sense that it will actually be possible to obtain an accurate picture of evolutionary history' (Donoghue & Sandeson, 1992), others are highly sceptical. Here we present a very simplified account of the subject, avoiding as far as possible the specialised terms used by cladists.

Cladists produce trees based on the assumption that it is possible to recognise primitive and advanced character states in homologous structures, including DNA sequences. The aim is to determine monophyletic groups defined by possession of shared derived character states (Fig. 14.8). The models used consider evolution to be an ordered divergent step-wise transformation from primitive to advanced states. For a very full account of cladistic methods, see Simpson (1986), Forey, Humphries & Kitching (1993), Swofford *et al.* (1996) and Hillis, Mable & Moritz (1996).

In a cladistic study, polarity of change in morphology in a group of plants (each member is called an evolutionary unit or EU) is determined, wherever possible, from the fossil record, but it is often necessary to determine such polarities from comparisons of homologous structures within a group of living organisms. As with numerical taxonomic studies, a nagging difficulty must be faced. Are structures truly homologous or merely analogous, i.e. similar functionally, but of different origin? 'Non-homologous similarities (known collectively as *homoplasy*) can occur either by *convergence* ..., in which a similar feature evolves independently in two or more different lineages, or by a *reversal*, in which a derived feature is lost and replaced by the original ancestral condition' (Simpson, 1986).

In addition to the group being studied, one or more related out-groups having what are taken to be primitive character states are included in the study (Fig. 14.8). Whether primitive and advanced characters can be established in this way has been hotly disputed. As we shall see, sometimes in the study of DNA the primitive state can be deduced by studying the genetic fingerprints of a range of taxa. Advanced taxa may be characterised

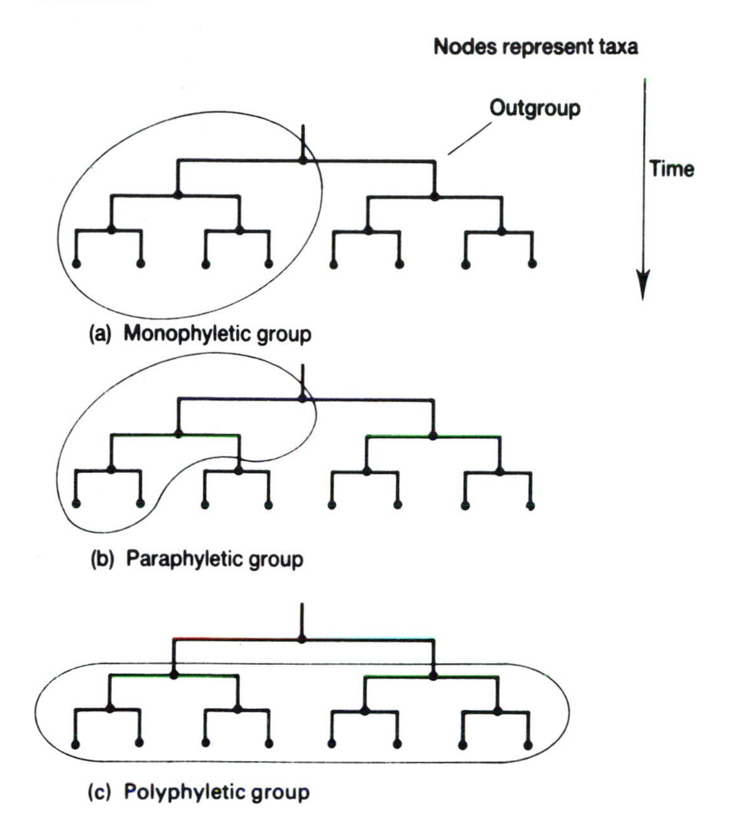

Fig. 14.8. Types of phyletic groups. (*a*) Indicates a monophyletic group in which the descendants are grouped with a common ancestor. (*b*) A paraphyletic group is indicated in which all the descendants of a common ancestor are not included. (*c*) A polyphyletic group is shown which does not include an immediate ancestor. (From Pankhurst, 1991.)

by nucleotide substitutions, inversions or deletions. However, there is increasing evidence that in some cases such characters may be homoplastic (Doyle, Doyle & Palmer, 1995).

A number of different numerical models, technically algorithms, are available to construct the trees from tables of character states of EUs. Very simple trees can be constructed by hand (see Dahlgren & Rasmussen, 1983; Simpson, 1986), but computers are needed for larger data sets. Algorithms differ in their assumptions; for example, some allow reversals of character states. Three general approaches are available (for details see Forey, Humphries & Kitching, 1993). A group of related models use algorithms based on parsimony, i.e. that character changes (and, therefore, evolution) took place by the most direct route, involving the minimum number of evolutionary steps or mutational events. Thus, the analysis can be carried out with Wagner, Farris, Dollo or Camin-Sokal parsimony methods, each of which differs in its assumptions, e.g. whether reversions are possible. Other algorithms, based on character compatibility or maximum likelihood, are also used, but they are too technical to be described here (see Forey, Humphries & Kitching, 1993). Again, there have been disagreements about the techniques of producing trees.

Cladistic methods were developed for analysing morphological character sets of EUs, but in the last few years they have been used to analyse DNA data sets. For example, using 29 restriction enzymes Sytsma & Gottlieb (see Gottlieb, 1986) studied the restriction site variation in chloroplast DNA of the genus *Clarkia* (Onagraceae). They used several algorithms in the construction of trees. The shortest and most parsimonious tree is shown in Fig. 14.9. The figures on the internodes refer to the number of mutations shared by EUs distal to that point on the tree. As with all trees generated by cladistic methods, branch lengths do not indicate the degree of divergence of the groups or a time scale.

Cladistic approaches have also been employed in the examination of the relationships of 'higher' groupings of plants. For example, Jansen *et al.* (1990, 1991) have examined many hypotheses concerning the phylogenetic relationships of tribes of the Compositae (Asteraceae). Representatives of 57 genera representing 15 tribes were studied. Chloroplast DNA from the different samples was treated with 11 restriction enzymes and a tree was constructed, using parsimony methods, on the basis of the cleavage sites. Included in the sample as an out-group were members of the related Barnadesiinae. There is evidence that this group has primitive chloroplast DNA, as all other members of the Compositae have a more 'advanced' structure with a particular inverted segment (22 kilobases in length). Thus, the authors feel confident in using the Barnadesiinae as an out-group. When trees were generated using the Wagner parsimony algorithm, 20 equally most parsimonious trees were produced. As in many cladistic studies, an attempt was made to produce a single consensus tree, using, in this case, Dollo parsimony followed by analysis of the data set with a statistical technique known as the 'bootstrap' method, which produces a single tree from the many equally most parsimonious trees (Fig. 14.10).

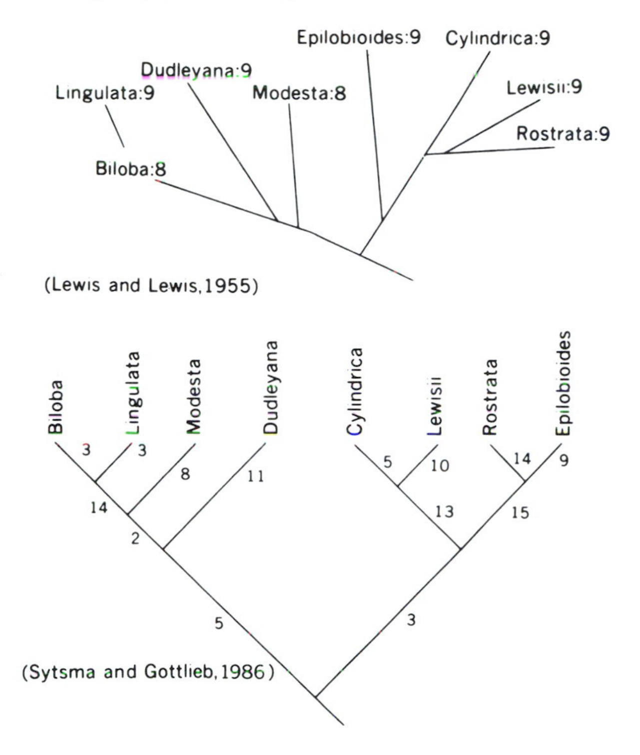

Fig. 14.9. The top diagram shows the phylogenetic relationships and gametic chromosome numbers of the eight species of section *Peripetasma* proposed by Lewis & Lewis (1955), primarily on the basis of morphological evidence. The bottom diagram shows a phylogenetic tree of the same section based on restriction endonuclease analysis of chloroplast DNA. The tree was generated by computer analysis from more than 100 restriction site mutations, and represents the shortest or most parsimonious tree. The numbers refer to mutations assigned to an internode in the tree. For example, *Clarkia epilobioides* and *C. rostrata* share 18 mutations of which 15 belong only to the branch leading to them. From their common ancestor, they differ by 9 and 14 mutations, respectively, and from each other by 23 mutations. (From Gottlieb, 1986.)

A critique of cladistic approaches

Many taxonomists have carried out research in cladistics: Stuessy (1990) and Soltis, Soltis & Doyle (1992) provide surveys of molecular systematics, including reviews of investigations in the Compositae (Asteraceae), Leguminosae (Fabaceae), Onagraceae and Orchidaceae. In a recent review, Sanderson *et al.* (1993) found 1140 papers in 79 journals that contained phylogenetic trees. Recently, many more have been published

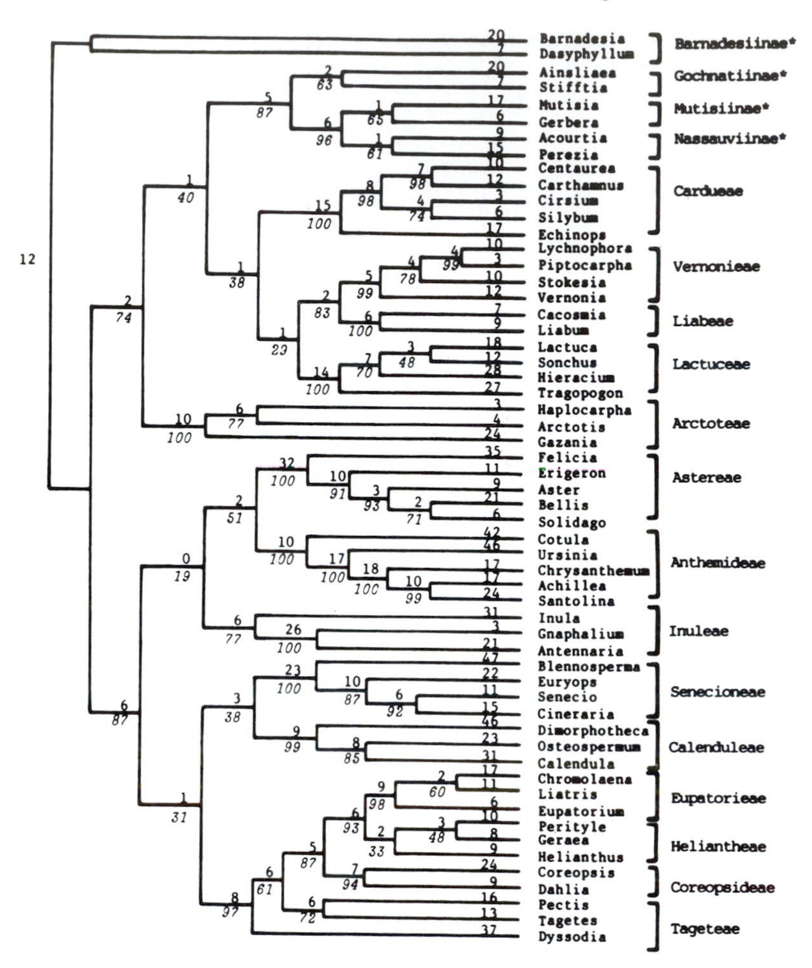

Fig. 14.10. Dollo tree summarising phylogenetic relationships in the Asteraceae (Compositae) using 328 phylogenetically informative chloroplast DNA restriction site mutations. This is a majority-rule consensus tree based on a bootstrap analysis. The numbers above and below the nodes indicate the number of restriction site changes and the number of times that a monophyletic group occurred in 100 bootstrap replicates, respectively. In the bootstrap method data points are sampled randomly, with replacement from the original data set until a new data set containing the original number of observations is obtained. Thus, some data points will not be included at all in a given bootstrap replication, others will be included only once, and still others twice or more from the original data set (Swofford & Olsen, 1990). High bootstrap values indicate parts of the tree found in many of the sets of trees; lower figures point to a higher degree of variability in the set for particular branches. Brackets show the current circumscription of 15 tribes, and the four subtribes of the Mutisieae (*sensu* Cabrera, 1977) are indicated by asterisks. (From Jansen, Michaels & Palmer, 1991.)

(see, for example, Schierwater *et al.*, 1994; Hoch & Stephenson, 1995; Sytsma & Hahn, 1997). Some researchers have studied the higher groupings. For instance, by sequencing the *rbc*L gene of the chloroplast, Chase and associates have attempted to determine the phylogeny of major seed-plant groups (see Baum, 1994). This research and investigations by Doyle, Donoghue & Zimmer (1994), studying both molecular and morphological data, although not conclusive, have shed light on the basal relationships of the flowering plants. Other broad-scale surveys have been attempted. Clark, Zang & Wendel (1995) have carried out a study of grasses using sequences of *ndh*F from the chloroplast genome, and Chase *et al.* (1995) have investigated the superorder Lilianae (which includes the Iridaceae, Orchidaceae, etc.). Darwin considered that carnivory in plants might have developed several times in evolution, and others, such as Croizat, tried to provide evolutionary links between all such plants. Phylogenetic studies of nucleotide sequences in the *rbc*L gene have provided a means of testing such hypotheses. The cladistic analysis indicates that carnivory and the different trap forms have arisen independently in different lineages of angiosperms (Albert, Williams & Chase, 1992). Thus, 'flypaper traps' have five separate origins, while pitcher traps have three. The evolution of breeding behaviour has been the focus for other studies. For example, Graham & Barrett (1995), using molecular cladistic methods, have studied the multiple losses of tristyly in the Pontederiaceae. Molecular systematists have also investigated micro-evolutionary relationships in species complexes. For instance, there is evidence in the *Streptanthus glandulosus* group that some endemic taxa growing on serpentine soils may be 'old', while others may have evolved more recently (Mayer & Soltis, 1994).

Clearly, the enormous growth of molecular systematics has provided many insights. However, a number of authors have pointed to problems and limitations of the subject. Some of the difficulties relate to empirical questions, whilst others are more concerned with conceptual issues, such as the uncertainty, in particular cases, of establishing ancestral and derived status of characters, and the problems caused by convergent and parallel evolution (Stuessy, 1990). Many taxonomists have discussed the difficulties of translating cladograms into 'classifications' (see, for example, Mayr in Duncan & Stuessy, 1985; Cronquist, 1987). Indeed, Darwin himself recognised that a 'genealogy by itself does not give classification' (Hull, 1979). A number of other problems and limitations have been identified in cladistic approaches (see, for example, Cronquist, 1987; Stuessy, 1990). Some are briefly considered here.

Many different trees can be generated from a large data set. Moreover, there are often a number of different equally parsimonious cladograms. Furthermore, while it is possible to test by modern cladistics the intuitive, so-called 'narrative' phylogenetic hypotheses generated by taxonomists, 'one cannot check them against the true phylogenies because these are not available' (Sneath, 1988). Some researchers – the 'transformed cladists' – , recognising the problems, have used cladistic methods as a means of studying relationships, without concerning themselves with evolutionary schemes (Platnick, 1979).

The trees generated by cladists are constrained by the number of samples. An immediate question must be faced: how representative are the samples chosen? Often only a single plant or a few specimens represent a particular group – genus, family, etc. – in an analysis, and often such groups are very diverse containing many species (Olmstead & Palmer, 1994). The selection of one or a small number of samples to represent the whole assumes that a group is monophyletic. How far is this assumption justified? Furthermore, it is not possible to work with very large data sets, for, as the number of EUs increases, so too does the time taken to run the software for the generation of the trees. Indeed, as the size of the data set increases, it is possible to provide only approximations, not a complete 'analysis' (Pankhurst, 1991; Olmstead & Palmer, 1994). Other problems are associated with the documentation of cladistic studies. Donoghue (1994) considers it important to preserve voucher specimens of the material studied, and stresses how vital it is to state the 'assumptions underlying the delimitation of taxa and characters'.

An important question must now be considered. Could cladistically generated trees of molecular data be provided with a time axis? Zuckerkandl & Pauling (1965) suggested that molecular change by nucleotide substitutions might occur at a rate proportional to clock time and, if such substitutions are selectively neutral (and as we have seen this possibility has been strongly debated) and are represented in the genomes of contemporary plants, then it would be possible to model the evolutionary branchings against a time axis. Chloroplast DNA is a conservative molecule because it is usually maternally transmitted and not, therefore, subject to the processes of recombination in meiosis. Thus, it might collect mutations at a stochastically regular rate. However, as molecular studies have proceeded, there is no evidence for a molecular clock: for example, in a comparative study of chloroplast DNA, annual grasses were found to have an eight-fold greater rate of molecular evolution than palms (Wilson, Gaut & Clegg, 1990). Thus, it is becoming clear that, for the determination of chronology, fossil evidence is the only reliable guide (see Hughes, 1994, for an emphatic statement of this view).

The question must now be raised as to whether the 'tree' is an appropriate model for the evolution of plants? Put simply, in cladistic models, a species produces two daughter species by divergent evolution and then becomes extinct. As Stuessy (1990) points out such a model is not likely to reflect the situation in many plants. For instance, a daughter species may evolve as a peripheral isolate, and both it and its parental species may survive. Also, because of the widespread occurrence of species-hybrids and polyploids in the flowering plants, evolution may be reticulate rather than simply divergent. Thus, to model the evolution of many groups, cross-connections between the branches would have to be shown. It has been suggested that hybrids could be excluded from the preliminary analysis. Then, later, they could be placed in their correct positions in the final tree. Such a procedure would appear to be fraught with difficulties. In cladistics parsimony methods are often applied. Pankhurst (1991) makes it clear that 'the choice is simply an expediency in order to make the problem soluble. It is not based on any assumption to the effect that evolution actually proceed[ed] by the most parsimonious or economical path'. This point needs stressing as studies of microevolution have sometimes provided evidence that variants arose by pathways with more than the minimum series of steps (see, for example, the studies of the origin of the pentaploid hybrids in *Holcus mollis* discussed in Chapter 12). Another example is provided by the fern *Asplenium plenum* (Gastony, 1986). Studies of isozymes have revealed that the species is not simply the allopolyploid derivative of the cross between the diploid species *A. cristatum* and *A. abscissum*. On the contrary, the evidence supports a much less parsimonious origin. *A. plenum* would appear to have arisen by a two-step process. A cross between *A. verecundum* ($4x$) × *A. abscissum* ($2x$) yielded a $3x$ hybrid. A triploid gamete from this plant then fused with a reduced gamete from *A. abscissum* to give the tetraploid *A. plenum*. Reflecting on the general issue raised by this and many other studies, Rieseberg (1995) considers that it is important 'to accept the reality of reticulate evolution ... and invest greater energy towards the development of phylogenetic algorithms that consider reticulate phylogeny'.

Next, we should consider the 'weighting' given to data sets from studies of DNA. Some molecular biologists consider that there is 'sufficient' content in DNA alone for the elucidation of phylogenies, and are inclined to see information from this source as more important than any other. This has led to a situation familiar to students of the history of science. The molecular biologists Donoghue & Sanderson (1992) clearly identify the problem and offer a strong recommendation for the future:

> Enthusiasm over a new source of evidence is understandable, as are exaggerated claims on its behalf. But too often in the history of systematics the rising popularity of one source of data takes place at the expense of another, which remains insufficiently explored ... it is felt that the best way to promote new data is to find fault with the old, and what could be better than to claim that the old data are worthless? But rhetoric of this sort, and the fads that it encourages, are unhealthy from the standpoint of our common goal, namely reconstructing the phylogeny of plants ... our efforts to construct phylogeny will be judged by their success in integrating *all* of our observations, which means that more attention should be devoted to combining molecular and morphological evidence.

It should not be anticipated that this will be an easy task, for where morphological and molecular data have been considered, the two data sets may be to an extent irreconcilable (see Kadereit, 1994). Some of these difficulties arise from procedural matters, but Kadereit considers the possibility 'that incongruence may result from the mutation of major morphogenetic genes leading to dramatic morphological divergence unaccompanied by equivalent change of the phylogenetic marker molecule(s) used'. As we saw in Chapter 6, there is evidence of mutations with large phenotypic effect, but Kadereit notes that 'it remains unclear how often large changes have contributed to evolutionary change'. Clearly, the role of macromutations in evolution is still a live issue for molecular taxonomists. It is important to see such debates in their historical perspective for, as we showed in Chapter 5, there were fierce arguments on this very subject between the Mendelians and Darwinians in the early years of the twentieth century.

Finally, we have to confront the central issues of cladistics yet again. In the opinion of Pankhurst (1991), cladistic methods involve 'many subjective decisions, and several major approximations, and do not have generally effective algorithms'. Perhaps radically different algorithms will be developed to analyse the molecular data now becoming available. Meanwhile, in the literature on cladistics, we note a major difference of opinion. Some biologists believe that evolution of the flowering plants will one day be elucidated, and Turrill's (1938, 1940) concept of a perfected taxonomy may then be realised. Other biologists, however, take the contrary view. Stuessy (1990) writes: 'We do not know the true phylogeny for any group of organisms, nor will we ever know it'. Thus, it is important to consider very carefully the relationship between the activities of traditional taxonomists and cladists. Cronquist (1987), in a powerful critique of cladism written before the full flowering of molecular systematics, presents a challenging view:

> If taxonomy is to serve its historical and continuing function as a general-purpose system, then it cannot be held in thrall to debates on arcane matters that bear little if any relationship to putting together the things that are most alike and separating them progressively from things progressively less like. If the participants find such discourse [on cladistics] interesting and mentally rewarding, well and good, but let them then admit they are working towards a special-purpose system that cannot replace the general-purpose taxonomic system.

Donoghue & Catino (1988) have challenged this view: 'Cronquist's critique of cladism illuminates a basic decision that systematists face. On the one hand, we might choose to maintain phenetically defined taxa, even when these are found to be at odds with our best estimates of phylogeny. On the other, we might continually update our system of classification so that it accurately reflects what we know of phylogenetic relationships, even if this means abandoning traditional groups. In our opinion, the first option represents subjectivity and stagnation, while the second offers objectivity and progress'. Recently, Brummitt (1996) has highlighted some of the key issues in the controversy. First, what is the relationship between a cladogram and a classification? Are they one and the same (as suggested for instance by Humphries & Funk, 1984) or different? Brummitt is of the opinion that 'neither a cladogram nor a phylogenetic tree is a classification, and subjective decisions must always be taken to impose the limits of taxa and their rank'. A second question involves the difficult issue of paraphyly. Paraphyletic groups are those which include an ancestor and *some, but not all*, of its descendants (Fig. 14.8). Brummitt notes that: 'In proposing his theory of phylogenetic systematics Willi Hennig argued that no paraphyletic group should be accepted as a formal taxon in a classification'. Cladistic research suggests that many recognised taxonomic groups are paraphyletic. Clearly this has practical consequences. Brummitt argues that: 'If paraphyletic taxa are not allowed, because every descendant taxon has to be of a lower rank than its ancestor and sunk into it, infinite regression occurs and the whole classification telescopes back into its original ancestral taxon'. Further, 'abandonment of Linnaean classification in favour of an attempt at a phylogenetic hierarchy must result in *loss* [author's italics] of ... information and some degree of predictivity, while causing very undesirable taxonomic and nomenclatural upsets'. He holds the view that 'ideally taxonomists should be able to present both a classification, which may allow paraphyletic taxa, and a putative phylogeny.... A phylogeny and a classification are both desirable but have different functions, and should be allowed to exist side by side, interrelated but not interdependent, to give maximum predictivity'.

Clearly, argument is set to continue. In discussing the relationship between phylogenetic studies, classification and practical taxonomy, we hope that the debate will be set in its proper historical perspective. At first sight many biologists may consider that they are faced with a set of 'new issues'. However, as we have shown, argument on essentially the same area has raged many times in the past. For instance, Winsor (1995) has provided an excellent account of the English debate about taxonomy and phylogeny in the period 1937–40. Most botanists 'asserted that taxonomy was a practical matter to be kept distinct from phylogenetic speculation,' but most zoologists insisted that 'taxonomists must strive to represent evolution if they wished to be scientific'.

Transgenic plants

In this book we have discussed many examples of microevolutionary change in man-disturbed habitats. In concluding this chapter on major evolutionary trends, two other themes must be introduced. The first concerns the threat of the loss of a major portion of the earth's biodiversity in the face of man's destruction of habitats, and this will be considered in the next chapter. The second topic, with which we close the present chapter, concerns the development of transgenic plants. It is always hazardous to predict the future, but in the case of biotechnology it is clear that there is the potential for the accelerated evolution of very different new variants of cultivated plants (Tiedje *et al.*, 1989).

Until recently the only route for introducing genes from one crop variety or species to another was through sexual hybridisation, and this was only possible between close relatives, either cultivars, or related species (Raybould & Gray, 1993). Moreover, only genes present in the group could be transferred. Seen in a modern perceptive, crossing to introduce 'new' genes is inefficient. Along with the genetic information for the desired trait, many unwanted genes are also introduced and have to be eliminated by selection in back-crossed progenies – a long, costly and difficult process. Now the molecular technology exists to introduce specific gene sequences (see Watson *et al.*, 1992, for a review of the methods). With the development of biotechnology it is possible, at least in principle, for DNA from any source to be used. Such transfers are made by a number of means, for example, by the bombardment of cells with metal particles coated with the DNA of the desired gene and an associated marker gene, for instance an antibiotic resistance gene. The presence of a marker enables the researcher to detect cells in which the DNA has been successfully transferred. Various traits are being introduced into familiar crops, e.g. herbicide resistance as well as genes for morphological traits, and resistance to pests and diseases. It is possible that, in the future, engineered plants will increasingly be used to produce pharmacological and industrial chemicals.

Many concerns have been expressed about these developments. Questions concerning the possible toxicity of transgenic plants to humans, domestic stock, pets, wildlife, etc. will have to be faced (Raybould & Gray, 1994). Others are more concerned that transgenic plants may themselves become weeds (see, for example, Tiedje *et al.*, 1989; Raybould & Gray, 1993, 1994; Abbott, 1994). They point to the presence of feral Oilseed Rape (*Brassica napus* subsp. *oleifera*) populations in many areas. Some consider that the proportion of introduced organisms that have become pests may be a guide to the potential for transgenic plants to become serious weeds. Thus, considering introduced plants and animals in Britain, Williamson & Brown (1986) discovered that about 10% of introductions become established and roughly 10% of those establishing themselves were generally regarded as pests. Given the huge potential for producing transgenes in a wider and wider range of plant species, it seems possible then that new 'weeds' may arise.

The most immediate concern is whether the genes will escape from transgenic individuals by hybridisation with crops of the same species and weedy relatives. Gene flow between crops and their weedy relatives is well known (see, for example, Langevin, Clay & Grace, 1990; Wilson, Lira & Rodriguez, 1994) and estimates of risk have been made for particular groups. Based on present knowledge it has been predicted that there is a minimal risk in Britain for such groups as Wheat and Cucumber, but a high probability of gene flow is predicted for Sugarbeet, Cabbage, Carrot, etc.

In an attempt to prevent hybridisation between a transgenic crop and untransformed varieties, and between crop and weedy relatives, the effectiveness of surrounding transgenic plants with bare areas and trap crops has been examined by Morris, Kareiva & Raymer (1994). In a particular situation, trap crops proved to be more effective than barren zones. However, if transgenic crops become more widely grown, it may be difficult to organise trap crops in all situations. Whether genes escaping via hybrids will persist in populations will depend upon the fitness of hybrids. In a study of transgenic Oilseed Rape in Britain, there was no evidence that genetic engineering had increased the invasive potential of this crop (Crawley *et al.*, 1993). However, an investigation of *Raphanus sativus* in the

USA has revealed that hybrids between non-transgenic crop and weedy variant had greater or equal fitness when compared with their wild siblings (Klinger & Ellstrand, 1994). These results suggest that, at least in this system, 'neutral or advantageous transgenes introduced into natural populations will tend to persist'.

The consequences of advances in biotechnology are profound. It seems highly likely that transgenic plants will be a major feature in the landscape of the twenty-first century, first as crops – both agricultural and horticultural – and then as garden plants. Recently, genes have been identified for pest resistance and flower characteristics, and these have been engineered into ornamental plants (Geneve, Preece & Merkle, 1996). It is appropriate to mention such plants in a review of evolution in the flowering plants, as these 'designed' plants will represent something quite new in evolution. Also, many of the new transgenic crops are being developed to grow in what are now marginal habitats – such as salted ground – some of which are at present occupied by semi-natural or natural communities. It seems probable, therefore, that further pressures could be placed on rare and endangered species living in areas at present unexploited for agriculture.

15

Conservation: confronting the extinction of species

In the *Origin*, Darwin (1859) considered that extinction was an almost inevitable consequence of evolution. Thus, he wrote: 'The theory of natural selection is grounded on the belief that each new variety and ultimately each new species, is produced and maintained by having some advantage over those with which it comes into competition; and the consequent extinction of the less-favoured forms almost inevitably follows'. Also, he notes: 'No fixed law seems to determine the length of time during which any single species or any single genus endures. There is reason to believe that the extinction of a whole group of species is generally a slower process than their production'. In his day little was known of the process of extinction, for he writes: 'Whenever we can precisely say why this species is more abundant in individuals than that; why this species and not another can be naturalised in a given country; then, and not until then, we may justly feel surprise why we cannot account for the extinction of any particular species or group of species'.

Evolutionists, in general, have paid more attention to the evolution of species than their extinction. However, recently, enormous advances in our understanding of extinction have come through investigating the devastating impact of man on contemporary organisms (Frankel & Soulé, 1981; Primack, 1993; Meffe & Carroll, 1994; Frankel, Brown & Burdon, 1995; Szaro & Johnston, 1996). In this short book it is not possible to consider the subject of extinction in detail, but some key questions will be examined. What are the threats to biodiversity and what proportion of the world's biodiversity is threatened? What factors operating in populations of rare and endangered plants lead to extinction? What are the possibilities of preventing extinction of endangered species by *ex situ* conservation in Botanic Gardens? How may effective nature reserves be created and managed (*in situ* conservation)? What is the future of 'restoration ecology' which aims, by manipulating environments and encouraging or transferring species, to restore damaged habitats or recreate vegetation types? And finally, given that extinction is probably the destiny of all species in geologic time, why should efforts be made to conserve threatened species?

What are the threats to biodiversity?

It is now clear that a significant proportion of the world's contemporary biodiversity is at risk, for it has been estimated that, globally, 10% of the world's flora may be threatened, while on isolated oceanic islands and in those regions of the world with a 'Mediterranean' climate, 25% or more may be endangered (Cody, 1986).

Man is endangering species through: the loss, degradation and fragmentation of natural habitats; the over-exploitation of natural resources; the catastrophic effects of introduced plants and animals, including pests and diseases; and the effects of environmental pollution, resulting not only in localised loss of biodiversity but also in the wider endangerment of plants and animals through, for instance, acid rain and the possibility of global climate change. Furthermore, the likely increase of human populations – from 5.3 billion in 1990 to a projected 6.5 billion by the year 2000 – will lead to increased demand for land, water, and resources of all kinds.

Assessments of the scale of the threat to biodiversity have been published in the scientific literature as well as in the popular press and on television. Some scientists consider that estimates of loss, or potential loss, of species following destruction of natural communities, such as forests, have often been exaggerated (Mann, 1991). For any particular situation, therefore, it is clearly important to consider the strength of the available evidence.

What classes of evidence are available for assessing claims concerning threats of extinction?

1. Important clues may sometimes be obtained by using archaeological techniques or those of the quaternary specialist, which examine vegetation change as it is revealed in pollen and other plant and animal remains preserved in anaerobic deposits of peat, lake sediments, etc. Where such studies have been carried out on datable deposits, they have provided information on forest clearance and the history of agricultural and urban-industrial development of the land in Europe, North America and elsewhere.
2. Settlements attract place names and these, used carefully, often indicate

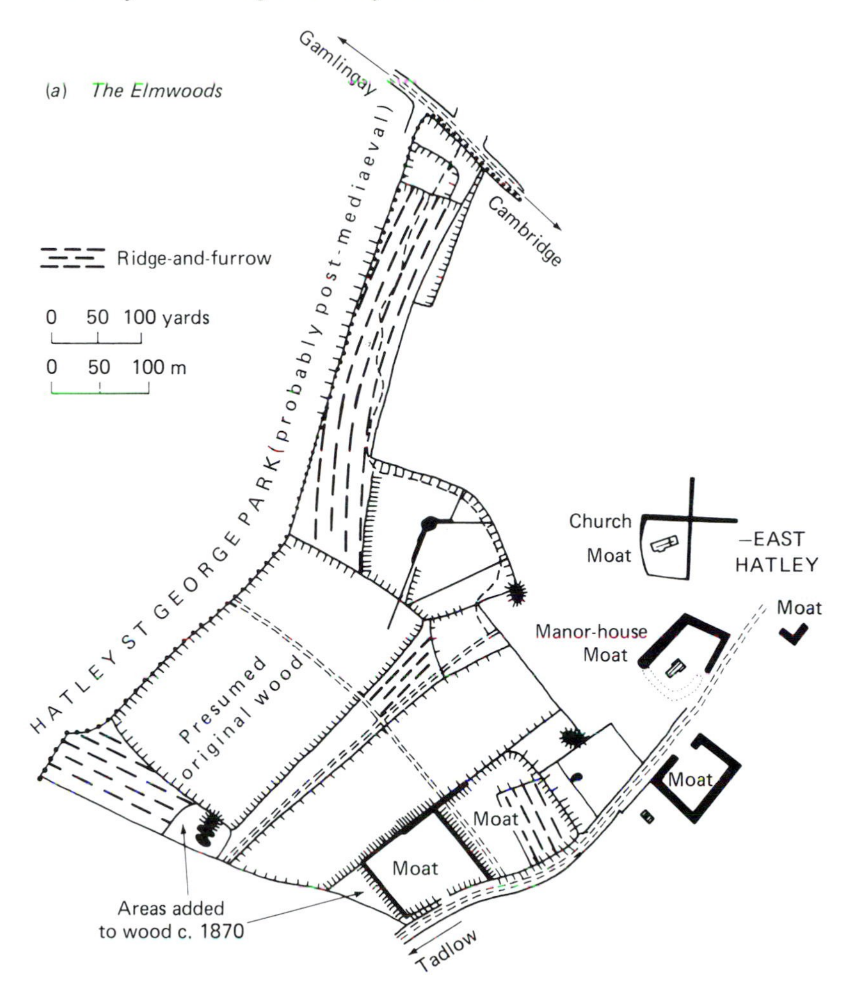

Fig. 15.1. Historical ecology of Buff Wood, Cambridgeshire, England. This small wood of *c.* 16 ha is situated on a calcareous boulder clay plateau west of Cambridge. Its history is complex; there is an original core of presumed ancient woodland surrounded by areas which have been allowed to revert to woodland from agricultural use or after the decline of human settlement in various periods from 1350 onwards. General plan showing historical evidence of earthworks and ridge-and-furrow ploughing. (From Rackham, 1980.)

something of past history, e.g. in Britain those names containing the Old English elements lēah and thveit suggest the clearing of woodland (Mills, 1993).

3. Documents, maps, charts and published evidence of all kinds are crucial for the interpretation of habitat change (Fig 15.1).

4. In the past, there was much disagreement on the extent of deforestation in different areas of the world. Aerial photographs and satellite images have provided very important new lines of evidence for testing hypotheses concerning change, in particular in estimating the losses of moist tropical forests in such places as Brazil (see Meffe & Carroll, 1994).
5. The study of present-day landmarks (see, for example, Fig. 15.1) and the form of plants, (including determining the age of trees by tree rings) may also provide important information on habitat change, with implications for the conservation of rare and endangered species.

The threats induced by changes in land use

Where forest has been cleared and the land settled by man, it is important to stress the obvious point that, while species might be endangered as the natural vegetation is destroyed, subsequent changes in land use may also endanger species. For instance, a number of rare and endangered species may be found in areas subject to traditional agricultural practices. Evidence suggests that natural selection has favoured certain weedy species or ecotypic variants that are able to survive and reproduce in the agricultural landscape, for example, in traditionally managed hay meadows and arable fields. Technological agriculture is replacing traditional practices throughout the world, with profound consequences for many rare species. Thus, traditional hay meadows in the Alps are famous for their floristic richness, and this diversity is threatened in some areas by the sowing of quick-growing grasses followed by fertiliser applications (Briand *et al.*, 1989). Also, throughout the world, the use of herbicides in arable agriculture has endangered many rare weed species (see, for example, the *British red data book*, Perring & Farrell, 1983).

Threats to native biota from introduced plants and animals

It is now recognised that the introduction by man of invasive plants and animals has important consequences for conservation (Cronk & Fuller, 1995; Williamson, 1996). For instance, Huenneke & Thomson (1995) have discussed the possibility that the introduced invasive Eurasian *Dipsacus sylvestris* is endangering the threatened endemic *Cirsium vinaceum* in New Mexico. The Everglades National Park in Florida is being invaded by *Casuarina* species and *Melaleuca quinquenervia* from Australia, and the Brazilian shrub, *Schinus terebinthifolius*. These species colonise disturbed sites faster than the native species, in part due to the diversion of water for agricultural and other purposes, and are very difficult to eradicate (Meffe & Carroll, 1994). The natural 'fynbos' vegetation of the southwestern Cape of South Africa and perhaps as much as 56% of its endangered species are seriously threatened by invasive species of *Acacia*, *Hakea*, *Pinus*, etc. (Cowling, 1992). Also, it has been estimated that *c.* 20% of species of the fynbos are mymecochorous plants, i.e. they provide a food 'reward' on the seed – the elaiosome – and their seeds are dispersed and buried by ants. Many of these species are being threatened by the introduced Argentinean ant species *Iridomyrmex humilis*, which consumes the 'reward' but does not bury the seed, leaving it exposed to predation (Johnson, 1992).

Many species have apparently been accidentally introduced. Clement & Foster (1994) give a number of examples. Seeds of alien species may arrive in ships' ballast and become established near ports and docks. Grain imported by brewers into Britain from Russia and eastern Europe has been found to contain alien species and these may grow on waste dumps near breweries. Wool contains seeds of alien plants and these have been introduced in the vicinity of woollen mills (Hayward & Druce, 1919).

There is also abundant evidence for the deliberate introduction of plant species. For instance, in the eighteenth century the Jardín de Aclimaticíon de la Orotava, Puerto de la Cruz, Tenerife was established to grow plants from overseas to see if they could be used commercially. Also, Mack (1991) points out that in the nineteenth century American seed catalogues offered gardeners a very wide range of species from around the world, and with hindsight we now see that some of these species were destined to become noxious weeds in the USA, e.g. *Eichhornia crassipes* and *Lysimachia nummularia*.

With regard to animal species, domestic animals have been deliberately introduced, and sometimes societies were established in colonial territories with the purpose of releasing wild animals for hunting. Introductions were made under such schemes in Australia (Groves & di Castri, 1991). In total, for various purposes, more than 60 species of vertebrates were released in Australia during 1860–80, and many have become serious pests and a threat to native plants and animals (Myers, 1986).

Not all introduced species flourish in their new environments, and a number of botanists have concerned themselves with the important question: what factors are important in the success of introduced plants (Crawley, 1987)?

1. Studies have revealed that success is more likely if the reception area has a similar climate and vegetation type to the source area. Many studies

support this generalisation: for instance, there have been reciprocal introductions of species between the Mediterranean proper and other areas – South Africa, South Australia, California and Chile – which share a Mediterranean climate (Groves & di Castri, 1991).
2. Generally, but not always, introduced plants invade disturbed habitats.
3. Given that the invader may be introduced as a single or small number of individuals, plants able to reproduce by vegetative means and/or be pollinated by generalist insects or the wind are favoured. Also alien species with a self-compatible breeding system and efficient seed dispersal are the most likely to succeed.
4. One of the most important factors favouring alien species invading new territory is that their pests and diseases have not been introduced. In their absence, new introductions flourish. In an attempt to provide biological control of alien species, the authorities in different countries have often introduced appropriate predators, pests and/or diseases.

The effects of pollution

Man's activities release a wide array of pollutants into the air, soil, and fresh water and marine environments. Vegetation is damaged, some species being more affected than others. In the present context, we stress the importance of testing hypotheses relating pollutant to specific plant damage. Thus, in field studies of damage, the levels of alleged pollutants must be measured. Also, while experimenters often investigate the effect of known quantities of single pollutants on plants, it must be appreciated that many pollutants react with each other, giving a range of primary and derivative toxic or damaging compounds (Klein & Perkins, 1988; Strauss & Mainwaring, 1984).

The widespread and damaging effects of pollution have been examined world-wide. For example, it is claimed that more than 50% of the conifers in Switzerland exhibit slight to severe defoliation, a phenomenon paralleled in the forest trees of Austria, Switzerland and Germany (Briand *et al.*, 1989). Newspaper reports on 'forest decline' in the Alps and elsewhere often blame atmospheric pollution and, indeed, sulphur dioxide (SO_2) pollution, produced by power stations, industrial plants and by domestic installations, damages plants directly. But SO_2 gas also reacts to produce weak sulphuric acid in rain, which not only causes immediate damage to plants, but also produces other effects, such as promoting the release of toxic aluminium ions in the soil. Vehicle emissions also contribute greatly to atmospheric pollution. They are a major source of damaging nitrogen oxides. In sunlight, nitrogen dioxide reacts to produce ozone, and experiments have shown that this gas damages the photosynthetic processes of plants. Thus, conifers exposed to a high dose of ozone in the summer may be more frost sensitive in the autumn. Furthermore, other pollutants, and their interaction products, also injure trees. The 'acid rain', damaging forests and acidifying lakes in many part of the world, is a complex phenomenon, containing not only sulphuric acid, but also hydrogen chloride and nitrogen oxides with their acidic oxidation products (Strauss & Mainwaring, 1984; Mason, 1992). There is evidence, therefore, that pollution may damage trees, but a review of all the factors contributing to 'forest decline' reveals a more complex picture, for climatic stresses and tree diseases may also be important (Klein & Perkins, 1988; Mueller-Dombois, 1988).

Pollutants are also likely to cause long-term effects, and there is broad agreement amongst experts that in the next decades pollution is likely to cause significant global warming, which will have profound implications for endangered species.

The earth is warmed by incoming solar radiation and some of the energy is re-radiated. A proportion of this energy is absorbed by gases in the air (including carbon dioxide; CO_2), while the rest escapes to space. It is possible that the CO_2 content of the air, which has been increasing since the industrial revolution, may double in the next 100 years. Increased CO_2 will result in the trapping of additional re-irradiated energy resulting in global warming, – the so-called 'greenhouse effect'. Other pollutant gases are known to cause the same effect – methane, chlorofluorocarbons (CFCs), carbon monoxide and nitrous oxide – and even if international controls are put in place, these other pollutants are expected to contribute to global warming over the next few decades. Models predict that a rise of + 1.5 to + 4.5°C global mean surface temperature is likely, if CO_2 were to double in the next century and other greenhouse gases increase. Thus, a temperature change is predicted that is 10–100 times faster than the warming following the last Ice Age (Huntley, 1991).

The effects of global warming are not uniform across the earth. The surface warming at high latitudes will be greater than the global average. The larger the warming the greater will be the increase in precipitation. The area of sea ice and seasonal snow cover will diminish (see Schneider, 1993 for a fuller account of the predictions).

Biologists have considered the likely effects of global climate change on plants and animals (see Meffe & Carroll, 1994). It seems probable that plants of restricted range – narrow endemics, specialised species and plants with poor seed/fruit dispersal – may have difficulty in migrating sufficiently to keep within the limits of their climatic tolerance. Furthermore, in montane areas the major community types are likely to move to higher and higher elevation and, therefore, it is possible that many plants of conservation interest, growing at the present time at the highest altitudinal levels, may be lost. In the Arctic, plant communities will be stressed by temperature increases, whilst in coastal areas sea levels will rise, flooding many plant communities containing rare and endangered species. Moreover, the distribution of weeds, pests and diseases may change: for instance, southern species in Europe may spread to higher latitudes.

It is clear from these comments that human activities are threatening many species with extinction. There are questions to be faced, however, concerning the scale of the threat.

How many species are there in the world?

In evaluating the claims of conservationists that 10% or more of the world's flora may be at risk of extinction, we return, yet again, to the species concept of taxonomists, for to estimate the risk of species extinction we must know how many species there are in a particular territory.

In defining species, taxonomists examine the variation in samples of plants, and look for patterns of correlated variation and 'gaps' – absence of intermediates – in the pattern of variation. Thus, the taxa so described reflect, to some degree, the breeding behaviour of the plants. As we have seen in earlier chapters, defining species in some plant genera has proved difficult, in particular those involving groups in which there is much hybridisation, polyploidy and/or apomixis. Thus, the species in different groups are equivalent only by designation. A decision about which taxa are ranked at the level of species is based on skill and judgement of the taxonomist. Judgements are not made in a vacuum, but are influenced by the taxonomic tradition in which the taxonomist works. This may differ in different countries. For instance, some taxonomists do not use the rank of subspecies, and taxa which would be given that rank by other taxonomists are treated as species (see Davis & Heywood, 1963).

Decisions of taxonomists with regard to species are not binding and, in areas where the flora is well known, the groups may have been 'reworked' many times. In contrast, only preliminary taxonomic surveys have been made of many areas and our knowledge of the flora is seriously incomplete. For instance, the clearance of tropical forests, with their rich biodiversity, makes it certain that 'many species are being eradicated before they have been discovered by botanists' (Stace, 1989). Therefore, there is an urgent need to complete a full inventory of the *c.* 250 000 vascular plants found in the world. Also it is necessary to evaluate the status of the one million species names currently in use. Some of these names reflect the attitudes of different taxonomists. Many names reflect the division of large genera into a number of smaller ones. Monographic treatments of groups world-wide are urgently required. Such approaches reveal another reason why there are so many species names. Some are synonyms, a species having been given different names by taxonomists working in isolation in different parts of the species range. The problem of synonyms is very troublesome. For instance, in the preparation of *Flora Europaea*, the total number of binomials considered was five times greater than the number accepted as correct names.

Our knowledge of vascular plants should be seen in the light of the numbers of species in other groups of organisms (see Minelli, 1993; May, 1994). Certain groups are well known. For instance, Diamond (1985), reviewing the situation in birds, discovered that, since 1934, only 134 new bird species have been discovered, resulting in a total of just over 9000 species. In contrast, we are only able to make educated guesses as to the number of species of lower plants and fungi. Amongst animal groups, the situation in insects has proved extremely controversial. Erwin (1982, 1983) injected insecticidal fog into the canopies of *Luehea seemannii* trees in Panama and discovered more than 1100 beetle species in the canopy. Using this figure, he then estimated the total number of beetle and insect species restricted to the *Luehea* trees, the likely number of specialised species per tropical tree species, and finally, on the basis of these estimates, he concluded that there could be as many as 30 million insect species in the world. This figure, an educated guess, has been challenged by some (see May, 1994) as it greatly exceeds the 790 000 species recorded up to 1970. May (1994) echoes the views of Stace, in calling for an inventory of the whole of the world's biodiversity. Such an undertaking has its difficulties. For instance, 80% of insect taxonomists are based in North America or Europe, and the financial support for traditional taxonomy has fallen in many countries (see for example, Small, 1993). However, some progress is being made. For example, a consortium of botanical institutions, Kew, Geneva, Leiden, New York, Missouri, Paris and the Smithsonian (USA) have inaugurated the 'Species Plantarum Project' (Prance, 1992). The aim of the first phase is to produce a computer-based checklist of the current taxonomic and nomenclatural knowledge of the world's plant species, and then, if funds permit, synoptic descriptions and keys for identification will be prepared. The task is urgent, the time limited, and the funding uncertain.

May (1994) clearly articulates the concern of many : 'The clock ticks faster and faster as human numbers continue to grow, and each year 1–2% of the tropical forests are destroyed. Future generations will, I believe, find it incomprehensible that Linnaeus still lags so far behind Newton, and that we continue to devote so little money and effort to understanding and conserving the other forms of life with which we share this planet'.

How many species are threatened with extinction?

In order to assess the conservation status of plant species in a region it is necessary to have information on their distribution and abundance (Morse & Henifin, 1981). Such information is available for well-studied areas but, for many regions of the world, very little is known, information being found scattered in the literature, in unpublished field notes or on the labels of herbarium specimens. Where there are old records or preserved specimens, it is often not known whether the species still exists at a given site. Furthermore, for all classes of data there are often nomenclatural problems.

In order to stimulate the collecting of records, provide a basis for scientific study and make available information for influencing public opinion, records of the distribution and status of rare and endangered species and details of nature reserves have been prepared by many individuals and conservation agencies. Much of this information is held nationally but, to provide global coverage, the World Conservation Monitoring Centre (WCMC) has been set up in Cambridge, UK. Computerised data on more than 80 000 species are stored on databases and new information is constantly being provided by countries around the world. Information on reserves is also held, and there is a computerised Biodiversity Map Library (GIS), which makes it possible to map the distribution of plants and reserves in relation to selected geographical and topographical features. On the WCMC databases, the threatened species are classified using the International Union for the Conservation of Nature (IUCN) categories (Table 15.1), which have been used to produce national red data books of rare and endangered species. Besides the IUCN categories, red data books often give threat numbers (e.g. the *British red data book*, Perring & Farrell, 1983). Species are ranked in relation to past and present distribution, attractiveness, accessibility, how many of the extant sites are in reserves, etc. A scale of threat is thereby produced, the higher the threat number the more endangered the species.

Our brief survey reveals the complexities of determining the number of species endangered in a region and categorising the precise degree of risk.

Table 15.1. *IUCN categories. The categories listed below have been used by conservationists for the last 30 years for a whole range of purposes, including the preparation of red data books, etc. Recently, after much debate, a new listing has been published (Anon., 1994). The aim is to define a number of categories – Extinct, Extinct in the wild, Critically endangered, Endangered, Vulnerable, and Lower risk (including Conservation dependent, Near threatened, and Least concern) – in a more objective fashion by introducing some quantitative measures. These include estimates of present and future: (a) population size; (b) area occupied; and (c) numbers of populations. Estimates of future risk are to be made within a specified time frame. It remains to be seen how far these new categories prove workable given our imperfect knowledge of endangered species and the complexities of clonal growth, seed banks, etc. found in many plants*

EX: Extinct species
E: Endangered species
Taxa in danger of extinction and whose survival is unlikely if the causal factors continue operating. Included here are taxa whose numbers have been reduced to a critical level or whose habitats are so drastically reduced that they are deemed to be in danger of extinction.
V: Vulnerable species
Taxa believed to move into the Endangered category in the near future if the causal factors continue operating. Included are taxa of which most or all the populations are decreasing, or with populations seriously depleted, or those with still abundant populations but under threat from several factors.
R: Rare
Taxa with small populations that are not at present Endangered or Vulnerable, but are at risk.
I: Indeterminate
Species which are Endangered, Vulnerable or Rare but where there is not enough information to say which of these categories is appropriate.
nt: Species which are not now rare or/and threatened

Clearly, the data sets are incomplete and imperfect. Estimates of the number of species currently in the different categories of risk are based on educated guesses.

The data sets held by WCMC allow us to ask the question: how many recorded extinctions of plants have occurred in the past? Smith *et al.*, (1993) have calculated that only 0.3% of vascular plants have become extinct since *c.* 1600. However, looking to the future, a higher percentage (9%) of vascular plants are threatened, with higher figures for gymnosperms (32%) and palms (33%).

Assessing future risk over the whole plant and animal kingdoms, is a problem fraught with difficulty. Many statements are found in the literature, e.g. Myers (1979) has written of an extinction spasm of one million species by the year 2000. Professor E. O. Wilson, the distinguished entomologist and renowned conservationist, addressing a meeting of the American Association for the Advancement of Science in 1995, calculated that, even with cautious assumptions, 'the number of species doomed each year is 27 000. Each day it is 74, and each hour three'.

While many biologists agree with this assessment, others have been critical (e.g. Mann, 1991). The counter-argument may be briefly stated as follows. Studies of the biota on oceanic islands have revealed that large islands contain more species than small islands. Man's activities, in causing the loss of natural habitats, results in the formation of islands of natural vegetation of different sizes in a sea of developed land. Figure 15.2 gives a general model of species area relationships from *The theory of island biogeography* (MacArthur & Wilson, 1967). If the area of original native

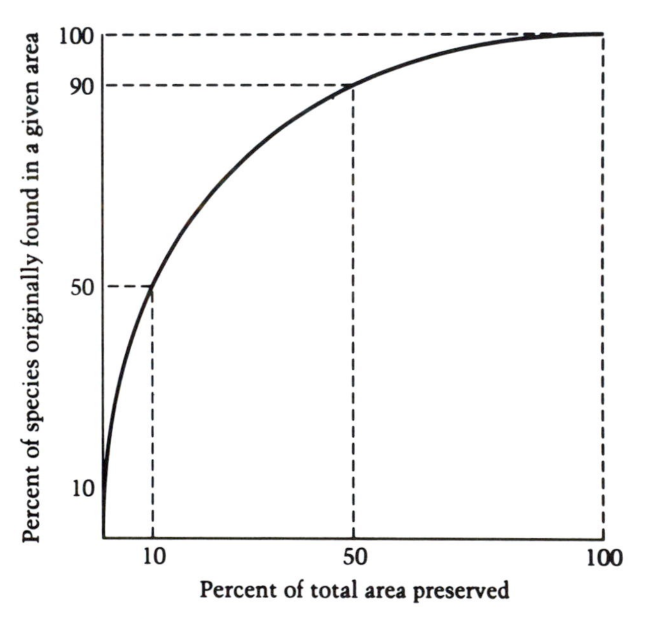

Fig. 15.2. The number of species present in an area increases asymptotically to a maximum value. As a result, if the area of habitat is reduced by 50%, the number of original endemic species going extinct may be 10%; if the habitat is reduced by 90%, the number of endemic species going extinct may be 50%. The shape of the curve is different for each region of the world and each group of species, but it gives a general indication of the impact of habitat destruction on species extinction and the persistence of species in the remaining habitat. (From Primack, 1993.)

habitat is reduced by 50%, then this will result in the extinction of a number of species restricted to that habitat. If the size of the islands of natural habitat is further reduced, then additional extinctions will occur. Mann (1991) criticises the use of such models, posing the question: do mainland islands behave in the same way as oceanic islands? He considers that as species in mainland territories may survive somewhere in the mosaic of derived habitats, a simple relationship between area and species lost must not be assumed. Mann points to the situation in eastern North America. In the colonisation and development of this area by colonial settlers, millions of acres of woodland were destroyed (see Williams, 1989). How many species became extinct in the process? Mann is also concerned about the fact that a very high proportion of the proposed losses in the estimates by Wilson are 'hypothetical species' not yet described by taxonomists. Thus, he ponders whether ecologists could be 'crying wolf' and could stand accused of overstating their case.

So far we have considered the complex question of assessing threats to species and the numbers at risk of extinction. What is abundantly clear is that many are at risk, and we now turn to consider what happens to a species as it is threatened with extinction. While Darwin could say little about mechanisms and time scales, theoretical and experimental investigations have recently transformed our understanding.

Processes involved in the extinction of populations

Plant species occur in populations, and recent studies are beginning to reveal the factors that predispose populations to extinction and point to the action needed to secure their future. It has frequently been noted that small populations have a tendency to become extinct. Many major factors influence population size, including the effects of human activities in damaging and changing habitats. The effect of all these factors has been likened to a vortex (Gilpin & Soulé, 1986). Once a population of an endangered species has become small by a bottleneck effect, problems in seedling establishment and/or reproduction may occur (Fig. 15.3). Decline in numbers in response to these factors is exacerbated by population subdivision and consequent curtailment of gene flow; populations often become totally isolated genetically. Reduction in population size then opens the door, in the longer term, to inbreeding depression and genetic drift that may result in genetic depauperisation and an inability to respond to further changes in the habitat. Thus, there is a tendency for small populations to be driven to extinction.

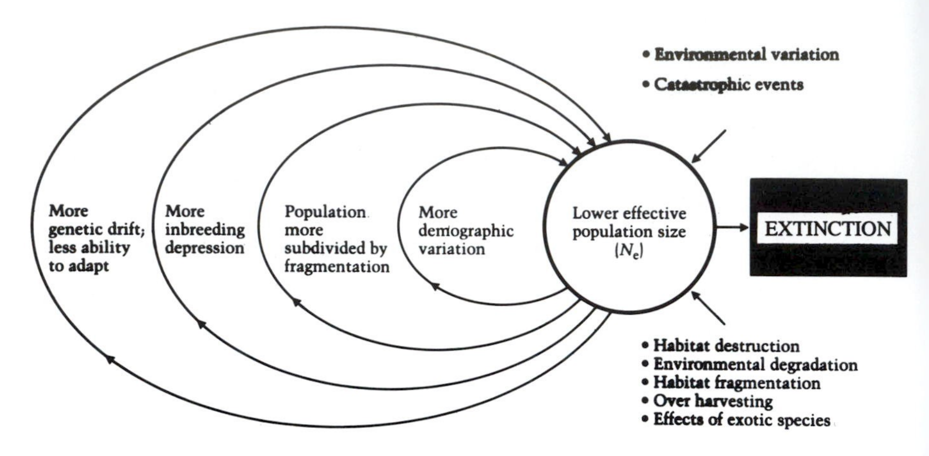

Fig. 15.3. Extinction vortices progressively lower population sizes, leading to local extinctions of species. Once a species enters a vortex, its population size becomes progressively lower, which in turn enhances the negative effects of the vortex. (Adapted from Guerrant, 1992, and Gilpin & Soulé, 1986, from Primack, 1993.)

While there is a good deal of observational and experimental information for some rare and endangered animal populations (Meffe & Carroll, 1994), as yet the interlocking effects of the vortex model have not been investigated in any particular rare or endangered plant species. However, some facets of the model have been examined in different investigations.

Demographic stochasticity

In a healthy population, birth rate is often more or less equivalent to the death rate (Silvertown & Lovett Doust, 1993). As we have seen, it is postulated that small fragmented populations may encounter demographic problems, and support for this hypothesis is offered by a number of investigations. For example, in a study of *Gentiana pneumonanthe* in The Netherlands, Oostermeijer *et al.* (1992) discovered that some populations were dynamic, with a high number of seedlings and low percentage of adult plants. Other populations, however, subject to increased nutrient status, ground water depletion and inappropriate management practices, were at a greater risk of extinction, for they were static or senile, consisting mainly of flowering individuals and hardly any seedlings. Clearly, a census of the number of adult flowering individuals may not be a good guide to the health of a population in the longer term. To understand the conservation status of populations it is important to carry out demographic monitoring of all life cycle stages (Davy & Jeffries, 1981). Such studies have revealed some important complications in monitoring certain rare plants. For example, studies of *Gentiana* (Oostermeijer *et al*., 1992) and certain Orchid species (e.g. *Ophrys sphegodes*: Hutchings, 1989) have revealed that individuals may remain underground in a dormant state for one or more years.

In the study of the demography of plant populations, the presence of persistent banks of seed in the soil is a complication found in some species. The above-ground population may become very small or even disappear, but because there is a persistent bank of seed in the soil, the population may reappear should the right conditions prevail. For instance, the Fen Violet, *Viola persicifolia*, lost to Wicken Fen, Cambridgeshire for the last 60 years, was 'rediscovered' in 1980 when a seedling was produced in a sample of fen soil taken from under scrub and placed in an unheated greenhouse (Rowell, Walters & Harvey, 1982). Subsequent investigations revealed the presence of a large population in a different part of the Fen, where it had not been detected in earlier visits. It is significant that scrub had recently been removed from this area and the surface soil disturbed (Rowell, 1984). It seems very likely that this population also arose from seed buried in the soil. In the past, when the Fen Violet was more abundant, large areas of Wicken Fen were dug for peat. Rowell argues that the key to the conservation of this species is to ensure that the soil is periodically disturbed. Two important points emerge from this example. It is clear that conservationists may declare a species to be extinct locally when invisible populations of seed may still be present. The converse is also true. Conservationists may cling to the erroneous belief that, where a rare species has disappeared, viable seed of the species may still be found in the soil. Clearly, such an assumption should be tested in any attempt to manage an endangered species.

In small populations there is also evidence that the reproduction of the plants may be adversely affected. For instance, Jennersten (1988) studied two populations of *Dianthus deltoides* in an area of South Sweden. Site A contained a large, more or less continuous population in a 1 hectare meadow surrounded by grasslands and forest. Site B, on the other hand, contained a much smaller population in two habitat fragments in a sea of arable fields. Site B had fewer flowering individuals and seed set was lower. A thorough investigation revealed that the number of ovules per flower did not differ between sites A and B and, at neither site, did development of seeds appear to be resource-limited. However, fewer insects visited the plants in site B and, as hand pollination of flowers increased the level of seed set in site B but not site A, it was concluded that low seed set was explained by the low level of pollination in the small fragmented population.

Studies of the rare perennial herb *Tephroseris integrifolia* (*Senecio integrifolius*) in Sweden also revealed that in small populations reproduction fails through lack of insect visits, but the investigation also suggested that population size is not the only factor, the density of flowering individuals being also important (Widén, 1993).

Effects of fragmentation

The vortex model suggests the great importance of population fragmentation in endangering species, by reducing population size, and potentially disrupting patterns of gene flow and demographic functioning. The time-scale of such effects will depend upon the longevity of the plants concerned. For instance, Turner *et al.* (1994) studied plant extinctions in the Republic of Singapore. Forest clearance gave rise to fragmentation in the last century. Many plants are long-lived and population turnover is very slow, and it may take decades or centuries for extinction to occur in long-lived species in remnant forest patches. In contrast, short-lived plants may become extinct quite quickly. For example, the epiphytic flora has declined with the reduction in the numbers of big old trees and the drier microclimate that accompanies fragmentation. Also, for certain epiphytic Orchids, already under pressure from habitat change, the final *coup de grâce* has sometimes been delivered by plant hunters.

Genetics of small populations

The vortex model predicts that small populations may become genetically depauperate. A number of studies of rare and endangered plants have examined variation in isozymes (Young, Boyle & Brown, 1996). No variation was detected in electrophoretic patterns of isozymes in populations of the endemic streamside species *Pedicularis furbishiae* of the St John Valley, Maine, USA (Waller, O'Malley & Gawler, 1987). Recently, DNA markers have been employed to study variation in rare species: for example, low variation was detected in the endangered island endemic *Malacothamnus fasciculatus* var. *nesioticus* of Santa Cruz Island, California. (Swensen *et al.*, 1995) and in Swedish populations of the rare species *Vicia pisiformis* (Gustafsson & Gustafsson, 1994).

With regard to the vortex model, the most informative of the studies of rare plants compare variation in small and large populations to test the hypothesis that genetic erosion occurs in populations with small numbers of individuals. Evidence consistent with this hypothesis has been detected in studies of variation in isozymes: in the New Zealand endemic *Halocarpus bidwillii* (Billington, 1991); *Salvia pratensis* and *Scabiosa columbaria* in The Netherlands (van Treuren *et al.*, 1991); *Eucalyptus albens* in southwestern Australia (Prober & Brown, 1994); and, to some degree, in studies of *Gentiana pneumonanthe* in The Netherlands (Raijmann *et al.*, 1994). However, Ellstrand & Elam (1993) in a review, note that in some cases small populations had the same level of genetic variation as large populations. In these cases, historical factors may have been more important; in particular the number of years since the bottleneck effect, in relation to the length of the life cycle of the plants.

While many investigations have examined variation in isozymes, a few studies have considered the question: is there evidence for the erosion of morphological traits in small populations? For instance, Ouborg & van Treuren (1995) collected seeds from two small and two large populations of *Salvia pratensis* in The Netherlands. These were grown in a common garden experiment. They studied a wide range of characters and discovered differences between populations, none of which were correlated with population size. They concluded that while there was some evidence for a decrease in isozyme diversity in the smaller populations, the genetic variation underlying morphological traits of consequence to fitness had not been affected and the small populations were likely to be at an early stage of genetic erosion. Similar conclusions were drawn from studies of large and small populations of *Scabiosa columbaria* (van Treuren *et al.*, 1993) and *Lychnis flos-cuculi* (Hauser & Loeschcke, 1994).

Evidence for loss of variability in small populations has also come from studies of breeding systems. For example, Demauro (1993, 1994) discovered that in the Great Lakes endemic *Hymenoxys acaulis* var. *glabra*, a remnant population in Illinois, produced no seed. Crossing experiments with plants from other sites revealed that plants of the Illinois population were self-incompatible and of the same mating type, and, thus, prevented from reproducing by seed. The author concludes: 'small populations of self-incompatible species are vulnerable to extinction if the number of self-incompatibly genes, either as a result of a bottleneck or of genetic drift, falls below the number needed for the breeding system to function.' However, investigations of populations of *Aster furcatus*, a rare, primarily self-incompatible species from Wisconsin, USA, have revealed that another outcome is possible (Reinartz & Les, 1994). Small populations of this species were found to have a small amount of variation at isozyme loci, and seed set proved to be limited by low numbers of *S* alleles. However, in some populations, self-compatible individuals were detected and, as they are able to reproduce by seed, it is very likely that such plants will be at a selective advantage and their numbers increase in small populations.

Because the level of inbreeding increases as the number of reproducing individuals decreases (Falconer, 1981), inbreeding depression is another genetic factor to be considered in small populations. A number of investigations have recently attempted to study the difficult question of the degree and effects of inbreeding in populations of rare species. For instance van Treuren *et al.* (1993) grew seed samples of *Scabiosa columbaria* from small and large populations in The Netherlands, and examined progenies obtained by self-fertilisation, within population crosses and between-population crosses. The performance of these progenies was followed throughout their life cycle. On average, the within-population progeny showed a four-fold advantage and the between-population crosses almost a ten-fold advantage over selfed progeny, indicating that the species is highly susceptible to inbreeding. Three points emerge from these findings. First, the enhanced fitness of between population progenies point to the importance, in conservation management, of improving genetic exchange between small isolated populations. Secondly, crosses within small populations may be between close relatives and such crosses may lead to inbreeding depression, albeit less severe than repeated selfing. This may explain the difference between the within-population and the between population progenies. Thirdly, it has been suggested that inbreeding for several generations in small populations might purge the genetic load and the difference between selfed and outcrossed progenies might then be less in small compared to large populations. The fact that van Treuren *et al.* (1993) found no clear relationship between population size and level of inbreeding depression suggests that the genetic load has not been substantially reduced in the smaller populations. Thus, it seems possible that the small *Scabiosa* populations studied have only comparatively recently been reduced in size. Clearly the history of the population will effect the level of inbreeding depression detected in experiments. Ouborg & van Treuren (1995) may be consulted for details of experiments on other species with small populations.

Minimum viable populations

While there is support for the vortex model from isolated pieces of research on different species, a thorough study has yet to be made of all the interlocking factors contributing to the endangerment of particular species. Even though our understanding remains incomplete, it is clear from these studies, and those of many animals, that, if the future of endangered species is to be secured, conservationists must pay attention to the size of populations, in particular the number of individuals actually reproducing – the so-called effective population size (Ne). In a population with a large number of non-breeding individuals, Ne is smaller, perhaps a great deal smaller, than the census population number (N). How large should the effective population be to ensure population survival? This question was first considered by zoologists, who formulated the concept of the Minimum Viable Population (MVP), which led to the 50/500 rule for animal populations (Franklin, 1980; Soulé, 1980). Thus, it was suggested that an effective population size of 50 was necessary in the short-term to protect against inbreeding depression; while an effective population size of 500 was needed to counter genetic drift and to provide the population with a capacity to evolve in the longer term.

The general application of the 50/500 rule has been questioned (Lande, 1988), for it is clear that species differ radically, and each needs a separate individual assessment or Population Viability Analysis (PVA). This has led to a modified concept of MVP which incorporates a probabilistic element (Shaffer, 1981): '[A] minimum viable population for any given species in any given habitat is the smallest isolated population having a 99% chance of survival for 1000 years'. Survival probabilities may be adjusted in the model to give different probabilities and for different time frames, for say 95% and 500 years (Primack, 1993). Having arrived at an estimate of an MVP, zoologists have then calculated a Minimum Critical (or Dynamic) Area necessary for such a population. Then, if necessary, active management efforts could reverse the decline in the habitat and provide sufficient area for the species to flourish, taking into account other organisms, whose activities or presence are essential for the survival of the species in question.

Clearly, these ideas, developed for animal populations, could be very important in the conservation of plants (Fig 15.4). However, it has to be acknowledged that plants are not the same as animals. The conservation strategies for endangered plant species must take account, where appropriate, of persistent seed banks, clonal growth, and the great variety of plant breeding systems, including self-incompatibility, dioecy, gynodioecy, apomixis, etc. Thus, we emphasise again the problems faced by small populations of self-incompatible species, where the lack of variation at the S locus might limit or prevent reproduction by seed. Reproduction in small populations may be prevented or impaired in other situations.

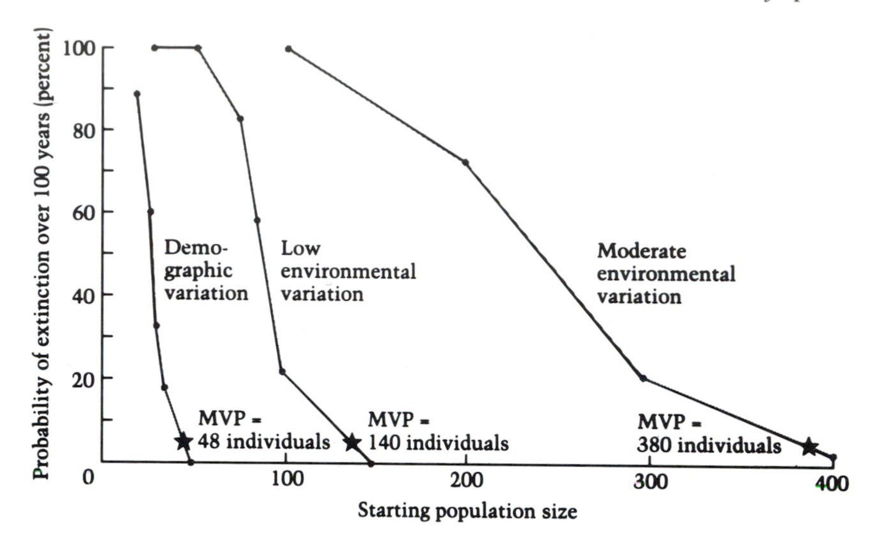

Fig. 15.4. The effects of demographic variation, low environmental variation and moderate environmental variation on the probability of extinction of a population of the Mexican palm, *Astrocaryum mexicanum*. In this study, the minimum viable population size, shown as stars, was defined as the population size at which there is a less than 5% chance of the population going extinct within 100 years. (After Menges, 1991, from Primack, 1993.)

1. In a dioecious species only one sex may be present (e.g. some populations of *Juniperus communis* in southern Britain).
2. Populations of rare gynodioecious species may contain only the hermaphrodite variant (e.g. small populations of *Salvia pratensis* in Oxfordshire: personal communication E. Arnold). In most species the genetic control of gynodioecy remains to be determined but, while some hermaphrodite plants may be of an appropriate genotype to produce the male sterile variant, the lack of such plants in contemporary populations may restrict the level of outbreeding.
3. Small populations of a facultatively apomictic species may contain only a single genotype and the production of new variants by sexual reproduction may be prevented.

Also, it is important to acknowledge that interspecific hybridisation is more widespread in plants than animals (Stace, 1975). Finally, we stress the implications for conservation of the polyploid nature of many flowering plants. Models on the effects of inbreeding have been devised for diploid organisms. Clearly, as we discussed in Chapter 12, the deleterious effects of inbreeding might take longer to appear in polyploids containing several different genomes.

What priorities should be set in attempting to reverse the decline of endangered species?

It is clear that many thousands of species are at present vulnerable or endangered by man's activities. Two conservation options are available. Plants may either be conserved *ex situ* in such places as Botanic Gardens or *in situ* in their native habitats. Many conservationists consider that all species should be conserved. Others believe that, as resources for conservation are limited, some priorities will have to be established. As we shall see the establishment of priorities is often the result of a critical assessment of the degree of endangerment in particular cases.

Our knowledge of the distribution and population size for plants growing in many parts of the world is seriously incomplete. However, where the flora is well-known, it is of very great interest that different kinds of rarity have been distinguished. (Rabinowitz, Cairns & Dillon, 1986). In their study of the British flora, three factors are considered: geographic range (broad or narrow); habitat specificity (wide or specialised); and local population size (large or small). For example, considering the species selected from the British flora by Rabinowitz and associates. Scottish Bird's Eye Primrose (*Primula scotica*) an endemic plant in the north of Scotland, has a narrow geographic range, a broad habitat specificity and somewhere in its range its populations are large. In contrast, *Lloydia serotina*, known as the Snowdon Lily in Britain, has a narrow geographic range, restricted habitat specificity and everywhere its populations are small. Thus, the *Primula* is restricted in one dimension (geographic range), while *Lloydia* is restricted in all three dimensions.

The logic of Rabinowitz's paper, is that we should pay more attention to the Snowdon Lily than the Primrose. We should note, however, that *Lloydia* has a wide circumpolar distribution, while *Primula scotica* is endemic to Britain. Conservationists, almost without exception, consider that endemic taxa, which occur only in a particular localised area, should be given priority. Not only species, but also genera and even families, may be narrowly endemic. They are often represented by small endangered populations in the wild. On the other hand, they may be common in the limited areas where they occur.

Major (1988) notes that high endemism is found on oceanic islands, roughly in proportion to their degree of isolation. In montane areas, the proportion of endemism increases with altitude. Tropical areas have the highest number of endemic species, genera and even families. Edaphic endemism also occurs where there are outcrops of specialised rocks, e.g. gabbro, serpentine, or heavy metal rich areas.

A case should be made for conserving not only the endemics developed as a result of long periods of speciation in geographic isolation but also the newly formed polyploid endemic species we discussed in Chapter 12 – namely, *Tragopogon miris* and *T. miscellus* in western USA, and *Senecio cambrensis* in Britain.

Holsinger & Gottlieb (1991) add to the list of priorities taxonomically distinctive groups, for example insectivorous plants. The relatives of agricultural and horticultural plants are also priority candidates for conservation. They cite as an example *Oryctes nevadensis*, small populations of which are found in a few localities in California and Nevada. This is the sole species in the genus (the so-called monotypic condition) and, as a member of the Solanaceae, the plant is a highly distinctive, distant relative of the Potato, Pepper, Tomato, Aubergine, Tobacco, etc. However, as *Orcytes* is a small, insignificant species, its conservation has been opposed. Holsinger & Gottlieb (1991) also suggest that the phylogenetic diagrams produced by cladists might be helpful in deciding priorities. In their view, it would be more important to conserve species or genera on distant branches of phylogenetic trees rather than selecting only closely related taxa on adjacent branches.

In assigning priorities to the conservation of rare and endangered species, some botanists have assessed the present status and future prospects of populations, and have evoked the concept of triage. In war-time, those casualties most likely to survive have often been selected for medical treatment. Thus, in conservation terms it might be possible to classify endangered species into three groups:

1. 'Stretcher cases' – which are unlikely to survive in the wild despite intervention by conservationists. These species could be conserved in gardens.
2. Those whose prospects of survival in the wild are such that remedial action is not immediately required.
3. Those whose populations are likely to thrive if remedial action is taken. There are, however, many conservationists who reject triage, considering that we should try to conserve all endangered species. The debate about triage will continue.

***Ex situ* conservation**

Endangered species, including those at the very edge of extinction, are often conserved *ex situ* in Botanic or other gardens. Many plant species, with their capacity for vegetative propagation and self-fertility, are potentially easier to conserve in gardens than are animals in zoos, where it often proves difficult to persuade the inmates to breed. From specialist and Botanic Gardens, plants can be made available to nurseries and may then be widely grown. Botanic Gardens, by providing interpretative displays of living endangered material, are also well-placed to educate the public on conservation issues.

The scale of the holdings of Botanic Gardens is impressive. Perhaps 30% of all vascular plants are represented in the collections of the *c.* 1500 Botanic Gardens world-wide. Some gardens hold very large general collections (e.g. Kew with *c.* 10% of the world's plants), while others have specialist collections (e.g. The Arnold Arboretum, Boston, USA, has a collection of several hundred species of temperate trees). The Botanic Gardens Conservation International (BGCI), Richmond, England, organises and co-ordinates the conservation efforts of gardens. Regrettably most Botanic Gardens are located in temperate areas of the world and it is costly to grow tropical plants in the glasshouses of Europe and North America. There are some notable Botanic Gardens in the tropics (Heywood & Wyse Jackson, 1991), but if the number could be increased it would be possible to conserve many tropical species of plants cheaply out-of-doors, and provide important centres for economic development and the exploitation of plant biodiversity.

While gardens obviously have an important role to play in providing a last refuge for endangered species, problems of long-term conservation in gardens have to be faced. First, there is the question of continuity of policy in gardens. The enthusiastic conservation efforts of one director might be difficult to fund by his/her successor in times of financial stringency. Also, many of the plants in existing collections are of unknown provenance. The move made by many Botanic Gardens towards growing plants of known wild origin is to be welcomed. Further, while technology provides the means to grow even the most exacting species, there are a number of practical difficulties to be considered. The number of plants that can be grown of any one species is strictly limited. The genetic purity of the conservation stocks may be compromised, hybrid seed being produced if plants of the same genus are grown together. Furthermore, if stocks are renewed repeatedly from seed, then the genetic variation represented in the original sample introduced into the garden may be reduced, and unconscious selection may be practised by the gardeners leading to insidious domestication of the material. At a more mundane level, labels may be lost and there are also problems of keeping good records. In the longer term, some collections of plants at present maintained out-of-doors in Botanic and other gardens may be threatened, as global climate change could make their cultivation more difficult, or impossible.

Botanic Gardens may increase their contribution to conservation if they maintain modern seed storage facilities or have links with such centres. It has been discovered that the viability of seeds of many species may be extended if they are kept cold and dry. The recommended conditions are a minimum of 8% water content and a temperature between − 20 and − 5 °C (Given, 1994). However, not all species may be stored in this way, as about 15% of higher plants – many from tropical forests – have recalcitrant seeds that lack seed dormancy and/or cannot withstand desiccation. Such seeds germinate immediately in the wild.

For conservation purposes it is important to collect appropriate seed samples for storage, samples which adequately reflect the genetic variation of the original population. Falk & Holsinger (1991) present a theoretical framework for their recommendation to collect seed from up to five populations per species, and from 10–50 individuals per population. Furthermore, if all the seed cannot be safely collected in one year without affecting the reproductive success and size of the populations, they recommend that seed should be collected over several seasons. Because stored seed, even at low temperature storage, eventually loses its viability, it has been proposed that stocks be rejuvenated by growing samples in gardens to obtain new seed, steps having been taken to make sure that the genetic variability is not lost in the rejuvenation process, with the plots so placed as to be free of any possibility of crossing with other stocks. However, rejuvenation is very expensive and it is cheaper, in some cases, to collect new samples in the wild, especially if several species are to be collected in the same area.

Seed from seed banks and propagated plant material are often made available by Botanic Gardens to reintroduce a species back into the wild. To facilitate this many Botanic Gardens, exploiting the capacity of plants to regenerate from small pieces in appropriate culture conditions, have set up micropropagation units. Several examples have been published in *Botanic Gardens Micropropagation News* from the Royal Botanic Gardens at Kew. For instance, the endangered endemic *Dianthus arenarius* subsp. *bohemicus* – of which there is only a single population of about 30 clumps in the wild – has been successfully grown in nutrient solutions to provide rooted plants (Kovác, 1995). In the development of micropropagation technology, it has also been found possible to freeze pieces of plant tissue (called cryopreservation). This could prove an important alternative to cold storage of seeds. For example, González-Benito & Pérez (1994) successfully preserved nodal segments of the Spanish threatened endemic species *Centaurium rigualii.*

So far our discussion has focused largely on species considered individually, to determine the factors leading to extinction and priorities for conservation. Given the difficulties of *ex situ* approaches, we stress the overwhelming importance of *in situ* conservation. Thus, legally protected reserve areas have been set up to secure the future of all the interacting species of plants and animals in functioning ecosystems.

The role of protected areas in countering the threat of extinction

About 2.8% of world's surface is occupied by reserves of various sorts, offering a degree of protection to biodiversity (Western, 1989). Historically, nature reserves and national parks were set up for a range of purposes, often to protect areas of outstanding natural beauty and landscape features. Some reserves were designated to protect particular groups of organisms, for example The National Trust has been involved with the Wicken Fen nature reserve, Cambridgeshire, England since 1899, primarily because entomologists were interested in continuing their collecting of fenland Lepidoptera in the area.

Over the years, reserves have been established in many different areas. For instance, some are remnants of semi-natural communities – such as ancient coppiced woodland, species-rich grasslands and heathlands – on low productivity land of marginal agricultural value (Meffe & Carroll, 1994). Such reserves are very rarely designed. Conservationists have purchased or negotiated management agreements to look after existing islands of distinctive biodiversity, found where traditional management practices – coppicing, haymaking, peat cutting, etc. – are still practised or more usually, have declined or have been abandoned altogether.

At the other extreme, reserves are sometimes set up in areas of pristine natural vegetation, such as tropical rain forests. In this case, conservationists may be able to design the reserve. In such circumstances, what would be the ideal size and shape to choose?

Ideas about the design of nature reserves have been influenced by theoretical and experimental studies of island biogeography, which, as we have seen above, have considered the diversity found on areas of land of different size surrounded by water. In this account we indicate some of the main ideas formulated by these studies; for a more comprehensive treatment of the ideas, as they affect nature reserves, see Shafer (1990). The general principle emerging from these approaches, which receives some support from theoretical and experimental studies, is that 'as area increases, so does the diversity of physical habitats and resources, which in turn support a larger number of species' (Meffe & Carroll, 1994). Taking these ideas as a starting point, and noting that they have proved controversial and very difficult to test, it is generally agreed that, within a territory, large reserves are better for conserving biodiversity than small reserves (Fig. 15.5), and are

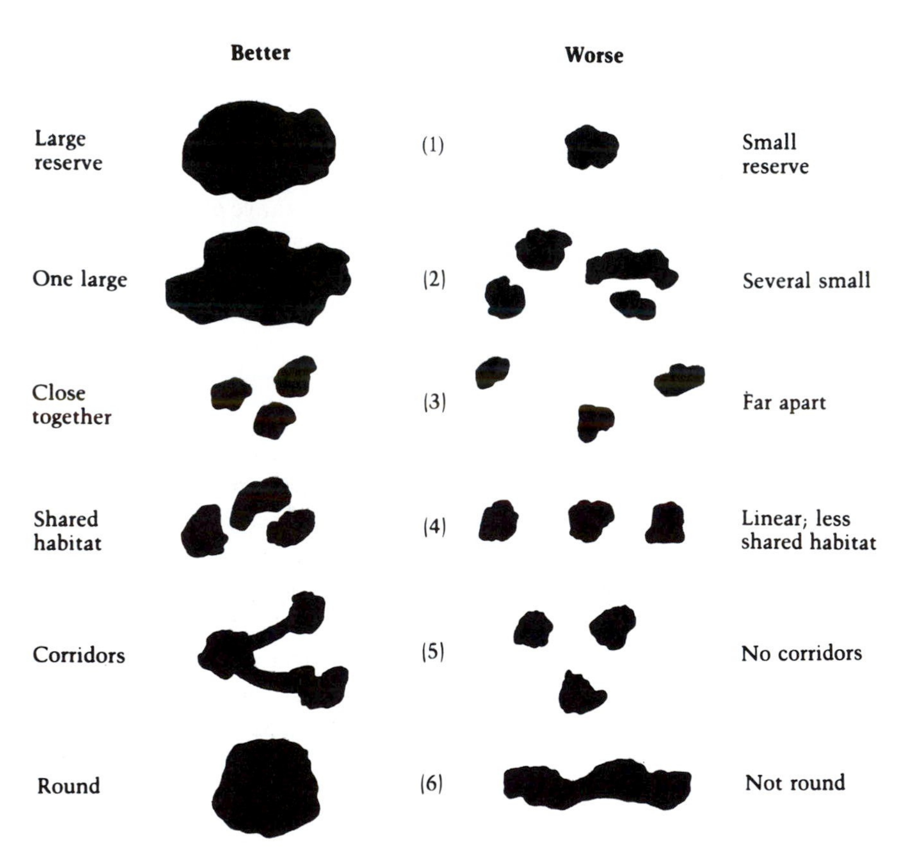

Fig. 15.5. Principles of reserve design that have been proposed based on theories of island biogeography. Imagine that the reserves are 'islands' of natural habitat surrounded by land that has been totally changed by human activity. The practical application of these principles is still under study. Principles 2 and 5 in particular have been the subjects of much debate. (After Diamond, 1975, from Primack, 1993.)

perhaps better able to accommodate the problems of introduced species than small areas. Modelling and experimental work have also attempted to resolve another question. Is it better to have a single large reserve or several small reserves of equivalent total area (the so-called SLOSS debate)? Clearly, one of the key issues is the extent to which reserves of different size, and also different shape, are affected by edge effects. For instance, where a reserve abuts on agricultural land, edge effects could include drift of herbicides and fertilisers, as well as invasions of weedy species into the reserve. The degree to which edge effects affect the main communities contained in the reserves is influenced by area/perimeter relationships. A single, large reserve suffers less edge effect than several smaller reserves of equal total area. Also, more or less circular reserves have much less edge than long, narrow reserves of the same area, which could be exposed to edge effects across their entire width.

In our view the SLOSS debate should not be used as an argument against setting up reserves that do not conform to a supposed 'ideal' size and shape. Well-positioned small reserves may have a vital role in conserving rare species, e.g. butterflies. And even long, thin reserves have an important local role to play in conservation in regions dominated by modern intensive agriculture. For example, Devil's Dyke, near Cambridge, England, is an important chalk grassland reserve (Walters, 1979). This archaeological site, dating from the *c.* 4th–7th centuries, consists of a 7.5 mile long, linear, massive defensive earthwork and an associated ditch. Also, the degree to which edge effects should be minimised in reserves is a complex issue, as edges of reserves may contain their own suite of interesting rare species.

Another key concept emerging from island biogeography concerns the distance between reserves (Fig. 15.5). Theory suggests that reserves close enough to permit any gene flow between the populations of particular species would be preferred to a collection of widely scattered reserves. However, given the limited gene flow in many plant species, it is clear that gene flow may be impossible even between reserves very close together. Taking the question one stage further, consideration should also be given as to whether corridors of appropriate vegetation, between two reserves, might permit gene flow in particular species. A first point to note is that narrow corridors are likely to suffer from extreme edge effects and, taking a specific case, one might ask: is there any evidence that strips of woodland and hedgerows provide effective routes for gene flow for woodland species, say between two ancient woodland 'island' nature reserves in eastern England? The characteristic species of ancient woodland are not very 'mobile', but there is some evidence that they are unable to migrate between island woods along hedgerows (Peterken, 1981). Some conservationists have suggested that corridors might be set up between reserves, for instance, to allow species to migrate in response to global climate change to the new areas appropriate to their climatic tolerances. The effectiveness of such corridors – both now and in the future – was called into question by Simberloff *et al.* (1992), for while the movement of birds and mammals might be facilitated by corridors, it should not be assumed that this would allow plant migration.

Reserves in particular areas are often very small, and it has recently been emphasised that what happens outside reserves is critical to the survival of much biodiversity. It has been suggested that, where blocks of privately held land, and public forests, etc. exist adjacent to reserves, they might be managed to an agreed common plan. Such agreements could provide increased habitats for wildlife and in some cases effectively enlarge reserves or provide a buffer zone of sympathetically managed ground around them.

Managing reserves to prevent extinction of species

Historically many reserves were established as preserves. The idea was to isolate them from human activities to allow 'the balance of nature' to be restored (Pickett, Parker & Fiedler, 1992). This outlook has been displaced by the realisation that man has decisively influenced the ecology of communities throughout most of the world, and that to conserve particular sites requires active intervention by man in the form of reserve management. For example, to conserve species-rich chalk grassland in Europe, which is the product of centuries of grazing (by sheep and rabbits in Britain), it is necessary to prevent the natural climax forest of the area developing by controlling the invasion of the site by shrubs and trees. Thus, active management is necessary, by using grazing animals, cutting or physical removal of plants.

The range of management techniques used in conservation is enormous (Sutherland & Hill, 1995). A few examples will indicate the range of options:

1. Site management and soils (e.g. control of water table, disturbance, watering, alteration of the nutrient status, etc).
2. Plants (selective action involving sowing, planting, cutting, burning or weeding, etc.).
3. Animals (employing fences to restrict or intensify grazing with fences, culling, introductions, etc.).
4. Management of human activities (legal protection of the site and its biota, preventing or limiting access, provision of educational and other facilities for visitors, etc.).

This fourth category presents some of the most difficult and politically complex issues. For instance, in tropical areas indigenous peoples live within or close to areas which have become reserves, and in their justified attempts to survive, often in conditions of abject poverty, the biodiversity of the area may be put at risk. Such situations provoke moral as well as scientific questions. How far should traditional land use be permitted? To what degree is increased use of resources of the reserve compatible with conservation? Is it possible to devise sustainable means of using resources, so that the long-term future of biodiversity is assured? In times past, local people were sometimes removed from newly set-up reserves. Often they received no material benefits from having a reserve on their traditional lands and in some cases became antagonistic to conservation efforts. Many conservationists now take the view that the people who live and work in or near nature reserves or national parks – be they in India, Europe or elsewhere – must be more fully involved in decision making and receive some income from conservation activities, perhaps from the sustainable development of resources or various kinds of ecotourism (Meffe & Carroll, 1994). The political, ecological and social conflicts involved in conservation and development have been studied in many places (see, for example, Anderson & Grove, 1987, for a discussion of the situation in different parts of Africa).

Restoration ecology

Man's activities have caused great damage to plant communities. From small beginnings, a major effort is being made, not only to restore and enlarge existing fragments, but also to create or recreate *new* areas of these vegetation types on former agricultural land or even in derelict areas. For example, species-rich grasslands (Crofts & Jefferson, 1994), woodland (Ferris-Kaan, 1995), wetlands (Wheeler & Shaw, 1995), and the use in Britain of agricultural set-aside schemes (Firbank *et al.*, 1993).

At the moment, such ventures often concentrate on the establishment of common species, sown as wild flower mixtures. Other rarer species may then be transplanted into the sward as small plants. Given that common species of plants show ecotypic variation in relation to climatic and edaphic factors, it is essential to plant material appropriate to the site. For instance, while the 'correct' species were sown in a project to restore chalk grassland in 1991 on the Gog Magog Hills, Cambridge, England, the 'wrong' variants were unwittingly used. The *Achillea millefolium* and *Centaurea cyanus* were most likely of horticultural origin, while the *Medicago lupulina*, *Trifolium pratense* and *Sanguisorba minor* subsp.

muricatus were most probably agricultural strains. There is evidence that 'seed impurities' were also included in the wild flower mixture, which was obtained from commercial sources: for instance, *Ranunculus marginatus* var. *marginatus*, native of the Balkans and Crimea to Iran, appeared on the site (Akeroyd, 1994).

Manipulating and creating populations of endangered species in an attempt to prevent extinction

Zoologists have led the way in many practical conservation techniques. Transfers are often made to replenish existing populations, to re-establish populations which have become extinct and to establish populations outside their normal range (Bowles & Whelan, 1994; Meffe & Carroll, 1994). Plant conservationists have also become interested in the possibilities of restocking, reintroductions and transfers in the management of both existing and lost populations. For instance, the English Nature Recovery Programme has been set up with the aim of securing the future of very rare native species (Maunder & Ramsay, 1994). Not only are populations of rare plants being manipulated as part of the management of reserves, but introductions of rare and endangered species are also being carried out. as part of restoration ecology. For example, the rare *Orchis laxiflora* has been successfully propagated in the laboratory (with its symbiotic fungus), and subsequently transferred to a new site in the grounds of Wakehurst Place, West Sussex, managed by the Royal Botanic Gardens, Kew (see Maunder & Ramsay, 1994).

While some naturalists accept habitat management, they condemn the setting up of new populations, as misusing valuable resources, devaluing natural populations, obscuring true and native distributions and likely to damage populations at source and maybe at point of transfer (Falk & Olwell, 1992). They draw attention to the fact that past transfers were often poorly documented and carried out without permission. Also, they are concerned about mitigation. If it becomes accepted by the general public that new habitats may be produced which are acceptable 'replicas' of present reserves, then a developer wishing to damage an existing site could offer to produce a 'new identical' area in mitigation. Conservationists are rightly suspicious of such a bargain.

While acknowledging the force of these arguments, do we rule out restoration and transfer of the populations of rare species under all circumstances, given the obvious plight of many species? For instance, transfers have been suggested to conserve species in the face of global climate change. Many endangered species are likely to be genetically depauperate and unable to respond *in situ* to the new selection pressures imposed by such changes (Bradshaw & McNeilly, 1991). As we have seen, the natural migration of species to more suitable areas, in response to climate change, may be effectively impossible as their natural gene flow is insufficient to enable them to cross, unaided, huge barriers of agricultural and urban/industrial land. Moreover, natural migration may be inadequate even in areas less influenced by man's activities. Zabinski & Davis (1989) modelled the future range of various tree species in North America under two-fold CO_2 global warming, taking into account their current capacity for migration (Fig. 15.6). As some potential ranges will not be exploited for decades by natural gene flow by seed, should seed/plants be transferred to these areas?

It seems to us that the debate about the manipulation of populations of rare and endangered species could be conducted with more rigor, if there were more case histories on which to make a more informed assessment of the potential and the limitations of the techniques. Indeed, we may very well ask the question: do manipulations and transfers succeed in the long term? How far are attempts aimed at achieving a minimum viable population?

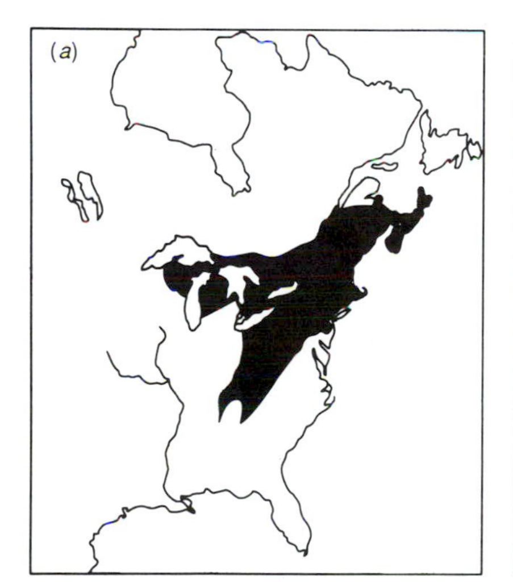

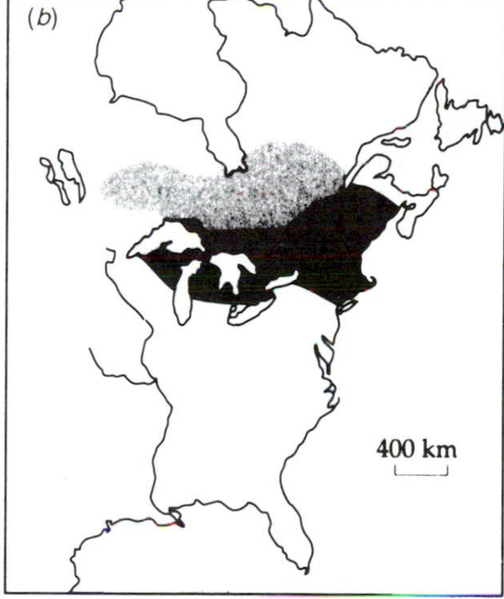

Fig. 15.6. Present and future geographic range for Hemlock (*Tsuga canadensis*). (*a*) Present range. (*b*) Range in AD 2090 under the Goddard Institute of Space Studies model with two-fold CO_2 scenario. The black area is the projected occupied range considering the rate of migration. The grey area is the potential projected range with climate change. (From Zabinski & Davis, 1989.)

Can the process of extinction be prevented by such means? Here, we give a few case studies. (For some additional examples and a comprehensive review of strategies for reintroduction of endangered species, see Falk, Millar & Olwell, 1996.)

Restocking an existing population

In 1975, the population of the rare mountain species *Saxifraga cespitosa* on Cwm Idwal, North Wales, had been reduced to 4 plants. From material collected from the population, plants were raised at Liverpool Botanic Garden and, in 1978, the population was restocked with 130 mature plants, 195 small seedlings and 1300 seeds (Parker, 1982). In 1980, there were 48 mature plants in the population. In this case, the intention was to produce a population equal to the estimated population size in 1796. Perhaps it would have been more appropriate to aim for a minimum viable population.

Re-establishment of an extinct population

Stephanomeria malheurensis is known only from one locality in Oregon, USA, where it became extinct less than 20 years after its discovery, its original habitat having been damaged by fire followed by invasion by the aggressive European introduced weed, *Bromus tectorum.* Using plants generated from stocks maintained *ex situ* at the University of Davis, California, 1000 seedlings, protected inside four rodent-proof exclosures, were set out in an artificially watered area cleared of *B. tectorum.* A total of 40 000 seeds were produced in the first year and population numbers fluctuated thereafter (Guerrant, 1992).

Founding of a new population

Demauro (1994) founded three new populations of the rare endemic self-incompatible *Hymenoxys acaulis* var. *glabra* in the Great Lakes area. As the species requires a mixture of *S* allele genotypes for the self-incompatible system to function, the new populations were set up using plants of different mating types from different populations, together with F_1 and open-pollinated F_1 progenies. Planting arrangements were designed to maximise the functioning of the self-incompatibility system. The plan for the restoration was based on an MVP size of 1000, a figure large enough, it was thought, to conserve variability. In the event, 95% of the first transplants were lost through drought within two months, and a much larger initial population might have been more appropriate. An interesting technical point concerns the use of composite stocks rather than material from a single source. Generally, it has been recommended that restocking and reintroduction be attempted with stocks native to the site in question. But, in this investigation, hybrid material was used. Such a procedure has been advocated in some cases by Barrett & Kohn (1991). It will result in a widening of the range of genetic variation in a population and the action of selection may result in a population more fitted to changing circumstances. However, other conservationists, anticipating the possibility that wide crosses may sometimes be made – involving variants distinct at the ecotypic level – have pointed out the procedure may lead to serious outbreeding depression, a phenomenon discovered in animals, where many or all of the hybrid progeny may lack fitness.

Our example of the founding of a new population involved the planting of seedlings. Given the high level of preparation necessary for such a transfer, it is interesting to consider whether sowing seeds might be a cheaper, quicker and more reliable method. Such an approach was made by Primack & Miao (1992), who set up experimental populations, in Massachusetts, USA, by sowing seeds of four annual species, 40–600 m from their existing populations. At three sites, new populations established and persisted for four generations, while at seven sites small initial populations were established but then died out. At 24 sites no establishment occurred. The authors consider that, perhaps, the sites were unsuitable in some way, or not enough seeds were sown. Given the very specific habitat requirements of many species, especially for germination and establishment, it should not be assumed that sowing seeds is a more reliable means of transferring species.

There is every reason to believe that more manipulations and transfers of endangered species will be attempted. However, legislation protecting wildlife may make experiments difficult. Thus, in Britain, the Wildlife and Countryside Act, 1981 makes it illegal to uproot any plant without the owner's permission, and consent is also required to remove or introduce organisms at any designated Site of Special Scientific Interest (SSSI). It is to be hoped that, when experiments are attempted with rare and endangered species, full permission is obtained from the owners of sites, and that the investigations are well-planned and fully documented. Often details of introductions are referred to briefly in reserve and other reports. Progress in this area would be more rapid if details of such introductions were published in the scientific literature, so that success and failure can be properly analysed. Akeroyd & Wyse Jackson (1995) have provided an important guide to all aspects of the reintroduction of plant species using stocks from Botanic Gardens and other sources.

Having set out the case for well-designed experiments on habitat management and reintroductions, we should acknowledge the difficulties faced by those who are trying to carry out practical conservation. Very often *ad hoc* decisions have to be made in situations of crisis, and there is neither the time nor the resources to carry out long-term research. Action has to be taken quickly on the basis of the best available information. However, sometimes, long-term funding for conservation does become available and then, we suggest, it is imperative to devise strategies to maximise the possibility of long-term success through well thought out experiments.

Arguments for conservation

Considering the millions of plants and animals presently found on the earth, Raup (1991) speculates that 'somewhere between five and fifty *billion* species' may have existed during geological history, and points out that 'only about one in a thousand species is still alive'. Given that, as far as we know, extinction is the fate of all species, why should action be taken to counter threats of extinction? What arguments can a conservationist produce to convince the sceptic that species on the edge of extinction should be conserved at the present day, and are such representations likely to produce the necessary financial support?

Heslop-Harrison (1973) has reviewed the arguments used to support conservation. First, some take the view that animals and plants have a 'right to exist' and that it is 'wrong to hound to extinction organisms which share the planet with us'. There are many fervent advocates of this view, which is essentially a moral stance involving value judgements. While some are prepared to accept this argument, others remain sceptical.

A second line of argument considers the 'quality of life'. A world rich in organisms is better to live in or indeed is the right place to live in. Aesthetic appreciation has become a powerful force in the conservation of species and communities of plants and animals, especially if there is linkage between conservation of biodiversity and the preservation of cherished landscapes, be they nearby or in remote pristine wilderness areas. However, an appreciation of the finer points of the environment cannot be relied upon. Tudge (1989) suspects that when politicians use the word 'environment ... all they have in mind is a generalised green-ness, a golf-course, a bit of Repton-style landscaping, or even a Disney-style theme park'.

The third, and by far the strongest argument for conservation, is that plants are 'essential for human survival'. With regard to plants 'we, like all other animals, depend for our existence on the biochemistry of plants, which in all essentials is complementary to our own' (Heslop-Harrison, 1973). With regard to food plants, Given (1994) notes that at least 75 000 species are edible and *c.* 3000 have been used for food, but only a handful are crop plants. He takes the view that thousands of species have desirable qualities and could be useful in the future. Extending the argument about the value of biodiversity to include both plants and animals, Oldfield (1989) provides a very detailed review of the uses man makes of genetic resources in terms of crop plants, domesticated animals, medicinal plants, tree resources, natural resources of rubber, oil, waxes, the fashion trade, curios, pets and garden plants. The argument that man should conserve species because they might be useful is very strong, but it does not apply with equal force to all organisms? What about all the myriad species of beetles? It has been argued that, for the health and survival of the planet, the stability and proper functioning of entire ecosystems, with *all* their plants and animals, is very important (Given, 1994). Thus, Tudge (1989) writes: 'It has been suggested that we are sowing the seeds of our own destruction by destroying so many species: [and] ... that we need a planet that is in ecological 'balance': and that balance depends upon the multitude of other species, perhaps between 10 and 30 million, that the earth is thought to contain'.

In the face of habitat destruction on a very large scale and the likelihood of extinction of many species, it is easy to become depressed. However, in recent years, the environmental movement has come of age, and many pressure groups and national and international agencies are concerned with the need for changes in attitude, calling for sustainable use of resources and the safeguarding of biodiversity. Tudge (1991) admirably captures the problem and the challenges when he writes:

> Since the Neolithic revolution, the transition from hunter-gatherer into farming, which began 10 000 years ago, human beings have been weaned, and have weaned their children, on the notion that it was the destiny of humans ... simply to take over the world ... we do need a post-Neolithic society, which in its attitudes, in its moral philosophy, in its religion, in its politics and economics and in its way of working would be quite different from all that we have been developing for the past 10 000 years. What exactly those attitudes should be, and how they should be expressed; what form the politics and economics should take; those, I suggest, are the most interesting questions now facing the human species.'

Glossary of selected terms

Definitions, some of which have been simplified, have been taken from Heslop-Harrison (1953), Davis & Heywood (1963), Whitehouse (1973), Rieger, Michaelis & Green (1976), Heywood (1976), Stace (1980), Futuyma (1986) and Asker & Jerling (1992).

Agamospermy The production of seeds by asexual means.

Alleles (*allelomorphs*) Alternative forms of a gene, presumably differing in DNA sequence, which, on account of their corresponding positions on homologous chromosomes, are subject to Mendelian inheritance.

Allopatry Of species or populations originating in or occurring in different geographical regions.

Allopolyploidy The doubling (allotetraploidy or amphidiploidy) or higher multiplication (allohexaploidy, etc.) of chromosome sets from different species or genera either spontaneously or by experimental means.

Aneuploid Individuals having one or more whole chromosomes of the normal complement absent or in addition to that complement.

Apomixis Replacement in sexual reproduction by various types of asexual reproduction, in which meiosis and fertilisation are suppressed.

Apospory The phenomenon shown by some higher plants in which a diploid embryo-sac is formed directly from a somatic cell of the nucellus or chalaza; an embryo is then formed without fertilisation.

Artificial classification A system of ordering based upon one or a few characters, which gives a convenient arrangement of plants for some specific purpose. In such systems closely related taxa may be placed in different groupings.

Autogamy Self-fertilisation; persistent autogamy may result in an increase in homozygosity and division of a population into a number of 'pure lines'.

Autopolyploidy Polyploids having three, four, etc. homologous chromosome sets.

B chromosomes Small chromosomes (frequently, but not always, heterochromatic), which are additional to the normal complement of A chromosomes.

Biological species Groups of actually or potentially interbreeding natural populations, which are reproductively isolated from other such groups. A complex concept: see Chapters 10 and 13.

Bivalent The associated pair of homologous chromosomes observed at prophase I in meiosis.

Chiasma (*pl. chiasmata*) An interchange occurring only at meiosis between chromatids derived from homologous chromosomes. Chiasmata are the visible evidence of genetic crossing-over.

Cladistics The concepts and methods used in the determination of branching patterns of evolution.

Classification The ordering of plants into a hierarchy of classes to produce an arrangement which serves both to express the interrelationships of plants and to act as an information retrieval system.

Cleistogamy Self-fertilisation within closed flowers.

Cline A variational trend in space found in a population or series of populations of a species.

Clone A group of individuals (or lineage of individuals) derived vegetatively (by mitosis) from a single plant.

Crossing-over The occurrence of new combinations of linked characters following the process of exchange between homologous chromosomes at meiosis.

'*-deme*' *terminology* Devised to bring clarity to a confusion of terms for units below, at and about the species level. (For a discussion of terms, including Turesson's experimental categories, see Stace, 1980.) In cases where precise usage is necessary, the term species, and its prefixed derivatives, is to be reserved for taxonomic categories. Experimentalists could use the '-deme' terminology based on the neutral suffix-deme, which denotes a group of individuals of a specified taxon. The term 'deme' should not stand by itself. Terms are made by adding the appropriate prefixes.

For example:

Groups of individuals of a specified taxon:

- in a particular area (topodeme);
- in a particular habitat (ecodeme);
- with a particular chromosome condition (cytodeme);
- within which free exchange of genes is possible (in a local area) (gamodeme: equivalent to the term Mendelian population);
- which are believed to interbreed with a high level of freedom under a specified set of conditions (hologamodeme: approximately equivalent to the term biological species).

Despite the clear advantages of the '-deme' terminology it has been little used by biologists (Briggs & Block, 1981); indeed the most frequent use of the terminology is the incorrect employment of 'deme', without prefix, in the sense of gamodeme, by many zoologists.

Diploid With two chromosome compliments; the condition is initiated with the fusion of two haploid gametes at fertilisation.

Diplospory The phenomenon shown by some higher plants in which a diploid embryo-sac is formed directly from a megaspore mother-cell; an embryo is then formed without fertilisation.

Ecocline A variational trend correlated with an ecological gradient.

Euploid A polyploid possessing a chromosome number which is an exact multiple of the basic number of the series.

Facultative apomicts Apomicts that retain some sexuality.

Founder effect The principle that a new colony established by one or a few individuals contains only a fraction of the genetic variation present in a large source population.

Gametophyte The haploid gamete-producing phase of the life cycle of plants.

Gene flow The dispersal of genes, in both gametes and zygotes, within and between breeding populations.

Genome A single complete set of chromosomes. One such set is present in the gametes of diploid species; two genomes are found in the somatic cells. Polyploid cells contain more than two genomes.

Genotype The totality of the genetic constitution of an individual.

Haploid With a single set of chromosomes (one genome), such as occurs at gamete formation.

Heterochromatin Parts of chromosomes, or whole chromosomes, which exhibit an abnormal degree of staining and/or contraction at nuclear divisions.

Heterostyly A combination of morphological, physiological and genetic mechanisms which promote outbreeding in certain angiosperms. Heterostyly refers to the presence in a species of two (distyly) or three (tristyly) genetically determined different types of individuals, distinguished by the relative positions of stigma(s) and anthers in the flower. Breakdown of heterostyly – by gene mutation, recombination or following polyploidy – may give rise to homostyles, which are self-compatible (e.g. in the genus *Primula* homostylous species occur, and, populations of heterostylous species may contain homostyles). The term homostyly may also be used in cases where one of the heterostylous variants has survived and reproduction is apomictic (e.g. in the genus *Limonium* species are generally homostylous (Ingrouille & Stace, 1985), with either one or other of two heteromorphic flower variants).

Heterozygote A zygote or individual carrying two different alleles of a gene (e.g. *Aa*)

Homozygote A zygote or individual formed from the fusion of gametes carrying the same allele of a gene (e.g. *AA* or *aa*).

Introgressive hybridisation (*introgression*) Genetic modification of one species by another through the intermediacy of hybrids.

Isolating mechanisms Genetically determined differences between populations that restrict or prevent interbreeding between them.

Karyotype The appearance and characteristics (shape, size, etc.) of the somatic chromosomes at mitotic metaphase.

Matroclinous (*maternal*) *inheritance* Condition found where a hybrid is closely similar to its seed parent.

Meiosis A special nuclear division in which the chromosome number is halved.

Meristic variation Variation in numbers of parts or of organs.

Mitosis The nuclear division typical of somatic plant tissues in which a nucleus divides to produce two identical complements of chromosomes (and hence genes).

Multivalent Association of more than two homologous chromosomes at meiosis, e.g. 3 = trivalent, 4 = qudrivalent.

Natural classification A classification based upon overall resemblance, and serving a variety of purposes.

Outgroup A taxon that diverged from a group of other taxa before they diverged from each other.

Phenetic classification A classification based upon present-day resemblances and differences between plants.

Phenotype The totality of characteristics of an individual; its appearance as a result of the interaction between genotype and environment.

Phenotypic plasticity The capacity of a genotype to respond to different environments by producing different phenotypes.
Phylogenetic classification A classification showing the supposed evolution of groups.
Polyhaploid An organism with the gametic chromosome number arising by parthenogenesis in a polyploid, e.g. a diploid (2*x*) plant arising from the parthenogenetic development of an embryo of a tetraploid (4*x*).
Polymorphism The occurrence of two or more distinct genetic variants of a species in a single habitat.
Polyploid Having three (triploid), four (tetraploid), five (pentaploid), six (hexaploid) or more complete chromosome sets, instead of two as in diploids.
Pseudogamy The phenomenon found in some apomictic plants, whereby pollination is necessary for seed development, even though no fertilisation of the egg-cell takes place.
Ramet An individual belonging to a clone.
Ratio-cline Clinal variation occurring in polymorphic species, in which successive populations show progressive change in the proportion of the variants.
Sporophyte The diploid spore-producing phase of the life cycle of plants arising from the fertilisation of haploid gametes.
Supergene A group of two or more loci between which recombination is so reduced that they are usually inherited as a single entity.
Sympatry Of species or populations, originating in or occupying the same geographical area, so that the opportunity to interbreed is presented.
Taxon (*pl. taxa*) A classificatory unit of any rank, e.g. Daisy: *Bellis perennis* (species); *Bellis* (genus); Compositae (family).
Topocline A geographical variational trend which is not necessarily correlated with an ecological gradient.
Univalent An unpaired chromosome at meiosis.
Variant Any definable individual or group of individuals. A valuable neutral term.

References

Abbott, R. J. (1994). Ecological risks of transgenic crops. *Trends in Ecology & Evolution*, **9**, 280–2.
Abbott, R. J., Curnow, D. J. & Irwin, J. A. (1995). Molecular systematics of *Senecio squalidus* L. and its close diploid relatives. In *Advances in Compositae systematics*, ed. D. J. N. Hind, C. Jeffrey & G. V. Pope, pp. 223–37. London: Royal Botanic Gardens, Kew.
Abbott, R. J., Ingram, R. & Noltie, H. J. (1983a). Discovery of *Senecio cambrensis* Rosser in Edinburgh. *Watsonia*, **14**, 407–8.
Abbott, R. J., Noltie, H. J. & Ingram, R. (1983*b*). The origin and distribution of *Senecio cambrensis* Rosser in Edinburgh. *Transactions of the Botanical Society of Edinburgh*, **44**, 103–6.
Ab-Shukor, N. A., Kay, Q. O. N., Stevens, D. P. & Skibinski, D. O. F. (1988). Salt tolerance in natural populations of *Trifolium repens*. *New Phytologist*, **109**, 483–9.
Ainsworth, C. C., Parker, J. S. & Horton, D. M. (1983). Chromosome variation and evolution in *Scilla autumnalis*. In *Kew Chromosome Conference II*, ed. P. E. Brandham & M. D. Bennett, pp. 261–8. London: Allen & Unwin.
Akeroyd, J. R. (1994). Some problems with introduced plants in the wild. In *The common ground of wild and cultivated plants*, ed. A. R. Perry & R. G. Ellis, pp. 31–40. Cardiff: National Museum of Wales.
Akeroyd, J. R. & Wyse Jackson, P. (1995). *A handbook for botanic gardens on the reintroduction of plants to the wild*. London: Botanic Gardens Conservation International.
Albert, V. A., Williams, S. E. & Chase, M. W. (1992). Carnivorous plants: phylogeny and structural evolution. *Science*, **257**, 1491–5.
Alexander, H. M., Thrall, P. H., Antonovics, J., Jarosz, A. M. & Oudemans, P. V. (1996). Population dynamics and genetics of plant disease: a case study of anther-smut disease. *Ecology*, **77**, 990–6.
Al-Hiyaly, S. A. K., McNeilly, T. & Bradshaw, A. D. (1988). The effect of zinc contamination from electricity pylons – evolution in a replicated situation. *New Phytologist*, **110**, 571–80.
Al-Hiyaly, S. A. K., McNeilly, T. & Bradshaw, A. D. (1990). The effect of zinc contamination from electricity pylons – contrasting patterns of evolution in five grass species. *New Phytologist*, **114**, 183–90.
Al-Hiyaly, S. A. K., McNeilly, T., Bradshaw, A. D. & Mortimer, A. M. (1993). The effect of zinc contamination from electricity pylons. Genetic constraints on selection for zinc tolerance. *Heredity*, **70**, 22–32.

Alston, R. E. & Turner, B. L. (1963*a*). *Biochemical systematics*. London & New York: Prentice Hall.
Alston, R. E. & Turner, B. L. (1963*b*). Natural hybridization among four species of *Baptisia* (Leguminosae). *American Journal of Botany*, **50**, 159–73.
Alvarez, L. W., Alvarez, W., Asaro, F. & Michel, H. V. (1980). Extraterrestrial cause for the Cretaceous–Tertiary extinction. *Science*, **208**, 1095–108.
Amann, J. (1896). Application du calcul des probabilités a l'étude de la variation d'un type végétal. *Bulletin de l'Herbier Boissier*, **4**, 578–90.
Anderson, D. & Grove, R. (1987). *Conservation in Africa: people, policies and practice*. Cambridge: Cambridge University Press.
Anderson, E. (1949). *Introgressive hybridisation*. London: Chapman & Hall; and New York: Wiley.
Anderson, E. (1953). Introgressive hybridization. *Biological Reviews*, **28**, 280–307.
Anon. (1965). *Iconographia Mendeliana*. Brno: The Moravian Museum.
Anon. (1994). *IUCN red list categories*. Gland: IUCN Council.
Antonius, K. & Nybom, H. (1994). DNA fingerprinting reveals significant amounts of genetic variation in a wild Raspberry *Rubus idaeus* population. *Molecular Ecology*, **3**, 177–80.
Antonovics, J. (1968). Evolution in closely adjacent plant populations. V. Evolution of self-fertility. *Heredity*, **23**, 219–38.
Antonovics, J. (1976). The input from population genetics: 'the new ecological genetics'. *Systematic Botany*, **1**, 233–45.
Antonovics, J. & Bradshaw, A. D. (1970). Evolution in closely adjacent plant populations. VIII. Clinal patterns at a mine boundary. *Heredity*, **25**, 349–62.
Antonovics, J., Bradshaw, A. D. & Turner, R. G. (1971). Heavy metal tolerance in plants. *Advances in Ecological Research*, **7**, 1–85.
Armstrong, H. E., Armstrong, F. & Horton, E. (1912). Herbage studies. 1. *Lotus corniculatus*, a cyanophoric plant. *Proceedings of the Royal Society, Series B*, **84**, 471–84.
Armstrong, J. A., Powell, J. M. & Richards, A. J. (1982). *Pollination and evolution*. Sydney: Royal Botanic Gardens.
Arnheim, N., White, T. & Rainey, W. E. (1990). Application of PCR: organismal and population biology. *Bioscience*, **40**, 174–82.
Arnold, M. L. (1992). Natural hybridisation as an evolutionary process. *Annual Review of Ecology & Systematics*, **23**, 237–61.
Arnold, M. L. (1993). *Iris nelsonii* (Iridaceae): origin and genetic composition of a homoploid hybrid species. *American Journal of Botany*, **80**, 577–83.
Arnold, M. L. & Bennett, B. D. (1993). Natural hybridization in Louisiana Irises: genetic variation and ecological determinants. In *Hybrid zones and the evolutionary process*, ed. R. G. Harrison, pp. 115–39. New York: Oxford University Press.
Arnold, M. L. & Hodges, S. A. (1995). Are natural hybrids fit or unfit relative to their parents? *Trends in Ecology & Evolution*, **10**, 67–71.
Ashton, P. A. & Abbott, R. J. (1991). Multiple origins and genetic diversity in the newly arisen allopolyploid species, *Senecio cambrensis* Rosser (Compositae). *Heredity*, **68**, 25–32.
Asker, S. E. & Jerling, L. (1992). *Apomixis in plants*. Boca Raton: CRC Press.
Assouad, M. W., Dommée, B., Lumaret, R. & Valdeyron, G. (1978). Reproductive capacities in the sexual forms of the gynodioecious species *Thymus vulgaris* L. *Botanical Journal of the Linnean Society*, **77**, 29–39.
Aston, J. L. & Bradshaw, A. D. (1966). Evolution in closely adjacent plant populations. II. *Agrostis stolonifera* in maritime habitats. *Heredity*, **21**, 649–64.

Atwood, S. S. & Sullivan, J. T. (1943). Inheritance of a cyanogenic glucoside and its hydrolysing enzyme in *Trifolium repens*. *Journal of Heredity*, **34**, 311–20.

Avers, C. J. (1989). *Process and pattern in evolution*. New York: Oxford University Press.

Avise, J. C. (1994). *Molecular markers, natural history and evolution*. New York: Chapman & Hall.

Ayazloo, M. & Bell, J. N. B. (1981). Studies on the tolerance to sulphur dioxide of grass populations in polluted areas. I. Identification of tolerant populations. *New Phytologist*, **88**, 203–22.

Bachmann, K. (1994). Molecular markers in plant ecology. *New Phytologist*, **126**, 403–18.

Bailey, J. P., Bennett, S. T., Bennett, M. D. & Stace, C. A. (1993). Genomic *in situ* hybridization identifies parental chromosomes in the wild grass hybrid × Festulpia hubbardii. *Heredity*, **71**, 413–20.

Baker, H. G. (1947). Criteria of hybridity. *Nature*, **159**, 1–5.

Baker, H. G. (1951). Hybridization and natural gene-flow between higher plants. *Biological Reviews*, **26**, 302–37.

Baker, H. G. (1954). Report of meeting of British Ecological Society, April 1953. *Journal of Ecology*, **42**, 570–2.

Baker, H. G. (1955). Self-compatibility and establishment after 'long-distance' dispersal. *Evolution*, **9**, 347–9.

Baker, H. G. (1965). Characteristics and modes of origin of weeds. In *The genetics of colonizing species*, ed. H. G. Baker & G. L. Stebbins, pp. 147–72. New York: Academic Press.

Baker, H. G. (1966). The evolution, functioning and breakdown of heteromorphic incompatibility systems. I. The Plumbaginaceae. *Evolution*, **20**, 349–68.

Baker, H. G. (1967). Support for Baker's Law – as a rule. *Evolution*, **21**, 853–6.

Baker, H. G. (1974). The evolution of weeds. *Annual Review of Ecology & Systematics*, **5**, 1–24.

Baker, H. G. (1991). The continuing evolution of weeds. *Economic Botany*, **45**, 445–9.

Bannister, M. H. (1965). Variation in the breeding system of *Pinus radiata*. In *The genetics of colonizing species*, ed. H. G. Baker & G. L. Stebbins, pp. 353–72. New York & London: Academic Press.

Bannister, P. (1976). *Introduction to physiological plant ecology*. Oxford: Blackwell.

Barber, H. N. & Jackson, W. D. (1957). Natural selection in action in *Eucalyptus*. *Nature*, **179**, 1267–9.

Barkley, T. M. (1966). A review of the origin and development of the florists' Cineraria, *Senecio cruentus*. *Economic Botany*, **20**, 386–95.

Barling, D. M. (1955). Some population studies in *Ranunculus bulbosus* L. *Journal of Ecology*, **43**, 207–18.

Barlow, B. A. & Wiens, D. (1977). Host parasite resemblance in Australian Mistletoes: the case for cryptic mimicry. *Evolution*, **31**, 69–84.

Barlow. P. W. & Nevin, D. (1976). Quantitative karyology of some species of *Luzula*. *Plant Systematics & Evolution*, **125**, 77–86.

Barrett, P. H., Gautrey, P. J., Herbert, S., Kohn, D. & Smith, S. (1988). *Charles Darwin's notebooks. 1836–1844*. Cambridge: Cambridge University Press; and London: British Museum (Natural History).

Barrett, S. C. H. (1980*a*). Sexual reproduction in *Eichhornia crassipes* (Water Hyacinth). I. Fertility of clones from diverse regions. *Journal of Applied Ecology*, **17**, 101–12.
Barrett, S. C. H. (1980*b*). Sexual reproduction in *Eichhornia crassipes* (Water Hyacinth). II. Seed production in natural populations. *Journal of Applied Ecology*, **17**, 113–24.
Barrett, S. C. H. (1983). Crop mimicry in weeds. *Economic Botany*, **37**, 255–82.
Barrett, S. C. H. (1985). Floral trimorphism and monomorphism in continental and island populations of *Eichhornia paniculata* (Spreng.) Solms. (Pontederiaceae). *Biological Journal of the Linnean Society*, **25**, 41–60.
Barrett, S. C. H. (1988). The evolution, maintenance and loss of self-incompatibility systems. In *Plant reproductive ecology*, ed. J. Lovett Doust & L. Lovett Doust, pp. 84–124. New York: Oxford University press.
Barrett, S. C. H. (1989*a*). *The evolutionary breakdown of heterostyly*. In *The evolutionary ecology of plants*, ed. J. H. Bock & Y. B. Linhart, pp. 151–69. Boulder: Westview Press.
Barrett, S. C. H. (1989*b*). Mating system evolution and speciation in heterostylous plants. In *Speciation and its consequences*, ed. D. Otte & J. A. Endler, pp. 257–83. Sunderland: Sinauer.
Barrett, S. C. H. (1992). Heterostylous genetic polymorphisms: model systems for evolutionary analysis. In *Evolution and function of heterostyly*, ed. S. C. H. Barrett, pp. 1–29. Heidelberg: Springer Verlag.
Barrett, S. C. H. & Harder, L. D. (1996). Ecology and evolution of plant mating. *Trends in Ecology & Evolution*, **11**, 73–9.
Barrett, S. C. H. & Kohn, J. R. (1991). Genetic and evolutionary consequences of small population size in plants: implications for conservation. In *Genetics and conservation of rare plants*, ed. D. A. Falk & K. E. Holsinger, pp. 3–30. New York: Oxford University Press.
Barrett, S. C. H. & Richardson, B. J. (1986). Genetic attributes of invading species. In *Ecology of biological invasions*, ed. R. H. Groves & J. J. Burdon, pp. 21–33. Canberra: Australian Academy of Science.
Barrett, S. C. H. & Shore, J. S. (1990). Isozyme variation in colonizing plants. In *Isozymes in plant biology*, ed. D. E. Soltis & P. S. Soltis, pp. 106–26. London: Chapman & Hall.
Barros, M. D. C. & Dyer, T. A. (1988). Atrazine resistance in the grass *Poa annua* is due to a single base change in the chloroplast gene for the D_1 protein of photosystem II. *Theoretical & Applied Genetics*, **75**, 610–16.
Bateson, W. (1895*a*). The origin of the cultivated *Cineraria*. *Nature*, **51**, 605–7.
Bateson, W. (1895*b*). The origin of the cultivated *Cineraria*. *Nature*, **52**, 29, 103–4.
Bateson, W. (1897). Notes on hybrid Cinerarias produced by Mr Lynch and Miss Pertz. *Proceedings of the Cambridge Philosophical Society*, **9**, 308–9.
Bateson, W. (1909). *Mendel's principles of heredity*. London: Cambridge University Press; and New York: Macmillan.
Bateson, W. (1913). *Problems of genetics*. London: Oxford University Press; and New Haven, CN: Yale University Press.
Bateson, W. & Punnett, R. C. (1911). On gametic series involving reduplication of certain terms. *Journal of Genetics*, **1**, 293–302.
Bateson, W. & Saunders, E. R. (1902). Experimental studies in the physiology of heredity. *Report to the Evolution Committee of the Royal Society*, **1**, 1–160.

Bateson, W., Saunders, E. R. & Punnett, R. C. (1905). Experimental studies in the physiology of heredity. *Report to the Evolution Committee of the Royal Society*, **2**, 1–55, 80–99.
Battaglia, E. (1963). Apomixis. In *Recent advances in the embryology of angiosperms*, ed. P. Maheshwari, Chapter 8, pp. 221–64. Delhi: University of Delhi Press.
Battjes, J., Bachmann, K. & Bouman, F. (1992). Early development of capitula in *Microseris pygmaea* D. Don strains C96 and A92 (Asteraceae: Lactuceae). *Botanische Jahrbücher für Systematik*, **113**, 461–75.
Battjes, J., Vischer, N. O. E. & Bachmann, K. (1993*a*). Capitulum phyllotaxis and numerical canalization in *Microseris pygmaea* (Asteraceae: Lactuceae). *American Journal of Botany*, **80**, 419–28.
Battjes, J., Vischer, N. O. E. & Bachmann, K. (1993*b*). Meristem geometry and heritable variation in numbers of florets and involucral bracts in *Microseris pygmaea* (Asteraceae: Lactuceae). *Acta Botanica Neerlandica*, **42**, 255–68.
Baum, D. (1994). *rbc*L and seed-plant phylogeny. *Trends in Ecology & Evolution*, **9**, 39–41.
Bawa, K. S., Perry, D. R. & Beach, J. H. (1985). Reproductive biology of tropical lowland rain forest trees. I. Sexual systems and incompatibility mechanisms. *American Journal of Botany*,**72**, 331–45.
Bayer, R. J., Ritland, K. & Purdy, B. G. (1990). Evidence of partial apomixis in *Antennaria media* (Asteraceae: Inuleae) detected by segregation of genetic markers. *American Journal of Botany*, **77**, 1078–83.
Beach, J. H. & Kess, W. J. (1980). Sporophyte versus gametophyte: a note on the origin of self-incompatibility in flowering plants. *Systematic Botany*, **5**, 1–5.
Beattie, A. (1978). Plant–animal interactions affecting gene flow in *Viola*. In *The pollination of flowers by insects*, ed. A. J. Richards, pp. 151–64, Linnean Society Symposium Series 6. London: Academic Press.
Beddall, B. G. (1957). Historical notes on avian classification. *Systematic Zoology*, **6**, 129–36.
Bell, G. (1982). *The masterpiece of nature: the evolution and genetics of sexuality*. London: Croom Helm.
Belzer, N. F. & Ownbey, M. (1971). Chromatographic comparison of *Tragopogon* species and hybrids. *American Journal of Botany*, **58**, 791–802.
Bennett, J. H. (1983). *Natural selection, heredity and eugenics*. Oxford: Oxford University Press.
Bennett, K. D. (1997). *Evolution and ecology. The pace of life*. Cambridge: Cambridge University Press.
Bennett, M. D. (1995). The development and use of genomic *in situ* hybridization (GISH) as a new tool in plant biosystematics. In *Kew Chromosome Conference IV*, ed. P. E. Brandham & M. D. Bennett, pp. 167–83. London: Royal Botanic Gardens, Kew.
Bennett, M. D. & Smith, J. B. (1991). Nuclear DNA amounts in angiosperms. *Philosophical Transactions of the Royal Society of London, B*, **334**, 309–45.
Bennett, S. T., Kenton, A. Y. & Bennett, M. D. (1992). Genomic *in situ* hybridization reveals the allopolyploid nature of *Milium montianum* (Graminae). *Chromosoma*, **101**, 420–4.
Benson, L. (1962). *Plant taxonomy*. New York: Ronald Press.
Bergman, B. (1935). Zytologische Studien über sexuelles und asexuelles *Hieracium umbellatum*. *Hereditas*, **20**, 47–64.

Bergman, B. (1941). Studies on the embryo sac mother cell and its development in *Hieracium* subg. *Archieracium*. *Svensk Botanisk Tidsskrift*, **35**, 1–42.

Berlin, B., Breedlove, D. E. & Raven, P. H. (1974). *Principles of Tzeltal plant classification*. London & New York: Academic Press.

Berry, P. E., Tobe, H. & Gómez, J. A. (1991). Agamospermy and the loss of distyly in *Erythroxylum undulatum* (Erythroxylaceae) from northern Venezuela. *American Journal of Botany*, **78**, 595–600.

Berry, R. J. (1977). *Inheritance and natural history*. London: Collins.

Bessey, C. E. (1897). Phylogeny and taxonomy of the angiosperms. *Botanical Gazette*, **24**, 145–78.

Bessey, C. E. (1915). The phylogenetic taxonomy of flowering plants. *Annals of the Missouri Botanical Garden*, **2**, 109–64.

Bharthan, G., Lambert, G. & Galbraith, D. W. (1994). Nuclear DNA content of monocotyledons and related taxa. *American Journal of Botany*, **81**, 381–6.

Billington, H. L. (1991). Effect of population size on genetic variation in a dioecious conifer. *Conservation Biology*, **5**, 115–19.

Bishop, J. A. & Cook, L. M. (1981) *Genetic consequences of man made change*. London: Academic Press.

Bishop, J. A. & Korn, M. E. (1969). Natural selection and cyanogenesis in White Clover, *Trifolium repens*. *Heredity*, **24**, 423–30.

Bishop, O. (1971). *Statistics for biology. A practical guide for the experimental biologist*, 2nd edn. London: Longmans.

Bittrich, V. & Kadereit, J. (1988). Cytogenetical and geographical aspects of sterility in *Lysimachia nummularia*. *Nordic Journal of Botany*, **8**, 325–8.

Blakeslee, A. F. & Avery, A. G. (1937). Methods of inducing chromosome doubling in plants. *Journal of Heredity*, **28**, 393–411.

Bøcher, T. W. (1949). Racial divergences in *Prunella vulgaris* in relation to habitat and climate. *New Phytologist*, **48**, 285–314.

Bøcher, T. W. (1963). The study of ecotypical variation in relation to experimental morphology. *Regnum Vegetabile*, **27**, 10–16.

Bøcher, T. W. & Larsen, K. (1958). Geographical distribution of initiation of flowering, growth habit and other factors in *Holcus lanatus*. *Botaniska Notiser*, **3**, 289–300.

Bøcher, T. W. & Lewis, M. C. (1962). Experimental and cytological studies on plant species. 7. *Geranium sanguineum*. *Biologiske Skrifter*, **11**, 1–25.

Bohm, W. (1979). *Methods of studying root systems*. Berlin, Heidelberg & New York: Springer Verlag.

Bolkhovskikh, Z., Grif, V., Matvejeva, T. & Zakharyeva, O. (1969). *Chromosome numbers of flowering plants*. Leningrad: Academy of Sciences of the USSR.

Bonnier, G. (1890). Cultures expérimentales dans les Alpes et les Pyrénées. *Revue Générale de Botanique*, **2**, 513–46.

Bonnier, G. (1895). Recherches expérimentales sur l'adaptation des plantes au climat Alpin. *Annales des Sciences Naturelles* (*Botanique*), **20**, 217–360.

Bonnier, G. (1920). Nouvelles observations sur les cultures expérimentales à diverses altitudes et cultures par semis. *Revue Générale de Botanique*, **32**, 305–26.

Borg, S. J. (1972). *Variability of* Rhinanthus serotinus (*Schönh.*) *Oborny in relation to environment*. Thesis, Rijksuniversiteit te Groningen, 158 pp.

Borgström, G. (1939). Formation of cleistogamic and chasmogamic flowers in Wild Violets as a photoperiodic response. *Nature*, **144**, 514–15.

Bosemark, N. O. (1954). On accessory chromosomes in *Festuca pratensis*. I. Cytological investigations. *Hereditas*, **40**, 346–76.
Boswell Syme, J. T. (ed.) (1866). *English botany; or coloured figures of British plants*, 3rd edn, vol. 6. London: Hardwicke.
Bowlby, J. (1990). *Charles Darwin. A biography*. London: Hutchinson.
Bowler, P. J. (1989*a*). *The Mendelian revolution*. London: The Athlone Press.
Bowler, P. J. (1989*b*). *Evolution: the history of an idea*. Revised edn. Berkeley: University of California Press.
Bowles, M. L. & Whelan, C. J. (1994). *Restoration of endangered species, conceptual issues, planning and implementation*. Cambridge: Cambridge University Press.
Box, J. F. (1978). *R. A. Fisher. The life of a scientist*. New York, Chichester, Brisbane & Toronto: Wiley.
Boyer, S. H. (1972). Extraordinary incidence of electrophoretically silent genetic polymorphisms. *Nature*, **239**, 453–4.
Brackman, A. C. (1980). *A delicate arrangement: the strange case of Charles Darwin and Alfred Russel Wallace*. New York: Times Books.
Bradshaw, A. D. (1959*a*). Population differentiation in *Agrostis tenuis* Sibth. I. Morphological differentiation. *New Phytologist*, **58**, 208–27.
Bradshaw, A. D. (1959*b*). Population differentiation in *Agrostis tenuis* Sibth. II. The incidence and significance of infection by *Epichloë typhina*. *New Phytologist*, **58**, 310–15.
Bradshaw, A. D. (1959*c*). Studies of variation in bent grass species. II. Variation within *Agrostis tenuis*. *Journal of the Sports Turf Research Institute*, **10**, 1–7.
Bradshaw, A. D. (1960). Population differentiation in *Agrostis tenuis* Sibth. III. Populations in varied environments. *New Phytologist*, **59**, 92–103.
Bradshaw, A. D. (1965). Evolutionary significance of phenotypic plasticity in Plants. *Advances in Genetics*, **13**, 115–55.
Bradshaw, A. D. (1972). Some of the evolutionary consequences of being a plant. *Evolutionary Biology*, **5**, 25–47.
Bradshaw, A. D. (1976). Pollution and evolution. In *Effects of air pollution on plants*, ed. T. A. Mansfield, pp. 135–59. London, New York & Melbourne: Cambridge University Press.
Bradshaw, A. D. (1989). Is evolution fettered or free? *Transactions of the Botanical Society of Edinburgh*, **45**, 303–11.
Bradshaw, A. D. & McNeilly, T. (1981). *Evolution and pollution*. London: Arnold.
Bradshaw, A. D. & McNeilly, T. (1991). Evolutionary response to global climate change. *Annals of Botany*, 67 (Suppl.), 5–14.
Bradshaw, M. E. (1963*a*). Studies on *Alchemilla filicaulis* Bus., *sensu lato* and *A. minima* Walters. Introduction and I. Morphological variation in *A. filicaulis, sensu lato*. *Watsonia*, **5**, 304–20.
Bradshaw, M. E. (1963*b*). Studies on *Alchemilla filicaulis* Bus., *sensu lato* and *A. minima* Walters. II. Cytology of *A. filicaulis, sensu lato*. *Watsonia*, **5**, 321–6.
Bradshaw, M. E. (1964). Studies on *Alchemilla filicaulis* Bus., *sensu lato* and *A. minima* Walters. III. *Alchemilla minima*. *Watsonia*, **6**, 76–81.
Brand, C. J. & Waldron, L. R. (1910). Cold resistance of Alfalfa and some factors influencing it. *U. S. Department of Agriculture, Bureau of Plant Industry. Bulletin*, No. 185, 1–80.
Brannigan, A. (1979). The reification of Mendel. *Social Studies of Science*, **9**, 423–54.

Brehm, B. G. & Ownbey, M. (1965). Variation in chromatographic patterns in the *Tragopogon dubius-pratensis-porrifolius* complex (Compositae). *American Journal of Botany*, **52**, 811–18.
Breiman, A. & Graur, D. (1995). Wheat evolution. *Israel Journal of Plant Sciences*, **43**, 85–98.
Breitenbach-Dorfer, M., Pinsker, W. Hacker, R. & Müller, F. (1992). Clone identification and clinal allozyme variation in populations of *Abies alba* from the eastern Alps (Austria). *Plant Systematics & Evolution*, **181**, 109–20.
Brenchley, W. E. & Warington, K. (1969). *The park grass plots at Rothamsted*, 1856–1949. Harpenden: Rothamsted Experimental Station.
Bretagnolle, F. & Thompson, J. D. (1995). Gametes with the somatic chromosome number: mechanisms of their formation and role in the evolution in autoployploid plants. *New Phytologist*, **129**, 1–22.
Briand, F., Dubost, M., Pitt, D. & Rambaud, D. (1989). *The Alps. A system under pressure*. Chambéry: IUCN.
Briggs, D. & Block, M. (1981). An investigation into the use of the '-deme' terminology. *New Phytologist*, **89**, 729–35.
Brochmann, C., Soltis, P. S. & Soltis, D. E. (1992*a*). Recurrent formation and polyphyly of Nordic polyploids in *Draba* (Brassicaceae). *American Journal of Botany*, **79**, 673–88.
Brochmann, C., Soltis, P. S. & Soltis, D. E. (1992*b*). Multiple origins of the octoploid Scandinavian endemic *Draba cacuminum*: electrophoretic and morphological evidence. *Nordic Journal of Botany*, **12**, 257–72.
Brooks, J. L. (1983). *Just before the Origin: Alfred Russel Wallace's theory of evolution*. New York: Columbia University Press.
Brougham, R. W. & Harris, W. (1967). Rapidity and extent of changes in genotypic structure induced by grazing in a Ryegrass population. *New Zealand Journal of Agricultural Research*, **10**, 56–65.
Brown, A. D. H. & Marshall, D. R. (1981). Evolutionary changes accompanying colonization in plants. In *Evolution today*, ed. G. C. C. Sudder & J. L. Reveal, pp. 351–63. Pittsburgh: Hunt Institute for Botanical Documentation.
Brown, A. D. H. & Schoen, D. J. (1992). Plant population genetic structure and biological conservation. In *Conservation of biodiversity for sustainable development*, ed. O. T. Sandlund, K. Hindar & A. D. H. Brown, pp. 88–104. Oslo: Scandinavian University Press.
Brown, A. H. D. & Burdon, J. J. (1983). Multilocus diversity in an outbreeding weed, *Echium plantagineum* L. *Australian Journal of Biological Sciences*, **36**, 503–9.
Brown, A. H. D. (1979). Enzyme polymorphism in plant populations. *Theoretical Population Biology*, **15**, 1–42.
Brown, R. C. & Schaack, C. G. (1972). Two new species of *Tragopogon* for Arizona. *Madroño*, **21**, 304.
Brown, V. K. & Lawton, J. H. (1991). Herbivory and the evolution of leaf size and shape. *Philosophical Transactions of the Royal Society of London, B*, **333**, 265–72.
Browne, J. (1995). *Charles Darwin: voyaging*. London: Jonathan Cape.
Bruhin, A. (1950). Beiträge zur Zytologie und Genetik schweizerischer *Crepis-Arten. Arbeiten aus dem Institut für Allgemeine Botanik der Universität Zürich, Serie B*, **1**, 1–101.
Bruhin, A. (1951). Auslösung von Mutationen in ruhenden Samen durch hohe Temperaturen. *Naturwissenschaften*, **38**, 565–6.

Brummitt, R. K. (1996). In defence of paraphyletic taxa. In *The biodiversity of African plants*, ed. L. J. G. van der Maesen *et al.*, pp. 371–84. Dordrecht: Kluwer Academic Publishers.

Bulmer, M. G. (1967). *Principles of statistics*, 2nd edn. London & Edinburgh: Oliver & Boyd.

Burchfield, J. D. (1975). *Lord Kelvin and the age of the earth.* London: Macmillan.

Burdon, J. J. (1980). Intraspecific diversity in a natural population of *Trifolium repens. Journal of Ecology*, **68**, 717–35.

Burdon, J. J. (1987). *Disease and plant population biology.* Cambridge: Cambridge University Press.

Burdon, J. J., Marshall, D. R. & Groves, R. H. (1980). Isozyme variation in *Chondrilla juncea* in Australia. *Australian Journal of Botany*, **28**, 193–8.

Burkhardt, F. & Smith, S. (1985). *The correspondence of Charles Darwin.* Cambridge: Cambridge University Press.

Burkhardt, F. & Smith, S. (1991). *The correspondence of Charles Darwin.* vol. 7, 1858–9. Cambridge: Cambridge University Press.

Burkill, I. H. (1895). On the variations in number of stamens and carpels. *Journal of the Linnean Society* (*Botany*), **31**, 216–45.

Cabrera, A. L. (1977). Mutisieae – systematic review. In *The biology and chemistry of the Compositae*, ed. V. H. Heywood, J. B. Harborne & B. L. Turner, pp. 1039–66. London: Academic Press.

Cahn, M. A. & Harper, J. L. (1976*a*). The biology of the leaf mark polymorphism in *Trifolium repens.* 1. Distribution of phenotypes at a local scale. *Heredity*, **37**, 309–25.

Cahn, M. A. & Harper, J. L. (1976*b*). The biology of the leaf mark polymorphism in *Trifolium repens.* 2. Evidence for the selection of marks by rumen fistulated sheep. *Heredity*, **37**, 327–33.

Cain, A. J. (1958). Logic and memory in Linnaeus's system of taxonomy. *Proceedings of the Linnean Society of London*, **169**, 144–63.

Callender, L. A. (1988). Gregor Mendel: an opponent of descent with modification. *History of Science*, **26**, 41–75.

Camp, W. H. & Gilly, C. L. (1943). The structure and origin of species. *Brittonia*, **4**, 323–85.

Campbell, R. C. (1967). *Statistics for biologists.* London & New York: Cambridge University Press. (2nd edn, 1974.)

Chadwick, M. J. & Salt, J. K. (1969). Population differentiation within *Agrostis tenuis* L. in response to colliery spoil substrate factors. *Nature*, **224**, 186.

Challice, J. & Kovanda, M. (1978). Chemotaxonomic survey of the genus *Sorbus* in Europe. *Naturwissenschaften*, **65**, 111–12.

Chanway, C. P., Holl, F. B. & Turkington, R. (1989). Effect of *Rhizobium leguminosarum* biovar. *trifolii* genotype on specificity between *Trifolium repens* and *Lolium perenne. Journal of Ecology*, **77**, 1150–60.

Charlesworth, B. & Charlesworth, D. (1978). A model for the evolution of dioecy and gynodioecy. *American Naturalist*, **112**, 975–97.

Charlesworth, D. & Charlesworth, B. (1979). The evolutionary genetics of sexual systems in flowering plants. *Proceedings of the Royal Society London, B*, **205**, 513–30.

Charlesworth, D. & Charlesworth, B. (1987). Inbreeding depression and its evolutionary consequences. *Annual Review of Ecology & Systematics*, **18**, 237–68.

Charnov, E. L. (1988). Foreword. In *Plant reproductive ecology*, ed. J. Lovett Doust & L. Lovett Doust, pp. ix–x. New York: Oxford University Press.
Chase, M. W., Duvall, M. R., Hills, H. G., Conran, J. G., Cox, A. V., Eguiarte, L. E., Hartwell, J., Fay, M. F., Caddick, L. R., Cameron, K. M. & Hoot, S. (1995). Molecular phylogenetics of Lilianae. In *Monocotyledons: systematics and evolution*, ed. P. J. Rudall, P. J. Cribb, D. F. Cutler & C. J. Humphries, pp. 109–37. London: Royal Botanic Gardens, Kew.
Chen, Q. F. & Armstrong, K. (1994). Genomic *in situ* hybridization in *Avena sativa*. *Genome*, **37**, 607–12.
Cheplick, G. P. & Quinn, J. A. (1982). *Amphicarpum purshii* and the 'pessimistic strategy' in amphicarpic annuals with subterranean fruit. *Oecologia*, **52**, 327–32.
Cheplick, G. P. & Quinn, J. A. (1983). The shift in aerial/subterranean fruit ratio in *Amphicarpum purshii*: causes and significance. *Oecologia*, **57**, 374–9.
Cherfas, J. (1991). Ancient DNA: still busy after death. *Science*, **253**, 1354–6.
Chung, M. G., Hamrick, J. L., Jones, S. B. & Derda, G. S. (1991). Isozyme variation within and among populations of *Hosta* (Liliaceae) in Korea. *Systematic Botany*, **16**, 667–84.
Clapham, A. R., Tutin, T. G. & Warburg, E. F. (1981). *Excursion flora of the British Isles*. London: Cambridge University Press.
Clark, L. G., Zang, W. & Wendel, J. F. (1995). A phylogeny of the grass family Poaceae based on *ndh*F sequence data. *Systematic Botany*, **20**, 436–60.
Clarke, G. M. (1980). *Statistics and experimental design*, 2nd edn. London: Arnold.
Clausen. J. (1951). *Stages in the evolution of plant species*. London: Oxford University Press; and New York: Cornell University Press.
Clausen, J. & Hiesey, W. M. (1958). *Experimental studies on the nature of species. IV. Genetic structure of ecological races*. Carnegie Institution of Washington Publication no. 615, Washington DC.
Clausen, J., Keck, D. D. & Hiesey, W. M. (1939). The concept of species based on experiment. *American Journal of Botany*, **26**, 103–6.
Clausen, J., Keck, D. D. & Hiesey, W. M. (1940). *Experimental studies on the nature of species. I. The effect of varied environments on western North American plants*. Carnegie Institution of Washington Publication no. 520, pp. 1–452, Washington DC.
Clausen, J., Keck, D. D. & Hiesey, W. M. (1941). Experimental taxonomy. *Carnegie Institution of Washington Year Book*, **40**, 160–70.
Clausen, R. E. & Goodspeed, T. H. (1925). Interspecific hybridization in *Nicotiana*. II. A tetraploid *glutinosa-tabacum* hybrid, an experimental verification of Winge's hypothesis. *Genetics*, **10**, 279–84.
Clement, E. J. & Foster, M. C. (1994). *Alien plants of the British Isles*. London: Botanical Society of the British Isles.
Clements, F. E., Martin, E. V. & Long, F. L. (1950). *Adaptation and origin in the plant world. The role of environment in evolution*. Waltham, MA: Chronica Britanica Co.
Cochran, W. G. (1963). *Sampling techniques*, 2nd edn. New York: Wiley.
Cockburn, A. (1991). *An introduction to evolutionary ecology*. Oxford: Blackwell.
Cody, M. L. (1986). Diversity, rarity, and conservation in Mediterranean climate regions. In *Conservation biology: the science of scarcity & diversity*, ed. M. E. Soulé, pp. 122–52. Sunderland: Sinauer.
Coen, E. S. (1991). The role of homeotic genes in flower development and evolution. *Annual Review of Plant Physiology & Plant Molecular Biology*, **42**, 241–79.

Coen, E. S. & Meyerwitz, E. M. (1991). The war of the whorls: genetic interactions controlling flower development. *Nature*, **353**, 31–7.
Collins, J. L. (1927). A low temperature type of albinism in Barley. *Journal of Heredity*, **33**, 82–6.
Cook, C. D. K. (1968). Phenotypic plasticity with particular reference to three amphibious plant species. In *Modern methods in plant taxonomy*, ed. V. H. Heywood, pp. 97–111. London: Academic Press.
Cook, S. A. (1962). Genetic system, variation and adaptation in *Eschscholzia californica. Evolution*, **16**, 278–99.
Cook, S. A. & Johnson, M. P. (1968). Adaptation to heterogeneous environments. I. Variation in heterophylly in *Ranunculus flammula* L. *Evolution*, **22**, 496–516.
Corcos, A. F. & Monaghan, F. V. (1990). Mendel's work and its rediscovery: a new perspective. *Critical Reviews in Plant Sciences*, **9**, 197–212.
Corkhill, L. (1942). Cyanogenesis in White Clover (*Trifolium repens* L.) V. The inheritance of cyanogenesis. *New Zealand Journal of Science & Technology, B*, **23**, 178–93.
Correns, C. (1909). Vererbungsversuche mit blass (gelb) grunen und buntblättrigen Sippen bei *Mirabilis jalapa*, *Urtica pilulifera*, und *Lunularia annua. Zeitschrift für Vererbungslehre*, **1**, 291–329.
Correns, C. (1913). Selbststerilität und Individualstoffe. *Biologisches Zentralblatt*, **33**, 389–443.
Cott, H. B. (1940). *Adaptive coloration in animals.* London: Methuen.
Coughtrey, P. J. & Martin, M. H. (1978). Tolerance of *Holcus lanatus* to lead, zinc and cadmium in factorial combination. *New Phytologist*, **81**, 147–54.
Cousens, R. & Mortimer, M. (1995). *Dynamics of weed populations.* Cambridge: Cambridge University Press.
Cowan, R. S. (1972). Francis Galton's statistical ideas: the influence of eugenics. *Isis*, **63**, 509–28.
Cowling, R. M. (1992). *The ecology of fynbos: nutrients, fire and diversity.* Capetown: Oxford University Press.
Cox , P. A. (1988). Monomorphic and dimorphic sexual strategies: a modular approach. In *Plant reproductive ecology*, ed. J. Lovett Doust & L. Lovett Doust, pp. 80–97. New York: Oxford University Press.
Cracraft, J. (1983). Species concepts and speciation analysis. *Current Ornithology*, **1**, 159–87.
Crane, P. R. (1987). Vegetational consequences of angiosperm diversification. In *The origins of angiosperms and their biological consequences*, ed. E. M. Friis, W. G. Chaloner & P. R. Crane, pp. 107–44. Cambridge: Cambridge University Press.
Crawford, D. J. (1990*a*). Enzyme electrophoresis and plant systematics. In *Isozymes in plant biology*, ed. D. E. Soltis & P. S. Soltis, pp. 146–64. London: Chapman & Hall.
Crawford,D. J. (1990*b*). *Plant molecular systematics.* New York: Wiley.
Crawford, D. J., Ornduff, R. & Vasey, M. C. (1985). Allozyme variation within and between *Lasthenia minor* and its derivative species, *Lasthenia maritima* (Asteraceae). *American Journal of Botany*, **72**, 1177–84.
Crawford, R. M. M. (1989). *Studies in plant survival.* Oxford: Blackwell.
Crawford, T. J. (1984). What is a population? In *Evolutionary ecology*, ed. B. Sharrocks, pp. 135–73. Oxford: Blackwell.
Crawford, T. J. & Jones, D. A. (1986). Variation in the colour of the keel petals

in *Lotus corniculatus* L. 2. Clines in Yorkshire and adjacent counties. *Watsonia*, **16**, 15–19.

Crawford-Sidebotham, T. J. (1971). *Studies of aspects of slug behaviour and the relation between molluscs and cyanogenic plants.* PhD Thesis, University of Birmingham.

Crawley, M. J. (1987). What makes a community invasible? In *Colonization, succession and stability*, ed. A. J. Gray, M. J. Crawley & P. J. Edwards, pp. 429–53. Oxford: Blackwell.

Crawley, M. J., Hails, R. S., Rees, M., Kohn, D. & Buxton, J. (1993). Ecology of transgenic oilseed rape in natural habitats. *Nature*, **363**, 620–3.

Crew, F. A. E. (1966). *Mendelism comes to England.* In G. *Mendel Memorial Symposium.* 1865–1965, ed. M. Sosna, pp. 15–30. Prague: Academia Publishing House of the Czechoslovak Academy of Sciences.

Crick, F. (1989). *What mad pursuit: a personal view of scientific discovery.* Harmondsworth: Penguin.

Crofts, A. & Jefferson, R. G. (1994). *The lowland grassland management handbook.* English Nature/ The wildlife Trusts.

Cronk, Q. C. B. & Fuller, J. L. (1995). *Plant invaders.* London: Chapman & Hall.

Cronquist, A. (1987). A botanical critique of cladism. *Botanical Review*, **53**, 1–52.

Cronquist, A. (1988). *The evolution and classification of flowering plants*, 2nd edn. New York: New York Botanical Garden.

Cruden, R. W. & Hermann-Parker, S. M. (1977). Temporal dioecism: an alternative to dioecism? *Evolution*, **31**, 863–6.

Curtis, O. F. & Clark D. G. (1950). *An introduction to plant physiology.* London, New York & Toronto: McGraw-Hill.

Daday, H. (1954*a*). Gene frequencies in wild populations of *Trifolium repens.* 1. Distribution by latitude. *Heredity*, **8**, 61–78.

Daday, H. (1954*b*). Gene frequencies in wild populations of *Trifolium repens.* 2. Distribution by altitude. *Heredity*, **8**, 377–84.

Daday, H. (1965). Gene frequencies in wild populations of *Trifolium repens* L. IV. Mechanism of natural selection. *Heredity*, **20**, 355–66.

Dafni, A. (1992). *Pollination ecology: a practical approach.* Oxford: Oxford University Press.

Dahlgren, K. V. O. (1922). Selbststerilität interhalb Klonen von *Lysimachia nummularia. Hereditas*, **3**, 200–10.

Dahlgren, R. & Rasmussen, F. N. (1983). Monocotyledon evolution; characters and phylogenetic estimation. In *Evolutionary Biology*, **16**, ed. M. K. Hecht, B. Wallace & G. T. Prance, pp. 255–395. New York: Plenum Press.

D'Amato, F. & Hoffmann-Ostenhof, O. (1956). Metabolism and spontaneous mutations in plants. *Advances in Genetics*, **8**, 1–28.

Darlington, C. D. (1937). *Recent advances in cytology*, 2nd edn. London: Churchill.

Darlington, C. D. (1939). *The evolution of genetic systems.* London: Cambridge University Press.

Darlington, C. D. (1956). *Chromosome botany.* London: Allen & Unwin.

Darlington, C. D. (1963). *Chromosome botany and the origins of cultivated plants.* London: Allen & Unwin.

Darlington, C. D. & Wylie, A. P. (1955). *Chromosome atlas of flowering plants*, 2nd edn. London: Allen & Unwin.

Darwin, C. (1859). *On the origin of species by means of natural selection.* London: Murray. (6th edn. 1872.)

Darwin, C. (1862). *On the various contrivances by which British and foreign Orchids are fertilised by insects and on the good effects of crossing.* London: Murray.

Darwin, C. (1868). *The variation of plants and animals under domestication.* London: Murray.

Darwin, C. (1871). Pangenesis. *Nature*, **3**, 502–3.

Darwin. C. (1876). *The effects of cross- and self-fertilisation in the vegetable kingdom.* London: Murray.

Darwin, C. (1877*a*). *The different forms of flowers of the same species.* London: Murray.

Darwin, C. (1877*b*). *The various contrivances by which Orchids are fertilised by insects*, 2nd edn. London: Murray.

Darwin, C. & Wallace, A. (1859). On the tendency of species to form varieties; and on the perpetuation of varieties and species by natural means of selection. *Proceedings of the Linnean Society of London*, **3**, 45–62.

Darwin. F. (ed.) (1909*a*). *The foundations of the origin of species. A sketch written in* 1842 *by Charles Darwin.* Cambridge: Cambridge University Press.

Darwin, F. (ed.) (1909*b*). *The foundations of the origin of species. Two essays written in* 1842 *and* 1844 *by Charles Darwin.* London: Cambridge University Press.

Davenport, C. B. (1904). *Statistical methods with special reference to biological variation*, 2nd edn. London: Chapman & Hall; and New York: Wiley.

David, F. N. (1971). *A first course in statistics*, 2nd edn. London: Griffin.

Davidson, J. F. (1947). The polygonal graph for simultaneous portrayal of several variables in population analysis. *Madroño*, **9**, 105–10.

Davies, E. (1956). Cytology, evolution, and origin of the aneuploid series in the genus *Carex*. *Hereditas*, **42**, 349–65.

Davies, E. W. (1955). The cytogenetics of *Carex flava* and its allies. *Watsonia*, **3**, 129–37.

Davies, M. S. (1975). Physiological differences among populations of *Anthoxanthum odoratum* collected from the Park Grass experiment. IV. Response to potassium and magnesium. *Journal of Applied Ecology*, **12**, 953–64.

Davies, M. S. (1993). Rapid evolution in plant populations. In *Evolutionary patterns and processes*, ed. D. R. Lees & D. Edwards, Linnean Society Symposium Series **14**, pp. 172–88. London: Published for the Linnean Society by Academic Press.

Davies, M. S. & Snaydon, R. W. (1973*a*). Physiological differences among populations of *Anthoxanthum odoratum* collected from the Park Grass experiment. I. Response to calcium. *Journal of Applied Ecology*, **10**, 33–45.

Davies, M. S. & Snaydon, R. W. (1973*b*). Physiological differences among populations of *Anthoxanthum odoratum* collected from the Park Grass experiment. II. Response to aluminium. *Journal of Applied Ecology*, **10**, 47–55.

Davies, M. S. & Snaydon, R. W. (1974). Physiological differences among populations of *Anthoxanthum odoratum* collected from the Park Grass experiment. III. Response to phosphate. *Journal of Applied Ecology*, **11**, 699–707.

Davies, M. S. & Snaydon, R. W. (1976). Rapid population differentiation in a mosaic environment. III. Measures of selection pressures. *Heredity*, **36**, 59–66.

Davies, T. M. & Snaydon, R. W. (1989). An assessment of the spaced-plant trial technique. *Heredity*, **63**, 37–45.

Davies, W. E. (1963). Leaf markings in *Trifolium repens*. In *Teaching genetics in school and university*, ed. C. D. Darlington & A. D. Bradshaw, pp. 94–8. Edinburgh: Oliver & Boyd.

Davis, P. H. & Heywood, V. H. (1963). *Principles of angiosperm taxonomy*. Edinburgh: Oliver & Boyd; and New York: Van Nostrand.

Davis, J. I. (1995). Species concepts and phylogenetic analysis – introduction. *Systematic Botany*, **20**, 555–9.

Davison, A. W. & Reiling, K. (1995). A rapid change in ozone resistance of *Plantago major* after summers with high ozone concentrations. *New Phytologist*, **131**, 337–44.

Davy, A. J. & Jefferies, R. L. (1981). Approaches to the monitoring of rare plant populations. In *The biological aspects of rare plant conservation*, ed. H. Synge, pp. 219–32. London: Wiley.

Dawson, C. D. R. (1941). Tetrasomic inheritance in *Lotus corniculatus* L. *Journal of Genetics*, **42**, 49–72.

De Beer, G. (ed.) (1960–61). *Darwin's notebooks on transmutation of species*. I–IV. Bulletin of British Museum (Natural History). Historical Series 2 nos. 2–6.

De Beer, G. (1963). *Charles Darwin*. London: Nelson.

De Beer, G. (1964). *Atlas of evolution*. London: Nelson.

De Haan, A., Maceira, N. O., Lumaret, R. & Delay, J. (1992). Production of 2*n* gametes in diploid subspecies of *Dactylis glomerata* L. 2. Occurrence and frequency of 2*n* eggs. *Annals of Botany*, **69**, 345–50.

de Litardière, R. (1939). Sur les caractères chromosomiques et la systématique des *Poa* du group du *P. annua* L. *Revue de Cytologie et de Cytophysiologie Végétales*, **4**, 82–5.

De Nettancourt, D. (1977). *Incompatibility in angiosperms*. Berlin, Heidelberg & New York: Springer Verlag.

de Vilmorin, P. (1910). Recherches sur l'héredité Mendélienne. *Compte Rendu Hebdomadaire des Séances de l Académie des Sciences, Paris*, **151**, 548–51.

de Vilmorin, P. (1911). Etude sur la caractère adhérence des grains entre eux chez 'le Pois, Chenille'. 4*th International Conference on Genetics, Paris*, 368–72.

de Vilmorin, P. & Bateson, W. (1911). A case of gametic coupling in *Pisum*. *Proceedings of the Royal Society, B*, **84**, 9–11.

De Vries, H. (1894). Uber halbe Galton-Kurven als Zeichnen diskontinurlichen Variation. *Bericht der Deutschen Botanischen Gesellschaft*, **12**, 197–207.

De Vries, H. (1897). Monstruosités héréditaires offertes en échange aux jardins botaniques. *Botanisch Jaarboek*, **9**, 80–93.

De Vries, H. (1905). *Species and varieties their origin by mutation*. Chicago: Open Court Publishing Co.

de Wet, J. M. J. (1968). Diploid-tetraploid-haploid cycles and the origin of variability in *Dichanthium* agamospecies. *Evolution*, **22**, 394–7.

de Wet, J. M. J. (1971). Reversible tetraploidy as an evolutionary mechanism. *Evolution*, **25**, 545–8.

de Wet, J. M. J. (1980). Origins of polyploids. In *Polyploidy*, ed. W. H. Lewis, pp. 3–15. New York & London: Plenum Press.

de Wet, J. M. J. & Harlan, J. R. (1970). Apomixis, polyploidy and speciation in *Dichanthium*. *Evolution*, **24**, 270–7.

de Wet, J. M. J. & Harlan, J. R. (1972). Chromosome pairing and phylogenetic affinities. *Taxon*, **21**, 67–70.

Delph, L. F. (1996). Flower size dimorphism in plants with unisexual flowers. In *Floral biology. Studies on floral evolution in animal-pollinated plants*, ed. D. G. Lloyd & S. C. H. Barrett, pp. 217–37. New York: Chapman & Hall.

Demauro, M. M. (1993). Relationship of breeding system to rarity in the Lakeside Daisy (*Hymenoxys acaulis* var. *glabra*). *Conservation Biology*, **7**, 542–50.

Demauro, M. M. (1994). Development and implementation of a recovery program for the federal threatened Lakeside Daisy (*Hymenoxys acaulis* var. *acaulis*). In *Restoration of endangered species*, ed. M. L. Bowles & C. J. Whelan, pp. 298–321. Cambridge: Cambridge University Press.

Desmond, A. & Moore, J. (1991). *Darwin*. London: Michael Joseph.

Dettner, K. & Liepert, C. (1994). Chemical mimicry and camouflage. *Annual Review of Entomology*, **39**, 129–54.

Diamond, J. M. (1975). The island dilemma: lessons of modern biogeographic studies for the design of natural reserves. *Biological Conservation*, **7**, 129–146.

Diamond, J. M. (1985). How many unknown species are yet to be discovered? *Nature*, **315**, 358–9.

Di Cesnola, A. P. (1904). Preliminary note on the protective value of colour in *Mantis religiosa*. *Biometrika*, **3**, 58–9.

Digby, L. (1912). The cytology of *Primula kewensis* and of other related *Primula* hybrids. *Annals of Botany*, **26**, 357–88.

Dirzo, R. & Harper, J. L. (1982*a*). Experimental studies of slug–plant interactions. III. Differences in the acceptability of individual plants of *Trifolium repens* in the field. *Journal of Ecology*, **70**, 101–17.

Dirzo, R. & Harper, J. L. (1982*b*). Experimental studies of slug–plant interactions. IV. The performance of cyanogenic and acyanogenic morphs of *Trifolium repens* in the field. *Journal of Ecology*, **70**, 119–38.

Dobzhansky, T. (1935). A critique of the species concept in biology. *Philosophy of Science*, **2**, 344–55.

Dobzhansky, T. (1937). *Genetics and the origin of species*. New York: Colombia University Press.

Dobzhansky, T. (1941). *Genetics and the origin of species*. New York: Columbia University Press.

Dommée, B., Assouad, M. W. & Valdeyron, G. (1978). Natural selection and gynodioecy in *Thymus vulgaris* L. *Botanical Journal of the Linnean Society*, **77**, 17–28.

Donoghue, M. J. (1994). Progress and prospects in reconstructing plant phylogeny. *Annals of the Missouri Botanical Garden*, **81**, 405–18.

Donoghue, M. J. & Catino, P. D. (1988). Paraphyly, ancestors, and the goals of taxonomy: a botanical defense of cladism. *Botanical Review*, **54**, 107–28.

Donoghue, M. J. & Kadereit, J. W. (1992). Walter Zimmermann and the growth of phylogenetic theory. *Systematic Biology*, **41**, 74–85.

Donoghue, M. J. & Sanderson, M. J. (1992). The suitability of molecular and morphological evidence in reconstructing plant phylogeny. In *Molecular systematics of plants*, ed. P. S. Soltis, D. E. Soltis & J. J. Doyle, pp. 340–68. New York: Chapman & Hall.

Doyle, J. A. (1978). Origin of angiosperms. *Annual Review of Ecology & Systematics*, **9**, 365–92.

Doyle, J. A., Donoghue, M. J. & Zimmer, E. A. (1994). Integration of morphological and ribosomal-DNA data on the origin of the angiosperms. *Annals of Missouri Botanical Garden*, **81**, 419–50.

Doyle, J. J., Doyle, J. L. & Palmer, J. D. (1995). Multiple independent losses of two genes and one intron from legume chloroplast genomes. *Systematic Botany*, **20**, 272–94.

Duncan, T. & Stuessy, T. F. (1985). *Cladistic theory and methodology*. New York: van Nostrand Reinhold Company.

Dunn, G. & Everitt, B. S. (1982). *An introduction to mathematical taxonomy*. Cambridge: Cambridge University Press.

Dunstan, W. R. & Henry, T. A. (1901). The nature and origin of the poison of *Lotus arabicus*. *Proceedings of the Royal Society of London*, **68**, 374–8.

East, E. M. (1913). Inheritance of flower size in crosses between species of *Nicotiana*. *Botanical Gazette*, **55**, 177–88.

East, E. M. & Mangelsdorf, A. J. (1925). A new interpretation of the hereditary behaviour of self-sterile plants. *Proceedings of the National Academy of Sciences, Washington*, **11**, 166–83.

Eckert, C. G. & Barrett, S. C. H. (1993). Clonal reproduction and patterns of genotypic diversity in *Decodon verticillatum* (Lythraceae). *American Journal of Botany*, **80**, 1175–82.

Edwards, A. W. F. (1986). Are Mendel's results really too close? *Biological Reviews*, **61**, 295–312.

Edwards, K. J. R. (1977). *Evolution in modern biology*. Studies in Biology no. 87. London: Arnold.

Ehrendorfer, F. (1959). Differentiation-hybridization cycles and polyploidy in *Achillea*. *Cold Spring Harbour Symposium of Quantitative Biology*, **24**, 141–52.

Ehrlich, P. R. & Raven, P. H. (1969). Differentiation of populations. *Science*, **165**, 1228–32.

Eigsti, C. J. & Dustin, P. (1955). *Colchicine in agriculture, medicine, biology and chemistry*. Ames, IA: Iowa State College Press.

Elkington, T. T. (1968). Introgressive hybridization between *Betula nana* L. and *B. pubescens* Ehrh. in Northwest Iceland. *New Phytologist*, **67**, 109–18.

Ellegren, H. (1991). DNA typing of museum birds. *Nature*, **354**, 113.

Elliot, E. (1914), *see* Lamarck.

Ellis, W. M., Keymer, R. J. & Jones, D. A. (1977*a*). The effect of temperature on the polymorphism of cyanogenesis in *Lotus corniculatus* L. *Heredity*, **38**, 339–47.

Ellis, W. M., Keymer, R. J. & Jones, D. A. (1977*b*). On the polymorphism of cyanogenesis in *Lotus corniculatus* L. VIII. Ecological studies in Anglesey. *Heredity*, **39**, 45–65.

Ellstrand, N. C. & Elam, D. R. (1993). Population genetic consequences of small population size: implications for plant conservation. *Annual Review of Ecology & Systematics*, **24**, 217–42.

Ellstrand, N. C. & Marshall, D. L. (1985). Interpopulation gene flow by pollen in Wild Radish *Raphanus sativus*. *American Naturalist*, **126**, 606–16.

Ellstrand, N. C. & Marshall, D. L. (1986). Patterns of multiple paternity in populations of *Raphanus sativus*. *Evolution*, **40**, 837–42.

Erikkson, G. (1983). Linnaeus the botanist. In *Linnaeus, the man and his work*, ed. T. Frängsmyr, pp. 63–109. Berkeley: University of California Press.

Ernst, A. (1955). Self-fertility in monomorphic primulas. *Genetica*, **27**, 391–448.

Erwin, T. L. (1982). Tropical forests: their richness in Coleoptera and other arthropod species. *Coleopterists' Bulletin*, **36**, 74–82.

Erwin, T. L. (1983). Beetles and other insects of tropical forest canopies at Manaus, Brazil sampled by insecticidal fogging. In *Tropical rain forest: ecology and management*, ed. S. L. Sutton, pp. 59–75. Oxford: Blackwell.

Evans, G. M. (1988). Genetic control of chromosome pairing in polyploids. In *Kew Chromosome Conference III*, ed. P. E. Brandham, pp. 253–60. London: HMSO.

Evans, R. C. & Turkington, R. (1988). Maintenance of morphological variation in a biotically patchy environment. *New Phytologist*, **109**, 369–76.

Evans, J. R., von Caemmerer, S. & Adams, W. W., III (1988). *Ecology of photosynthesis in sun and shade.* Melbourne: CSIRO.

Evenari, M. (1989). The history of research on white-green variegated plants. *Botanical Review* , **55**, 106–39.

Fagerlind, F. (1937). Embryologische, zytologische und bestäubungs-experimentelle Studien in der Familie Rubiaceae nebst Bemerkungen über einige Polyploiditätsprobleme. *Acta Horti Bergiani*, **11**, 195–470.

Falconer, D. S. (1981). *Introduction to quantitative genetics*, 2nd edn. London: Longmans.

Falk, D. A. & Holsinger, K. E. (1991). *Genetics and conservation of rare plants.* Oxford: Oxford University Press.

Falk, D. A. & Olwell, P. (1992). Scientific and policy considerations in restoration and reintroduction of endangered species. *Rhodora*, **94**, 287–315.

Falk, D. A., Millar, C. I. & Olwell, M. (1996). *Restoring diversity. Strategies for reintroduction of endangered plants.* Washington DC: Island Press.

Favarger, C. (1967). Cytologie et distribution des plantes. *Biological Review*, **42**, 163–206.

Favarger, C. & Villard, M. (1965). Nouvelles récherches cytotaxinomiques sur *Chrysanthemum leucanthemum* L. *sens. lat. Bericht der Schweizerischen Botanischen Gesellschaft*, **75**, 57–79.

Ferguson, A. (1980). *Biochemical systematics and evolution.* Glasgow & London: Blackie.

Ferris-Kaan, R. (1995). *The ecology of woodland creation.* Chichester: Wiley.

Fincham, J. R. S. (1983). *Genetics.* Bristol: Wright.

Firbank, L. G., Arnold, H. R., Eversham, B. C., Mountford, J. O., Radford, G. L., Tefler, M. G., Treweek, J. R., Webb, N. R. C. & Wells, T. C. E. (1993). *Managing set-aside land for wildlife.* ITE Research Publication 7, London: HMSO.

Fisher, R. A. (1929). *The genetical theory of natural selection*, 2nd edn, reprinted 1958. London: Constable; and New York: Dover Books.

Fisher, R. A. (1935). *The design of experiments.* Edinburgh & London: Oliver & Boyd.

Fisher, R. A. (1936). Has Mendel's work been rediscovered? *Annals of Science*, **1**, 115–37.

Fisher, R. A. & Yates, F. (1963). *Statistical tables for biological, agricultural and medical research*, 6th edn. Edinburgh: Longman (Oliver & Boyd).

Flake, R. H., Urbatsch, L. & Turner, B. L. (1978). Chemical documentation of allopatric introgression in *Juniperus. Systematic Botany*, **3**, 129–44.

Flake, R. H., von Rudloff, E. & Turner, B. L. (1969). Quantitative study of clinal variation in *Juniperus virginiana* using terpenoid data. *Proceedings of the National Academy of Sciences, USA*, **64**, 487–94.

Forey, P. L., Humphries, C. J. & Kitching, I. J. (1993). *Cladistics: a practical course in systematics*. Oxford: Oxford University Press.

Fowler, N. L. & Levin, D. A. (1984). Ecological constraints on the establishment of a novel polyploid in competition with its diploid progenitor. *American Naturalist*, **12**, 703–11.

Frankel, O. H., Brown, A. H. D. & Burdon, J. J. (1995). *The conservation of plant biodiversity*. Cambridge: Cambridge University Press.

Frankel, O. H. & Soulé, M. E. (1981). *Conservation and evolution*. London: Cambridge University Press.

Franklin, I. R. (1980). Evolutionary change in small populations. In *Conservation biology: an evolutionary-ecological perspective*, ed. M. E. Soulé & B. A. Wilcox, pp. 135–50. Sunderland: Sinauer.

Friday, A. & Ingram, D. S. (1985). *The Cambridge encyclopedia of life sciences*. Cambridge: Cambridge University Press.

Friedman, S. T. & Adams, W. T. (1985). Estimation of gene flow into two seed orchards of Loblolly-Pine (*Pinus taeda* L.). *Theoretical & Applied Genetics*, **69**, 609–15.

Friis, E. M. & Crepet, W. L. (1987). Time of appearance of floral features. In *The origins of angiosperms and their biological consequences*, ed. E. M. Friis, W. G. Chaloner & P. R. Crane, pp. 145–79. Cambridge: Cambridge University Press.

Friis, E. M., Chaloner, W. G. & Crane, P. R. (1987). Introduction to angiosperms. In *The origins of angiosperms and their biological consequences*, ed. E. M. Friis, W. G. Chaloner & P. R. Crane, pp. 1–15. Cambridge: Cambridge University Press.

Fritsch, P. & Rieseberg, L. H. (1992). High outcrossing rates maintain male and hermaphrodite individuals in populations of the flowering plant *Datisca glomerata*. *Nature*, **359**, 633–6.

Frost, H. B. (1938). Nuclear embryony and juvenile characters in clonal varieties of *Citrus*. *Journal of Heredity*, **29**, 423–32.

Fröst, S. & Ising, G. (1968). An investigation into the phenolic compounds in *Vaccinium myrtillus* L. (Bilberries), *Vaccinium vitis-idaea* L. (Cowberries), and the hybrid between them V. *intermedium* Ruthe employing thin layer chromatography. *Hereditas*, **60**, 72–6.

Futuyma, D. J. (1986). *Evolutionary biology*, 2nd edn. Sunderland: Sinauer.

Futuyma, D. J. (1995). *Science on trial*. Sunderland: Sinauer.

Gajewski, W. (1957). A cytogenetic study on the genus *Geum*. *Monographiae Botanicae* no. 4.

Gale, M. D. & Miller, T. E. (1987). The introduction of alien genetic variation in wheat. In *Wheat breeding; its scientific basis*, ed. F. G. H. Lupton, pp. 173–210. London: Chapman & Hall.

Gallez, G. P. & Gottlieb, L. D. (1982). Genetic evidence for the hybrid origin of the diploid plant *Stephanomeria diegensis*. *Evolution*, **36**, 1158–67.

Galton, F. (1871). Experiments in pangenesis, by breeding from rabbits of a pure variety, into whose circulation blood taken from other varieties had previously been largely transfused. *Proceedings of the Royal Society of London*, **19**, 393–410.

Galton, F. (1876). The history of twins, as a criterion of the relative powers of nature and nurture. *Journal of the Royal Anthropological Institute*, **5**, 391–406.

Galton, F. (1889). *Natural inheritance*. London & New York: Macmillan.

Ganders, F. R. (1979). The biology of heterostyly. *New Zealand Journal of Botany*, **17**, 607–35.

Ganders, F. R. (1990). Altitudinal clines for cyanogenesis in introduced populations of White Clover near Vancouver, Canada. *Heredity*, **64**, 387–90.

Gastony, G. J. (1986). Electrophoretic evidence for the origin of fern species by unreduced gametes. *American Journal of Botany*, **73**, 1563–9.

Gates, R. R. (1909). The stature and chromosomes of *Oenothera gigas* de Vries. *Archiv für Zellforschung*, **3**, 525–52.

Gay, P. A. (1960). A new method for the comparison of populations that contain hybrids. *New Phytologist*, **59**, 219–26.

Geiger, R. (1965). *The climate near the ground*. Cambridge, MA: Harvard University Press.

Geneve, R. L., Preece, J. E. & Merkle, S. A. (1996). *Biotechnology of ornamental plants*. Biotechnology in Agriculture series no. 16. Wallingford: CAB International.

Gerstel, D. U. (1950). Self-incompatibility studies in Guayule. II. Inheritance. *Genetics*, **35**, 482–506.

Gibbs, P. E. (1986). Do homomorphic and heteromorphic self-incompatibility systems have the same sporophytic mechanism? *Plant Systematics & Evolution*, **154**, 285–323.

Gibby, M. (1981). Polyploidy and its evolutionary significance. In *The evolving biosphere*, ed. P. L. Forey, pp. 87–96. Cambridge: British Museum (Natural History) & Cambridge University Press.

Gibson, R. W. & Pickett, J. A. (1983). Wild Potato repels aphids by release of aphid alarm pheromone. *Nature*, **302**, 608–9.

Gill, B. S. & Kimber, G. (1974). Giemsa C-banding and the evolution of Wheat. *Proceedings of the National Academy of Sciences, USA*, **71**, 4086–90.

Gill, D. E., Chao, L., Perkins, S. L. & Wolf, J. B. (1995). Genetic mosaicism in plants and clonal animals. *Annual Review of Ecology & Systematics*, **26**, 423–44.

Gilmour, J. S. L. (1937). A taxonomic problem. *Nature*, **139**, 1040.

Gilmour, J. S. L. (1940). Taxonomy and philosophy. In *The new systematics*, ed. J. Huxley, pp. 461–74. Oxford: Clarendon Press.

Gilmour, J. S. L. (1951). The development of taxonomic theory since 1851. *Nature*, **168**, 400–2.

Gilmour, J. S. L. & Gregor, J. W. (1939). Demes: a suggested new terminology. *Nature*, **144**, 333–4.

Gilmour, J. S. L. & Heslop-Harrison, J. (1954). The deme terminology and the units of micro-evolutionary change. *Genetica*, **27**, 147–61.

Gilmour, J. S. L. & Walters, S. M. (1963). Philosophy and classification. *Vistas in Botany*, **4**, 1–22.

Gilpin, M. E. & Soulé, M. E. (1986). Minimum viable populations: processes of species extinction. In *Conservation biology: the science of scarcity and diversity*, (ed.) M. E. Soulé, pp. 19–34. Sunderland: Sinauer.

Given, D. R. (1994). *Principles and practice of plant conservation*. London: Chapman & Hall.

Givnish, T. (1979). On the adaptive significance of leaf form. In *Topics in plant population biology*, ed. O. T. Solbrig, S. Jain, G. B. Johnson & P. H. Raven, pp. 375–407. London: Columbia University Press.

Givnish, T. J. & Vermeij, G. J. (1976). Sizes and shapes of Liane leaves. *American Naturalist*, **110**, 743–76.

Glass, B. (1959). Heredity and variation in the eighteenth century concept of the species. In *Forerunners of Darwin* 1745–1859, ed. B. Glass, O. Temkin & W. I. Straus, pp. 144–72. London: Oxford University Press.
Gliddon, C. & Saleem, M. (1985). Gene flow in *Trifolium repens* – an expanding genetic neighbourhood. In *Genetic differentiation and dispersal in plants*, ed. P. Jacquard, G. Heim & J. Antonovics, pp. 293–309. Berlin: Springer Verlag.
Godward, M. B. E. (1985). The kinetochore. *International Review of Cytology*, **94**, 77–105.
Goebel, K. (1897). *Uber Jugendformen von Pflanzen und deren künstliche Wiederhervorrufung*. Sitzungsbericht der Mathematisch-Physikalischen Class der Königlich-Bayerischen Akademie der Wissenschaften, München, 26.
Goerke, H. (1989). *Carl von Linné*. Stuttgart: Wissenschaftliche Verlagsgesellschaft.
Goldblatt, P. (1980). Polyploidy in angiosperms: monocotyledons. In *Polyploidy* ed. W. H. Lewis, pp. 219–39. New York & London: Plenum Press.
Goldblatt, P. & Johnson, D. E. (1990). *Index to plant chromosome numbers* 1986–1987, 1988–1989, 1990–1991. Missouri: Missouri Botanical Garden.
Goldschmidt, R. B. (1940). *The material basis of evolution*. New Haven: Yale University Press.
Goldschmidt, R. B. (1955). *Theoretical genetics*. Berkeley & Los Angeles: University of California Press.
Golenberg, E. M., Giannasi, D. E., Clegg, M. T., Smiley, C. J., Durbin, M., Henderson, D. & Zurawski, G. (1990). Chloroplast DNA sequence from a Miocene Magnolia species. *Nature*, **344**, 656–8.
González-Benito, M. E. & Pérez, C. (1994). Studies on the cryopreservation of nodal explants of *Centaurium rigualli* Esteve, an endemic threatened species, through vitrification. *Botanic Gardens Micropropagation News*, **1**, 82–4.
Good, R. d'O. (1956). *Features of evolution in the flowering plants*. London & New York: Longmans.
Gottlieb, L. D. (1972). Levels of confidence in the analysis of hybridization in plants. *Annals of Missouri Botanical Garden*, **59**, 435–46.
Gottlieb, L. D. (1973). Genetic differentiation, sympatric speciation, and the origin of a diploid species of *Stephanomeria*. *American Journal of Botany*, **60**, 545–53.
Gottlieb, L. D. (1979). The origin of phenotype in a recently evolved species. In *Topics in plant population biology*, ed. O. T. Solbrig, S. Jain, G. B. Johnson & P. H. Raven, pp. 264–86. New York: Columbia University Press.
Gottlieb, L. D. (1981*a*). Electrophoretic evidence and plant populations. *Progress in Phytochemistry*, **7**, 1–46.
Gottlieb, L. D. (1981b). Gene number in species of Asteraceae that have different chromosome numbers. *Proceedings of the National Academy of Sciences, USA*, **78**, 3726–9.
Gottlieb, L. D. (1982). Conservation and duplication of isozymes in plants. *Science*, **216**, 373–80.
Gottlieb, L. D. (1984). Isozyme evidence and problem solving in plant systematics. In *Plant biosystematics*, ed. W. F. Grant, pp. 343–57. Toronto: Academic Press.

Gottlieb, L. D. (1986). Genetic differentiation, speciation and phylogeny in *Clarkia* (Onagraceae). In *Modern aspects of species*, ed. K. Iwatsuki, P. H. Raven & W. J. Bock, pp. 145–60. Tokyo: University of Tokyo Press.

Gottlieb, L. C. & Pilz, G. (1976). Genetic similarity between *Gaura longifolia* and its obligately outcrossing derivative *G. demareei*. *Systematic Botany*, **1**, 181–7.

Gould, S. J. (1979). Species are not specious. *New Scientist*, **83**, 374–6.

Gould, S. J. (1994). The evolution of life on the earth. *Scientific American*, **271**, 63–9.

Gould, S. J. & Eldredge, N. (1977). Punctuated equilibria: the tempo and mode of evolution reconsidered. *Paleobiology*, **3**, 115–51.

Gould, S. J. & Eldredge, N. (1993). Punctuated equilibrium comes of age. *Nature*, **366**, 223–7.

Gould, S. J. & Lewontin, R. G. (1979). The spandrels of San Marco and the panglossian paradigm: a critique of the adaptionist programme. *Proceedings of the Royal Society of London, B*, **205**, 581–8.

Govindaraju, D. R. (1988). Relationship between dispersal ability and levels of gene flow in plants. *Oikos*, **52**, 31–5.

Graham, B. F, Jr. & Bormann, F. H. (1966). Natural root grafts. *Botanical Review*, **32**, 255–92.

Graham, S. W. & Barrett, S. C. H. (1995). Phylogenetic systematics of the Pontederiales: implications for breeding-system evolution. In *Monocotyledons: systematics and evolution*, ed. P. Rudall, P. J. Cribb, D. F. Cutler & C. J. Humphries, pp. 415–41. Kew: Royal Botanic Gardens.

Grant, V. (1950). The protection of the ovules in flowering plants. *Evolution*, **4**, 179–201.

Grant, V. (1966). The selective origin of incompatibility barriers in the plant genus *Gilia*. *American Naturalist*, **100**, 99–118.

Grant, V. (1971). *Plant speciation*. New York & London: Columbia University Press.

Grant, V. (1975). *Genetics of flowering plants*. New York & London: Columbia University Press.

Grant, V. (1981). *Plant speciation*, 2nd edn. New York: Columbia University Press.

Grant, V. & Grant, K. A. (1965). *Flower pollination in the Phlox family*. New York: Columbia University Press.

Gray, A. J., Marshall, D. F. & Raybould, A. F. (1991). A century of evolution in *Spartina anglica*. *Advances in Ecological Research*, **21**, 1–62.

Green, R. H. (1979). *Sampling design and statistical methods for environmental biologists*. New York, Chichester, Brisbane & Toronto: Wiley.

Greene, E. L. (1909). Linnaeus as an evolutionist. *Proceedings of the Washington Academy of Sciences*, **11**, 17–26.

Gregor, J. W. (1930). Experiments on the genetics of wild populations. I. *Plantago maritima*. *Journal of Genetics*, **22**, 15–25.

Gregor, J. W. (1931). Experimental delimitation of species. *New Phytologist*, **30**, 204–17.

Gregor, J. W. (1938). Experimental taxonomy. 2. Initial population differentiation in *Plantago maritima* in Britain. *New Phytologist*, **37**, 15–49.

Gregor, J. W. (1939). Experimental taxonomy. 4. Population differentiation in North American and European Sea Plantains allied to *Plantago maritima* L. *New Phytologist*, **38**, 293–322.

Gregor, J. W. (1944). The ecotype. *Biological Reviews*, **19**, 20–30.
Gregor, J. W. (1946). Ecotypic differentiation. *New Phytologist*, **45**, 254–70.
Gregor, J. W., Davey, V. McM. & Lang, J. M. S. (1936). Experimental taxonomy. 1. Experimental garden technique in relation to the recognition of small taxonomic units. *New Phytologist*, **35**, 323–50.
Gregor, J. W. & Lang, J. M. S. (1950). Intra-colonial variation in plant size and habit in Sea Plantains. *New Phytologist*, **49**, 135–41.
Greig-Smith, P. (1964). *Quantitative plant ecology*, 2nd edn. London: Butterworth.
Groot, J. & Boschuizen, R. (1970). A preliminary investigation into the genecology of *Plantago major* L. *Journal of Experimental Botany*, **21**, 835–41.
Groves, R. H. & di Castri, F. (1991). *Biogeography of Mediterranean invasions*. Cambridge: Cambridge University Press.
Guerrant, E. O. (1992). Genetic and demographic considerations in the sampling and reintroduction of rare plants. In *Conservation biology*, ed. P. L. Fiedler & S. K. Jain, pp. 321–44. New York: Chapman & Hall.
Guignard, L. (1891). Nouvelles études sur la fécondation. *Annales des Sciences Naturelles* (*Botanique*), **14**, 163–296.
Gunther, R. W. T. (1928). *Further correspondence of John Ray*. London & New York: Oxford University Press.
Gustafsson, Å. (1946). Apomixis in higher plants. 1. The mechanism of apomixis. *Acta Universitatis lundensis*, **42**(3), 1–67.
Gustafsson, Å. (1947*a*). Apomixis in higher plants. 2. The causal aspect of apomixis. *Acta Universitatis lundensis*, **43**(3), 69–179.
Gustafsson, Å. (1947*b*). Apomixis in higher plants. 3. Biotype and species formation. *Acta Universitatis lundensis*, **43**(12), 183–370.
Gustafsson, L. & Gustafsson, P. (1994). Low genetic variation in Swedish populations of the rare species *Vicia pisiformis* (Fabaceae) revealed with RFLP (rDNA) and RAPD. *Plant Systematics & Evolution* **189**, 133–48.
Haase, P. (1993). Genetic variation, gene flow and the founder effect in pioneer populations of *Nothofagus menziesii* (Fagaceae) South Island, New Zealand. *Journal of Biogeography*, **20**, 79–85.
Habakkuk, H. J. (1960). Thomas Robert Malthus, F. R. S. (1766–1834). *Notes and Records of the Royal Society of London*, **14**, 99–108.
Haeckel, E. H. P. A. (1876). *The history of creation*. (Translation revised by E. R. Lankester.) London: Routledge.
Hadrys, H., Balick, M. & Schierwater, R. (1992). Applications of random amplified polymorphic DNA (RAPD) in molecular ecology. *Molecular Ecology*, **1**, 55–63.
Hagerup, O. (1931). Uber Polyploide in Beziehung zu Klima, Okologie und Phylogenie. *Hereditas*, **16**, 19–40.
Hair, J. B. (1956). Subsexual reproduction in *Agropyron*. *Heredity*, **10**, 129–60.
Haldane, J. B. S. (1932). *The causes of evolution*. London: Longmans.
Hall, M. T. (1952). A hybrid swarm in *Juniperus*. *Evolution*, **6**, 347–66.
Hallam, A. (1994). Theories in collision. *Nature*, **372**, 296.
Hall, O. (1972). Oxygen requirements of root meristems in diploid and autotetraploid Rye. *Hereditas*, **70**, 69–74.
Hamrick, J. L. (1990). Isozymes and the analysis of genetic structure in plant populations. In *Isozymes in plant biology*, ed. D. E. Soltis & P. S. Soltis, pp. 87–105. London: Chapman & Hall.

Hamrick, J. L. & Godt, M. J. W. (1989). Allozyme diversity in plant species. In *Plant population genetics, breeding and genetic resources*, ed. A. D. H. Brown, M. T. Clegg, A. L. Kahler & B. S. Weir, pp. 43–63. Sunderland: Sinauer.

Hamrick, J. L. & Loveless, M. D. (1989). The genetic structure of tropical tree populations: associations with reproductive biology. In *The evolutionary ecology of plants*, ed. J. H. Bock & Y. B. Linhart, pp. 129–46. Boulder: Westview Press.

Hamrick, J. L., Linhart, Y. B. & Mitton, J. B. (1979). Relationships between life history characteristics and electrophoretically detectable genetic variation in plants. *Annual Review of Ecology & Systematics*, **10**, 173–200.

Harberd, D. J. (1957). The within population variance in genecological trials. *New Phytologist*, **56**, 269–80.

Harberd, D. J. (1958). Progress and prospects in genecology. *Record of the Scottish Breeding Station*, 1958, 52–60.

Harberd, D. J. (1961). Observations on population structure and longevity of *Festuca rubra*. L. *New Phytologist*, **60**, 184–206.

Harberd, D. J. (1962). Some observations on natural clones in *Festuca ovina*. *New Phytologist*, **61**, 85–100.

Harberd, D. J. (1963). Observations on natural clones of *Trifolium repens* L. *New Phytologist*, **62**, 198–204.

Harborne, J. B. (1993). *Introduction to ecological biochemistry*, 4th edn. San Diego: Academic Press.

Harborne, J. B. & Turner, B. L. (1984). *Plant chemosystematics*. Orlando: Academic Press.

Harborne, J. B., Williams, C. A. & Smith, D. M. (1973). Species-specific kaempferol derivatives in ferns of the Appalachian *Asplenium* complex. *Biochemical Systematics*, **1**, 51–4.

Harder, L. D. & Barrett, S. C. H. (1996). Pollen dispersal and mating patterns in animal-pollinated plants. In *Floral biology. Studies on floral evolution in animal-pollinated plants*, ed. D. G. Lloyd & S. C. H. Barrett, pp. 140–90. New York: Chapman & Hall.

Hardin, G. (1966). *Biology. Its principles and implications*, 2nd edn. London & San Francisco: Freeman.

Harlan, H. V. & Martini, M. L. (1938). The effect of natural selection on a mixture of Barley varieties. *Journal of Agricultural Research*, **57**, 189–99.

Harlan, J. R. & de Wet, J. M. J. (1975). On Ö. Winge and a prayer: the origins of polyploidy. *Botanical Review*, **41**, 361–90.

Harley, J. L. & Harley, E. L. (1987). A check-list of mycorrhiza in the British flora. *New Phytologist* (*Suppl.*), **105**, 1–102.

Harper, J. L. (1977). *Population biology of plants*. London & New York: Academic Press.

Harper, J. L. (1978). The demography of plants with clonal growth. In *Structure and functioning of plant populations*, ed. A. H. J. Freyson & J. W. Waldendorp, pp. 27–48. Amsterdam, Oxford & New York: North Holland Publishing Co.

Harper, J. L. (1983). A Darwinian plant ecology. In *Evolution from molecules to men*, ed. D. S. Bendall, pp. 323–45. Cambridge: Cambridge University Press.

Harris, S. A. & Ingram, R. (1991). Chloroplast DNA and biosystematics: the effects of intraspecific diversity and plastid transmission. *Taxon*, **40**,

393–412.

Harris, S. A. & Ingram, R. (1991). Molecular systematics of the genus *Senecio* L. I. Hybridization in a British polyploid complex. *Heredity*, **69**, 1–10.

Harshberger, J. W. (1901). The limits of variation in plants. *Proceedings of the National Academy of Sciences, USA*, **53**, 303–19.

Hartman, H. T. & Kester, D. E. (1975). *Plant propagation: principles and practices.* Englewood Cliffs, NY: Prentice Hall.

Harvey, W. H. (1860). Darwin on the origin of species. *The Gardeners' Chronicle and Agricultural Gazette*, February 18 1860, 145–6.

Hathaway, W. H. (1962). Weighted hybrid index. *Evolution*, **16**, 1–10.

Haufler, C. H. & Zhongren, W. (1991). Chromosomal analyses and the origin of allopolyploid *Polypodium virginianum* (Polypodiaceae). *American Journal of Botany*, **78**, 624–9.

Haufler, C. H., Soltis, D. E. & Soltis, P. S. (1995). Phylogeny of the *Polypodium vulgare* complex: insights from chloroplast DNA restriction site data. *Systematic Botany*, **20**, 110–19.

Haufler, C. H., Windham, M. D. & Rabe, E. W. (1995). Reticulate evolution in the *Polypodium vulgare* complex. *Systematic Botany*, **20**, 89–109.

Hauser, T. P. & Loeschcke, V. (1994). Inbreeding depression and mating-distance dependent offspring fitness in large and small populations of *Lychnis flos-cuculi* (Caryophyllaceae). *Journal of Evolutionary Biology*, **7**, 609–22.

Hayes, W. (1964). *The genetics of bacteria and their viruses.* Oxford & Edinburgh: Blackwell.

Hayward, I. M. & Druce, G. C. (1919). *The adventive flora of Tweedside.* Arbroath: Buncle.

Hazarika, M. H. & Rees, H. (1967). Genotypic control of chromosome behaviour in Rye. X. Chromosome pairing and fertility in autotetraploids. *Heredity*, **22**, 317–32.

Heiser, C. B., Jr. (1949*a*). Study in the evolution of sunflower species *Helianthus annuus* and *H. bolanderi*. *University of California Publication in Botany*, **23**, 157–208.

Heiser, C. B., Jr. (1949*b*). Natural hybridization with particular reference to introgression. *Botanical Review*, **15**, 645–87.

Heiser, C. B., Jr. (1973). Introgression re-examined. *Botanical Review*, **39**, 347–66.

Heiser, C. B., Jr. (1979). Hybrid populations of *Helianthus divaricatus* and *H. microcephalus* after 22 years. *Taxon*, **28**, 71–5.

Hennig, W. (1950). Grundzüge einer Theorie der phylogenetischen Systematik. Berlin: Deutscher Zentralverlag.

Hermanutz, L. A., Innes, D. J. & Weis, I. M. (1989). Clonal structure of Arctic Dwarf Birch (*Betula glandulosa*) at its northern limit. *American Journal of Botany*, **76**, 755–61.

Heslop-Harrison, J. (1952). A reconsideration of plant teratology. *Phyton*, **4**, 19–34.

Heslop-Harrison, J. (1953). *New concepts in flowering-plant taxonomy.* London: Heinemann .

Heslop-Harrison, J. (1964). Forty years of genecology. *Advances in Ecological Research*, **2**, 159–247.

Heslop-Harrison, J. (1973). The plant kingdom: an exhaustible resource? *Transactions of the Botanical Society of Edinburgh*, **42**, 1–15.

Heslop-Harrison, J. (1978). *Cellular recognition systems in plants*. Studies in Biology no. 100. London: Arnold.

Hesselman, H. (1919). Iakttagelser øver skogsträdpollens spridningsførmagå. *Meddelanden från Statens Skogsførsoksanstalt*, **16**, 27.

Heywood, V. H. (1976). *Plant taxonomy*, 2nd edn. Studies in Biology no. 5. London: Arnold.

Heywood, V. H. (1980). The impact of Linnaeus on botanical taxonomy – past, present and future. *Veröffentlichungen der Joachim Jungius-Gesellschaft der Wissenschaften Hamburg*, **43**, 97–115.

Heywood, V. H. & Wyse Jackson, P. S. (1991). *Tropical botanic gardens: their role in conservation and development*. London: Academic Press.

Hickey, L. J. & Doyle, J. A. (1977). Early Cretaceous fossil evidence for angiosperm evolution. *Botanical Review*, **43**, 2–104.

Hiesey, W. M. & Milner, H. W. (1965). Physiology of ecological races and species. *Annual Review of Plant Physiology*, **16**, 203–16.

Higuchi, R., Bowman, B., Freiberger, M., Ryder, O. A., Wilson, A. C. (1984). DNA sequences from the quagga, an extinct member of the horse family. *Nature*, **312**, 282–4.

Hill, C. R. & Crane, P. R. (1982). Evolutionary cladistics and the origin of angiosperms. In *Problems of phylogenetic reconstruction*, ed. K. A. Joysey & A. E. Friday, pp. 269–361. London & New York: Academic Press.

Hillis, D. M., Mable, B. K. & Moritz, C. (1996). Applications of molecular systematics. In *Molecular systematics*, 2nd edn, ed. D. M. Hillis, C. Moritz & B. K. Mable, pp. 515–43. Sunderland: Sinauer.

Hillis, D. M., Moritz, C. & Mable, B. K. (1996). *Molecular systematics*, 2nd edn. Sunderland: Sinauer.

Hilu, K. W. (1983). The role of single-gene mutations in the evolution of flowering plants. *Evolutionary Biology*, **16**, 97–128.

Hinton, W. F. (1976). The evolution of insect-mediated self-pollination from an outcrossing system in *Calyptridium* (Portulacaceae). *American Journal of Botany*, **63**, 979–86.

Hoch, P. C. & Stephenson, A. G. (1995). Experimental and molecular approaches to plant biosystematics. *Monographs in Systematic Botany*, **53**, St Louis: Missouri Botanical Garden.

Hoffmann, H. (1881). Rückblick auf meine Variations-Versuche von 1855–80. *Botanische Zeitung*, 1881, 345–51, 361–7, 377–83, 393–9, 409–15, 424–31.

Holliday, R. J. & Putwain, P. D. (1977). Evolution of resistance to simazine in *Senecio vulgaris* L. *Weed Research*, **17**, 291–6.

Holliday, R. J. & Putwain, P. D. (1980). Evolution of herbicide resistance in *Senecio vulgaris*: variation in susceptibility to simazine between and within populations. *Journal of Applied Ecology*, **17**, 779–91.

Holsinger, K. E. (1984). The nature of biological species. *Philosophy of Science*, **51**, 293–307.

Holsinger, K. E. & Gottlieb, L. D. (1991). Conservation of rare and endangered plants. In *Genetics and conservation of rare plants*, ed. D. A. Falk & K. E. Holsinger, pp. 195–208. New York: Oxford University Press.

Holttum, R. E. (1967). Comparative morphology, taxonomy and evolution. *Phytomorphology*, **17**, 36–41.

Hooglander, N., Lumaret, R. & Bos, M. (1993). Inter-intraspecific variation of chloroplast DNA of European *Plantago* species. *Heredity*, **70**, 322–34.

Hopkins, W. G. (1995). *Introduction to plant physiology*. New York: Wiley.

Horsman, D. C., Roberts, T. M. & Bradshaw, A. D. (1979). Studies on the effect of sulphur dioxide on perennial ryegrass (*Lolium perenne* L.). II. Evolution of sulphur dioxide tolerance. *Journal of Experimental Botany*, **30**, 495–501.
Hort, A. (1938). *The Critica botanica of Linnaeus.* London: Ray Society, British Museum.
Hsiao, J. Y. & Li, H. L. (1973). Chromatographic studies on the Red Horsechestnut (*Aesculus* × *carnea*) and its putative parent species. *Brittonia*, **25**, 57–63.
Huenneke, L. F. & Thomson, J. K. (1995). Potential interference between a threatened endemic Thistle and an invasive non-native plant. *Conservation Biology*, **9**, 416–25.
Hughes, A. (1959). *A history of cytology.* London & New York: Abelard-Schuman.
Hughes, J. & Richards, A. J. (1988). The genetic structure of populations of sexual and asexual *Taraxacum* (Dandelions). *Heredity*, **60**, 161– 71.
Hughes, M. A. (1991). The cyanogenic polymorphism in *Trifolium repens L.* (White Clover). *Heredity*, **66**, 105–15.
Hughes, M. A. (1996). *Plant molecular genetics.* Harlow: Addison Wesley Longman Ltd.
Hughes, M. B. & Babcock, E. B. (1950). Self-incompatibility in *Crepis foetida* L. subsp. *rhoeadifolia. Genetics*, **35**, 570–88.
Hughes, N. F. (1994). *The enigma of angiosperm origins.* Cambridge: Cambridge University Press.
Hull, D. L. (1979). The limits of cladism. *Systematic Zoology*, **28**, 416–40.
Humphries, C. J. & Funk, V. A. (1984). Cladistic methodology. In *Current concepts in plant taxonomy*, ed. V. H. Heywood & D. M. Moore, pp. 323–62. London: Academic Press.
Huntley, B. (1991). How plants respond to climate change: migration rates, individualism and the consequences for plant communites. *Annals of Botany*, **67** (Suppl.), 15–22.
Huskins, C. L. (1930). The origin of *Spartina townsendii. Genetica*, **12**, 531–8.
Hutchings, M. J. (1989). Population biology and conservation of *Ophrys sphegodes.* In *Modern methods in orchid conservation*, ed. H. W. Pritchard, pp. 101–15. Cambridge: Cambridge University Press.
Hutchinson, A. H. (1936). The polygonal representation of polyphase phenomena. *Transactions of the Royal Society of Canada, Ser.* 3, *Sect. V*, **30**, 19–26.
Hutchinson, T. C. (1967). Ecotype differentiation in *Teucrium scorodonia* with respect to susceptibility to lime-induced chlorosis and to shade factors. *New Phytologist*, **66**, 439–53.
Huxley, A., Griffiths, M., & Levy, M. (1992). *The new Royal Horticultural Society dictionary of gardening.* London: Macmillan.
Huxley, C. R. (1991). Ants and plants: a diversity of interactions. In *Ant–plant interactions*, ed. C. R. Huxley & D. F. Cutler, pp. 1–11. Oxford: Oxford University Press.
Huxley, C. R. & Cutler, D. F. (1991). *Ant–plant interactions.* Oxford: Oxford University Press.
Huxley, J. S. (1938). Clines: an auxiliary taxonomic principle. *Nature*, **142**, 219–20.
Huxley, J. S. (ed.) (1940). *The new systematics.* Oxford: Clarendon Press.
Huxley, J. S. (1942). *Evolution: the modern synthesis.* London: Allen & Unwin.
Iltis, H. (1932). *Life of Mendel.* London: Allen & Unwin, reprinted 1966; and New York: Hafner.

Ingrouille, M. J. & Stace, C. A. (1985). Pattern of variation of agamospermous *Limonium* (Plumbaginaceae) in the British Isles. *Nordic Journal of Botany*, **5**, 113–25.

Jackson, R. C. (1962). Interspecific hybridization in *Haplopappus* and its bearing on chromosome evolution in the Blepharodon section. *American Journal of Botany*, **49**, 119–132.

Jackson, R. C. (1965). A cytogenetic study of a three-paired race of *Haplopappus gracilis*. *American Journal of Botany*, **52**, 946–53.

Jain, S. K. & Martins, P. S. (1979). Ecological genetics of the colonizing ability of Rose Clover (*Trifolium hirtum* All.) *American Journal of Botany*, **66**, 361–6.

Jameson, D. L. (ed.) (1977). (ed.) *Evolutionary genetics*. Stroudsburg, PA: Dowden, Hutchinson & Ross.

Jansen, R. K., Holsinger, K. E., Michaels, H. J. & Palmer, J. D. (1990). Phylogenetic analysis of chloroplast DNA restriction site data at higher taxonomic levels: an example from the Asteraceae. *Evolution*, **44**, 2089–105.

Jansen, R. K., Michaels, H. J. & Palmer, J. D. (1991). Phylogeny and character evolution in the Asteraceae based on chloroplast DNA restriction site mapping. *Systematic Botany*, **16**, 98–115.

Jardine, N. & Sibson, R. (1971). *Mathematical taxonomy*. London & New York: Wiley.

Jauhar, P. P. (1975). Genetic control of diploid-like meiosis in hexaploid Tall Fescue. *Nature*, **254**, 595–7.

Jenkin, F. (1867). Unsigned review of Darwin's 'On the origin of species'. *The North British Review*, June 1867, 277–318.

Jennersten, O. (1988). Pollination in *Dianthus deltoides* (Caryophyllaceae): effects of habitat fragmentation on visitation and seed set. *Conservation Biology*, **2**, 359–66.

Jensen, I. & Bogh, H. (1941). On conditions influencing the danger of crossing in the case of wind-pollinated cultivated plants. *Tidsskrift for Planteavl*, **46**, 238–66.

Johannsen, W. (1909). *Elemente der exakten Erblichkeitslehre*. Jena: Fischer.

Johannsen, W. (1911). The genotype concept in heredity. *American Naturalist*, **45**, 129–59.

Johnson, B. L. (1972). Protein electrophoretic profiles and the origin of the B genome of wheat. *Proceedings of the National Academy of Sciences, USA*, **69**, 1398–402.

Johnson, H. (1945). Interspecific hybridization within the genus *Betula*. *Hereditas*, **31**, 163–76.

Johnson, H. B. (1975). Plant pubescence: an ecological perspective. *Botanical Review*, **41**, 233–58.

Johnson, M. A. T., Kenton, A. Y., Bennett, M. D. & Bradham, P. E. (1989). *Voaniola gerardii* has the highest known chromosome number in the monocotyledons. *Genome*, **32**, 328–33.

Johnson, S. D. (1992). Plant–animal relationships. In *The ecology of fynbos: nutrients, fire and diversity*, ed. R. M. Cowling, pp. 175–205. Cape Town: Oxford University Press.

Jones, D. A. (1962). Selective eating of the acyanogenic form of the plant *Lotus corniculatus* L. by various animals. *Nature*, **193**, 1109–10.

Jones, D. A. (1966). On the polymorphism of cyanogenesis in *Lotus corniculatus*. Selection by animals. *Canadian Journal of Genetics & Cytology*, **8**, 556–67.

Jones, D. A. (1972). Cyanogenic glycosides and their function. In *Phytochemical ecology*, ed. J. B. Harborne, pp. 103–24. London & New York: Academic Press.

Jones, D. A. (1973). Co-evolution and cyanogenesis. In *Taxonomy and ecology*, ed. V. H. Heywood, pp. 213–42. Systematics Association Special Volume no. 5. London & New York: Academic Press.

Jones, D. A., Keymer, R. J. & Ellis, W. M. (1978). Cyanogenesis in plants and animal feeding. In *Biochemical aspects of plant and animal coevolution*, ed. J. B. Harborne, pp. 21–34. London, New York & San Francisco: Academic Press.

Jones, D. F. (1924). The attainment of homozygosity in inbred strains of Maize. *Genetics*, **9**, 405–18.

Jones, K. (1958). Cytotaxonomic studies in *Holcus*. I. The chromosome complex in *Holcus mollis* L. *New Phytologist*, **57**, 191–210.

Jones, K. (1964). Chromosomes and the nature and origin of *Anthoxanthum odoratum* L. *Chromosoma*, **15**, 248–74.

Jones, K. (1978). Aspects of chromosome evolution in higher plants. *Advances in Botanical Research*, **6**, 120–94.

Jones, K. & Borrill, M. (1961). Chromosomal status, gene exchange and evolution in *Dactylis*. 3. The role of the inter-ploid hybrids. *Genetica*, **32**, 296–322.

Jones, K. & Carroll, C. P. (1962). Cytotaxonomic studies in *Holcus*. II. Morphological relationships in *Holcus mollis* L. *New Phytologist*, **61**, 63–84.

Jones, M. D. & Brooks, J. S. (1952). *Effect of tree barriers on outcrossing in Corn*. Oklahoma Agricultural Experiment Station Bulletin no. T–45.

Jones, R. N. (1995). B chromosomes in plants. *New Phytologist*, **131**, 411–34.

Jonsell, B. (1978). Linnaeus's views on plant classification and evolution. *Botaniska Notiser*, **131**, 523–30.

Jonsell, B. (1984). The biological species concept reexamined. In *Plant biosystematics*, ed. W. F. Grant, pp. 159–68. Toronto: Academic Press.

Jordan, A. (1864). *Diagnoses d'espèces nouvelles ou méconnues pour servir de matériaux à une flore réformée de la France et des Contrées voisines.* Paris: Savy.

Jordanova, L. J. (1984). *Lamarck*. Oxford: Oxford University Press.

Kadereit, J. W. (1994). Molecules and morphology, phylogenetics and genetics. *Botanica Acta*, **107**, 369–73.

Kadereit, J. W. & Briggs, D. (1985). Speed of development of radiate and non-radiate plants of *Senecio vulgaris* L. from habitats subject to different degrees of weeding pressure. *New Phytologist*, **99**, 155–69.

Karpechenko, G. D. (1927). Polyploid hybrids of *Raphanus sativus* L. × *Brassica oleracea* L. *Bulletin of Applied Botany, Genetics & Plant Breeding* (*Leningrad*), **17**, 305–410.

Karpechenko, G. D. (1928). Polyploid hybrids of *Raphanus sativus* L. × *Brassica oleracea* L. *Zeitschrift für induktive AbstammungsVererbungslehre*, **39**, 1–7.

Kay, Q. O. N. (1978). The role of preferential and assortative pollination in the maintenance of flower colour polymorphisms. In *The pollination of flowers by insects*, ed. A. J. Richards, pp. 175–90. Linnean Society Symposium Series 6. London: Academic Press.

Kellerman, W. A. (1901). Variation in *Syndesmon thalictroides*. *Ohio Naturalist*, **1**, 107–11.

Kelvin, Lord. (Sir W. Thompson) (1871). On geological time. *Transactions of the Geological Society of Glasgow*, **3**, 1–28.

Kendall, M. G. & Plackett, R. L. (1977). *Studies in the history of statistics and probability*, vol. II. London: Griffin.

Kerner, A. (1895). *The natural history of plants, their forms, growth, reproduction and distribution*. Translated and edited by F. W. Oliver. London: Blackie.

Kernick, M. D. (1961). Seed production of specific crops. In *Agricultural and horticultural seeds*, pp. 181–547. FAO Agriculture Studies no. 55.

Keymer, R. & Ellis, W. M. (1978). Experimental studies on plants of *Lotus corniculatus* L. from Anglesey polymorphic for cyanogenesis. *Heredity*, **40**, 189–206.

Keynes, W. M. (1993). *Sir Francis Galton, FRS. The legacy of his ideas*. London: Macmillan.

Khandelwal, S. (1990). Chromosome evolution in the genus *Ophioglossum* L. *Botanical Journal of the Linnean Journal*, **102,** 205–17.

Kihara, H. & Ono, T. (1926). Chromosomenzahlen und systematische Gruppierung der *Rumex* – Arten. *Zeitschrift für Zellforshung und mikroskopische Anatomie*, **4**, 475–81.

Kimber, G. (1961). Basis of the diploid-like meiotic behaviour of polyploid Cotton. *Nature*, **191**, 98–100.

Kimber, G. & Athwal, R. S. (1972). A reassessment of the course of evolution of Wheat. *Proceedings of the National Academy of Sciences, USA*, **69**, 912–15.

Kimura, M. (1983). *The neutral theory of molecular evolution*. Cambridge: Cambridge University Press.

King, L. M. & Schaal, B. A. (1990). Genotypic variation within asexual lineages of *Taraxacum officinale*. *Proceedings of the National Academy of Sciences, USA*, **87,** 998–1002.

King, M. (1993). *Species evolution, the role of chromosome change*. Cambridge: Cambridge University Press.

Kirby, L. T. (1990). *DNA fingerprinting*. London: Freeman.

Kirk, J. T. O. & Tilney-Bassett, R. A. E. (1978). *The plastids: their chemistry, structure, growth and inheritance*, 2nd edn. Amsterdam: Elsevier.

Klein, R. M. & Perkins, T. D. (1988). Primary and secondary causes and consequences of contemporary forest decline. *Botanical Review*, **54**, 1–43.

Klinger, T. & Ellstrand, N. C. (1994). Engineered genes in wild populations: fitness of weed–crop hybrids of *Raphanus sativus*. *Ecological Applications*, **4**, 117–20.

Knight, G. R., Robertson, A. & Waddington, C. H. (1956). Selection for sexual isolation within a species. *Evolution*, **10**, 14–22.

Knox, R. B. & Heslop-Harrison, J. (1963). Experimental control of aposporous apomixis in a grass of the Andropogoneae. *Botaniska Notiser*, **116**, 127–41.

Kohn, D. (1981). On the origin of the principle of diversity. *Science*, **213**, 1105–8.

Kohn, D. (1985). *The Darwinian heritage*. New Jersey: Princeton University Press.

Kojima, A., Nagato, Y. & Hinata, K. (1991). Degree of apomixis in Chinese Chive (*Allium tuberosum*) estimated by esterase isozyme analysis. *Japanese Journal of Breeding*, **41,** 73–83.

Koopman, K. F. (1950). Natural selection for reproductive isolation between *Drosophila pseudoobscura* and *Drosophila persimilis*. *Evolution*, **4**, 135–48.

Koshy, T. K. (1968). Evolutionary origin of *Poa annua* L. in the light of karyotypic studies. *Canadian Journal of Genetics & Cytology*, **10**, 112–18.

Kovác, J. (1995). Micropropagation of *Dianthus arenarius* subsp. *bohemicus* – an endangered endemic from the Czech Republic. *Botanic Gardens Micropropagation News*, **1** , 106–8.

Kozlowski, T. T., Kramer, P. J. & Pallardy, S. G. (1991). *The physiological ecology of woody plants.* San Diego: Academic Press.

Kruckeberg, A. R. (1951). Intraspecific variability in the response of certain native plant species to serpentine soil. *American Journal of Botany*, **38**, 408–19.

Kruckeberg, A. R. (1954). The ecology of serpentine soils. III. Plant species in relation to serpentine soils. *Ecology*, **35**, 267–74.

Kyhos, D. W. (1965). The independent aneuploid origin of two species of *Chaenactis* (Compositae) from a common ancestor. *Evolution*, **19**, 26–43.

La Cour, L. F. (1953). The *Luzula* system analysed by X-rays. *Heredity*, **6** (Suppl.), 77–81.

Lack, A. J. & Kay, Q. O. N. (1988). Allele frequencies, genetic relationships and heterozygosity in *Polygala vulgaris* populations from contrasting habitats in southern Britain. *Biological Journal of the Linnean Society*, **34**, 119–47.

Lamarck, J. B. (1809). *Philosophie zoologique.* (English translation, Zoological philosophy, translated by H. Elliot, published 1914, London & New York: Macmillan.)

Lamb, H. H. (1970). Our changing climate. In *Flora of a changing Britain*, ed. F. H. Perring, pp. 11–24. Hampton, Middlesex: Botanical Society of the British Isles.

Lamprecht, H. (1961). Die Genenkarte von *Pisum* bei normaler Struktur der Chromosomen. *Agri Hortique Genetica*, **19**, 360–401.

Lamprecht, H. (1966). *Die Entstehung der Arten und höheren Kategorien.* New York & Vienna: Springer Verlag.

Lande, R. (1988). Genetics and demography in biological conservation. *Science*, **241**, 1455–60.

Landman, O. E. (1991). The inheritance of acquired characteristics. *Annual Review of Genetics*, **25**, 1–20.

Lane, C. (1962). Notes on the Common Blue (*Polyommatus icarus*) egg laying and feeding on the cyanogenic strains of the Bird's-Foot Trefoil (*Lotus corniculatus*). *Entomologist's Gazette*, **13**, 112–16.

Langevin, S. A., Clay, K. & Grace, J. B. (1990). The incidence and effects of hybridization between cultivated Rice and its related weed Red Rice (*Oryza sativa* L.) *Evolution*, **44**, 1000–8.

Langlet, O. (1934). Om variationen hos tallen *Pinus sylvestris* och dess samband med climatet. *Meddelanden från Statens Skogsførsøksanstalt*, **27**, 87–93.

Langlet, O. (1971). Two hundred years genecology. *Taxon*, **20**, 653–722.

Lankester, E. (1848). *The correspondence of John Ray.* London: Ray Society, British Museum.

Larsen, E. C. (1947). Photoperiodic responses of geographical strains of *Andropogon scoparius. Botanical Gazette*, **109**, 132–50.

Lawrence, W. E. (1945). Some ecotypic relations of *Deschampsia caespitosa. American Journal of Botany*, **32**, 298–314.

Lawrence, W. J. C. (1950). *Science and the glasshouse.* Edinburgh & London: Oliver & Boyd.

Lawrence, E. (1989). *A guide to modern biology, genetics, cells and systems.* Harlow: Longmans.

Lebaron, H. M. & Gressel, J. (eds) (1982). *Herbicide resistance in plants.* New York: Wiley.

Lee, A. (1902). Dr Ludwig on variation and correlation in plants. *Biometrika*, **1**, 316–19.

Lefèbvre, C. (1973). Outbreeding and inbreeding in a zinc–lead mine population of *Armeria maritima*. *Nature*, **243**, 96–7.

Levan, A. (1938). The effect of colchicine on root mitosis in *Allium*. *Hereditas*, **24**, 471–86.

Levin, D. A. (1973). The role of trichomes in plant defense. *Quarterly Review of Biology*, **48**, 3–15.

Levin, D. A. (1975). Minority cytotype exclusion in local plant populations. *Taxon*, **24**, 35–43.

Levin, D. A. (1978*a*). Pollinator behaviour and the breeding structure of plant populations. In *The pollination of flowers by insects*, ed. A. J. Richards, pp. 133–50. Linnean Society Symposium Series 6. London: Academic Press.

Levin, D. A. (1978*b*). The origin of isolating mechanisms in flowering plants. *Evolutionary Biology*, **11**, 185–317.

Levin, D. A. (1979). The nature of plant species. *Science*, **204**, 381–4.

Levin, D. A. (1984). Immigration in plants: an exercise in the subjunctive. In *Perspectives on plant population ecology*, ed. R. Dirzo & J. Sarukhán, pp. 242–60. Sunderland: Sinauer.

Levin, D. A. (1985). Reproductive character displacement in *Phlox*. *Evolution*, **39**, 1275–81.

Levin, D. A. (1988). Local differentiation and the breeding structure of plant populations. In *Plant evolutionary biology*, ed. L. D. Gottlieb & S. K. Jain, pp. 305–29. London: Chapman & Hall.

Levin, D. A. (1993). Local speciation in plants: the rule not the exception. *Systematic Botany*, **18**, 197–208.

Levin, D. A. & Kerster, H. W. (1967). An analysis of interspecific pollen exchange in *Phlox*. *American Naturalist*, **101**, 387–400.

Levin, D. A. & Kerster, H. W. (1974). Gene flow in seed plants. *Evolutionary Biology*, **7**, 139–220.

Lewin, B. (1990). *Genes IV*. Oxford: Oxford University Press.

Lewin, B. (1997). *Genes VI*. New York: Oxford University Press.

Lewin, R. A. (1993). *Origins of plastids: symbiogenesis, prochlorophytes, and the origins of chloroplasts*. New York: Chapman & Hall.

Lewis, D. (1979). *Sexual incompatibility in plants*. Studies in Biology no. 110. London: Arnold .

Lewis, D. & Crowe, L. K. (1956). The genetics and evolution of gynodioecy. *Evolution*, **10**, 115–25.

Lewis, H. (1973). The origin of diploid neospecies in *Clarkia*. *American Naturalist*, **107**, 161–70.

Lewis, W. H. (1976). Temporal adaptation correlated with ploidy in *Claytonia virginia*. *Systematic Botany*, **1**, 340–7.

Lewis, W. H. (ed.) (1980*a*). *Polyploidy*. London & New York: Plenum Press.

Lewis, W. H. (1980*b*). Polyploidy in species populations. In *Polyploidy*, ed. W. H. Lewis, pp. 103–44. New York & London: Plenum Press.

Lewis, W. H. (1980*c*). Polyploidy in angiosperms: dicotyledons. In *Polyploidy*, ed. W. H. Lewis, pp. 241–68. New York & London: Plenum Press.

Li, H. L. (1956). The story of the cultivated Horse-Chestnuts. *Morris Arboretum Bulletin*, **7**, 35–9.

Liljefors, A. (1953). Studies on propagation, embryology and pollination in

Sorbus. Acta Horti Bergiani, **16**, 227–329.

Liljefors, A. (1955). Cytological studies in *Sorbus. Acta Horti Bergiani*, **17**, 47–113.

Linhart, V. B. & Baker, I. (1973). Intra-population differentiation of physiological response to flooding in a population of *Veronica peregrina. Nature*, **242**, 275–6.

Linhart, Y. B., Mitton, J. B., Sturgeon, K. B. & Davis, M. L. (1981). Genetic variation in space and time in a population of Ponderosa Pine. *Heredity*, **46**, 407–26.

Linnaeus, C. (1737, but not distributed until 1738). *Hortus cliffortianus.* Amsterdam.

Linnaeus, C. (Carl von Linné). (1737). *Critica botanica.* (English translation by A. Hort, 1938. London: Ray Society, British Museum.)

Linnaeus, C. (1744). Peloria. In *Amoenitates academicae* (1749–90). (See Stearn, 1957, for details of the many editions.)

Linnaeus, C. (1749–90). *Amoenitates academicae.* (See Stearn, 1957, for details of the many editions.)

Linnaeus, C. (1751). *Philosophia botanica.* Stockholm.

Linnaeus, C. (1753). *Species plantarum.* (Facsimile edition 1957, London: Ray Society, British Museum.)

Linnaeus, C. (1762–3). *Species plantarum*, 2nd edn. Stockholm.

Linroth, S. (1983). The two faces of Linnaeus. In *Linnaeus, the man and his work*, ed. T. Frangsmyr, pp. 1–62. Berkeley: University of California Press.

Lloyd, D. G. (1975). The maintenance of gynodioecy and androdioecy in angiosperms. *Genetica*, **45**, 325–39.

Lloyd, D. G. (1979*a*). Some reproductive factors affecting the selection of self-fertilisation in plants. *American Naturalist*, **113**, 67–79.

Lloyd, D. G. (1979*b*). Evolution towards dioecy in heterostylous populations. *Plant Systematics & Evolution*, **131**, 71–80.

Lloyd, D. G. (1980). Demographic factors and mating patterns in angiosperms. In *Demography and evolution in plant populations*, ed. O. T. Solbrig, pp. 67–88. Oxford: Blackwell.

Lloyd, D. G. (1992). Self- and cross-fertilization in plants. II. The selection of self-fertilization. *International Journal of Plant Science*, **153**, 370–80.

Lloyd, D. G. & Webb, C. J. (1992). The evolution of heterostyly. In *Evolution and function of heterostyly*, ed. S. C. H. Barrett, pp. 151–78. Heidelberg: Springer Verlag.

Lönn, M., Prentice, H. C. & Tegelström, H. (1995). Genetic differentiation in *Hippocrepis emerus* (Leguminosae): allozyme and DNA fingerprint variation in disjunct Scandinavian populations. *Molecular Ecology*, **4**, 39–48.

Lord, E. M. (1981). Cleistogamy: a tool for the study of floral morphogenesis, function and evolution. *Botanical Review*, **47**, 421–49.

Löve, A. (1962). The biosystematic species concept. *Preslia*, **34**, 127–39.

Löve, A. & Löve, D. (1961). Chromosome numbers of Central and Northwest European plant species. *Opera Botanica*, **5**, 1–581.

Löve, A. & Löve, D. (1974). Origin and evolution of the arctic and alpine floras. In *Arctic and alpine environments*, ed. J. D. Ives & R. G. Barry, pp. 571–603. London: Methuen.

Lovejoy, A. O. (1966). *The great chain of being*. Cambridge, MA: Harvard University Press.

Lovett Doust, J. & Lovett Doust, L. (1988). *Plant reproductive ecology, patterns and strategies.* New York: Oxford University Press.
Lovis, J. D. (1977). Evolutionary patterns and processes in ferns. *Advances in Botanical Research*, **4**, 229–415.
Lövkvist, B. (1956). The *Cardamine pratensis* complex. *Symbolae Botanicae Upsalienses*, **14**(2), 1–131.
Lövkvist, B. (1962). Chromosome and differentiation studies in flowering plants of Skåne, South Sweden. 1. General aspects. Type species with coastal differentiation. *Botaniska Notiser*, **115**, 261–87.
Lowe, A. J. & Abbott, R. J. (1996). Origins of the new allopolyploid species *Senecio cambrensis* (Asteraceae) and its relationship to the Canary Islands endemic *Senecio teneriffae. American Journal of Botany*, **83**, 1365–72.
Luceño, M. & Castroviejo, S. (1991). Agmatoploidy in *Carex laevigata* (Cyperaceae). Fusion and fission of chromosomes as the mechanism of cytogenetic evolution in Iberian populations. *Plant Systematics & Evolution*, **177**, 149–59.
Luckow, M. (1995). Species concepts: assumptions, methods and applications. *Systematic Botany*, **20** , 589–605.
Ludwig, F. (1895). Uber Variationskurven und Variationsflächen der Pflanzen. *Botanisches Zentralblatt*, **64**, 1–8, 33–41, 65–72, 97–105.
Ludwig, F. (1901). Variationsstatistische Probleme und Materialen. *Biometrika*, **1**, 11–29.
Lumaret, R. (1984). The role of polyploidy in the adaptive significance of polymorphism at the GOT I locus in the *Dactylis glomerata* complex. *Heredity*, **52**, 153–69.
Lumaret, R. (1988). Cytology, genetics and evolution in the genus *Dactylis. Critical Reviews in Plant Sciences*, **7**, 55–91.
Lynch, R. I. (1900). Hybrid Cinerarias. *Journal of the Royal Horticultural Society*, **24**, 269–74.
Mabberley, D. J. (1987). *The plant-book.* Cambridge: Cambridge University Press.
MacArthur, R. H. & Wilson, E. O. (1967). *The theory of island biogeography.* Princeton: Princeton University Press.
Maceira, N. O., De Haan, A. A., Lumaret, R., Billon, M. & Delay, J. (1992). Production of 2*n* gametes in diploid subspecies of *Dactylis glomerata* L. I. Occurrence and frequency of 2*n* pollen. *Annals of Botany*, **69**, 335–43.
Mack, R. N. (1991). The commercial seed trade: an early disperser of weeds in the United States. *Economic Botany*, **45**, 257–73.
Mackenzie, D. A. (1981). *Statistics in Britain*, 1865–1930. Edinburgh: Edinburgh University Press.
Macnair, M. R. (1983). The genetic control of copper tolerance in the Yellow Monkey Flower, *Mimulus guttatus. Heredity*, **50**, 283–93.
Macnair, M. R. (1993). The genetics of metal tolerance in vascular plants. *New Phytologist*, **124**, 541–59.
Macnair, M. R. & Christie, P. (1983). Reproductive isolation as a pleiotropic effect of copper tolerance in *Mimulus guttatus. Heredity*, **50**, 295–302.
Macnair, M. R., Cumbes, Q. J. & Meharg, A. A. (1992). The genetics of arsenate tolerance in Yorkshire Fog, *Holcus lanatus* L. *Heredity*, **69**, 325–35.
Maddox, D. G., Cook, R. E., Wimberger, P. H. & Gardescu, S. (1989). Clone structure in four *Solidago altissima* populations: rhizome connections

within genotypes. *American Journal of Botany*, **76**, 318–26.

Major, J. (1988). Endemism; a botanical perspective. In *Analytical biogeography*, ed. A. A. Myers & P. S. Giller, pp. 117–46. London: Chapman & Hall.

Mann, C. C. (1991). Extinction – are ecologists crying wolf? *Science*, **253**, 736–8.

Manton, I. (1950). *Problems of cytology and evolution in the Pteridophyta.* London & New York: Cambridge University Press.

Marchant, C. J. (1963). Corrected chromosome numbers for *Spartina* × *townsendii* and its parent species. *Nature*, **199**, 299.

Marchant, C. J. (1967). Evolution in *Spartina* (Gramineae). 1. The history and morphology of the genus in Britain. *Journal of the Linnean Society* (*Botany*), **60**, 1–24.

Marchant, C. J. (1968). Evolution in *Spartina* (Gramineae). 2. Chromosomes, basic relationships and the problem of *S.* × *townsendii* agg. *Journal of the Linnean Society* (*Botany*), **60**, 381–409.

Marchi, P., Illuminati, O., Macioce, A., Capineri, R. & D'Amato, G. (1983). Genome evolution and polyploidy in *Leucanthemum vulgare* Lam. aggr. (Compositae). Karyotype analysis and DNA microdensitometry. *Caryologia*, **36**, 1–18.

Marks, G. E. (1966). The origin and significance of intraspecific polyploidy: experimental evidence from *Solanum chacoense*. *Evolution*, **20**, 552–7.

Marsden-Jones, E. (1930). The genetics of *Geum intermedium* Willd. haud Ehrh. and its back-crosses. *Journal of Genetics*, **23**, 377–95.

Marsden-Jones, E. M. & Turrill, W. B. (1945). Report of the transplant experiments of the British Ecological Society. *Journal of Ecology*, **33**, 59–81. (See also earlier reports in the *Journal of Ecology*: **18**, 352; **21**, 268; **23**, 443; **25**, 189; **26**, 359 & 380.)

Marshall, D. R. & Brown, A. H. D. (1981). The evolution of apomixis. *Heredity*, **47**, 1–15.

Marshall, D. R. & Weiss, P. W. (1982). Isozyme variation within and among Australian populations of *Emex spinosa* (L.). Campd. *Australian Journal of Biological Sciences*, **35**, 327–32.

Mason, B. J. (1992). *Acid rain: its causes and its effects on island waters.* Oxford: Clarendon Press.

Massart, J. (1902). L'accomodation individuelle chez le *Polygonum amphibium. Bulletin de Jardin Botanique de l'Etat à Bruxelles*, **1**, 73–95.

Mather, K. (1966). Breeding systems and response to selection. In *Reproductive biology and taxonomy of vascular plants*, ed. J. G. Hawkes, pp. 13–19. Conference report of Botanical Society of the British Isles. Oxford: Pergamon Press.

Matthew, P. (1831). Ideas on evolution, published in an Appendix to *On naval timber and arboriculture*. London.

Matthews, R. E. F. (1991). *Plant virology*, 3rd edn. New York: Academic Press.

Maunder, M. & Ramsay, M. (1994). The reintroduction of plants into the wild: an integrated approach to the conservation of native plants. In *The common ground of wild and cultivated plants*, ed. A. R. Perry & R. G. Ellis, pp. 81–8. Cardiff: National Museum of Wales.

Maurice, S., Charlesworth, D., Desfeux, C., Couvet, D., & Gouyon, P. -H. (1993). The evolution of gender in hermaphrodites of gynodioecious populations with nucleo-cytoplasmic male-sterility. *Proceedings of the Royal Society of London, B*, **251**, 253–61.

May, R. M. (1994). Past efforts and future prospects towards understanding how many species there are. In *Biodiversity and global change*, ed. O. T. Solbrig, H. M. van Emden & P. G. W. J. van Oordt, pp. 71–84. Wallingford: CAB International.

Mayer, M. S. & Soltis, P. S. (1994). The evolution of serpentine endemics: a chloroplast DNA phylogeny of the *Streptanthus glandulosus* complex (Cruciferae). *Systematic Botany*, **19**, 557–74.

Mayer, M. S., Soltis, P. S. & Soltis, D. E. (1994). The evolution of the *Streptanthus glandulosus* complex (Cruciferae): genetic divergence and gene flow in serpentine endemics. *American Journal of Botany*, **81**, 1288–99.

Maynard Smith, J. (1989). *Evolutionary genetics*. New York: Oxford University Press.

Mayr, E. (1942). *Systematics and the origin of species*. New York: Columbia University Press.

Mayr, E. (1963). *Animal species and evolution*. London: Oxford University Press; and Cambridge, MA: Harvard University Press.

Mayr, E. (1969). *Principles of systematic zoology*. New York: McGraw-Hill.

Mayr, E. (1982). *The growth of biological thought; diversity, evolution and inheritance*. Cambridge, MA: Harvard University Press.

McCauley, D. E. (1994). Contrasting the distribution of chloroplast DNA and allozyme polymorphism among local populations of *Silene alba*: implications for studies of gene flow in plants. *Proceedings of the National Academy of Sciences, USA*, **91**, 8127–31.

McFadden, E. S. & Sears, E. R. (1946). The origin of *Triticum spelta* and its free-threshing hexaploid relatives. Hybrids of synthetic *T. spelta* with cultivated hexaploids. *Journal of Heredity*, **37**, 81–9, 107–16.

McLean, R. C. & Ivimey-Cook, W. R. (1956). *Textbook of theoretical botany*. London, New York & Toronto: Longmans.

McLeish, J. & Snoad, B. (1962). *Looking at chromosomes*. London & New York: Macmillan. (2nd edn, 1972).

McMillan, C. (1970). Photoperiod in *Xanthium* populations from Texas and Mexico. *American Journal of Botany*, **57**, 881–8.

McMillan, C. (1971). Photoperiod evidence in the introduction of *Xanthium* (Cocklebur) to Australia. *Science*, **171**, 1029–31.

McMullen, C. K. (1987). Breeding systems of selected Galapagos Islands angiosperms. *American Journal of Botany*, **74**, 1694–1705.

McNeill, C. I. & Jain, S. K. (1983). Genetic differentiation studies and phylogenetic inference in the plant genus *Limnanthes* (Section *Inflexae*). *Theoretical & Applied Genetics*, **66**, 257–69.

McNeill, J. (1976). The taxonomy and evolution of weeds. *Weed Research*, **16**, 399–413.

McNeilly, T. (1968). Evolution in closely adjacent plant populations. III. *Agrostis tenuis* on a small copper mine. *Heredity*, **23**, 99–108.

McNeilly, T. & Antonovics, J. (1968). Evolution in closely adjacent plant populations. IV. Barriers to gene flow. *Heredity*, **23**, 205–18.

Mead, R. (1988). *The design of experiments*. Cambridge: Cambridge University Press.

Meagher, T. R. (1986). Analysis of paternity within a natural population of *Chamaelirium luteum*. I. Identification of most-likely parents. *American Naturalist*, **128**, 199–215.

Meagher, T. R. & Thompson, E. (1987). Analysis of parentage for naturally established seedlings of *Chamaelirium luteum* (Liliacae). *Ecology*, **68**, 803–12.

Meeuse, A. D. J. (1966). *Fundamentals of phytomorphology*. New York: Ronald Press.

Meeuse, A. D. J. (1987). *All about angiosperms*. Delft: Eburon.

Meffe, G. K. & Carroll, C. R. (1994). *Principles of conservation biology*. Sunderland: Sinauer.

Meharg, P. A., Cumbes, Q. J. & Macnair, M. R. (1993). Pre-adaptation of Yorkshire Fog *Holcus lanatus* L. (Poaceae) to arsenate tolerance. *Evolution*, **47**, 313–16.

Mendel, G. (1866). Versuche über Planzenhybriden. *Verhandlungen des Naturforschenden Vereins in Brünn*, **4**, 3–44. (English translation in Bateson, W., 1909, *Mendel's principles of heredity*. London: Cambridge University Press; also Bennett, J. H. (ed.), 1965. *Experiments in plant hybridisation*. Edinburgh & London: Oliver & Boyd.)

Mendel, G. (1869). Uber einige aus künstlicher Befruchtung gewonnenen *Hieracium*-Bastarde. *Verhandlungen des naturforschen den Vereines in Brünn*, **8**, 26pp.

Menges, E. S. (1991). The application of minimum viable population theory to plants. In *Genetics and conservation of rare plants*, ed. D. A. Falk & K. E. Holsinger, pp. 45–61. New York: Oxford University Press.

Mergen, F. (1963). Ecotypic variation in *Pinus strobus*. *Ecology*, **44**, 716–27.

Merrell, D. J. (1962). *Evolution and genetics: the modern theory of evolution*. New York: Holt, Rinehart & Winston.

Merxmüller, H. (1970). 1. Biosystematics: still alive? Provocation of biosystematics. *Taxon*, **19**, 140–5.

Midgley, J. J. & Bond, W. J. (1991). Ecological aspects of the rise of angiosperms: a challenge to the reproductive superiority hypotheses. *Biological Journal of the Linnean Society*, **44**, 81–92.

Millener, L. H. (1961). Day length as related to vegetative development in *Ulex europaeus*. 1. The experimental approach. *New Phytologist*, **60**, 339–54.

Miller, T. E. (1987). Systematics and evolution. In *Wheat breeding; its scientific basis*, ed. F. G. H. Lupton, pp. 1–30. London: Chapman & Hall.

Mills, A. D. (1993). *English place names*. Oxford: Oxford University Press.

Minelli, A. (1993). *Biological systematics: the state of the art*. London: Chapman & Hall.

Mitchell, R. S. (1968). Variation in the *Polygonum amphibium* complex and its taxonomic significance. *University of California Publications in Botany*, **45**, 1–54.

Mivart, St. G. (1871). *The genesis of species, 2nd edn*. London: Macmillan.

Mogie, M. (1985). Morphological, developmental and electrophoretic variation within and between obligately apomictic *Taraxacum* species. *Biological Journal of the Linnean Society*, **24**, 207–16.

Mølgaard, P. (1976). *Plantago major* ssp. *major* and ssp. *pleiosperma*. Morphology, biology and ecology in Denmark. *Botanik Tidsskrift*, **71**, 31–56.

Mooney, H. A. & Billings, W. D. (1961). Comparative physiological ecology of arctic and alpine populations of *Oxyria digyna*. *Ecological Monographs*, **31**, 1–29.

Mooney, H. A., Winner, W. E. & Pell, E. J. (1991). *Reponse of plants to multiple stresses*. San Diego: Academic Press.

Moore, D. M. (1959). Population studies on *Viola lactea* Sm. and its wild hybrids. *Evolution*, **13**, 318–32.

Moore, D. M. (1976). *Plant cytogenetics*. London: Chapman & Hall; and New York: Wiley & Sons.
Moore, D. M. (1982). *Flora europaea check-list and chromosome index*. London: Cambridge University Press.
Moore, D. M. & Harvey, M. J. (1961). Cytogenetic relationships of *Viola lactea* Sm. and other West European arosulate Violets. *New Phytologist*, **60**, 85–95.
Moore, R. J. & Mulligan, G. A. (1956). Natural hybridization between *Carduus acanthoides* and *Carduus nutans* in Ontario. *Canadian Journal of Botany*, **34**, 71–85.
Moore, R. J. & Mulligan, G. A. (1964). Further studies on natural selection among hybrids of *Carduus acanthoides* and *Carduus nutans*. *Canadian Journal of Botany*, **42**, 1605–13.
Morison, R. (1672). *Plantarum umbelliferarum distributio nova*. Oxford.
Morisset, P. & Boutin, C. (1984). The biosystematic importance of phenotypic plasticity. In *Plant systematics*, ed. W. F. Grant, pp. 293–306. Toronto: Academic Press.
Morris, M. G. & Perring, F. H. (eds.) (1974). *The British oak: its history and natural history*. Published for The Botanical Society of the British Isles. Classey: Faringdon.
Morris, W. F., Kareiva, P. M. & Raymer, P. L. (1994). Do barren zones and pollen traps reduce gene escape from transgenic crops? *Ecological Applications*, **4**, 157–65.
Morse, L. E. & Henifin, M. S. (1981). *Rare plant conservation. Geographical data organization*. New York: The New York Botanic Garden.
Morton, A. G. (1981). *History of botanical science*. London: Academic Press.
Morton, J. K. (1966). The role of polyploidy in the evolution of a tropical flora. In *Chromosomes today*, vol. 1, ed. C. D. Darlington & K. R. Lewis, pp. 73–6, Edinburgh: Oliver & Boyd.
Mueller-Dombois, D. (1988). Forest decline and dieback – a global ecological problem. *Trends in Ecology & Evolution*, **3**, 310–12.
Muenchow, G. (1982). A loss-of-alleles model for the evolution of distyly. *Heredity*, **49**, 81–93.
Muller, G. (1977). Cross-fertilization in a Conifer stand inferred from enzyme gene-markers in seeds. *Silvae Genetica*, **26**, 223–6.
Müntzing, A. (1930*a*). Uber Chromosomenvermehrung in *Galeopsis*-kreuzungen und ihre phylogenetische Bedeutung. *Hereditas*, **14**, 153–72.
Müntzing, A. (1930*b*). Outlines to a genetic monograph of the genus *Galeopsis* with special reference to the nature and inheritance of partial sterility. *Hereditas*, **13**, 185–341
Müntzing, A. (1961). *Genetic research*. Stockholm: L. T. Førlag.
Müntzing, A., Tedin, O. & Turesson, G. (1931). Field studies and experimental methods in taxonomy. *Hereditas*, **15**, 1–12.
Myers, K. (1986). Introduced vertebrates in Australia, with emphasis on the mammals. In *Ecology of biological invasions*, ed. R. H. Groves & J. J. Burdon, pp. 120–36. Cambridge: Cambridge University Press.
Myers, N. (1979). *The sinking ark: a new look at the problem of disappearing species*. Oxford: Pergamon Press.
Nannfeldt, J. A. (1937). The chromosome numbers of *Poa*, Sect. *Ochlopoa* A. and Gr. and their taxonomical significance. *Botaniska Notiser*, 1937, 238–57.

Naumova, T. N. (1993). *Apomixis in angiosperms; nucellular and integumental embryology*. Boca Raton: CRC Press.
Navashin, M. (1926). Variabilität des Zellkerns bei *Crepis*-Arten in Bezug auf die Artbildung. *Zeitschrift für Zellforschung und mikroskopische Anatomie*, **4**, 171–215.
Nei, M. (1972). Genetic distance between populations. *American Naturalist*, **106**, 283–92.
Nelson, A. P. (1967). Racial diversity in Californian *Prunella vulgaris*. *New Phytologist*, **66**, 707–46.
Neuhaus, D., Kühl, H., Kohl, J. G., Dörfel, P. & Börner, T. (1993). Investigations on the genetic diversity of *Phragmites* stands using genomic fingerprinting. *Aquatic Botany*, **45**, 357–64.
New, J. K. (1958). A population study of *Spergula arvensis* 1. *Annals of Botany, New Series*, **22**, 457–77.
New, J. K. (1959). A population study of *Spergula arvensis* 2. *Annals of Botany, New Series*, **23**, 23–33.
New, J. K. (1978). Change and stability of clines in *Spergula arvensis* L. (Corn Spurrey) after 20 years. *Watsonia*, **12**(2), 137–43.
New, J. K. & Herriott, J. C. (1981). Moisture for germination as a factor affecting the distribution of the seedcoat morphs of *Spergula arvensis* L. *Watsonia*, **13**(4), 323–4.
Newton, W. C. F. & Pellew, C. (1929). *Primula kewensis* and its derivatives. *Journal of Genetics*, **20**, 405–66.
Niemela, P. & Tuomi, J. (1987). Does the leaf morphology of some plants mimic caterpillar damage? *Oikos*, **50**, 256–7.
Nieuwhof, M. (1963). Pollination and contamination of *Brassica oleracea* L. *Euphytica*, **12**, 17–26.
Nilsson-Ehle, E. (1909). Kreuzungsuntersuchungen an Hafer und Weizen. *Acta Universitatis lundensis, Ser. 2*, **5**(2), 1–122.
Nitecki, M. H. (1983). *Coevolution*. Chicago: University of Chicago Press.
Njoku, E. (1956). Studies on the morphogenesis of leaves. II. The effect of light intensity on leaf shape in *Ipomoea caerulea*. *New Phytologist*, **55**, 91–110.
Nogler, G. A. (1984). Gametophytic apomixis. In *Embryology of angiosperms*, ed. B. M. Johri, pp. 475–518. Berlin: Springer Verlag.
Nordenskiøld, H. (1949). The somatic chromosomes of some *Luzula* species. *Botaniska Notiser*, 1949, 81–92.
Nordenskiøld, H. (1951). Cyto-taxonomical studies in the genus *Luzula*. 1. Somatic chromosomes and chromosome numbers. *Hereditas*, **37**, 325–55.
Nordenskiøld, H. (1956). Cyto-taxonomical studies in the genus *Luzula*. 2. Hybridization experiments in the *campestris-multiflora* complex. *Hereditas*, **42**, 7–73.
Nordenskiøld, H. (1961). Tetrad analysis and the course of meiosis in three hybrids of *Luzula campestris*. *Hereditas*, **47**, 203–38.
Norton, B. J. (1983). Fisher's entrance into evolutionary science: the role of eugenics. In *Dimensions of Darwinism*, ed. M. Grene, pp. 19–30. Cambridge: Cambridge University Press.
Novak, S. J. & Mack, R. N. (1995). Allozyme diversity in the apomictic vine *Bryonia alba* (Cucurbitaceae): potential consequences of multiple introductions. *American Journal of Botany*, **82**, 1153–62.
Novak, S. J., Soltis, D. E. & Soltis, P. S. (1991). Ownbey's Tragopogons: 40 years later. *American Journal of Botany*, **78**, 1586–1600.

Nyárády, E. I., Beldie, A. L., Morariu, I. & Nyárády, A. (1972). Flora Republicii Socialiste România XII. Bucharest: Editura Academiei Republicii Socialiste România.
Nybom, H. & Schaal, B. A. (1990). DNA 'fingerprints' reveal genotypic distributions in natural populations of Blackberries and Raspberries (*Rubus*, Rosaceae). *American Journal of Botany*, **77**, 883–88.
Ohno, S. (1970). *Evolution by gene duplication*. London: Allen & Unwin.
Olby, R. C. (1974). *The path to the double helix*. London: Macmillan.
Olby, R. C. (1979). Mendel no Mendelian? *History of Science*, **17**, 53–72.
Olby, R. C. (1985). *Origins of Mendelism*, 2nd edn. Chicago: University of Chicago Press.
Olby, R. C. & Gautry, P. (1968). Eleven references to Mendel before 1900. *Annals of Science*, **24**, 7–20.
Oldfield, M. L. (1989). *The value of conserving genetic resources*. Sunderland: Sinauer.
Olmstead, R. G. & Palmer, J. D. (1994). Chloroplast DNA systematics: a review of methods and data analysis. *American Journal of Botany*, **81**, 1205–24.
Oostermeijer, G., Hvatum, H., Den Nijs, H., Borgen, L. (1966). Genetic variation, plant growth strategy and population structure of the rare, disjunctly distributed *Gentiana pneumonanthe* (Gentianaceae) in Norway. *Symbolae Botanicae Upsaliensis*, **31**(3), 185–203.
Oostermeijer, J. G. B., Den Nijs, J. C. M., Raijmann, L. E. L. & Menken, S. B. J. (1992). Population biology and management of the Marsh Gentian (*Gentiana pneumonanthe* L.), a rare species in The Netherlands. *Botanical Journal of the Linnean Society*, **108**, 117–30.
Orel, V. (1984). *Mendel*. Past Masters Series. Oxford: Oxford University Press.
Orel, V. & Matalová, A. (1983). *Gregor Mendel and the foundations of genetics*. Brno: Mendelianum of the Moravian Museum.
Ornduff, R. (1966). The origin of dioecism from heterostyly in *Nymphoides* (Menyanthaceae). *Evolution*, **20**, 309–14.
Ornduff, R. (1969). Reproductive biology in relation to systematics. *Taxon*, **18**, 121–33.
Ornduff, R. (1970). Pathways and patterns of evolution – a discussion. *Taxon*, **19**, 202–4.
Osawa, J. (1913). Studies on the cytology of some species of *Taraxacum*. *Archiv für Zellforschung*, **10**, 450–69.
Osborn, H. F. (1894). *From the Greeks to Darwin: an outline of the development of the evolution idea*. London & New York: Macmillan.
Ouborg, N. J. & van Treuren, R. (1995). Variation in fitness-related characters among small and large populations of *Salvia pratensis*. *Journal of Ecology*, **83**, 369–80.
Ownbey, M. (1950). Natural hybridisation and amphidiploidy in the genus *Tragopogon*. *American Journal of Botany*, **37**, 487–99.
Ownbey, M. & McCollum, G. D. (1953). Cytoplasmic inheritance and reciprocal amphiploidy in *Tragopogon*. *American Journal of Botany*, **40**, 788–96.
Ownbey, M. & McCollum, G. D. (1954). The chromosome of *Tragopogon*. *Rhodora*, **56**, 7–21.
Palmer, J. D. (1988). Intraspecific variation and multicircularity in *Brassica* mitochondrial DNAs. *Genetics*, **118**, 341–51.
Pankhurst, R. J. (1991). *Practical taxonomic computing*. Cambridge: Cambridge University Press.

Pantin, C. F. A. (1960). Alfred Russel Wallace, FRS and his essays of 1858 and 1855. *Notes and Records of the Royal Society of London*, **14**, 67–84.

Parker, D. M. (1982). The conservation, by restocking, of *Saxifraga cespitosa* in North Wales. *Watsonia*, **14**, 104–5.

Parker, J. S., Jones, G. H., Edgar, L. A. & Whitehouse, C. (1991). The population cytogenetics of *Crepis capillaris*. IV. The distribution of B-chromosomes in British populations. *Heredity*, **66**, 211–18.

Parker, J. S., Wilby, A. S. & Taylor, S. (1988). Chromosome stability and instability in plants. In *Kew Chromosome Conference III*, ed. P. E. Brandham, pp. 131–40. London: HMSO.

Parker, R. E. (1973). *Introductory statistics for biology*. London: Arnold.

Parks, J. C. & Werth, C. R. (1993). A study of spatial features of clones in a population of Bracken Fern, *Pteridium aquilinum* (Dennstaedtiaceae). *American Journal of Botany*, **80**, 537–44.

Parokonny, A. S., Kenton, A. Y., Meridith, L., Owens, S. J. & Bennett, M. D. (1992). Genomic divergence of allopatric sibling species studied by molecular cytogenetics of their F_1 hybrids. *The Plant Journal*, **2**, 665– 704.

Parokonny, A. S., Kenton, A. Y., Gleba, Y. Y. & Bennett, M. D. (1994). The fate of recombinant chromosomes and genome interaction in *Nicotiana* asymmetric hybrids and their sexual progeny. *Theoretical & Applied Genetics*, **89**, 488–97.

Parsons, P. A. (1959). Some problems in inbreeding and random mating in tetrasomics. *Agronomy Journal*, **51**, 465–7.

Paterniani, E. (1969). Selection for reproductive isolation between two populations of Maize, *Zea mays* L. *Evolution*, **23**, 534–47.

Pazy, B. & Zohary, D. (1965). The process of introgression between *Aegilops* polyploids: natural hybridization between *A. variabilis*, *A. ovata* and *A. biuncialis*. *Evolution*, **19**, 385–94.

Pearson, E. S. & Kendall, M. G. (1970). *Studies in the history of statistics and probability*. London: Griffin.

Pearson, K. (1900). *The grammar of science*, 2nd edn. London: Black.

Pearson, K. (1924). *The life, letters and labours of Francis Galton*, vol. II. Cambridge: Cambridge University Press.

Pearson, K. *et al.* (1903). Cooperative investigation on plants. 2. Variation and correlation in Lesser Celandine from diverse localities. *Biometrika*, **2**, 145–64.

Pearson, K., Lee, A., Warren, E., Fry, A., Fawcett, C. D. *et al.* (1901). Mathematical contributions to the theory of evolution. IX. On the principle of homotyposis and its relation to heredity, to the variability of the individual, and to that of the race. Part I – homotyposis in the vegetable kingdom. *Philosophical Transactions of the Royal Society*, *A*, **197**, 285–379.

Pearson, K. & Yule, G. U. (1902). Variation in ray-flowers of *Chrysanthemum leucanthemum*, 1133 heads gathered at Keswick during July 1895. *Biometrika*, **1**, 319.

Peckham, M. (1959). *The origin of species by Charles Darwin. A variorum text.* London: Oxford University Press; and Philadelphia: University of Pennsylvania Press.

Pellew, C. (1913). Note on gametic reduplication in *Pisum*. *Journal of Genetics*, **3**, 105–6.

Pellmyr, O. & Thien, L. B. (1986). Insect reproduction and floral fragrances: keys to the evolution of the angiosperms? *Taxon*, **35**, 76–85.

Pennington, W. (1974). *The history of British vegetation*, 2nd edn. London: English Universities Press.
Perring, F. H. & Farrell, L. (1983). *British red data books. Vol.* 1, *vascular plants*, 2nd edn. Lincoln: The Society for the Promotion of Nature Conservation with the financial support of the World Wildlife Fund.
Perring, F. H. & Walters, S. M. (1976). *Atlas of the British flora*, 2nd edn. Wakefield: EP Publishing.
Peterken, G. F. (1981). *Woodland conservation and management*. London: Chapman & Hall.
Pharis, R. P. & Ferrell, W. K. (1966). Differences in drought resistance between coastal and inland sources of Douglas Fir. *Canadian Journal of Botany*, **44**, 1651–9.
Pichersky, E., Soltis, D. E. & Soltis, P. S. (1990). Defective chlorophyll a/b-binding protein genes in the genome of a homosporous fern. *Proceedings of the National Academy of Sciences, USA*, **87**, 195–9.
Pickett, S. T. A., Parker, V. T. & Fiedler, P. L. (1992). The new paradigm in ecology: implications for conservation biology above the species level. In *Conservation biology: the theory and practice of nature conservation, preservation and management*, ed. P. L. Fiedler & S. K. Jain, pp. 65–88. New York: Chapman & Hall.
Platnick, N. I. (1979). Philosophy and the transformation of cladistics. *Systematic Zoology*, **28**, 537–46.
Pollard, A. J. (1980). Diversity of metal tolerances in *Plantago lanceolata* L. from the southeastern United States. *New Phytologist*, **86**, 109–17.
Pope, O. A., Simpson, D. M., & Duncan, E. N. (1944). Effect of Corn barriers on natural crossing in Cotton. *Journal of Agricultural Research*, **68**, 347–61.
Porter, T. M. (1986). *The rise of statistical thinking*, 1820–1900. Princeton: Princeton University Press.
Portugal, F. H. & Cohen, J. S. (1977). *A century of DNA*. Cambridge, MA & London: The MIT Press.
Poulton, J. E. (1990). Cyanogenesis in plants. *Plant Physiology*, **94**, 401–5.
Powers, L. (1945). Fertilization without reduction in Guayule (*Parthenium argentatum* Gray) and a hypothesis as to the evolution of apomixis and polyploidy. *Genetics*, **30**, 323–46.
Prance, G. T. (1992). The *Species plantarum* Project. *Botanical Journal of the Linnean Society*, **109**, 569–74.
Prentice, H. C. (1986). Climate and clinal variation in seed morphology of the White Campion, *Silene latifolia* (Caryophyllaceae). *Biological Journal of the Linnean Society*, **27**, 179–89.
Primack, R. B. (1993). *Essentials of conservation biology*. Sunderland: Sinauer.
Primack, R. B. & Miao, S. L. (1992). Dispersal can limit local plant distribution. *Conservation Biology*, **6**, 513–19.
Prime, C. T. (1960). *Lords and ladies*. London: Collins.
Prober, S. M. & Brown, A. H. D. (1994). Conservation of the Grassy White Box woodlands: population genetics and fragmentation of *Eucalyptus albens*. *Conservation Biology*, **8**, 1003–13.
Proctor, J. (1971*a*). The plant ecology of serpentine. II. Plant response to serpentine soils. *Journal of Ecology*, **59**, 397–410.
Proctor, J. (1971*b*). The plant ecology of serpentine. III. The influence of a high magnesium/calcium ratio and high nickel and chromium levels in some British and Swedish serpentine soils. *Journal of Ecology*, **59**, 827–42.

Proctor, M. C. F., Proctor, M. E. & Groenhof, A. C. (1989). Evidence from peroxidase polymorphism on the taxonomy and reproduction of some *Sorbus* populations in Southwest England. *New Phytologist*, **112**, 569–75.

Proctor, M. C. F. & Yeo, P. F. (1973). *The pollination of flowers*. London: Collins.

Proctor, M. C. F., Yeo, P. F. & Lack, A. J. (1996). *The natural history of pollination*. London: Harper Collins.

Provine, W. B. (1971). *The origins of theoretical population genetics*. Chicago & London: University of Chicago Press.

Provine, W. B. (1986). *Sewall Wright and evolutionary biology*. Chicago: University of Chicago.

Provine, W. B. (1987). *The origins of theoretical population genetics*, 2nd edn. Chicago: University of Chicago Press.

Putnam, A. R. & Tang, C. S. (1986). *The science of allelopathy*. New York: Wiley.

Quarin, C. L. & Hanna, W. W. (1980). Effect of three ploidy levels on meiosis and mode of reproduction in *Paspalum hexastachyum*. *Crop Science*, **20**, 69–75.

Quetelet, M. A. (1846). *Lettres à S. A. R. Ie Duc Régnant de Saxe-Coburg et Gotha, sur la théorie des probabilités, appliquée aux sciences morales et politiques*. Brussels. Translation by O. G. Downes (1849): *Letters addressed to H. R. H. the Grand Duke of Saxe-Coburg and Gotha on the theory of probabilities as applied to the moral and political sciences*. London: Charles & Edwin Layton.

Quinn, J. A. (1978). Plant ecotypes: ecological or evolutionary units. *Bulletin of the Torrey Botanical Club*, **105**, 58–64.

Rabinowitz, D., Cairns, S. & Dillon, T. (1986). Seven forms of rarity and their frequency in the flora of the British Isles. In *Conservation biology*, ed. M. E. Soulé, pp. 182–204. Sunderland: Sinauer.

Rackham, O. (1975). *Hayley Wood. Its history and ecology*. Cambridge: Cambridgeshire and Isle of Ely Naturalists Trust.

Rackham, O. (1980). *Ancient woodland: its history, vegetation and uses in England*. London: Arnold.

Radford, A. E., Dickison, W. C., Massey, J. R. & Bell, R. C. (1974). *Vascular plant systematics*. New York: Harper & Row.

Rafinski, J. N. (1979). Geographic variability of flower colour in *Crocus scepusiensis* (Iridaceae). *Plant Systematics & Evolution*, **131**, 107–25.

Raijmann, L. E. L., van Leeuwen, N. C., Kersten, R., Oostermeijer, J. G. B., Den Nijs, H. C. M. & Menken, S. B. J. (1994). Genetic variation and outcrossing rate in relation to population size in *Gentiana pneumonathe* L. *Conservation Biology*, **8**, 1014–26.

Rajhathy, T. & Thomas, H. (1972). Genetic control of chromosome pairing in hexaploid Oats. *Nature*, **239**, 217–19.

Ramsbottom, J. (1938). Linnaeus and the species concept. *Proceedings of the Linnean Society of London*, **150**, 192–219.

Ramsey, M. W., Cairns, S. C. & Vaughton, G. V. (1994). Geographic variation in morphological and reproductive characters of coastal and tableland populations of *Blandfordia grandiflora*. *Plant Systematics & Evolution*, **192**, 215–30.

Randolph, L. F., Nelson, I. S. & Plaisted, R. L. (1967). Negative evidence of introgression affecting the stability of Louisiana Iris species. *Cornell University Agriculture Experimental Station Memoir* no. 398.

Ranker, T. A., Floyd, S. K. & Trapp, P. G. (1994). Multiple colonizations of *Asplenium adiantum-nigrum* onto the Hawaiian archipelago. *Evolution*, **48**, 1364–7.

Raup, D. M. (1991). *Extinction*. Oxford: Oxford University Press.

Raven, C. E. (1950). *John Ray: naturalist*, 2nd edn., reissued 1986. Cambridge: Cambridge University Press.

Raven, P. H. (1976). Systematics and plant population biology. *Systematic Botany*, **1**, 284–316.

Raven, P. H. & Thompson, H. J. (1964). Haploidy and angiosperm evolution. *American Naturalist*, **98**, 251–2.

Raybould, A. F. & Gray, A. J. (1993). Genetically modified crops and hybridization with wild relatives: a UK perspective. *Journal of Applied Ecology*, **30**, 199–219.

Raybould, A. F. & Gray, A. J. (1994). Will hybrids of genetically modified crops invade natural communities? *Trends in Ecology & Evolution*, **9**, 85–9.

Raybould, A. F., Gray, A. J., Lawrence, M. J. & Marshall, D. F. (1990). The origin and taxonomy of *Spartina* × *neyrautii* Foucaud. *Watsonia*, **18**, 207–9.

Raybould, A. F., Gray, A. J., Lawrence, M. J. & Marshall, D. F. (1991*a*). The evolution of *Spartina anglica* C. E. Hubbard (Gramineae): origin and genetic variability. *Biological Journal of the Linnean Society*, **43**, 111–26.

Raybould, A. F., Gray, A. J., Lawrence, M. J. & Marshall, D. F. (1991*b*). The evolution of *Spartina anglica* C. E. Hubbard (Gramineae): genetic variation and status of the parental species in Britain. *Biological Journal of the Linnean Society*, **44**, 369–80.

Rayner, A. A. (1969). *A first course in biometry for agricultural students*. Pietermaritzburg: University of Natal Press.

Rees, H. & Jones, R. N. (1977). *Chromosome genetics*. London: Arnold.

Reiling, K. & Davison, A. W. (1992). Spatial variation in ozone resistance of British populations of *Plantago major* L. *New Phytologist*, **122**, 699–708.

Reinartz, J. A. & Les, D. H. (1994). Bottleneck-induced dissolution of self-incompatibility and breeding system consequences in *Aster furcatus*. *American Journal of Botany*, **81**, 446–55.

Rice, W. R. & Hostert, E. E. (1993). Laboratory experiments on speciation: what have we learned in 40 years? *Evolution*, **47**, 1637–53.

Rice, E. L. (1984). *Allelopathy*, 2nd edn. London: Academic Press.

Richard, M., Jubier, M. F., Bajon, R., Gouyon, P. H. & Lejeune, B. (1995). A new hypothesis for the origin of pentaploid *Holcus* from diploid *Holcus lanatus* L. and tetraploid *Holcus mollis* L. in France. *Molecular Ecology*, **4**, 29–38.

Richards, A. J. (1979). Reproduction in flowering plants. *Nature*, **278**, 306.

Richards, A. J. (1986). *Plant breeding systems*. London: George Allen & Unwin.

Richards, A. J. (1993). *Primula*. London: Batsford.

Richards, A. J. & Ibrahim, H. (1978). Estimation of neighbourhood size in two populations of *Primula veris*. In *The pollination of flowers by insects*, ed. A. J. Richards, pp. 165–74. Linnean Society Symposium Series 6. London: Academic Press.

Ridge, I. (1991). *Plant physiology*. Sevenoaks: Hodder & Stoughton and the Open University.

Ridgman, W. J. (1975). *Experimentation in biology*. Glasgow: Blackie.

Rieger, R., Michaelis, A. & Green, M. M. (1976). *Glossary of genetics and cytogenetics*, 4th edn. Berlin, Heidelberg & New York: Springer Verlag.

Rieseberg, L. H. (1995). The role of hybridization in evolution: old wine in new skins. *American Journal of Botany*, **82**, 944–53.

Rieseberg, L. H. & Ellstrand, N. C. (1993). What can molecular and morphological markers tell us about plant hybridization? *Critical Reviews in Plant Sciences*, **12**, 213–41.

Rieseberg, L. H., Soltis, D. E. & Palmer, J. D. (1988). A molecular re-examination of introgression between *Helianthus annuus* and *H. bolanderi*, (Compositae). *Evolution*, **42**, 227–38.

Rieseberg, L. H., Soltis, D. E. & Soltis, P. S. (1988). Genetic variation in *Helianthus annuus* and *H. bolanderi. Biochemical Systematics & Ecology*, **16**, 393–9.

Rieseberg, L. H. & Wendel, J. F. (1993). Introgression and its consequences in plants. In *Hybrid zones and the evolutionary process*, ed. R. G. Harrison, pp. 70–109. New York: Oxford University Press.

Riley, H. P. (1938). A character analysis of colonies of *Iris fulva* and *Iris hexagona* var. *giganticaerulea* and natural hybrids. *American Journal of Botany*, **25**, 727–38.

Riley, R. (1965). Cytogenetics and the evolution of Wheat. In *Essays on crop plant evolution*, ed. J. Hutchinson, pp. 103–22. London: Cambridge University Press.

Riley, R. & Chapman, V. (1958). Genetic control of the cytologically diploid behaviour of hexaploid Wheat. *Nature*, **183**, 713–15.

Riley, R. & Law, C. N. (1965). Genetic variation in chromosome pairing. *Advances in Genetics*, **13**, 57–114.

Riley, R., Unrau, J. & Chapman, V. (1958). Evidence on the origin of the B genome of Wheat. *Journal of Heredity*, **49**, 91–8.

Ritchie, J. C. (1955*a*). A natural hybrid in *Vaccinium*. 1. The structure, performance and chorology of the cross *Vaccinium intermedium* Ruthe. *New Phytologist*, **54**, 49–67.

Ritchie, J. C. (1955*b*). A natural hybrid in *Vaccinium*. 2. Genetic studies in *Vaccinium intermedium* Ruthe. *New Phytologist*, **54**, 320–35.

Ritchie, J. C. (1987). *Postglacial vegetation of Canada*. Cambridge: Cambridge University Press.

Rizvi, S. J. H. & Rizvi, V. (1992). *Allelopathy: basic and applied aspects*. London: Chapman & Hall.

Roach, D. A. & Wulff, R. D. (1987). Maternal effects in plants. *Annual Review of Ecology & Systematics*, **18**, 209–35.

Roberts, H. F. (1929). *Plant hybridisation before Mendel*. Princeton: Princeton University Press; and London: Oxford University Press.

Roberts, N. (1989). *The Holocene: an environmental history*. Oxford: Blackwell.

Rogers, S. O. (1994). Phylogenetic and taxonomic information from herbarium and mummified DNA. In *Conservation of Plant Genes II: utilization of ancient and modern DNA*, ed. R. P. Adams, J. S. Miller, E. M. Golenberg & J. E. Adams, pp. 47–67. Monographs in Systematic Botany from the Missouri Botanical Garden **48**.

Rogstad, S. H., Nybom, H. & Schaal, B. A. (1991). The tetrapod 'DNA fingerprinting' M13 repeat probe reveals genetic diversity and clonal growth in Quaking Aspen (*Populus tremuloides*, Salicaceae). *Plant Systematics & Evolution*, **175**, 115–23.

Roles, S. J. (1960). Illustrations (Part II) to *Flora of the British Isles*, Clapham, A. R., Tutin, T. G. & Warburg, E. F. Cambridge: Cambridge University Press.

Roose, M. L. & Gottlieb, L. D. (1976). Genetic and biochemical consequences of polyploidy in *Tragopogon*. *Evolution*, **30**, 818–30.

Rosen, F. (1889). Systematische und biologische Beobachtungen über *Erophila verna*. *Botanische Zeitung*, **47**, 565–80, 581–91, 597–608, 613–20.

Ross-Craig, S. (1948–1973). *Drawings of British plants*. London: Bell & Sons Ltd.

Rosser, E. M. (1955). A new British species of *Senecio*. *Watsonia*, **3**, 228–32.

Rothwell, N. V. (1993). *Understanding genetics*. New York: Wiley-Liss.

Rowell, T. A. (1984). Further discoveries of the Fen Violet (*Viola persicifolia* Schreber) at Wicken Fen, Cambridgeshire. *Watsonia*, **15**, 122–3.

Rowell, T. A., Walters, S. M. & Harvey, H. J. (1982). The rediscovery of the Fen Violet, *Viola persicifolia* Schreber, at Wicken Fen, Cambridgeshire. *Watsonia*, **14**, 183–4.

Roy, B. A. & Rieseberg (1989). Evidence for apomixis in *Arabis*. *Journal of Heredity*, **80**, 506–8.

Rückert, J. (1892). Zur Entwicklungs Geschichte des Ovarioleies bei Selachiern. *Anatomischer Anzeiger*, **7**, 107.

Ruse, M. (1987). Biological species: natural kinds, individuals, or what? *British Journal for the Philosophy of Science*, **38**, 225–42.

Rushton, B. S. (1978). *Quercus robur* L. and *Quercus petraea* (Matt.) Liebl: a multivariate approach to the hybrid problem. 1. Data acquisition, analysis and interpretation. *Watsonia*, **12**, 81–101.

Rushton, B. S. (1979). *Quercus robur* L. and *Quercus petraea* (Matt.) Liebl.: a multivariate approach to the hybrid problem. 2. The geographical distribution of population types. *Watsonia*, **12**, 209–24.

Russell, B. (1931). *The scientific outlook*. London: Allen & Unwin.

Sakai, A. & Larcher, W. (1987). *Frost survival in plants. Responses and adaptation to freezing stress*. Berlin: Springer Verlag.

Salisbury, F. B. & Ross, C. W. (1992). *Plant physiology*, 4th edn. Belmont, CA: Wadsworth.

Salmon, S. C. & Hanson, A. A. (1964). *The principles and practice of agricultural research*. London: Leonard Hill.

Sanderson, M. J., Baldwin, B. G., Bharathan, G., Campbell, C. S., Von Dohlen, C., Ferguson, D., Porter, J. M., Wojciechowski, M. F. & Donoghue, M. J. (1993). The growth of phylogenetic information, and the need for a phylogenetic data base. *Systematic Biology*, **42**, 562–8.

Sarkar, P. & Stebbins, G. L. (1956). Morphological evidence concerning the origin of the B genome in Wheat. *American Journal of Botany*, **43**, 297–304.

Sauer, J. D. (1988). *Plant migration*. Berkeley: University of California Press.

Saunders, E. R. (1897). On discontinuous variation occurring in *Biscutella laevigata*. *Proceedings of the Royal Society of London, B*, **62**, 11–26.

Savidan, Y. (1978). Genetic control of facultative apomixis and application in breeding *Panicum maximum*. *Communication to XIVth International Congress of Genetics*. Moscow.

Schaal, B. A. (1980). Measurement of gene flow in *Lupinus texensis*. *Nature*, **284**, 450–1.

Schaal, B. A. (1988). Somatic variation and genetic structure in plant populations. In *Plant population ecology*, ed. A. J. Davy, M. J. Hutchings & A. R. Watkinson, pp. 47–58. Oxford: Blackwell.

Schaal, B. A., Leverich, W. J. & Rogstad, S. H. (1991). A comparison of methods for assessing genetic variation in plant conservation biology. In

Genetics and conservation of rare plants, ed. D. A. Falk & K. E. Holsinger, pp. 123–34. New York: Oxford University Press.

Schaal, B. A., O'Kane, S. L. & Rogstad, S. H. (1991). DNA variation in plant populations. *Trends in Ecology & Evolution*, **6**, 329–33.

Schierwater, B., Steit, B., Wagner, G. P. & De Salle, R. (1994). *Molecular ecology and evolution: approaches and applications*. Basel: Birkhäuser.

Schlichting, C. D. (1986). The evolution of phenotypic plasticity in plants. *Annual Review of Ecology & Systematics*, **17**, 667–93.

Schlichting, C. D. & Levin, D. A. (1984). Phenotypic plasticity of annual Phlox: tests of some hypotheses. *American Journal of Botany*, **71**, 252–60.

Schlising, R. A. & Turpin, R. A. (1971). Hummingbird dispersal of *Delphinium cardinale* pollen treated with radioactive iodine. *American Journal of Botany*, **58**, 401–6.

Schmalhausen, I. I. (1949). *Factors of evolution*. Blakiston Press: New York.

Schmidt, J. (1899). Om ydre faktorers indflydelse paa løvbladets anatomiske bygning hos en af vore strandplanter. *Botanisk Tidsskrift*, **22**, 145–65.

Schneider, S. H. (1993). Scenarios of global warming. In *Biotic interactions and global change*, ed. P. M. Kareiva, J. G. Kingsolver & R. B. Huey, pp. 9–23. Sunderland: Sinauer.

Schoen, D. J. & Lloyd, D. G. (1984). The selection of cleistogamy and heteromorphic diaspores. *Biological Journal of the Linnean Society*, **23**, 303–22.

Schrödinger, E. (1944). *What is life*? London: Cambridge University Press; and New York: Macmillan.

Schwaegerle, K. E. & Schaal, B. A. (1979). Genetic variability and founder effect in the Pitcher Plant *Sarracenia purpurea* L. *Evolution*, **33**, 1210–18.

Schweber, S. S. (1977). The origin of the origin revisited. *Journal of the History of Biology*, **10**, 229–316.

Seavey, S. R. & Bawa, K. S. (1986). Late-acting self-incompatibility in angiosperms. *Botanical Review*, **52**, 195–219.

Shafer, C. L. (1990). *Nature reserves*. Washington: Smithsonian Institution Press.

Shaffer, M. L. (1981). Minimum population sizes for species conservation. *Bioscience*, **31**, 131–4.

Sharma, C. B. S. R. & Panneerselvam, N. (1990). Genetic toxicology of pesticides in higher plant systems. *Critical Reviews in Plant Sciences*, **9**, 409–42.

Sheffield, E., Wolf, P. G., Rumsey, F. J., Robson, D. J., Ranker, T. A. & Challinor, S. M. (1993), Spatial distribution and reproductive behaviour of a triploid Bracken (*Pteridium aquilinum*) clone in Britain. *Annals of Botany*, **72**, 231–37.

Sherman, M. (1946). Karyotype evolution: a cytogenetic study of seven species and six interspecific hybrids of *Crepis*. *University of California Publications in Botany*, **18**, 369–408.

Shivas, M. G. (1961*a*). Contributions to the cytology and taxonomy of species of *Polypodium* in Europe and America. 1. Cytology. *Journal of the Linnean Society*, **58**, 13–25.

Shivas, M. G. (1961*b*). Contributions to the cytology and taxonomy of species of *Polypodium* in Europe and America. 2. Taxonomy. *Journal of the Linnean Society*, **58**, 27–38.

Shore, J. S. & Barrett, S. C. H. (1985). The genetics of distyly and homostyly in *Turnera ulmifolia* L. (Turneraceae). *Heredity*, **55**, 167–74.

Shorrocks, B. (1984). *Evolutionary ecology*. Oxford: Blackwell.
Silvertown, J. W. (1984). Phenotypic variety in seed germination behaviour: the ontogeny and evolution of somatic polymorphism in seeds. *American Naturalist*, **124**, 1–16.
Silvertown, J. W. & Lovett Doust, J. (1993). *Introduction to plant population biology*. Oxford: Blackwell.
Simberloff, D. S., Farr, J. A., Cox, J. & Mehlman, D. W. (1992). Movement corridors: conservation bargains or poor investments? *Conservation Biology*, **6**, 493–504.
Simmonds, N. W. (ed.) (1976). *Evolution of crop plants*. London: Longmans.
Simpson, D. M. (1954). Natural cross-pollination in Cotton. *U. S. Department of Agriculture Technical Bulletin* no. 1094.
Simpson, G. G. (1944). *Tempo and mode in evolution*. New York: Columbia University Press.
Simpson, G. G. (1961). *Principles of animal taxonomy. The species and lower categories*. New York: Columbia University Press.
Simpson, M. G. (1986). Phylogeny and structural evolution of plants. In *Fundamentals of plant systematics*, ed. A. E. Radford, pp. 217–48. New York: Harper & Row.
Sims, T. L. (1993). Genetic regulation of self-incompatibility. *Critical Reviews in Plant Sciences*, **12**, 129–167.
Sindu, A. S. & Singh, S. (1961). Studies on the agents of cross pollination of Cotton. *Indian Cotton Growing Review*, **15**, 341–53.
Small, E. (1993). The economic value of plant systematics in Canadian agriculture. *Canadian Journal of Botany*, **71**, 1537–51.
Smartt, J. & Simmonds, N. W. (1995). *Evolution of crop plants*, 2nd edn. Harlow: Longmans.
Smith, A. (1965). The assessment of patterns of variation in *Festuca rubra* L. in relation to environmental gradients. *Scottish Plant Breeding Station Record*, 1965, 163–95.
Smith, A. (1972). The pattern of distribution of *Agrostis* and *Festuca* plants of various genotypes in a sward. *New Phytologist*, **71**, 937–45.
Smith, A. C. (1957). Fifty years of botanical nomenclature. *Brittonia*, **9**, 2–8.
Smith, D. C., Nielsen, E. L. & Ahlgren, H. L. (1946). Variation in ecotypes of *Poa pratensis*. *Botanical Gazette*, **108**, 143–66.
Smith, D. M. & Levin, D. A. (1963). A chromatographic study of reticulate evolution in the Appalachian *Asplenium* complex. *American Journal of Botany*, **50**, 952 –8.
Smith, F. D. M., May, R. M., Pellew, R., Johnson, T. H. & Walters, K. R. (1993). How much do we know about the current extinction rate? *Trends in Ecology & Evolution*, **8**, 375–8.
Smith, G. L. (1963*a*). Studies in *Potentilla* L. 1. Embryological investigations into the mechanism of agamospermy in British *P. tabernaemontani* Aschers. *New Phytologist*, **62**, 264–82.
Smith, G. L. (1963*b*). Studies in *Potentilla* L. 2. Cytological aspects of apomixis in *P. crantzii* (Cr.) Beck ex Fritsch. *New Phytologist*, **62**, 283–300.
Smith, G. L. (1971). Studies in *Potentilla* L. 3. Variation in British *P. tabernaemontani* Aschers. and *P. crantzii* (Cr.) Beck ex Fritsch. *New Phytologist*, **70**, 607–18.
Smith, J. (1841). Notice of a plant which produces perfect seeds without any apparent action of pollen. *Transactions of the Linnean Society of London*, **18**, 509–12.

Smith, P. M. (1976). *The chemotaxonomy of plants.* London: Arnold.
Smith, S. (1960). The origin of the Origin. *Advancement of Science*, **16**, 391–401.
Smyth, C. A. & Hamrick, J. L. (1987). Realised gene flow via pollen in artificial populations of Musk Thistle, *Carduus nutans*. *Evolution*, **41**, 613–19.
Snaydon, R. W. (1970). Rapid population differentiation in a mosaic environment. 1. The response of *Anthoxanthum odoratum* populations to soils. *Evolution*, **24**, 257–69.
Snaydon, R. W. (1976). Genetic change within species. In *The Park Grass experiment on the effect of fertilisers and liming on the botanical composition of permanent grassland and on the yield of hay*, J. M. Thurston, G. V. Dyke & E. D. Williams, Appendix. Harpenden: Rothamsted Experimental Station.
Snaydon, R. W. (1978). Genetic changes in pasture populations. In *Plant relations in pastures*, ed. J. R. Wilson, pp. 253–69. Melbourne: CSIRO.
Snaydon, R. W. & Davies, M. S. (1972). Rapid population differentiation in a mosaic environment. II. Morphological variation in *Anthoxanthum odoratum*. *Evolution*, **26**, 390–405.
Sneath, P. H. A. (1962). The construction of taxonomic groups. In *Microbial classification*, ed. G. C. Ainsworth & P. H. A. Sneath, pp. 289–332. Cambridge: Cambridge University Press.
Sneath, P. H. A. (1988). The phenetic and cladistic approaches. In *Prospects in systematics*. The Systematic Association Special Volume no. 36, ed. D. L. Hawksworth, pp. 252–73. Oxford: Clarendon Press.
Sneath, P. H. A. & Sokal, R. R. (1973). *Numerical taxonomy*. San Francisco: Freeman.
Snedecor, G. W. & Cochran, W. G. (1980). *Statistical methods*, 7th edn. Ames, IA: Iowa State University Press.
Snyder, L. A. (1950). Morphological variability and hybrid development in *Elymus glaucus*. *American Journal of Botany*, **37**, 628–35.
Snyder, L. A. (1951). Cytology of inter-strain hybrids and the probable origin of variability in *Elymus glaucus*. *American Journal of Botany*, **38**, 195–202.
Sokal, R. R. & Crovello, T. J. (1970). The biological species concept: a critical evaluation. *American Naturalist*, **104**, 127–53.
Sokal, R. R. & Rohlf, F. J. (1969). *Biometry. The principles and practice of statistics in biological research*. San Francisco: Freeman. [2nd edn, 1981].
Sokal, R. R. & Sneath, P. H. A. (1963). *Principles of numerical taxonomy*. London & San Francisco: Freeman.
Solbrig, O. T. & Solbrig, D. J. (1979). *Introduction to population biology and evolution*. London: Addison-Wesley Publishing Co.
Soltis, D. E. & Soltis, P. S. (1989). Allopolyploid speciation in *Tragopogon*: insights from chloroplast DNA. *American Journal of Botany*, **76**, 1119–24.
Soltis, D. E. & Soltis, P. S. (1990). *Isozymes in plant biology*. London: Chapman & Hall.
Soltis, D. E. & Soltis, P. S. (1993). Molecular data and the dynamic nature of polyploidy. *Critical Reviews in Plant Sciences*, **12**, 243–73.
Soltis, D. E. & Soltis, P. S. (1995). The dynamic nature of polyploid genomes. *Proceedings of the National Academy of Sciences, USA*, **92**, 8089–91.
Soltis, P. S., Plunkett, G. M., Novak, S. J. & Soltis, D. E. (1995). Genetic variation in *Tragopogon* species: additional origins of the allotetrapolyploids *T. mirus* and *T. miscellus* (Compositae). *American Journal of Botany*, **82**, 1329–41.

Soltis, P. S. & Soltis, D. E. (1991). Multiple origins of the allotetraploid *Tragopogon mirus* (Compositae): r DNA evidence. *Systematic Botany*, **16**, 407–13.

Soltis, P. S., Soltis, D. E. & Doyle, J. J. (1992). *Molecular systematics of plants*. New York: Chapman & Hall.

Solymosi, P. & Lehoczki, E. (1989). Characterization of a triple (atrazine-pyrazon-pyridate) resistant biotype of Common Lambs Quarters (*Chenopodium album* L.) *Journal of Plant Physiology*, **134**, 685–90.

Song, K., Lu, P., Tang, K. & Osborn, T. C. (1995). Rapid genome change in synthetic polyploids of *Brassica* and its implications for polyploid evolution. *Proceedings of the National Academy of Sciences, USA*, **92**, 7719–23.

Song, K. M., Osborn, T. C. & Williams, P. H. (1988). *Brassica* taxonomy based on nuclear restriction length polymorphisms (RFLPs). 1. Genome evolution of diploid and amphidiploid species. *Theoretical & Applied Genetics*, **75**, 784–94.

Sørenson, T. & Gudjónsson, G. (1946). Spontaneous chromosome-aberrants in apomictic *Taraxaca. Biologiske Skrifter, K. Danske Videnskabernes Selskab*, **4**, no. 2.

Soulé, M. E. (1980). Thresholds for survival: maintaining fitness and evolutionary potential. In *Conservation biology: an evolutionary–ecological perspective*, ed. M. E. Soulé & B. A. Wilcox, pp. 151–69. Sunderland: Sinauer.

Spencer, K. C. (1988). Chemical mediation of coevolution. San Diego: Academic Press.

Sprengel, C. K. (1793). *Das entdeckte Geheimniss der Natur im Bau und in der Befruchtung der Blumen*. Berlin.

Srb, A. M. & Owen, R. D. (1958). *General genetics*. San Francisco: Freeman.

Stace, C. A. (ed.) (1975). *Hybridization and the flora of the British Isles*. London: Academic Press.

Stace, C. A. (1980). *Plant taxonomy and biosystematics*. London: Arnold. (2nd edn, 1989.)

Stace, C. A. (1993). The importance of rare events in polyploid evolution. In *Evolutionary patterns and processes* ed. D. R. Lees & D. Edwards, Linnean Society Symposium Series **14**, pp. 157–69. London: Published for the Linnean Society by Academic Press.

Stadler, L. J. (1942). *Some observations on gene variability and spontaneous mutation*. The Spragg Memorial Lectures, Michigan State University.

Stapledon, R. G. (1928). Cocksfoot grass (*Dactylis glomerata* L.): ecotypes in relation to the biotic factor. *Journal of Ecology*, **16**, 72–104.

Stearn, W. T. (1957). *Introduction to facsimile edition of Linnaeus' Species plantarum*. London: Ray Society, British Museum.

Stearns, S. C. (1992). *The evolution of life histories*. Oxford: Oxford University Press.

Stebbins, G. L. (1947). Types of polyploids: their classification and significance. *Advances in Genetics*, **1**, 403–29.

Stebbins, G. L. (1950). *Variation and evolution in plants*. London: Oxford University Press; and New York: Columbia University Press.

Stebbins, G. L. (1957). Self-fertilization and population variability in the higher plants. *American Naturalist*, **41**, 337–54.

Stebbins, G. L. (1966). *Processes of organic evolution*. Englewood Cliffs, NJ:

Prentice Hall. [2nd edn, 1971.]
Stebbins, G. L. (1971). *Chromosomal evolution in higher plants*. London: Arnold.
Stebbins, G. L. (1974). *Flowering plants. Evolution above the species level.* London: Arnold.
Stebbins, G. L. (1984). Polyploidy and the distribution of the arctic-alpine flora: new evidence and a new approach. *Botanica Helvetica*, **94**, 1–13.
Stebbins, G. L., Harvey, B. L., Cox, E. L., Rutger, J. N., Jelencovic, G. & Yagil, E. (1963). Identification of the ancestry of an amphiploid *Viola* with the aid of paper chromatography. *American Journal of Botany*, **50**, 830–9.
Stelleman, P. (1978). The possible role of insect visits in pollination of reputedly anemophilous plants, exemplified by *Plantago lanceolata*, and syrphid flies. In *The pollination of flowers by insects*, ed. A. J. Richards, pp. 41–6. Linnean Society Symposium Series 6. London: Academic Press.
Stern, C. & Sherwood, E. R. (1966). *The origin of genetics. A Mendel source book*. London & San Francisco: Freeman.
Sternberg, L. (1976). Growth forms of *Larrea tridentata*. *Madroño*, **23**, 408–17.
Stewart, R. N. (1947). The morphology of somatic chromosomes in *Lilium*. *American Journal of Botany*, **34**, 9–26.
Stewart, W. N. & Rothwell, G. W. (1993). *Palaeobotany and the evolution of plants*, 2nd edn. Cambridge: Cambridge University Press.
Stone, J. L., Thomson, J. D. & Dent-Acosta, S. J. (1995). Assessment of pollen viability in hand-pollination experiments: a review. *American Journal of Botany*, **82**, 1186–97.
Stowe, M. K. (1988). Chemical mimicry. In *Chemical mediation of coevolution*, ed. K. C. Spencer, pp. 513–80. San Diego: Academic Press.
Strasburger, E. (1910). Chromosomenzahl. *Flora*, **100**, 398–446.
Strauss, W. & Mainwaring, S. J. (1984). *Air pollution*. London: Arnold.
Strickberger, M. W. (1976). *Genetics*, 2nd edn. New York & London: Macmillan.
Strickberger, M. (1985). *Genetics*, 3rd edn. London: Prentice Hall, Collier Macmillan.
Strid, A. (1970). Studies in the Aegean flora. XVI. Biosystematics of the *Nigella arvensis* complex. With special reference to the problem of non-adaptive radiation. *Opera Botanica* no. 28. Lund: Gleerup.
Stuart, A. (1984). *The ideas of sampling*. High Wycombe: Charles Griffin.
Stuessy, T. F. (1990). *Plant taxonomy: the systematic evaluation of comparative data*. New York: Columbia University Press.
Sturtevant, A. H. (1965). *A history of genetics*. New York: Harper Row.
Sultan, S. E. (1987). Evolutionary implications of phenotypic plasticity in plants. *Evolutionary Biology*, **21**, 127–78.
Sutherland, W. J. & Hill, D. A. (1995). *Managing habitats for conservation*. Cambridge: Cambridge University Press.
Sutton, W. S. (1902). On the morphology of the chromosome group in *Brachystola magna*. *Biological Bulletin, Marine Biological Laboratory, Woods Hole, MA*, **4**, 24–39.
Sutton, W. S. (1903). The chromosomes in heredity. *Biological Bulletin, Marine Biological Laboratory, Woods Hole, MA*, **4**, 231–48.
Swensen, S. M., Allan, G. J., Howe, M., Elisens, W. J., Junak, S. A. & Rieseberg, L. H. (1995). Genetic analysis of the endangered island endemic *Malacothamnus fasciculatus* (Nutt.) Green var. *nesioticus* (Rob.) Kearn (Malvaceae). *Conservation Biology*, **9**, 404–15.

Swofford, D. L. & Olsen, G. J. (1990). Phylogenetic reconstruction. In *Molecular systematics*, ed. D. K. Hillis & C. Moritz, pp. 411–501. Sunderland: Sinauer.

Swofford, D. L., Olsen, G. J., Waddell, P. J. & Hillis, D. M. (1996). Phylogenetic inference. In *Molecular systematics*, 2nd edn., ed. D. M. Hillis, C. Moritz & B. K. Mable, pp. 407–514. Sunderland: Sinauer.

Syomov, A. B., Ptitsyna, S. N., & Sergeeva, S. A. (1992). Analysis of DNA breaks, induction and repair in plants from the vicinity of Chernobyl. *Journal of the Total Environment*, **112,** 1–8.

Sytsma, K. J. (1994). DNA extraction from recalcitrant plants: long, pure and simple? In *Conservation of plant genes II: utilization of ancient and modern DNA*, ed. R. P. Adams, J. S. Miller, E. M. Golenberg & J. E. Adams, pp. 69–81. Monographs in Systematic Botany from the Missouri Botanical Garden **48**.

Sytsma, K. J. & Hahn, W. J. (1997). Molecular systematics: 1994–1995. *Progress in Botany*, **58**, 470–99.

Sytsma, K. J. & Schaal, B. A. (1985). Phylogenetics of the *Lisianthius skinneri* (Gentianaceae) species complex in Panama utilizing DNA restriction fragment analysis. *Evolution*, **39**, 594–608.

Szaro, R. C. & Johnston, D. W. (1996). *Biodiversity in managed landscapes*. New York: Oxford University Press.

Täckholm, G. (1922). Zytologische studien über die Gattung *Rosa*. *Acta Horti Bergiani*, **7**, 97–381.

Taggart, J. B., McNally S. F. & Sharp, P. M. (1990). Genetic variability and differentiation among founder populations of the Pitcher Plant (*Sarracenia purpurea* L.) in Ireland. *Heredity*, **64**, 177–83.

Tahara, M. (1915). Cytological studies on *Chrysanthemum*. *Botanical Magazine (Tokyo)*, **29**, 48–50.

Taiz, L. & Zeiger, E. (1991). *Plant physiology*. New York: Benjamin Cummings.

Takhtajan, A. (1969). *Flowering plants – origin and dispersal*. Authorised translation from the Russian by C. Jeffrey. Edinburgh: Oliver & Boyd.

Tardif, B. & Morisset, P. (1991). Chromosomal C-Band variation in *Allium schoenoprasum* (Lilaceae) in eastern – North America. *Plant Systematics & Evolution*, **174**, 125–37.

Taylor, G. E., Jr. & Murdy, W. H. (1975). Population differentiation of an annual plant species, *Geranium carolinianum* in response to sulfur dioxide. *Botanical Gazette*, **136**, 212–15.

Taylor, G. E., Pitelka, L. F. & Clegg, M. T. (1991). *Ecological genetics and air pollution*. New York: Springer Verlag.

Taylor, K. & Markham, B. (1978). *Ranunculus ficaria* L. Biological flora of the British Isles. *Journal of Ecology*, **66**, 1011–31.

Theaker, A. J. & Briggs, D. (1993). Genecological studies of Groundsel (*Senecio vulgaris L.*). IV. Rate of development in plants from different habitat types. *New Phytologist*, **123**, 185–94.

Thiselton-Dyer, W. T. (1895*a*). Variation and specific stability. *Nature*, **51**, 459–61.

Thiselton-Dyer, W. T. (1895*b*). Origin of the cultivated *Cineraria*. *Nature*, **52**, 3–4, 78–9, 128–9.

Thoday, J. M. (1972). Disruptive selection. *Proceedings of the Royal Society of London, B*, **182**, 109–43.

Thomas, D. A. & Barber, H. N. (1974). Studies of leaf characteristics of a cline of *Eucalyptus urnigera* from Mount Wellington, Tasmania. II. Reflection, transmission and absorption of radiation. *Australian Journal of Botany*, **22**, 701–7.

Thomas, H. H. (1947). The rise of geology and its influence on contemporary thought. *Annals of Science*, **5**, 325–41.

Thomson, J. D., Herre, E. A., Hamrick, J. L. & Stone, J. L. (1991). Genetic mosaics in Strangler Fig trees: implications for tropical conservation. *Science*, **254**, 1214–6.

Thompson, J. D. (1991). Phenotypic plasticity as a component of evolutionary change. *Trends in Ecology & Evolution*, **6**, 246–9.

Thompson, J. D. & Lumaret, R. (1992). The evolutionary dynamics of polyploid plants: origins, establishment and persistence. *Trends in Ecology & Evolution*, **7**, 302–6.

Thurston, J. M., Dyke, G. V. & Williams, E. D. (1976). *The Park Grass experiment on the effect of fertilisers and liming on the botanical composition of permanent grassland and on the yield of hay*. Harpenden: Rothamsted Experimental Station.

Tiedje, J. M., Colwell, R. K., Grossman, Y. L., Hodson, R. E., Lenski, R. E., Mack, R. N. & Regal, P. J. (1989). The planned introduction of genetically engineered organisms: ecological considerations and recommendations. *Ecology*, **70**, 298–315.

Till, I. (1987). Variability of expression of cyanogenesis in White Clover (*Trifolium repens* L.) Heredity, **59**, 265–71.

Tindall, K. R. & Kunkel, T. A. (1988). Fidelity of DNA synthesis by the *Thermus aquaticus* DNA polymerase. *Biochemistry*, **27**, 6008–13.

Tischler, G. (1950). Die Chromosomenzahlen der Gefässpflanzen Mitteleuropas. 'S-Gravenhage: Junk.

Tobgy, H. A. (1943). A cytological study of *Crepis fuliginosa*, *C. neglecta* and their F_1 hybrid, and its bearing on the mechanism of phylogenetic reduction in chromosome number. *Journal of Genetics*, **45**, 67–111.

Tower, W. L. (1902). Variation in the ray-flowers of *Chrysanthemum leucanthemum* L. at Yellow Springs, Green County, O, with remarks upon the determination of the modes. *Biometrika*, **1**, 309–15.

Townsend, C. R. & Calow, P. (1981). *Physiological ecology: an evolutionary approach to resource use*. Oxford: Blackwell.

Tralau, H. (1968). Evolutionary trends in the genus *Ginkgo*. *Lethaia*, **1**, 63–101.

Tudge, C. (1989). The rise and fall of *Homo sapiens sapiens*. *Philosophical Transactions of the Royal Society, London, B*, **325**, 479–88.

Turesson, G. (1922*a*). The species and variety as ecological units. *Hereditas*, **3**, 100–13.

Turesson, G. (1922*b*). The genotypical response of the plant species to the habitat. *Hereditas*, **3**, 211–350.

Turesson, G. (1925). The plant species in relation to habitat and climate. *Hereditas*, **6**, 147–236.

Turesson, G. (1927*a*). Erbliche Transpirationsdifferenzen zwischen Ökotypen derselben Pflanzen Art. *Hereditas*, **11**, 193–206.

Turesson, G. (1927*b*). Untersuchungen über Grenzplasmolyse und Saugkraftwerte in verschiedenen Ökotypen derselben Art. *Jahrbücher für wissenschaftliche Botanik*, **66**, 723–47.

Turesson, G. (1930). The selective effect of climate upon the plant species. *Hereditas*, **14**, 99–152.

Turesson, G. (1943). Variation in the apomictic microspecies of *Alchemilla vulgaris L. Botaniska Notiser*, 1943, 413–27.

Turesson, G. (1961). Habitat modifications in some widespread plant species. *Botaniska Notiser*, **114**, 435–52.

Turkington, R. (1989). The growth, distribution and neighbour relationships of *Trifolium repens* in a permanent pasture. V. The co-evolution of competitors. *Journal of Ecology*, **77**, 717–33.

Turkington, R. & Harper, J. L. (1979*a*). The growth, distribution and neighbour relationships of *Trifolium repens* in a permanent pasture. I. Ordination, pattern and contact. *Journal of Ecology*, **67**, 201–18.

Turkington, R. & Harper, J. L. (1979b). The growth, distribution and neighbour relationships of *Trifolium repens* in a permanent pasture. II. Fine scale biotic differentiation. *Journal of Ecology*, **67**, 245–54.

Turner, I. M., Tan, H. T. W., Wee, Y. C., Ibrahim, A. B., Chew, P. T. & Corlett, R. T. (1994). A study of plant species extinction in Singapore: lessons for the conservation of tropical biodiversity. *Conservation Biology*, **8**, 705–12.

Turrill, W. B. (1938). The expansion of taxonomy with special reference to Spermatophyta. *Biological Reviews*, **13**, 342–73.

Turrill, W. B. (1940). Experimental and synthetic plant taxonomy. In *The new systematics*, ed. J. S. Huxley, pp. 47–71. Oxford: Clarendon Press.

Tutin, T. G. (1957). A contribution to the experimental taxonomy of *Poa annua* L. *Watsonia*, **4**, 110.

Tutin, T. G., Heywood, V. H., Burges, N. A., Moore, D. M., Valentine, D. H., Walters, S. M. & Webb, D. A. (1964–80). *Flora europaea*. London: Cambridge University Press.

Tzuri, G., Hillel, J., Lavi, U., Haberfeld, A. & Vainstein, A. (1991). DNA fingerprint analysis of ornamental plants. *Plant Science*, **76**, 91–7.

Uhl, C. H. (1978). Chromosomes of Mexican *Sedum*. II. Section *Pachysedum*. *Rhodora*, **80**, 491–512.

Upcott, M. (1940). The nature of tetraploidy in *Primula kewensis*. *Journal of Genetics*, **39**, 79–100.

Uphof, J. C. Th. (1938). Cleistogamic flowers. *Botanical Review*, **4**, 21–49.

Valentine, D. H. (1941). Variation in *Viola riviniana* Rchb. *New Phytologist*, **40**, 189–209.

Valentine, D. H. (1956). Studies in British Primulas. V. The inheritance of seed incompatibility. *New Phytologist*, **55**, 305–18.

Valentine, D. H. (1975). *Primula*. In *Hybridization and the flora of the British Isles*, ed. C. A. Stace, pp. 346–8. London & New York: Academic Press.

van der Pijl, L. (1960). Ecological aspects of flower evolution. I. Phyletic evolution. *Evolution*, **14**, 403–16.

van der Pijl, L. & Dodson, C. H. (1966). *Orchid flowers: their pollination and evolution*. Coral Gables: University of Miami Press.

van Dijk, P. & Bijlsma, R. (1994). Simulations of flowering time displacement between two cytotypes that form inviable hybrids. *Heredity*, **72**, 522–35.

van Groenendael, J. M. (1986). Life history characteristics of two ecotypes of *Plantago lanceolata*. *Acta Botanica Neerlandica*, **35**, 71–86.

van Tienderen, P. H. & van der Toorn, J. (1991*a*). Genetic differentiation between populations of *Plantago lanceolata*. I. Local adaptation in three contrasting habitats. *Journal of Ecology*, **79**, 27–42.

van Tienderen P. H. & van der Toorn, J. (1991*b*). Genetic differentiation between populations of *Plantago lanceolata*. II. Phenotypic selection in a transplant experiment in three contrasting habitats. *Journal of Ecology*, **79**, 43–59.

van Treuren, R., Bijlsma, R., Ouborg, N. J. & van Delden, W. (1993). The significance of genetic erosion in the process of extinction. IV. Inbreeding depression and heterosis effects caused by selfing and outcrossing in *Scabiosa columbaria*. *Evolution*, **47**, 1669–80.

van Treuren, R., Bijlsma, R., van Delden, W. & Ouborg, N. J. (1991). The significance of genetic erosion in the process of extinction. I. Genetic differentiation in *Salvia pratensis* and *Scabiosa columbaria* in relation to population size. *Heredity*, **66**, 181–89.

van Valen, L. (1976). Ecological species, multispecies, and Oaks. *Taxon*, **25**, 223–39.

Vasek, F. C. (1980). Creosote Bush: long-lived clones in the Mojave Desert. *American Journal of Botany*, **67**, 246–55.

Venable, D. L. & Levin, D. A. (1985). Ecology of achene dimorphism in *Heterotheca latifolia*. I. Achene structure, germination and dispersal. *Journal of Ecology*, **73**, 133–45.

Vernon, H. M. (1903). *Variation* in *animals and plants*. London: Kegan Paul.

Verschaffelt, E. (1899). Galton's regression to mediocrity bij ongeslachtelijke verplanting. In *Livre Jubilaire dédié à Charles wan Bambeke*, pp. 1–5. Brussels: Lamerton.

Vickery, R. K. (1964). Barriers to gene exchange between members of the *Mimulus guttatus* complex (Scrophulariaceae). *Evolution*, **18**, 52–69.

Viosca, P., Jr. (1935). The Irises of southeastern Louisiana: a taxonomic and ecological interpretation. *Bulletin of the American Iris Society*, **57**, 3–56.

von Bothmer, R., Engstrand, L., Gustafsson, M., Persson, J., Snogerup, S. & Bentzer, B. (1971). Clonal variation in populations of *Anemone nemorosa* L. *Botaniska Notiser*, **124**, 505–19.

von Nägeli, C. (1865). Die Bastardbindung im Pflanzenreiche. *Sitzungsbericht der Königlich-Bayerischen Akademie der Wissenschaften zu München Botanische Mitteilungen*, **2**, 159–87.

Vorzimmer, P. J. (1972). *Charles Darwin: the years of controversy*. London: University of London Press.

von Tubeuf, K. F. (1923). *Monographie der Mistel*. Munich & Berlin: Oldenbourg.

von Wettstein, R. (1895). Der Saison-Dimorphismus als Ausgangpunkt für die Bildung neuer Arten im Pflanzenreich. *Berichte der Deutschen botanischen Gesellschaft*, **13**, 303–13.

von Wettstein, R. (1907). *Handbuch der systematischen Botanik*, 2nd edn. Leipzig & Wien: Franz Deuticke.

Waddington, C. H. (1966). Mendel and evolution. In *G. Mendel Memorial Symposium*, 1865–1965, ed. M. Sosna, pp. 145–50. Prague: Academia Publishing House of the Czechoslovak Academy of Sciences.

Wagner, M. (1868). *Die Darwin'sche Theorie und das Migrationgesetz der Organismen*. Leipzig: Duncker & Humblot.

Wagner, W. H., Jr (1970). Biosystematics and evolutionary noise. *Taxon*, **19**, 146–51.

Walbot, V. (1996). Sources and consequences of phenotypic and genotypic plasticity in flowering plants. *Trends in Plant Sciences*, **1**, 27–32.

Waldron, L. R. (1912). Hardiness in successive Alfalfa generations. *American Naturalist*, **46**, 463–9.

Waller, D. M. (1979). The relative costs of self- and cross-fertilized seeds in *Impatiens capensis* (Balsaminaceae). *American Journal of Botany*, **66,** 313–20.

Waller, D. M. (1984). Differences in fitness between seedlings derived from cleistogamous and chasogamous flowers in *Impatiens capensis. Evolution*, **38,** 427–40.

Waller, D. M., O'Malley, D. M. & Gawler, S. C. (1987). Genetic variation in the extreme endemic *Pedicularis furbishiae* (Scrophulariaceae). *Conservation Biology*, **1**, 335–40.

Walters, S. M. (1961). The shaping of angiosperm taxonomy. *New Phytologist*, **60**, 74–84.

Walters, S. M. (1962). Generic and specific concepts in the European flora. *Preslia*, **34**, 207–26.

Walters, S. M. (1970). Dwarf variants of *Alchemilla* L. *Fragmenta Floristica et Geobotanica*, **16**, 91–8.

Walters, S. M. (1972). Endemism in the genus *Alchemilla* in Europe. In *Taxonomy, phytogeography and evolution*, ed. D. H. Valentine, pp. 301–5. London & New York: Academic Press.

Walters, S. M. (1979). Progress in biological conservation in Cambridge. In *Landscape towards 2000: conservation or desolation*, ed. D. Smith, pp. 56–8. London: Lanscape Institute.

Walters, S. M. (1986*a*). *Alchemilla*: a challenge to biosystematists. *Acta Universitatus Upsaliensis, Symbolae Botanicae Upsalienses*, **XXVII**, 193–8.

Walters, S. M. (1986*b*). The name of the rose: a review of ideas on the European bias in angiosperm classification. *New Phytologist*, **104** , 527–46.

Walters, S. M. (1989*a*). Obituary of John Scott Lennox Gilmour. *Plant Systematics & Evolution*, **167**, 93–5.

Walters, S. M. (1989*b*). Experimental and orthodox taxonomic categories and the deme terminology. *Plant Systematics & Evolution*, **167**, 35–41.

Ward, D. B. (1974). The 'ignorant man' technique of sampling plant populations. *Taxon*, **23**, 325–30.

Warwick, S. I. (1990*a*). Allozyme and life history variation in five northwardly colonizing North American weed species. *Plant Systematics & Evolution*, **169**, 41–54.

Warwick, S. I. (1990*b*). Genetic variation in weeds – with particular reference to Canadian agricultural weeds. In *Biological approaches and evolutionary trends in plants*, ed. S. Kawano, pp. 3–18. London: Academic Press.

Warwick, S. I. (1991). Herbicide resistance in weedy plants: physiology and population biology. *Annual Review of Ecology & Systematics*, **22**, 95–114.

Warwick, S. I. & Briggs, D. (1978a). The genecology of lawn weeds. I. Population differentiation in *Poa annua* L. in a mosaic environment of bowling green lawns and flower beds. *New Phytologist*, **81**, 711–23.

Warwick, S. I. & Briggs, D. (1978*b*). The genecology of lawn weeds. II. Evidence for disruptive selection in *Poa annua* L. in a mosaic environment of bowling green lawns and flower beds. *New Phytologist*, **81**, 725–37.

Warwick, S. I. & Briggs, D. (1979). The genecology of lawn weeds. III. Cultivation experiments with *Achillea millefolium* L., *Bellis perennis* L., *Plantago lanceolata L.*, *Plantago major* L. and *Prunella vulgaris* L. collected from lawns and contrasting grassland habitats. *New Phytologist*, **83**, 509–36.

Warwick, S. I. & Briggs, D. (1980a). The genecology of lawn weeds. IV.

Adaptive significance of variation in *Bellis perennis* L. as revealed in a transplant experiment. *New Phytologist*, **85**, 275–88.

Warwick, S. I. & Briggs, D. (1980*b*). The genecology of lawn weeds. V. The adaptive significance of different growth habit in lawn and roadside populations of *Plantago major* L. *New Phytologist*, **85**, 289–300.

Warwick, S. I. & Briggs, D. (1980*c*). The genecology of lawn weeds. VI. The adaptive significance of variation in *Achillea millefolium* L. as investigated by transplant experiments. *New Phytologist*, **85**, 451–60.

Warwick, S. I. & Gottlieb, L. D. (1985). Genetic divergence and geographic speciation in *Layia* (Compositae). *Evolution*, **39**, 1236–41.

Warwick, S. I., Bain, J. F., Wheatcroft, R. & Thompson, B. K. (1989). Hybridization and introgression in *Carduus nutans* and *C. acanthoides* re-examined. *Systematic Botany*, **14**, 476–94.

Warwick, S. I., Thompson, B. K. & Black, L. D. (1987). Genetic variation in Canadian and European populations of the colonising weed species *Apera spica-venti*. *New Phytologist*, **106**, 301–17.

Watson, J. D. & Crick, F. H. C. (1953). A structure of deoxyribose nucleic acid. *Nature*, **171**, 737–8.

Watson, J. D. (1968). *The double helix*. London: Weidenfeld & Nicolson.

Watson, J. D. (1965). *Molecular biology of the gene. New York: Benjamin. (2nd edn*, 1970.)

Watson, J. D., Gilman, M., Witkowski, J. & Zoller, M. (1992). *Recombinant DNA*, 2nd edn. New York: Freeman & Co.

Watson, P. J. (1969). Evolution in closely adjacent plant populations. VI. An entomophilous species, *Potentilla erecta*, in two contrasting habitats. *Heredity*, **24**, 407–22.

Webster, S. D. (1988). *Ranunculus penicillatus* (Dumort.) Bab. in Great Britain and Ireland. *Watsonia*, **17**, 1–22.

Weeden, N. F. & Gottlieb, L. D. (1979). Isolation of cytoplasmic enzymes from pollen. *Plant Physiology*, **66**, 400–3.

Weeden, N. F. & Wendel, J. F. (1990). Genetics of plant isozymes. In *Isoenzymes in plant biology*, ed. D. E Soltis & P. S. Soltis, pp. 46–72. London: Chapman & Hall.

Weimarck, H. (1945). Experimental taxonomy in *Aethusa cynapium. Botaniska Notiser*, 1945, 351–80.

Weir, J. & Ingram, R. (1980). Ray morphology and cytological investigations of *Senecio cambrensis* Rosser. *New Phytologist*, **86**, 237–41.

Weising, K., Nybom, H., Wolff, K. & Meyer, W. (1995). *DNA fingerprinting in plants and fungi*. London: CRC Press.

Weismann, A. (1883). *Uber die Vererbung*. English translation, *On heredity* (1889), by A. E. Shipley. Oxford: Clarendon Press.

Weldon, W. F. R. (1895*a*). The origin of the cultivated *Cineraria. Nature*, **52**, 54, 104, 129.

Weldon, W. F. R. (1895*b*). Remarks on variation in animals and plants. *Proceedings of the Royal Society of London*, B, **57**, 379–82.

Weldon, W. F. R. (1898). Presidential address. Section D. Zoology. *Nature*, **58**, 499–506.

Weldon, W. F. R. (1902*a*). On the ambiguity of Mendel's categories. *Biometrika*, **2**, 44–55.

Weldon, W. F. R. (1902*b*). Seasonal changes in the characters of *Aster prenanthoides* Muhl. *Biometrika*, **2**, 113–14.

Wells, W. C. (1818). An account of a white female, part of whose skin resembles that of a Negro. Paper given at the Royal Society, 1813. In *Two essays upon dew and single vision*. London.

Wendel, J. F. & Weeden, N. F. (1990). Visualization and interpretation of plant isozymes. In *Isozymes in plant biology*, ed. D. E. Soltis & P. S. Soltis, pp. 5–45. London: Chapman & Hall.

Werth, C. R., Guttman, S. I. & Eshbaugh, W. H. (1985*a*). Electrophoretic evidence of reticulate evolution in the Appalachian *Asplenium* complex. *Systematic Botany*, **10**, 184–92.

Werth, C. R., Guttman, S. I. & Eshbaugh, W. H. (1985*b*). Recurring origins of allopolyploid species in *Asplenium*. *Science*, **228**, 731–3.

Werth, C. R., Riopel, J. L. & Gillespie, N. W. (1984). Genetic uniformity in an introduced population of Witchweed (*Striga asiatica*) in the United States. *Weed Science*, **32**, 645–8.

Western, D. (1989). Conservation without parks: wildlife in the rural landscape. In *Conservation for the twenty-first century*, ed. D. Western & M. C. Pearl, pp. 158–65. New York: Oxford University Press.

Whatley, J. M. (1993). Membranes and plastid origins. In *Origins of plastids: symbiogenesis, protochlorophytes, and the origins of chloroplasts*, ed. R. A. Lewin, pp. 77–106. New York: Chapman & Hall.

Wheeler, B. D. & Shaw, S. C. (1995). *Restoration of damaged peatlands*. London: HMSO.

Wheeler, N. C. & Guries, R. P. (1987). A quantitative measure of introgression between Lodge Pole and Jack Pines. *Canadian Journal of Botany*, **65**, 1876– 85.

White, G. (1789). *The natural history of Selborne*. World Classics Edition (1951). London: Oxford University Press.

White, M. J. D. (1978). *Modes of speciation*. San Francisco: Freeman & Co.

White, O. E. (1917). Inheritance studies in *Pisum*. 2. The present state of knowledge of heredity and variation in Peas. *Proceedings of the American Philosophical Society*, **56**, 487–588.

Whitehouse, H. L. K. (1950). Multiple-allelomorph incompatibility of pollen and style in the evolution of the angiosperms. *Annals of Botany*, **14**, 199–216.

Whitehouse, H. L. K. (1959). Cross- and self-fertilisation in plants. In *Darwin's biological work*, ed. P. R. Bell, pp. 207–61. London: Cambridge University Press.

Whitehouse, H. L. K. (1965). *Towards an understanding of the mechanism of heredity*. London: Arnold. [3rd edn, 1973.]

Wickler, W. (1968). *Mimicry in plants and animals*. New York: McGraw-Hill.

Widén, B. (1991). Phenotypic selection on flowering phenology in *Senecio integrifolius*, a perennial herb. *Oikos*, **61**, 205–15.

Widén B. (1993). Demographic and genetic effects on reproduction as related to population size in a rare, perennial herb, *Senecio integrifolius*. *Biological Journal of the Linnean Society*, **50**, 179–95.

Widén, M. (1992). Sexual reproduction in a clonal, gynodioecious herb *Glechoma hederacea*. *Oikos*, **63**, 430–8.

Wiens, D. (1978). Mimicry in plants. *Evolutionary Biology*, **11**, 365–403.

Wiley, E. O. (1981). *Phylogenetics: the theory and practice of phylogenetic systematics*. New York: Wiley.

Wilkins, D. A. (1959). Sampling for genecology. *Record of the Scottish Plant Breeding Station*, 1959, 92–6.

Wilkins, D. A. (1960). Recognising adaptive variants. *Proceedings of the Linnean Society of London*, **171**, 122–6.

Williams, G. C. (1966). *Adaptation and natural selection. A critique of some current evolutionary thought*. Princeton, NJ: Princeton University Press.

Williams, M. (1989). *Americans and their forests*. Cambridge: Cambridge University Press.

Williamson, M. (1996). *Biological invasions*. London: Chapman & Hall.

Williamson, M. H. & Brown, K. C., (1986). The analysis and modelling of British invasions. *Philosophical Transactions of the Royal Society, London*, B, **314**, 505–22.

Williamson, P. G. (1981). Morphological stasis and developmental constraint: real problems for Neo-Darwinism. *Nature*, **294**, 214–15.

Willis, J. C. (1922). *Age and area*. London: Cambridge University Press; and New York: Macmillan.

Willis, J. C. (1940). *The course of evolution*. London: Cambridge University Press.

Willis, J. C. (1949). *The birth and spread of plants*. Geneva: Conservatoire et Jardin Botaniques.

Willson, M. F. (1983). *Plant reproductive ecology*. New York: Wiley.

Wilmott, A. J. (1949). Intraspecific categories of variation. In *British flowering plants and modern systematic methods*, ed. A. J. Wilmott, pp. 28–45. London: Botanical Society of the British Isles.

Wilson, G. B. & Bell, J. N. B. (1986). Studies on the tolerance to sulphur dioxide of grass populations in polluted areas. IV. The spatial relationship between tolerance and a point source of pollution. *New Phytologist*, **102**, 563–74.

Wilson, H. D., Lira, R. & Rodríguez, I. (1994). Crop/weed gene flow – *Cucurbita argyrosperma* Huber and *C. fraterna* L. H. Bailey (Cucurbitaceae). *Economic Botany*, **48**, 293–300.

Wilson, M. A., Gaut, B. & Clegg, M. T. (1990). Chloroplast DNA evolves slowly in the Palm family (Arecaceae). *Molecular Biology & Evolution*, **7**, 303–14.

Winchester, A. M. (1966). *Genetics*. Boston: Houghton.

Wing, S. L. & Tiffney, B. H. (1987). Interactions of angiosperms and herbivorous tetrapods through time. In *The origins of angiosperms and their biological consequences*, ed. E. M. Friis, W. G. Chaloner & P. R. Crane, pp. 203–24. Cambridge: Cambridge University Press.

Winge, Ø. (1917). The chromosomes, their numbers and general importance. *Comptes Rendus des Travaux du Laboratoire Carlsberg*, **13**, 131–275.

Winge, Ø. (1940). Taxonomic and evolutionary studies in *Erophila* based on cytogenetic investigations. *Comptes Rendus des Travaux du Laboratoire Carlsberg* (Ser. Physiol.), **23**, 41–74.

Winkler, H. (1908). Uber Parthenogenesis und Apogamie im Pflanzenreich. *Progressus rei Botanicae*, **2**, 293–454.

Winkler, H. (1916). Uber die experimentelle Erzeugung von Pflanzen mit abweichenden Chromosomenzahlen. *Zeitschrift für Botanik*, **8**, 417–531.

Winsor, M. P. (1995). The English debate on taxonomy and phylogeny, 1937–1940. *History and Philosophy of Life Sciences*, **17**, 227–52.

Wolff, K., Rogstad, S. H. & Schaal, B. A. (1994). Population and species variation of minisatellite DNA in *Plantago*. *Theoretical & Applied Genetics*, **87**, 733–40.

Woodell, S. R. J. (1965). Natural hybridization between the Cowslip (*Primula veris* L.) and the Primrose (P. *vulgaris* Huds.) in Britain. *Watsonia*, **6**, 190–202.

Woodson, R. E., Jr. (1964). The geography of flower color in Butterflyweed. *Evolution*, **18**, 143–63.

Wright, J. W. (1953). Pollen dispersion studies: some practical applications. *Journal of Forestry*, **51**, 114–18.

Wright, S. (1931). Evolution in Mendelian populations. *Genetics*, **16**, 97–159.

Wright, S. (1943). Isolation by distance. *Genetics*, **28**, 114–38.

Wright, S. (1946). Isolation by distance under diverse systems of mating. *Genetics*, **31**, 39–59.

Wright, S. (1966). Mendel's ratios. In *The origin of genetics*, ed. C. Stern & E. R. Sherwood, pp. 173–5. London & San Francisco: Freeman.

Wright, S. (1977). *Evolution and the genetics of populations*, vol. 3, *Experimental results and evolutionary deductions.* Chicago & London: The University of Chicago Press.

Wu, L., Bradshaw, A. D. & Thurman, D. A. (1975). The potential for evolution of heavy metal tolerance in plants. III. The rapid evolution of copper tolerance in *Agrostis stolonifera. Heredity*, **34**, 165–87.

Wyatt, R. (1988). Phylogenetic aspects of the evolution of self-pollination. In *Plant evolutionary biology*, ed. L. D. Gottlieb & S. K. Jain, pp. 109–31. London: Chapman & Hall.

Yahara T., Ito, M., Watanabe, K. & Crawford, D. J. (1991). Very low genetic heterozygosities in sexual and agamospermous populations of *Eupatorium altissimum* (Asteraceae). *American Journal of Botany*, **78**, 706–10.

Yates, F. (1960). *Sampling methods for censuses and surveys*, 3rd edn. London: Griffin. (4th edn, 1981.)

Yeo, P. F. (1975). Some aspects of heterostyly. *New Phytologist*, **75**, 147–53.

Young, A., Boyle, T. & Brown, T. (1996). The population genetic consequences of habitat fragmentation for plants. *Trends in Ecology & Evolution*, **11**, 413–8.

Youngner, V. B. (1960). Environmental control of initiation of the inflorescence, reproductive structures and proliferations in *Poa bulbosa. American Journal of Botany*, **47**, 753–7.

Yule, G. U. (1902). Mendel's laws and their probable relations to intra-racial heredity. *New Phytologist*, **1**, 193–207, 222–38.

Zabinski, C. & Davis, M. B. (1989). Hard times ahead for Great Lakes forests: a climate threshold model predicts responses to CO_2-induced climate change. In *The potential effects of global climate change on the United States*, ed. J. B. Smith & D. Tirpak, pp. 5.1–5.19 Appendix D. Washington: U. S. Environmental Protection Agency.

Zander, B. & Wiegleb, G. (1987). Biosystematische Untersuchungen an Populationen von *Ranunculus* subgen. *Batrachium* in Nordwest-Deutschland. *Botanische Jahrbücher für Systematik, Planzengeschichte und Planzengeographie*, **109**, 81–130.

Zeyl, C. & Bell, G. (1996). Symbiotic DNA in eukaryotic genomes. *Trends in Ecology & Evolution*, **11**, 10–15.

Zirkle, C. (1941). Natural selection before the 'Origin of species'. *Proceedings of the American Philosophical Society*, **84**, 71–123.

Zirkle, C. (1966). Some anomalies in the history of Mendelism. In *G. Mendel Memorial Symposium* 1865–1965, ed. M. Sosna, pp. 31–7. Prague: Academia Publishing House of the Czechoslovak Academy of Sciences.

Zohary, D. & Feldman, M. (1962). Hybridisation between amphidiploids and the evolution of polyploids in the Wheat (*Aegilops Triticum*) group. *Evolution*, **16**, 44–61.

Zohary, D. & Hopf, M. (1993). *Domestication of plants in the old world: the origin and spread of cultivated plants in West Asia, Europe and the Nile Valley*, 2nd edn. Oxford: Clarendon.

Zohary, D. & Nur. V. (1959). Natural triploids in the Orchard Grass *Dactylis glomerata* polyploid complex and their significance for gene flow from diploid to tetraploid levels. *Evolution*, **13**, 311–17.

Zuckerkandl, E. & Pauling, L. (1965). Molecules as documents of evolutionary history. *Journal of Theoretical Biology*, **8**, 357–66.

Index

Plant names. If a genus is represented only by a particular species, the full binomial is indexed (e.g. *Acer pseudoplatanus*). Generic names standing alone (e.g. *Achillea*) may include references to the genus only, and to individual species of the genus. In the case of the large genera *Potentilla*, *Ranunculus* and *Viola*, subsidiary references are given to individual species.
Animal names. The occasional animal name has not normally been indexed, though important references to animal groups (e.g. butterfly) have been included.
Names. In the captions of some Figures and Tables names containing many references have not been separately indexed.
Authors cited. Only those authors are included whose works are cited in the main text (as opposed to bracketed references). The full citations are given in the References.
Page numbers. The numbers in **bold type** refer to Figures.

Abbott, 336
abscisic acid, 118
Acacia, 403
Acer pseudoplatanus, 41–2
Achillea, 44, 180, **181**, 346, 352, 427
acid rain, 400
acquired characters, 19 *et seq.*
acyanogenesis, 229–31, **232**, 233–7
Adams, 219
Adanson, 16, 29
adaptive characters, 252–5, 371–2
adaptive radiation, 372
Aegilops, 322–4
Aesculus, 331
Aethusa cynapium, 251
AFLP, *see* amplified fragment length polymorphism
agamospermy, 124–66, 134–7, **137**, 138 *et seq.*, 365
agamotype, 183
agmatoploidy, 356
agricultural experiments, *see* experiments
Agropyron, 159, 277, 363
Agrostemma githago, 250
Agrostis, 149, 185, **242**, 247–8, 287, 326
Akeroyd, 431
Alchemilla, 136, 154–5, **156**, 157, 183, **184**, 342
Alchornea ilicifolia, 136
Alder, 280
Alfalfa, 83
allele (allelomorph), 64, 66, 71, **72**, 78, 95, 106–7, 115, 128–33, 143, 145, 192, 211, 218 *et seq.*, 223, 263, 332, 341, 416, 430
Allium 149, 159
allopatric differentiation, 357, 365
allopolyploid(y), 219 *et seq.*, 265–8, 312 *et seq.*
allozyme, 106, 218–19, 226, 333, 336, 339
Alnus glutinosa, 280
alpha taxonomy, *see* taxonomy
Alston, 297
Alvarez, 374
Amentiferae, 369
American Association for the Advancement of Science, 410
Amphicarpum purshii, 218
amplified fragment length polymorphism, 111
Amyema, 371
analysis of variance, 33–51, 201–6
Anderson, 283, 290–1, 295, **301**, 304, 307, 322
Andropogon scoparius, 180–1
Anemone, 382

Anethum graveolens, 40
aneuploid(y), 98, 318, 353–7, 359
aneuspory, 141
Antennaria, 136, 159, 165
Anthoxanthum odoratum, 199, **240**, 241–2, 287, 326
Antirrhinum majus, 97
Antonius, 158
Antonovics, 244, 287
Apera spica-venti, 225
Aphanes, 165
apogamy, 138
apomixis, 134–42, 154–9, 169, 212, 340–4, **343**, 351, 365, 417
  embryology of, 137–42
  facultative, 154–9, 183, 365, 418
  obligate, 154–5, 365
  polyploids and, 342–4, **343**
  vegetative, 134
apospory, 141, **142**
Aquilegia, 63
Arabidopsis thaliana, 97, 189–90
Arabis holbellii, 158
arable weeds, *see* weeds
archaesporium, 141
arithmetic mean, 36
Armeria, 161, 287
Arnold, **304–6**, 307–8, 358
Arrhenatherum elatius, 326
Arum maculatum, **254**
Ash, 1–5
Asker, 342
Asplenium, 219, **330**, 393
Aster furcatus, 415
Astrocaryum mexicanum, 418
Atwood, 231
Aubergine, 420
autogamy, 161, **163**, 212, 344
autopolyploidy, 265–8, 314–18, 328, 331–46
Avena barbata, 225
average deviation, 37
Avers, **375**, **377**, **379**
Avery, 314
Ayazloo, 249

Bachmann, 114
Bailey, 360
Baker, 163, 295
Banana, 322
Bannister, 182
Barber, 301
Barley, 244, **246**
Barling, 206
Barrett, 133, 160, 226, 250–1, 340, 391, 431
Bateson, 59, 66, 77–8, 80–2, 84–5
Bath Information and Data Services, 189
Battaglia, 137–8, 141
Bauplan, 376
Bawa, 133
Bayer, 159
Beach, 160
Beech, 2
Bell, 249, 374
Bellis perennis, 49
Bennett, **304–5**, 347, 359–60
Benson, **290**
Berlin, 362
Beta, 12, **13**
Betula, 345
Bialowieza Nature Reserve, 283
BIDS, *see* Bath Information and Data Services
BGCI, *see* Botanic Gardens Conservation International
Biffen, 66
Bijlsma, 285
Billings, 182
bildiversity, 6, 399–433
  Map Library, 408
  threats to, 400–6
biological evolution, *see* evolution
biological species concept, *see* species
biological unit, 361–2
Biometrika, 48
biometry, 33–51, 52
biosystematics, 263 *et seq.*
biotechnology, 396–8
Biscutella laevigata, 50
Bishop, 198
Bladder Campion, 344
Blakeslee, 314
Bloody Cranesbill, 8
Bobart, 7
Bocher, 171
Bolkhovskikh, 309
Bond, 372
Bonnier, 55–8, **56**, 207, 237
Borg, 83
Borrill, 344–5
botanical institutions, 407
Botanic Gardens, 124, 190, 399, 419, 421–2, 432
  Arnold Arboretum, 421
  Cambridge, 82, 203, **204**, **205**
  Jardin de Aclimaticion de la Orotava, 403
  Innsbruck, 57
  Kew, 136, 312–13, 407, 421–2, 428
  Liverpool, 430
  Munich, 57
  Oxford, 337
  St Petersburg, 16
  University of California, 180
  Vienna, 57

Botanic Gardens Conservation International, 421
Botanic Gardens Micropropagation News, 422
Bothriochloa–Dichanthium complex, 351
Boutin, 122
Boveri, 74
Bowler, 20
Bracken, 192
Bradshaw, A.D., 185–6, 244, 376
Bradshaw, M.E., 183, 242, 247
Brand, 83–4
Brassica, **215**, 265, 310, 322, 348, 397
Breedlove, 362
Breiman, 323, **325**
Bretagnolle, 339
Briggs, 205, 237, 240
Broadbalk Wheat experiment, *see* experiments
Brochmann, 337
Bromus, 225, 430
Brougham, 245
Brown, 29, 255, 397
Brummitt, 395
Bryum cirrhatum, 40
Buffon, 17–19
Buff Wood, Cambridgeshire, **401**
bulbil, 134, **135**
Bumble Bee, 278
Burdon, 235
Burkill, 47–8, 352
Buttercup, 42
Butterfly, 234

Cabbage, 215, 265, 310
Caltha palustris, 41–2, 170, 352
Calytridium, 148
Camin, 388
Campanula rotundifolia, 98, 171, **172**
Campbell, 198
Cardamine, 126, 150, 356
Carduus, 303, 307
Carex, 356
Carrot, 3
Castanea, 2
Castroviejo, 356
Casuarina, 402
Catino, 395
Centaurea cyanus, 427
Centaurium, 345, 423
centromere, 353, 356
Chaenactis, **355**
Chain of Being, 9, 12
Chamaelirium luteum, 219
Chanway, 244
Chapman, 325–6
Charlesworth, 145
Charnov, 165
Chase, 391
Chenopodium album, 251
Cheplick, 218
Chestnut, 2
chloroplast, 103
Chondrilla juncea, 225
chromatogram, **330**, 334
chromosome, 74 *et seq*., **75**, **77**, **79**, 103, 309–60 *passim*
  accessory, 100
  A chromosome, 101
  atlas, 309
  B chromosome, **100**, 101
  breakage, **102**
  doubling, 338
  fragmentation, 356
  numbers, 352–3
  pairing, 322, 325–7, 331
  repatterning, 347–8, 353, 357
  satellited, 327
  sex, 100
  supernumerary, 100
  translocation, **355**, 357
Chrysanthemum, 310, 352
Cineraria cruenta, 80–1
Cirsium, 63, 402
Citrus, 155, 165
cladism, 260, 379
cladistics, 385–96, **386**, **389**, **390**, 420
Clark, 391
Clarke, 198
Clarkia, 357, **358**, 359, 388, **389**
classification, 1–3, 29, 379–81
  artificial, 3, 29, 379–81
  general purpose, 380
  hierarchical, 2, 380
  natural, 1–3, 29, 380–1
  phylogenetic, 379–80
  special purpose, 379, 381
Clausen, 57, 120, 174–5, **176**, **178**, 260, 262, 271, **272**
Claytonia virginiana, 355
cleistogamy, 133
Clement, 403
Clements, 57, 207, 237
Clifford, 11
climatic change, 400
cline, 183–9
clone, 5, 55–6, 58, 120, 175, 185, 192–5, 199, 207, 236–8
Cochran, 191, 198
Cockburn, 373
codon, 90, 91
coefficient of variation, 36 *et seq*., 42 *et seq*., **44**, 51
Cohen, 88

colchicine, 266, 322, 336–7
cold treatment, 119
comparative morphology, 367
computers in taxonomy, *see* taxonomy
Comte, 20
Conium, 3
conservation, 399–433
  arguments for, 432–3
  cryopreservation 423
  ex situ, 421–3
  in Botanic Gardens, 421–2
  in situ, 423
  micropropagation, 422–3
continental drift, 374
Conyza bonariensis, 251
Cook, 256
Corkhill, 231
correlation, 49–50
Correns, 66, 74, 102, 126
Correspondence of Charles Darwin, 20
Coryanthes speciosa, 25
Cotton, 213, 322, 326
Cowan, 33
Cowslip, 131, **132**
Cox, 164
Crab, 82
Cracraft, 260
Crane, 370
Crawford, 93
Crepis, 96, 98, 129, 355, 359
Crick, 88–9
Critica botanica, 10–12, 17
Crocus scepusiensis, 223
Croizat, 391
Cronquist, 374, 386, 394
cross-fertilisation, 125–33, 192, 261
cross(ing), 30, 58–79
  experiments, 363–4
  polygon, 363–4
  single factor, **61**
  two-factor, 61, **62**
  three-factor, 63
Crovello, 363
cryopreservation, *see* conservation
Cumbes, 249
Cuvier, 17
cyanogenesis, 229–30, **231**, **232**, 233–7, 256–7
Cymadotheca, 235
cytometry, 339

Dactylis glomerata, 171, **173**, 249, 326, 331–2, 339, 344, 352
Daday, 231, **232**, **233**, 234
Dafni, 217
Dandelion, 136, **137**, 141, **142**, 158–9
Darwin, C., 3, 6, 7–32, 33, 53, 64, 80 *et seq.*, **81**, 88, 125–6, **127**, 131, 133, 154, 226, 257, 264, 284, 361, 368–9, 376, **377**, 391, 399, 411
Darwin, E., 19
database, 366
Daucus, 3
D'Amato, 96
Davenport, 34
Davidson, 293
Davies, 199, 237, 239–40, **241**
Davis, **429**
Dawson, 231
De Candolle, 27, 29
defence mechanism, 371
De Haan, 339
de Jussieu, 29
Demauro, 415, 430
deme terminology, 381
demographic monitoring, 412–13
dendrogram, 383, **385**
De Nettancourt, 160
deoxyribonucleic acid, 88–123, **89**, **93**, 145, 158, 208, 323, 331–4, 338–9, 347, 353, 356, 360, 373, 386–9, 393, 396–7, 414
Deschampsia cespitosa, 149, 175, **179**, 180
designed experiment, *see* experiments
de Vilmorin, 58
Devils's Dyke, near Cambridge, 425
De Vries, 30, 41, 66, 84, 95, 102, 264–5
de Wet, 339, 351
Diamond, 407
Dianthus, 413, 422
Di Cesnola, 83
Dichanthium aristatum, 157
Digitalis purpurea, 8, 35, 126
dimorphism, 132, 162
dioecy 125, *et seq.*, 163, 219, 417–18
diplospory, 141, **142**
Dipsacus sylvestris, 402
Dirzo, 235
distribution,
  complex, 43–6
  frequency, **37**, 38–42, 51
  leptokurtic, 42, 217
  normal, 38–41, **40**, **41**
  platykurtic, 42
  skewed, 42
distyly, 124 *et seq.*, **131**, 161, 212
DNA, *see* deoxyribonucleic acid
Dobzhansky, 260–1, 301
Dollo, 388, 390
dominance, 59, 60, **61**, 62–79 *passim*
Donoghue, 391–2, 394–5
Doyle, 369, 374, 391
Draba, 338
Drosophila melanogaster, 77–8, 91, 95, 284
Duchesnes, 16

Duhamel du Monceau, 53
Dustin, 315

East, 67, 70, 126, 143
Echinochloa, 225, 250
ecosystem, 433
ecotone, 242
ecotype, 167–207 *passim*, 431
  biotic, 171
  edaphic, 180–1
  occurrence, 183
Edwards, 64
Ehrhart, 279
Ehrlich, 212
Eichhornia, 161, 162, 225, 403
Eigsti, 315
Elam, 415
Eldredge, 378
electrophoresis, 104–7, **109**, 159, 219, 312, 333, 337, 347, 357, 414
Elkington, 345
Ellis, 257
Ellstrand, 219, 415
Elodea canadensis, 134
Elymus, 276, **277**, 359, 363
embryo-sac, 138 *et seq*., 141
Emex spinosa, 225
Empedocles, 22
endangered species, 399–433
  attempts to prevent extinction, 423–32
  attempts to reverse decline, 419–23
  strategies for reintroduction of, 430–2
  transference of, 428–9
endemism, 419–20
enzyme, 104–9
  monomeric, 106
  polymorphic, 104–5
  restriction, 107, **108**, **109**
Epilobium, 56–7
Equisetum, 351
Erikkson, 13, 17
Erophila verna, 30, **31**
Erwin, 407
Erythroxylum, 165
EU, *see* evolutionary unit
euchromatin, 101, 354
Eucalyptus albens, 415
eugenics, 33–4
Eupatorium altissimum, 158
Euphrasia, 83, 345
Evans, 199
Evening Primrose, 30, 40, 84–5, 102–3, 264–5
evolution,
  biological, 367 *et seq*.
  co-evolution with animals, 371–2
  macroevolution, 372–8
  microevolution, 372–8
  organic, 368
  progressive, 368
evolutionary unit, 386–8, 392
experiments,
  agricultural, 209, 213–14, 244–6
  Broadbalk Wheat, 196, **197**
  cultivation, 196–200
  crossing, 363–4
  designed, 201–6
  genecological, 190–200
  interpretation of, 206–7
  Park Grass experiment, Rothamsted, 239–40
  transplant, 55–8, **56**, 206–7, 237–8
extinction, 368–78, 399–433

Fagaceae, 2
Fagerlind, 345
Fagus, 2, 41
Falk, 422
Farris, 388
Favarger, 348
Fee, 319
Fen Violet, 413
Ferguson, 297
Festuca, **100**, 149, 192, 249, 326, 345, 360
Fibonacci sequence, 44–6, 97
Ficus, 192
fingerprinting, 107–11
Fir, 84
Fisher, 34, 64, 85, 196–7, 228
Flora Europaea, 309
flow cytometry, 339
folk taxonomy, *see* taxonomy
Forey, 386
fossil record, 368–70, 377 *et seq*., **378**, 380
Foster, 403
founder effect, 223–6
Four O'Clock, 63, **86**, 102
Foxglove, 8, 35
Fragaria, 16
Fraxinus, 1–5
French Bean, 63
frequency diagram, 38–42, **42**
frequency distribution, *see* distribution
Friday, **92**
Friedman, 219
Frost, 298, **300**
Fruit Fly, 77–8, 284
Fundamenta fructificationis, 16

Gajewski, 279, **280**, **281**, 282
Gale, 326
Galeopsis, 313–14
Galium, 345
Galton, 33, 49–50, 52, 65, 66–7, 85

gamete, 98, 124
  unreduced, 339
gametophyte, 138 *et seq.*
gametophytic incompatibility, *see* incompatibility
gamodeme, 211
Ganders, 132–3
Gartner, 27, 59
Gates, 265
Gaura, 358
Gay, 293
Geiger, 274
gemmules, 64–5
gene, 66, 90, **91**, 103, 128, 396–8
  dominant, **68–9**
  flow, 113, 208, 211–23, 239–41, 344–6, 411, 414, 425, 429
  frequency, 223
  information, **93**
  marker, 396–7
  recessive, **68–9**
  resistant, 397
  silencing, 347
genecology, 53, 171 *et seq.*, 190–200, 206, 208–58
genetic code, 90, **91**
genetic drift, 223–5, 263, 373, 411
genetic erosion, 415
genetics, 58–79
genome, 108, 112–13, 134, 226, 266–7, 312–53 *passim*, 418
genomic *in situ* hybridisation, 347–8, 360
Gentiana pneumonanthe, 258, 412–13, 415
Geranium sanguineum, 8, 52, **188**
Gesner, 26
Geum, 8, 171, 277–8, **279**, **281**, 282–3
Gibasis, 360
gibberellic acid, 118
Gibbs, 160
Gibby, 317–18, 340
Giemsa dye, 101
Gilia, 288, **289**, 359
Gill, 324
Gilmour, 211, 379–81
Ginkgo, **378**
Giseke, 18
GISH, *see* genomic *in situ* hybridisation
global warming, 405–6, 429
Goatsbeard, 14–16
Godron, 30
Goebel, 116
Gog Magog Hills, Cambridge, 427
Goldblatt, 309–10
Goldschmidt, 357
Gonzalez-Benito, 423
Good, 378
Gorse, 116, **117**
Gossypium, 213, 303, 326
Gottlieb, 145, **275**, 312, 333, 388, **389**, 420
Gould, 362, 376, 378
Govindaraju, 215
Graham, 391
Grand Duke of Saxe-Coburg and Gotha, 33
Grant, 161, 288, **289**, 310, 312, 355, 372
Grape, 322
Grau, **311**
Grauer, 323, **325**
Gray, A., 20, 21, 23
Gray, A.J., 337
Green, 191, 264
Greene, 12
greenhouse effect, 405
Gregor, 43, 183–4, 186, 198
Greig-Smith, 191
gynodioecy, 150, **152**, 417–18
Groundsel, 35
Guinard, 98
Gustafsson, 137–8

habitat management, 167–8, 428
Hagerup, 348
Hakea, 403
Haldane, 85, 228
Half-Galton curve, 41–2
Hall, **301**
Halocarpus bidwillii, 415
Hanson, 198
haploidy, 128 *et seq.*, 140, 265–8, **358**
Haplopappus gracilis, 355
Harberd, 195
Harborne, 104
Harder, 160
Hardy-Weinberg Law, 211–12
Harlan, 244, 339, 351
Harper, 195, 235–6, 244, 257
Harris, 245
Harvey, 23, **322**
Hathaway, 293
heavy metal tolerance, 226, 228, 247, **248**, 249–50
Hedera helix, 118, **119**
Heiser, 299–300, 302, 359
Helianthus, 303, 308, 359
Hemlock, 3, **429**
Hennig, 385, 395
Herbert, 59
herbicide resistance, 251–2
Hereditary genius, 33
heredity, 58–79
heterochromatin, 101, **354**
heterosis, 143, **144**, 145
hermaphrodite, 125 *et seq.*
Heslop-Harrison, 182, 262, 432

heteroblastic development, 116
heterophylly, 116–18, **119**, 120
heterostyly, 131, **132**, 161, 166
Heterotheca latifolia, 218
heterozygosity, 66, 106–7, 122, 128 *et seq.*, **143**, 145, 147, 158–9, 341–2, 346
hexaploid(y), 322, 336, 340, 346, 348
  autohexaploid, 341
  segmental allohexaploid, 325
Heywood, 365, **383–5**
Hickey, 369
Heiracium umbellatum, 27, 63–4, 136, 168–70, 190, 341
Hiesey, 57, 174–5, **176**, **178**, 182, 271
Hillis, 386
Hinata, 159
Hippocrepis emerus, 258
histogram, 38–41, **39**
Hodges, 308
Hoffmann, 30
Hoffmann-Ostenhof, 96
Holl, 244
Holcus, 247, 249, 327, **328**, 393
Holliday, 252
Holly, 40
Holsinger, 363–4, 420, 422
Holttum, 380
homomorphism, 130 *et seq.*, 161
homozygosity, 66, 106–7, 143, 145–6, 159, 341–2
Hooker, 20
Hopkins, 119
Hordeum, 244, **246**, **277**, 363
Horse Chestnut, 331
Hortus cliffortianus, 11
Hosta, 276
Huenneke, 402
Hughes, 128, 159, 234–5, 369–70
Humphries, 386
Huskins, 337
Hutchinson, 293
Hutton, 26
Huxley, 184, 262, 381
hybrid vigour, 143, **144**, 145
hybridisation, 6, 27, 30, 58–79, 363, 418
  experimental, 58
  introgressive (introgression), 290–1, 358
  natural, 58
  speciation and, 358
Hymenoxys acaulis, 415, 430

idiogram, 98, **99**
Ilex aquifolium, 40
Impatiens capensis, 150
Imperial Academy of Sciences, 336
inbreeding, 143, **144**, 145, 159–60, 183, 411, 416, 418–19
incompatibility, 192
  gametophytic, **129**, 130
  self, 124–66, 192, 212, 370, 417, 430
  sporophytic, 129, **130**, 131–3
Index to plant chromosome numbers, 309
Ingram, **92**, 336
inheritance
  acquired characters and, 373
  blending, 23, **86**
  cytoplasmic, 103
  extranuclear, 103
  maternal (matroclinous), 136, **138**
  particulate, 60–79 *passim*
Institute for Scientific Information, 189
interbreeding, 363–5
  natural population, 364
International Botanical Congresses, 10
International Code of Botanical Nomenclature, 10
International Union for the Conservation of Nature, 408, **409**
  categories, **409**
Ipomoea purpurea, 116, **117**, 118, 125
Iris, **292**, 303–7, 359
Ising, 298, **300**
island bigeography, 423–5
isloating mechanisms, 261, **262**
isozyme, 106, 145, 158–9, 194, 208–9, 219, 225–6, 235, 248, 251, 312, 333, 336–8, 342, 346, 347, 358, 359, 414
IUCN, *see* International Union for the Conservation of Nature
Ivy, 118, **119**

Jackson, 301
Jacob's Ladder, 8
Jameson, 20
Jansen, 388, **390**
Jenkin, 23–4
Jennersten, 413
Jerling, 137, 141, 164, 342, 351
Johannsen, 53–5, 67, 85, 122
John Innes Horticultural Institution, 313
Johnson, 256, 309–10, 345
Jones, 234, 257, 326–7, 344–5
Jonsell, 363
Jordan, 30–1
Jordanova, 19
Juncus, 356
Juniperus, **301**, 301–2, 418

Kadereit, 394
Kareiva, 397
Karyotype, 98, 100–1, 323, 327–8, **328**, **329**, 333, 347, 357, 360
  banding patterns in, 101, 104, **105**, 106–7
Keck, 57, 174–5, **176**

Kelvin, Lord, 26
Kenton, 347
Kerner, 57 *et seq.*
Kerster, 217
Kess, 160
Keymer, 257
Khandewal, 310
Kimber, **324**
Kimura, 373
King, 158, 353, 363
Kitching, 386
Kohn, 20, 431
Kojima, 159
Kolreuter, 16, 27, 58–9
Koshy, 328, **329**
Kruckeberg, 180–1
Kyhos, **355**

Lady's Mantle, 154
Lamarck, 17–19, 20, 55
Lamprecht, 378
Lane, 234
Langlet, 52–3, 183
Larsen, 180
Lasthenia, 358–9
Lathyrus, 77, 167
Lawrence, 88, 175
Lawton, 255
Laxton, 58
Layia, **272**, **273**, 274–5, **275**
Lecooq, 59
Lee, 47
Lefebvre, 287
Leontodon autumnalis, 190
leptokurtic distribution, *see* distribution
Lesser Celandine, 46–7
Leucanthemum vulgare, 40, 44, **45**, 46, **56**, 352
leucine aminopeptidase, 218
Levin, 122, 217, **222**, 228, 261, **262**, 290, **330**, 339, 344, 353, 365
Lewin, 88
Lewis, 310, 312, 350, **357**, 359, 389
Lewontin, 376
Life, letters and labours of Francis Galton, 63
Lilac, 170
Lilium martagon, 98
Limnanthes, 276
Limonium, 165
Linaria vulgaris, 13–15, **14**, **15**, 97, 125
linkage, 212
Linnean Society, 21–2
Linnaeus, 2–3, 10–17, **13**, **18**, 19, 27, 43, 53, 58, 170, 318, 334, 361, 380, 408
Linroth, 17
Lisianthus, 276
Lithops, 371
Lloyd, 154, 218
Lloydia serotina, 419
local races, 46–8
Lolium, 244–5, 249, 326
Lotus, 56, 115, 229–35, 256–7
Love, **349**, 365
Lovejoy, 10
Lovett Doust, 165
Lovkvist, 169
Lowe, 336
Luceno, 356
Lucretius, 22
Ludwig, Professor, 43, 46–7
Luehea seemannii, 407
Lumaret, 331, 339
Lupinus texensis, 219
Luzula, 356
Lycopersicum esculentum, 264
Lyell, 20
Lynch, 82
Lysimachia nummularia, 149, 403
Lythrum salicaria, 133

Mable, 386
McCauley, 219
McClintock, 373
McFadden, 322
Mack, 403
McLeish, **75**, **79**
MacLeod, 47
McMillan, 182
Macnair, 247, 249
McNally, 225
McNeilly, 287
Maize, 63, 95, **96**, 126, **127**, 143, 215, **245**, 285, **286**, 339, 373
Major, 419
Malacothamnus fasciculatus, 414
Malthus, 20–1
Mangelsdorf, 126
Mann, 411
Manton, 319, **320**, **321**, 348, 350, 376
Marsden-Jones, 202, 279–80
Marsh Marigold, 41–2, 171
Marshall, 219, 337
Martini, 244
Matthew, 22
May, 407–8
Mayr, 260, 262, 362–4
Mead, 198
Medicago, 83, 427
Meeuse, 369
megaspore, 140 *et seq.*
Meharg, 249
meiosis, 74 *et seq.*, **76**, **78**, 93, **99**, 101, 113, 128 *et seq.*, 140 *et seq.*, 159, 266–7,

313, 316–18, 323–57 *passim*
Melaleuca quinquenervia, 402
Mendel, 30, 32, 50–2, 58–79, **61**, **62**, **68**, **69**, 84–5, 95, 208
Mendelian ratio, 65–6, 341
Mercurialis, 16
meristem, 115
Merxmüller, **311**
metaphase, **100**
Mexican Palm, 418
Miao, 431
Michaelis, 264
Michaels, **390**
micropropagation, *see* conservation
microsatellites, 112
microspore, 140 *et seq.*
Midgley, 372
migration, 264
Milium, 347–8
Miller, 324, 328
Milner, 182
mimicry, 371
Mimulus, 126, 247, 276, 288
Minimum Viable Population, 417
minisatellites, 112
Mirabilis, 63, **86**, 102, 103
Mistletoe, 84, 371
Mitchell, 120
mitochondria, 103
mitosis, 74 *et seq.*, 75, 93, 98, **99**, 101, 140 *et seq.*, 265, 353
Mivart, 22–3, 25, 26
Mogie, 159
monophylesis, 369–70
Mooney, 182
Moore, 293, **298**, 322, **354**, 355
Morgan, 78
Morison, 3–4
Morisset, 122
Moritz, 386
Morning Glory, 116, 117
morphological plasticity, 122–3
Morris, 397
Morton, 10, 348
Moth Mullein, 9
Muenchow, 160
Muller, 218
Muntzing, 313–14, **315**
Murbeck, 136
Musa, 322
mutagen, 96
mutation, 84–5, 92–3, **94**, **95**, **96**, 97, 102, 106, 108, 124, 134, 159, 212, 219, 263, 265, 346–7, 372–3, 376
MVP, *see* Minimum Viable Population
Myers, 410
Myosotis, **311**

Nagato, 159
Nannfeldt, 328–9
Nasturtium, **267**, **268**
national parks, 423
National Trust, 423
natural classification, *see* classification
Natural history of plants, their forms, growth, reproduction and distribution, 57
natural selection, *see* selection
Natural System, 3
nature reserves, 423
  design of, **424**, 425–6
  management of, 426–7
Naumova, 137
Nei, 275
neighbourhood, 222–3
Nelson, 198, **292**
Nemec, 314
neo-Darwinism, 85–7
Newton, 313, 408
Nicotiana, 16, 59, 67, **70**, 126, 265, 322, 326
Niemela, 255
Nigella, 163
Nightshade, 264
Nilsson-Ehle, 67, **71**
Njoku, 116–17
Nogler, 137, 341, 351
non-adaptive characters, 252–5
Nothofagus, 2, 225
Novak, 333
nucleotide, 90
null hypothesis, 203
nullisomic plant, 326
numerical taxonomy, *see* taxonomy
Nur, 345
Nybom, 158
Nymphoides, 164

Oak, 1 *et seq.*, 27, 299
Oat, 326
Odontites, 83
Oenothera, 30, 40, 84–5, 102–3, 264–5
Oilseed Rape, 397
Olby, 88
Oldfield, 433
Olea europaea, 2
Oleaceae, 2
Olive, 2
omega taxonomy, *see* taxonomy
On naval timber and arboriculture, 22
Oostermeijer, 412
operational taxonomic unit, 382–3
Ophioglossum reticulatum, 310
Ophrys, 372, 413
Orchid, 125
Orchis laxiflora, 428

organelle, 90, 103, 373
Ornduff, 164, 351
Oryctes nevadensis, 420
Osborn, 17
OTU, *see* operational taxonomic unit
Ouborg, 415–16
outbreeding, 124–33, 149–50 *et seq.*, 183, 344, 370, 418, 431
outcrossing, **153**, 154, 239, 416
Ownbey, 332–3, **334–5**
Ox-Eye Daisy, 44–5
Oxyria digyna, 182

Palmer, **390**
pangenesis, 64–5
Panicum, 226, 341
Pankhurst, **387**, 393–4
Papaver, 34–6, **35**, **36**, 38, 163, **164**, 344
Parker, 198
Park Grass Experiment, Rothamsted, *see* experiments
Parsnip, 3
Parthenium, 129
particulate inheritance, *see* inheritance
Paspalum, 340
Pastinaca, 3
Paterniani, 285, **286**
Pauling, 392
PCR, *see* polymerase chain reaction
Pea, 50, 58–79, **79**
Pearson, 34–5, **39**, 48–9, 66–7
Peckham, 22
Pedicularis, 8, 414
Pelargonium, 103
Persicaria amphibia, 120
Pellew, 77, 313
Peloria, 13–16, **14**, **15**, 97
Pepper, 420
Perez, 423
Phaseolus vulgaris, 53,**54**
Philosophia botanica, 12
Philosophie zoologique, 17, 19
Phleum, 57
Phlox, 161, 288, 290
photoperiodism, 119, 180–2
phylogenetic tree, **379**, 420
phylogeny, 29, 104, 113, 260–3
Pilosella, 155, 345
Pine, 53, 84, 217
Pinus, 53, 217–19, 308, 403
Pisum sativum, 50
Pitcher Plant, 225
Plaisted, **292**
Plantae hybridae, 14
Plantago, 43, 45, 145, 150, 183–4, **185**, **186**, 198, 201–6, 216, 237, **238**, 250, 285
plant cell, **92**
plasticity, 256–7
plastid, 103
Pledge, 42
Poa, **135**, 149, 183, 240, 251, 328, **329**, 341
Polemonium caeruleum, 8, 66, **162**
Polemoniaceae, **162**
polyhaploidy, 351
pollen dispersal, 124, 215, **216**, 217–19
pollen fertility, **139**
pollination, 125 *et seq.*, 140, 261
  animal, 216–17
  insect, 125, **216**
  wind, 125, 215–16
  mechanisms of, 124
pollution, 209, 247–50, 404–6
  acid rain, 405
  gases, other, 406, 429
  ozone, 404–5
  sulphur dioxide, 404
  vehicle emissions, 404
Polygala vulgaris, 225
Polygonum amphibium, 120
polymerase chain reaction, 109–11, **110**, 114
polymorphism, 27, 48, 111, 158–9, 212, 223, 227, 229–31, 235, 238, 254
polyploidy, 85, 98–100, 113, 264–9, **266**, 309–60, 418–20
  chemical compounds in, 330–1
  chemically induced, 314, 322
  cultivar evolution, 313
  distribution of, 348–51, **349**
  evolutionary change, 313
  latitude, 348–50, **349**
  lower plants, 312
  persistance of, 340–4
Polypodium, 318–21, **320**, 338
polytopic origins, 332–8, 346
Poppy, 34–5, 38
population(s), 124, 208–9, 211–12
  census, 417
  genetics of small, 414–16
  Mendelian, 211
  minimum viable, 416–18, 429–30
  panmictic, 211
  re-establishment of extinct, 430–1
  selection in, 226–8, 263
Population Viability Analysis, 417
Populus tremuloides, 194
Portugal, 88
post-zygotic mechanism, 261, **262**
Potato, 322, 371, 420
Potentilla, 137, **138**, 155, 165, 341
  anserina, 41–2
  arenaria, **138**
  crantzii, 156–7
  glandulosa, 120, 175, **178**, 271

gracilis, 183
neumanniana (*tabernaemontani*), **138**, 141
Powers, 342
protected areas, 423–6, 431
Praying Mantis, 83
pre-zygotic mechanism, 261, **262**, 363
Primack, **410**, **412**, **418**, **424**, 431
Primula, 131, **132**, 149, 161, 212, **267**, 292, **294**, 295, **296**, 300, 312–13, 338, 340, 419
probable error, 37–8
probe, radioactive, **109**
prothallus, 138
Prunella vulgaris, **56**, 171, **174**, 198
pseudogamy, 136 *et seq*., 141
Pseudopeziza, 235
Pteridium aquilinum, 192
punctuated equilibrium model, 378, **379**
Punnett, 77–8
Purdy, 159
pure line, **143**
Purple Loosestrife, 133
Putwain, 252
PVA, *see* Population Viability Analysis

Quercus, 1 *et seq*., 293, **299**, 308
Quetelet, 20, 33–4, 170
quinacrine, 101
Quinn, 218

Rabinowitz, 419
Rackham, **401**
Radford, 271
Radish, 265
Rafinski, 223
ramet, 55–6, 115, 120, 175
Ramsbottom, 16
Randolph, **292**
random amplified polymorphic DNA, 111, **112**, 114
Ranunculus, 11, 345, 382
aquatilis, 19
auricomus, 136, 351
Batrachium (subg.), 120, **121**
bulbosus, 206
ficaria, 46–9, **47**, 342, 352
flammula, 256
hederaceus, 19
marginatus var. *marginatus*, 428
repens, 42
RAPD, *see* random amplified polymorphic DNA
Raphanobrassica, 265
Raphanus, 212, 219, 265, 397
ratio-cline, 231, 253
Raup, 368, 374, 432
Raven, C.E., 7, 10
Raven, P., 212, 351, 362–3
Ray, 3, 7–32, 52, 66, 116, 260, 361
Raybould, 337
Raymer, 397
recessiveness, **61**, 62–79 *passim*
reciprocal transplant, 237–8, **238**, 244
recombination, 124
Red Rattle, 8
reduction division, 74
regression analysis, 49
reproductive isolation, 364
restoration ecology, 399–400, 427–8
restriction fragment length polymorphism, 111, **112**
RFLP, *see* restriction fragment length polymorphism
Rhinanthus, 83
Rhizobium, 224
ribonucleic acid, **91**
Richards, 125, 159, 160, 164
Ridgman, 198
Rieger, 264
Rieseberg, 291, 302, **303**, 308, 358, 393
Riley, **292**, **323**, 325–6
Ritland, 159
RNA, *see* ribonucleic acid
Roach, 199
Rogstad, 145
Rohlf, 198
Roose, 333
Rosa, 12, 27, 348
Rosen, **31**
Ross, 119
Rothwell, 369
Rowell, 413
Rubus, 27, 136, 155, 158
Ruckert, 74
Rumex, 310, 357
Ruse, 363
Rushton, **299**
rust, 235
Rye, 348

Salisbury, 119
Salix, 308
Salmon, 198
saltation, 23, 25, 264, 360
Salvia, 359, 415, 418
Sanguisorba minor, 427–8
sampling, 48, 189–96, 201, 208
Sanderson, 394
Sarkar, 322
Sarracenia purpurea, 225
Saunders, 59, 77
Saxifraga, **135**, 149, 430
Scabiosa columbaria, 415–16
Scala Naturae, 10

Schaal, 145, 158, 219, 225
Schinus terebinthifolius, 402
Schlichting, 122
Schmalhausen, 122
Schmidt, 167–8
Schoen, 218
Schrödinger, 88
Schwaegerle, 225
Scilla autumnalis, 357
Scirpus, 356
Scots Pine, 53
Scottish Bird's Eye Primrose, 419
Sea Plantain, 43
Sears, 322
Seavey, 133
Secale cereale, 348
seed dispersal, 124, 217, 222–3
  animal, 218
  wind, 217–18, 223
Sedum suaveolens, 310
seed banks, 422
segregation, 59–79
selection
  apostatic, 237
  catastrophic, 359
  directional, 252
  disruptive, 226, 238–44, 252
  natural, 6, 21, 25, 33, 212, 227 *et seq.*
  pollinators and, 371
  populations and, 226–8, 263
  stabilising, 226
  variation and, 376
  weighting, 380–1
self-compatibility, 134, **153**, 154, 160–5, 341, 358, 416
self-fertilisation, 58–79, 133–4, **143**, **144**, 145 *et seq.*, 261, 416
self-incompatibility, *see* incompatibility
self-sterility, 126–33
Senecio, 35, 80, 222, 226, 252, 336, 337, 414, 420
sequencing, 107
Sewall Wright effect, 85, 223
Sexual System, 3, 380
Shafer, 424
Sharp, 225
Sharrock, 7
Sherwood, 63
Shivas, 319, **320**, **321**
Shore, 226, 340
Shull, 143
Silene, 50–1, 219, 344
Silvertown, 218
Silverweed, 41–2
Simberloff, 426
Simpson, 260, 262, 386
Sims, 161
Sitanion, **277**, 363
Site of Special Scientific Interest, 431
SLOSS debate, 425
Slugs, 232–4, **236**
Smith, A., 195
Smith, Adam, 20
Smith, F.D.M., 297, **330**
Smith, J., 136
Snails, 232–4, **236**
Snaydon, 199, 237, 239, **240**
Sneath, **383**, **384**, **385**
Snedecor, 198
Snoad, **75**, **79**
Snowdon Lily, 419
Snyder, **277**, 363
Sokal, 198, 363
Solanum, **266**, 310, 322, 340, 344, 371
Soltis, 331, 333, 336–7, 389
Somnus plantarum, 14
Sorbus, 136, 342, **343**
Sorghum halepense, 226
Southern Beech, 2
Southern blotting technique, 108
Spartina, 337–9
speciation, 259 *et seq.*, 363
  abrupt, 262, 264–5, 309–60
  gradual, 263–4, 357
  hybridisation and, 358
  saltational, 359–60
  sympatric, 353
species, 1, 7, 29–32, 259–61
  biological, 260–4
  concept, 259 *et seq.*, 361–6
  elemental, 10
  elementary, 30
  nature of, 5–6
  numbers, 406–11
  origins of, 263 *et seq.*
  reproductive community, 260–1
  taxonomic, 259–60
  transmutation of, 8, 12, 20
Species plantarum, 12–13, 17, 29, 53
Spergula arvensis, 252, **253**
sporophyte, 138 *et seq.*
sports, 81, 84, 97
Sports Turf Research Institute, 249
Sprengel, 125
SSSI, *see* Site of Special Scientific Interest
Stace, 345, 364–5, 382, 407
standard deviation, 36 *et seq.*
Stapledon, 171, **173**
statistics, 33–51, 190
Stearn, 11
Stearns, 252
Stebbins, 163, 262, 310, 322, 330, **334–5**, 348–50, 370
Stephanomeria, 358, 430

Stern, 63
Stewart, 369
stochasticity, 412–14
Strangler Fig, 192
Strasburger, 74, 136, 264
Strawberry, 16
Streptanthus, 276, 391
Strickberger, 95, 145
Striga, 225
Stuart, 198
Stuessy, 383, 386, 389, 393–4
Sullivan, 231
Sultan, 122
Sutton, 74, **77**
Swede, 310
Sweet Pea, 77
Swofford, 386
Sycamore, 41
sympatry, 261, 264
Syndesmon thalictroides, 49
Syringa vulgaris, 170
Systema naturae, 17
Sytsma, 388

Täckholm, 348
Taggart, 225
Taiz, 119
Taq, 111
Taraxacum officinale, 136, **137**, 141, **142**, 158–9
taxa, reality of, 361–21
taxonomy, 2
  alpha, 262, 381
  computers in, 382–3, **383**
  folk, 2, 361–2
  numerical, 382–3, **384**, 385
  omega, 262, 381
Tephroseris integrifolia (*Senecio integrifolius*), 226, 414
Theory of the earth, 26
Thermophilus aquaticus, 111
Thiselton-Dyer, 80–1
Thomas, H. Hamshaw, 26
Thompson, 122, 339, 351
Thomson, 402
Thurman, 247
Thymus, 150, **152**
Toadflax, 13–15
Tobacco, 16, 265, 322, 326, 348, 420
Tomato, 264, 420
Tradescantia virginiana, 326
Tragopogon, 14–16, 332–7, **334–5**, 420
transmutation of species, *see* species
Tralau, **378**
transgenic plants, 396–8
transplant experiments, *see* experiments
transposon, 373–4
Trathan, 235
triage, 420
Trifolium, 66, 149, 192, 199, 225–6, 229, 231, 232, 233, 234–5, 246–7, 256
triploid(y), 314, 318, 321, 322, 327, 342, 344–5
tristyly, **131**, 161
Triticum, 66–7, **71**, 314, 322, **323**, **324**, **325**, 326
tropical rain forests, 423
Tsuga canadensis, **429**
Tudge, 432–3
Tuomi, 255
Turesson, 120, 167–73, 180, 183, 186, 189–90, 196, 199, 208, 260, 352
Turkington, 199, 244
Turner, 104, 297, 414
Turnera ulmifolia, 161, 340
Turnip, 310, 322
Turrill, 202, 381, 394
Tutin, 328

Ulex europaeus, 116, **117**
Umbelliferae, 2, **4**
University of Davis, California, 430
Uromyces trifolii, 235
Urtica dioica, **236**

Vaccinium, 298–9, **300**,
Van der Toorn, 237, **238**
Van Dijk, 285
Van Tienderen, 237, **238**
van Treuren, 415–16
van Valen, 260
variance, 36 *et seq.*
variation, 6, 34, **43–5**, 88–124, 367
  adaptive, 52
  basis of, 51
  biotic, 171
  chance, 223–5, 263
  continuous, 35, 39, 50, 66–7
  correlated, 48–50
  cytological, 98–102
  demographic, **418**
  developmental, 5, 114, 116
  discontinuous, 23, 34–5, 39, 50–1, 183–4, 191
  ecotypic, 208, **230**
  environmental, 5, 51–2, 114–15, 167 *et seq.*, **418**
  extreme habitat, 255–6
  genetic, 23, 51–2, 103–4, 114, 167, 208–9, 225
  genotypic, 114–16
  individual, 5, 52–79, **54**, **72**
  interspecific, 159

variation (*cont.*)
  intraspecific, 6–11, 19, 21, 33–51, 114, 159, 167–207
  limits of, 24–5
  meristic, 34, 39
  natural, 10, 33
  numerical, 379
  patterns of, 255–6
  phenotypic, 101, 114–16, 196, 199
  seasonal, 255–6
  selection and, 376
  small population, 415
  static, 367
Variation of plants and animals under domestication, 64
vegetative propagation, 140
Veitch, 313
Verbascum, 9, 63
Vestiges of the natural history of Creation, 20
Vicia, 90, 112, **414**
Vilmorin, 77
Viola, 63, 150, 218, 293, **322**, 344
  aurea, 331
  hirta, **151**
  lactea, **298**
  persicifolia, **322**, 413
  purpurea, 331
  quercetorum, 330
  riviniana, **151**, **298**
  stagnina, **322**
  tricolor, 30
Viscum album, 84
Vitis, 322
Voaniola gerardii, 310
von Nägeli, 30, 57, 63–4
von Niessl, 52
Von Tschermak, 66
von Tubeuf, 84
von Wettstein, 83
Vorzimmer, 22–3
Vulpia, 345, 360

Wagner, 348, 386, 388
Wakehurst, 428
Waldron, 83–4
Wallace, 20–2, 257, 284, 288–90
Walters, 183, 379
Ward, 190
Warwick, 205, 237, 240, 251, **275**, 307
Water Avens, 8
Watercress, **267**
Waterweed, 134
Watson, 88–9
weeds
  arable, 250–2
  crop mimicry by, 250–2
WCMC, *see* World Conservation Monitoring Centre
Wegener, 374
Weir, 336
Weising, 112
Weismann, 65
Weldon, 50, 81–3, 227
Wells, 22
Wendel, 291, 302, **303**, 308, 358, 391
What is life?, 88
Wheat, 66–7, **71**, 196–7, 314, **324**
White, **79**, 353, 359
Whitehouse, 65, **76**, 160, 370
Wichura, 59
Wicken Fen, 201 *et seq.*, **204**, **205**, 413, 423
Wild Beet, 12
wild flower mixtures, 427–8
Wildlife and Countryside Act, 431
Williams, 26
Williamson, 397
Willis, 378
Wilson, E.O., 410–11
Wilson, G.B., 249
Winge, 265, 312
Winkler, 264–5, **266**
Winsor, 395
Wittrock, 30
Wolf, 145
Woodell, 292, 296, 300
World Conservation Monitoring Centre, 408–9
Wright, 64, 223
Wu, 247
Wulff, 199
Wyse Jackson, 431

Xanthium strumarium, 182

Yahara, 158
Yarrow, 44
Yates, 191, 198
Yule, 67

Zabinski, **429**
Zang, 391
Zea mays, 63, 95, **96**, 126, **127**, 143, 215, **245**, 285, **286**
Zeiger, 119
Zeyl, 374
Zimmer, 391
Zimmermann, 386
Zirkle, 22
Zohary, 345
Zuckerlandl, 392
zygote, 124–33
zymogram, 104, **105**, 106